THE HORSE, THE WHEEL, AND LANGUAGE

THE
HORSE
THE
WHEEL
AND
LANGUAGE

HOW
BRONZE-AGE RIDERS
FROM THE
EURASIAN STEPPES
SHAPED THE
MODERN WORLD

DAVID W. ANTHONY

Princeton University Press
Princeton and Oxford

Published by Princeton University Press, 41 William Street, Princeton, New Jersey 08540

In the United Kingdom: Princeton University Press, 3 Market Place, Woodstock, Oxfordshire OX20 1SY

ISBN-13: 978-0-691-05887-0

Library of Congress Control Number: 2007932082

British Library Cataloging-in-Publication Data is available

This book has been composed in Adobe Caslon

Printed on acid-free paper ∞

press.princeton.edu

Printed in the United States of America

9   10

CONTENTS

Chapter Four
Language and Time 2:
Wool, Wheels, and Proto-Indo-European 59

Chapter Five
Language and Place:
The Location of the Proto-Indo-European Homeland 83

Chapter Six
The Archaeology of Language 102

PART TWO
The Opening of the Eurasian Steppes 121

Chapter Seven
How to Reconstruct a Dead Culture 123

ACKNOWLEDGMENTS

This book would not have been written without the love and support of my mother and father, David F. and Laura B. Anthony. Laura B. Anthony read and commented on every chapter. Bernard Wailes drew me into the University of Pennsylvania, led me into my first archaeological excavation, and taught me to respect the facts of archaeology. I am blessed with Dorcas Brown as my partner, editor, critic, fellow archaeologist, field excavation co-director, lab director, illustrator, spouse, and best friend through thick and thin. She edited every chapter multiple times. All the maps and figures are by D. Brown. Much of the content in chapters 10 and 16 was the product of our joint research, published over many years. Dorcas's brother, Dr. Ben Brown, also helped to read and edit the ms.

The bit-wear research described in chapter 10 and the field work associated with the Samara Valley Project (chapter 16) was supported by grants from Hartwick College, the Freedman and Fortis Foundations, the American Philosophical Society, the Wenner-Gren Foundation, the National Geographic Society, the Russian Institute of Archaeology (Moscow), the Institute for the History and Archaeology of the Volga (Samara), and the National Science Foundation (United States), with assistance for chapter 10 from the State University of New York at Cobleskill. We are particularly grateful to the National Science Foundation (NSF).

Support to write this book was provided by a fellowship from the National Endowment for the Humanities in 1999–2000 and a membership in the School of Historical Studies at the Institute for Advanced Study (IAS) at Princeton, New Jersey, in 2006, where Nicola DiCosmo and Patricia Crone made us welcome. The term at the IAS was crucial.

People who have helped me in numerous different ways include:

Near East and East Asia: Kathy Linduff, Victor Mair, Oscar Muscarella, Karen Rubinson, Chris Thornton, Lauren Zych, C. C. Lamberg-Karlovsky, Fred Hiebert, Phil Kohl, Greg Possehl, Glenn Schwartz, David Owen, Mitchell Rothman, Emmy Bunker, Nicola DiCosmo, and Peter Golden.

Horses and wheeled vehicles: Dexter Perkins and Pat Daly; Şandor Bökönyi, Sandra Olsen, Mary Littauer and Joost Crouwel (my instructors in ancient transport); and Peter Raulwing, Norbert Benecke, and Mindy Zeder.

Bit wear and the riding experiment: Mindy Zeder, Ron Keiper; the Bureau of Land Management, Winnemucca, Nevada; Cornell University Veterinary School; University of Pennsylvania New Bolton Center; the Assateague Island Wildlife Refuge; and, at the State University of New York at Cobleskill, Steve MacKenzie, Stephanie Skargensky, and Michelle Beyea.

Linguistics: Ward Goodenough, Edgar Polomé, Richard Diebold, Winfrid Lehmann, Alexander Lubotsky, Don Ringe, Stefan Zimmer, and Eric Hamp. A special thanks to Johanna Nichols, who helped edit chapter 5, and J. Bill Darden and Jim Mallory, who reviewed the first draft.

Eastern European archaeology: Petar Glumac (who made me believe I could read Russian-language sources), Peter Bogucki, Douglass Bailey (who reviewed chapter 11), Ruth Tringham (who gave me my first field experience in Eastern Europe), Victor Shnirelman (our first guide in Russia), Dimitri Telegin (my first source on steppe archaeology), Natalya Belan, Oleg Zhuravlev, Yuri Rassamakin, Mikhail Videiko, Igor Vasiliev, Pavel Kuznetsov, Oleg Mochalov, Aleksandr Khokhlov, Pavel Kosintsev, Elena Kuzmina, Sergei Korenevskii, Evgeni Chernykh, R. Munchaev, Nikolai Vinogradov, Victor Zaibert, Stanislav Grigoriev, Andrei Epimakhov, Valentin Dergachev, and Ludmila Koryakova. Of these I owe the deepest debts to Telegin (my first guide) and my colleagues in Samara: Vasiliev, Kuznetsov, Mochalov, Khokhlov, and (honorary Samaran) Kosintsev.

The errors I have made are mine alone; these people tried their best.

PART ONE

Language and Archaeology

Chapter One

The Promise and Politics
of the Mother Tongue

Ancestors

When you look in the mirror you see not just your face but a museum. Although your face, in one sense, is your own, it is composed of a collage of features you have inherited from your parents, grandparents, great-grandparents, and so on. The lips and eyes that either bother or please you are not yours alone but are also features of your ancestors, long dead perhaps as individuals but still very much alive as fragments in you. Even complex qualities such as your sense of balance, musical abilities, shyness in crowds, or susceptibility to sickness have been lived before. We carry the past around with us all the time, and not just in our bodies. It lives also in our customs, including the way we speak. The past is a set of invisible lenses we wear constantly, and through these we perceive the world and the world perceives us. We stand always on the shoulders of our ancestors, whether or not we look down to acknowledge them.

It is disconcerting to realize how few of our ancestors most of us can recognize or even name. You have four great-grandmothers, women sufficiently close to you genetically that you see elements of their faces, and skin, and hair each time you see your reflection. Each had a maiden name she heard spoken thousands of times, and yet you probably cannot recall any one of their maiden names. If we are lucky, we may find their birth names in genealogies or documents, although war, migration, and destroyed records have made that impossible for many Americans. Our four great-grandmothers had full lives, families, and bequeathed to us many of our most personal qualities, but we have lost these ancestors so completely that we cannot even name them. How many of us can imagine being so utterly forgotten just three generations from now by our own descendents that they remember nothing of us—not even our names?

In traditional societies, where life is still structured around family, extended kin, and the village, people often are more conscious of the debts they owe their ancestors, even of the power of their ghosts and spirits. Zafimaniry women in rural Madagascar weave complicated patterns on their hats, which they learned from their mothers and aunts. The patterns differ significantly between villages. The women in one village told the anthropologist Maurice Bloch that the designs were "pearls from the ancestors." Even ordinary Zafimaniry houses are seen as temples to the spirits of the people who made them.[1] This constant acknowledgment of the power of those who lived before is not part of the thinking of most modern, consumer cultures. We live in a world that depends for its economic survival on the constant adoption and consumption of new things. Archaeology, history, genealogy, and prayer are the overflowing drawers into which we throw our thoughts of earlier generations.

Archaeology is one way to acknowledge the humanity and importance of the people who lived before us and, obliquely, of ourselves. It is the only discipline that investigates the daily texture of past lives not described in writing, indeed the great majority of the lives humans have lived. Archaeologists have wrested surprisingly intimate details out of the silent remains of the preliterate past, but there are limits to what we can know about people who have left no written accounts of their opinions, their conversations, or their names.

Is there a way to overcome those limits and recover the values and beliefs that were central to how prehistoric people really lived their lives? Did they leave clues in some other medium? Many linguists believe they did, and that the medium is the very language we use every day. Our language contains a great many fossils that are the remnants of surprisingly ancient speakers. Our teachers tell us that these linguistic fossils are "irregular" forms, and we just learn them without thinking. We all know that a past tense is usually constructed by adding -t or -ed to the verb (kick-kicked, miss-missed) and that some verbs require a change in the vowel in the middle of the stem (run-ran, sing-sang). We are generally not told, however, that this vowel change was the older, original way of making a past tense. In fact, changing a vowel in the verb stem was the usual way to form a past tense probably about five thousand years ago. Still, this does not tell us much about what people were thinking then.

Are the words we use today actually fossils of people's vocabulary of about five thousand years ago? A vocabulary list would shine a bright light on many obscure parts of the past. As the linguist Edward Sapir observed, "The complete vocabulary of a language may indeed be looked upon as a complex inventory of all the ideas, interests, and occupations that take up the attention of the community."[2] In fact, a substantial vocabulary list has been reconstructed for one of the languages spoken about five thousand years ago. That language is the ancestor of modern English as well as many other modern and ancient languages. All the languages that are descended from this same mother tongue belong to one family, that of the Indo-European languages. Today Indo-European languages are spoken by about three billion people—more than speak the languages of any other language family. The vocabulary of the mother tongue, called "Proto-Indo-European", has been studied for about two hundred years, and in those two centuries fierce disagreements have continued about almost every aspect of Indo-European studies.

But disagreement produces light as well as heat. This book argues that it is now possible to solve the central puzzle surrounding Proto-Indo-European, namely, who spoke it, where was it spoken, and when. Generations of archaeologists and linguists have argued bitterly about the "homeland" question. Many doubt the wisdom of even pursuing it. In the past, nationalists and dictators have insisted that the homeland was in their country and belonged to their own superior "race." But today Indo-European linguists are improving their methods and making new discoveries. They have reconstructed the basic forms and meanings of thousands of words from the Proto-Indo-European vocabulary—itself an astonishing feat. Those words can be analyzed to describe the thoughts, values, concerns, family relations, and religious beliefs of the people who spoke them. But first we have to figure out where and when they lived. If we can combine the Proto-Indo-European vocabulary with a specific set of archaeological remains, it might be possible to move beyond the usual limitations of archaeological knowledge and achieve a much richer knowledge of these particular ancestors.

I believe with many others that the Proto-Indo-European homeland was located in the steppes north of the Black and Caspian Seas in what is today southern Ukraine and Russia. The case for a steppe homeland is stronger today than in the past partly because of dramatic new archaeological discoveries in the steppes. To understand the significance of an Indo-European homeland in the steppes requires a leap into the complicated and fascinating world of steppe archaeology. *Steppe* means "wasteland" in the language of the Russian agricultural state. The steppes resembled the prairies of North America—a monotonous sea of grass framed under a huge, dramatic sky. A continuous belt of steppes extends from eastern Europe on the west (the belt ends between Odessa and Bucharest) to the Great Wall of China on the east, an arid corridor running seven thousand kilometers across the center of the Eurasian continent. This enormous grassland was an effective barrier to the transmission of ideas and technologies for thousands of years. Like the North American prairie, it was an unfriendly environment for people traveling on foot. And just as in North America, the key that opened the grasslands was the horse, combined in the Eurasian steppes with domesticated grazing animals—sheep and cattle—to process the grass and turn it into useful products for humans. Eventually people who rode horses and herded cattle and sheep acquired the wheel, and were then able to follow their herds almost anywhere, using heavy wagons to carry their tents and supplies. The isolated prehistoric societies of China and Europe became dimly aware of the possibility of one another's existence only after the horse was domesticated and the covered wagon invented. Together, these two innovations in transportation made life predictable and productive for the people of the Eurasian steppes. The opening of the steppe—its transformation from a hostile ecological barrier to a corridor of transcontinental communication—forever changed the dynamics of Eurasian historical development, and, this author contends, played an important role in the first expansion of the Indo-European languages.

LINGUISTS AND CHAUVINISTS

The Indo-European problem was formulated in one famous sentence by Sir William Jones, a British judge in India, in 1786. Jones was already widely known before he made his discovery. Fifteen years earlier, in 1771, his *Grammar of the Persian Language* was the first English guide to the language of the Persian kings, and it earned him, at the age of twenty-five, the reputation as one of the most respected linguists in Europe. His translations of medieval Persian poems inspired Byron, Shelley, and the European Romantic movement. He rose from a respected barrister in Wales to a correspondent, tutor, and friend of some of the leading men of the kingdom. At age thirty-seven he was appointed one of the three justices of the first Supreme Court of Bengal. His arrival in Calcutta, a mythically alien place for an Englishman of his age, was the opening move in the imposition of royal government over a vital yet irresponsible merchant's colony. Jones was to regulate both the excesses of the English merchants and the rights and duties of the Indians. But although the English merchants at least recognized his legal authority, the Indians obeyed an already functioning and ancient system of Hindu law, which was regularly cited in court by Hindu legal scholars, or pandits (the source of our term *pundit*). English judges could not determine if the laws the pandits cited really existed. Sanskrit was the ancient language of the Hindu legal texts, like Latin was for English law. If the two legal systems were to be integrated, one of the new Supreme Court justices had to learn Sanskrit. That was Jones.

He went to the ancient Hindu university at Nadiya, bought a vacation cottage, found a respected and willing pandit (Rāmalocana) on the faculty, and immersed himself in Hindu texts. Among these were the *Vedas*, the ancient religious compositions that lay at the root of Hindu religion. The *Rig Veda*, the oldest of the Vedic texts, had been composed long before the Buddha's lifetime and was more than two thousand years old, but no one knew its age exactly. As Jones pored over Sanskrit texts his mind made comparisons not just with Persian and English but also with Latin and Greek, the mainstays of an eighteenth-century university education; with Gothic, the oldest literary form of German, which he had also learned; and with Welsh, a Celtic tongue and his boyhood language which he had not forgotten. In 1786, three years after his arrival in Calcutta, Jones came to a startling conclusion, announced in his third annual discourse to the Asiatic Society of Bengal, which he had founded when he first arrived. The key sentence is now quoted in every introductory textbook of historical linguistics (punctuation mine):

> The Sanskrit language, whatever be its antiquity, is of a wonderful structure: more perfect than the Greek, more copious than the Latin, and more exquisitely refined than either; yet bearing to both of them a stronger affinity, both in the roots of verbs and in the forms of grammar, than could possibly have been produced by accident; so strong indeed, that no philologer could examine them all three, without believing them to have sprung from some common source, which, perhaps, no longer exists.

Jones had concluded that the Sanskrit language originated from the same source as Greek and Latin, the classical languages of European civilization. He added that Persian, Celtic, and German probably belonged to the same family. European scholars were astounded. The occupants of India, long regarded as the epitome of Asian exotics, turned out to be long-lost cousins. If Greek, Latin, and Sanskrit were relatives, descended from the same ancient parent language, what was that language? Where had it been it spoken? And by whom? By what historical circumstances did it generate daughter tongues that became the dominant languages spoken from Scotland to India?

These questions resonated particularly deeply in Germany, where popular interest in the history of the German language and the roots of German traditions were growing into the Romantic movement. The Romantics wanted to discard the cold, artificial logic of the Enlightenment to return to the roots of a simple and authentic life based in direct experience and community. Thomas Mann once said of a Romantic philosopher (Schlegel) that his thought was contaminated too much by reason, and that he was therefore a poor Romantic. It was ironic that William Jones helped to inspire this movement, because his own philosophy was quite different: "The race of man . . . cannot long be happy without virtue, nor actively virtuous without freedom, nor securely free without rational knowledge."[3] But Jones had energized the study of ancient languages, and ancient language played a central role in Romantic theories of authentic experience. In the 1780s J. G. Herder proposed a theory later developed by von Humboldt and elaborated in the twentieth century by Wittgenstein, that language creates the categories and distinctions through which humans give meaning to the world. Each particular language, therefore, generates and is enmeshed in a closed social community, or "folk," that is at its core meaningless to an outsider. Language was seen by Herder and von Humboldt as a vessel that molded community and national identities. The brothers Grimm went out to collect "authentic" German folk tales while at the same time studying the German language, pursuing the Romantic conviction that language and folk culture were deeply related. In this setting the mysterious mother tongue, Proto-Indo-European, was regarded not just as a language but as a crucible in which Western civilization had its earliest beginnings.

After the 1859 publication of Charles Darwin's *The Origin of Species*, the Romantic conviction that language was a defining factor in national identity was combined with new ideas about evolution and biology. Natural selection provided a scientific theory that was hijacked by nationalists and used to rationalize why some races or "folks" ruled others—some were more "fit" than others. Darwin himself never applied his theories of fitness and natural selection to such vague entities as races or languages, but this did not prevent unscientific opportunists from suggesting that the less "fit" races could be seen as a source of genetic weakness, a reservoir of barbarism that might contaminate and dilute the superior qualities of the races that were more "fit." This toxic mixture of pseudo-science and

Romanticism soon produced its own new ideologies. Language, culture, and a Darwinian interpretation of race were bundled together to explain the superior biological–spiritual–linguistic essence of the northern Europeans who conducted these self-congratulatory studies. Their writings and lectures encouraged people to think of themselves as members of long-established, biological–linguistic nations, and thus were promoted widely in the new national school systems and national newspapers of the emerging nation-states of Europe. The policies that forced the Welsh (including Sir William Jones) to speak English, and the Bretons to speak French, were rooted in politicians' need for an ancient and "pure" national heritage for each new state. The ancient speakers of Proto-Indo-European soon were molded into the distant progenitors of such racial–linguistic–national stereotypes.[4]

Proto-Indo-European, the linguistic problem, became "the Proto-Indo-Europeans," a biological population with its own mentality and personality: "a slim, tall, light-complexioned, blonde race, superior to all other peoples, calm and firm in character, constantly striving, intellectually brilliant, with an almost ideal attitude towards the world and life in general".[5] The name *Aryan* began to be applied to them, because the authors of the oldest religious texts in Sanskrit and Persian, the *Rig Veda* and *Avesta*, called themselves Aryans. These Aryans lived in Iran and eastward into Afghanistan–Pakistan–India. The term *Aryan* should be confined only to this Indo-Iranian branch of the Indo-European family. But the *Vedas* were a newly discovered source of mystical fascination in the nineteenth century, and in Victorian parlors the name Aryan soon spread beyond its proper linguistic and geographic confines. Madison Grant's *The Passing of the Great Race* (1916), a best-seller in the U.S., was a virulent warning against the thinning of superior American "Aryan" blood (by which he meant the British–Scots–Irish–German settlers of the original thirteen colonies) through interbreeding with immigrant "inferior races," which for him included Poles, Czechs, and Italians as well as Jews—all of whom spoke Indo-European languages (Yiddish is a Germanic language in its basic grammar and morphology).[6]

The gap through which the word *Aryan* escaped from Iran and the Indian subcontinent was provided by the *Rig Veda* itself: some scholars found passages in the *Rig Veda* that seemed to describe the Vedic Aryans as invaders who had conquered their way into the Punjab.[7] But from where? A feverish search for the "Aryan homeland" began. Sir William Jones placed it in Iran. The Himalayan Mountains were a popular choice in the early nineteenth century, but other locations soon became the subject of animated debates. Amateurs and experts alike joined the search, many hoping to prove that their own nation had given birth to the Aryans. In the second decade of the twentieth century the German scholar Gustav Kossinna attempted to demonstrate on archaeological grounds that the Aryan homeland lay in northern Europe—in fact, in Germany. Kossinna illustrated the prehistoric migrations of the "Indo-Germanic" Aryans with neat black arrows that swept east, west, and south from his presumed Aryan homeland. Armies followed the pen of the prehistorian less than thirty years later.[8]

The problem of Indo-European origins was politicized almost from the beginning. It became enmeshed in nationalist and chauvinist causes, nurtured the murderous fantasy of Aryan racial superiority, and was actually pursued in archaeological excavations funded by the Nazi SS. Today the Indo-European past continues to be manipulated by causes and cults. In the books of the Goddess movement (Marija Gimbutas's *Civilization of the Goddess*, Riane Eisler's *The Chalice and the Blade*) the ancient "Indo-Europeans" are cast in archaeological dramas not as blonde heroes but as patriarchal, warlike invaders who destroyed a utopian prehistoric world of feminine peace and beauty. In Russia some modern nationalist political groups and neo-Pagan movements claim a direct linkage between themselves, as Slavs, and the ancient "Aryans." In the United States white supremacist groups refer to themselves as Aryans. There actually were Aryans in history—the composers of the *Rig Veda* and the *Avesta*—but they were Bronze Age tribal people who lived in Iran, Afghanistan, and the northern Indian subcontinent. It is highly doubtful that they were blonde or blue-eyed, and they had no connection with the competing racial fantasies of modern bigots.[9]

The mistakes that led an obscure linguistic mystery to erupt into racial genocide were distressingly simple and therefore can be avoided by anyone who cares to avoid them. They were the equation of race with language, and the assignment of superiority to some language-and-race groups. Prominent linguists have always pleaded against both these ideas. While Martin Heidegger argued that some languages—German and Greek—were unique vessels for a superior kind of thought, the linguistic anthropologist Franz Boas protested that no language could be said to be superior to any other on the basis of objective criteria. As early as 1872 the great linguist Max Müller observed that the notion of an Aryan skull was not just unscientific but anti-scientific; languages are not white-skinned or long-headed. But then how can the Sanskrit language be connected with a skull type? And how did the Aryans themselves define

"Aryan"? According to their own texts, they conceived of "Aryan-ness" as a *religious–linguistic* category. Some Sanskrit-speaking chiefs, and even poets in the *Rig Veda*, had names such as Balbūtha and Bṛbu that were foreign to the Sanskrit language. These people were of non-Aryan origin and yet were leaders among the Aryans. So even the Aryans of the *Rig Veda* were not genetically "pure"—whatever that means. The *Rig Veda* was a ritual canon, not a racial manifesto. If you sacrificed in the right way to the right gods, which required performing the great traditional prayers in the traditional language, you were an Aryan; otherwise you were not. The *Rig Veda* made the *ritual* and *linguistic* barrier clear, but it did not require or even contemplate racial purity.[10]

Any attempt to solve the Indo-European problem has to begin with the realization that the term *Proto-Indo-European* refers to a language community, and then work outward. Race really cannot be linked in any predictable way with language, so we cannot work from language to race or from race to language. Race is poorly defined; the boundaries between races are defined differently by different groups of people, and, since these definitions are cultural, scientists cannot describe a "true" boundary between any two races. Also, archaeologists have their own, quite different definitions of race, based on traits of the skull and teeth that often are invisible in a living person. However race is defined, languages are not normally sorted by race—all racial groups speak a variety of different languages. So skull shapes are almost irrelevant to linguistic problems. Languages and genes are correlated only in exceptional circumstances, usually at clear geographic barriers such as significant mountain ranges or seas—and often not even there.[11] A migrating population did not have to be genetically homogeneous even if it did recruit almost exclusively from a single dialect group. Anyone who *assumes* a simple connection between language and genes, without citing geographic isolation or other special circumstances, is wrong at the outset.

THE LURE OF THE MOTHER TONGUE

The only aspect of the Indo-European problem that has been answered to most peoples' satisfaction is how to define the language family, how to determine which languages belong to the Indo-European family and which do not. The discipline of linguistics was created in the nineteenth century by people trying to solve this problem. Their principal interests were comparative grammar, sound systems, and syntax, which provided the basis for classifying languages, grouping them into types, and otherwise

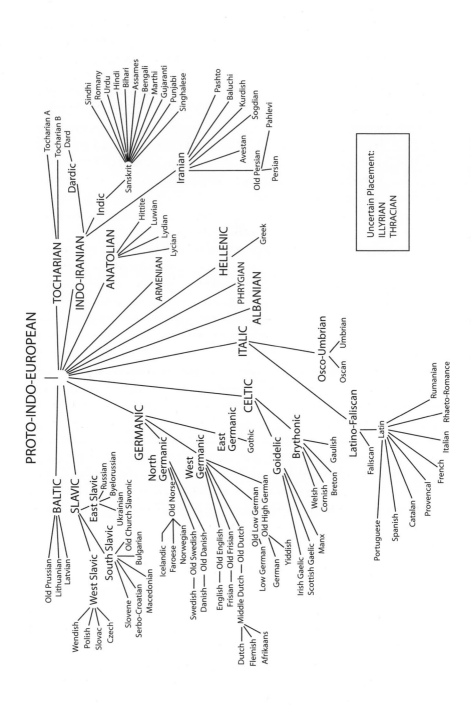

PROTO-INDO-EUROPEAN

Uncertain Placement:
ILLYRIAN
THRACIAN

defining the relationships between the tongues of humanity. No one had done this before. They divided the Indo-European language family into twelve major branches, distinguished by innovations in phonology or pronunciation and in morphology or word form that appeared at the root of each branch and were maintained in all the languages of that branch (figure 1.1). The twelve branches of Indo-European included most of the languages of Europe (but not Basque, Finnish, Estonian, or Magyar); the Persian language of Iran; Sanskrit and its many modern daughters (most important, Hindi and Urdu); and a number of extinct languages including Hittite in Anatolia (modern Turkey) and Tocharian in the deserts of Xinjiang (northwestern China) (figure 1.2). Modern English, like Yiddish and Swedish, is assigned to the Germanic branch. The analytic methods invented by nineteenth-century philologists are today used to describe, classify, and explain language variation worldwide.

Historical linguistics gave us not just static classifications but also the ability to reconstruct at least parts of extinct languages for which no written evidence survives. The methods that made this possible rely on regularities in the way sounds change inside the human mouth. If you collect Indo-European words for *hundred* from different branches of the language family and compare them, you can apply the myriad rules of sound change to see if all of them can be derived by regular changes from a single hypothetical ancestral word at the root of all the branches. The proof that Latin *kentum* (hundred) in the Italic branch and Lithuanian *shimtas* (hundred) in the Baltic branch are genetically related cognates is the construction of the ancestral root *$k'mtom$-. The daughter forms are compared sound by sound, going through each sound in each word in each branch, to see if they can converge on one unique sequence of sounds that could have evolved into all of them by known rules. (I explain how this is done in the next chapter.) That root sequence of sounds, if it can be found, is the proof that the terms being compared are genetically related cognates. A reconstructed root is the residue of a successful comparison.

Figure 1.1 The twelve branches of the Indo-European language family. Baltic and Slavic are sometimes combined into one branch, like Indo-Iranian, and Phrygian is sometimes set aside because we know so little about it, like Illyrian and Thracian. With those two changes the number of branches would be ten, an acceptable alternative. A tree diagram is meant to be a sketch of broad relationships; it does not represent a complete history.

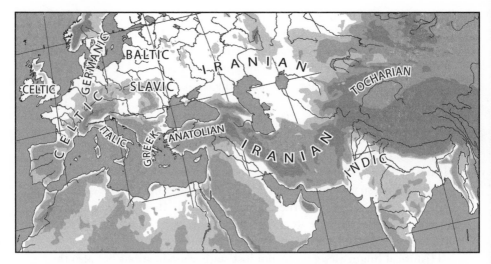

Figure 1.2 The approximate geographic locations of the major Indo-European branches at about 400 BCE.

Linguists have reconstructed the sounds of more than fifteen hundred Proto-Indo-European roots.[12] The reconstructions vary in reliability, because they depend on the surviving linguistic evidence. On the other hand, archeological excavations have revealed inscriptions in Hittite, Mycenaean Greek, and archaic German that contained words, never seen before, displaying precisely the sounds previously reconstructed by comparative linguists. That linguists accurately predicted the sounds and letters later found in ancient inscriptions confirms that their reconstructions are not entirely theoretical. If we cannot regard reconstructed Proto-Indo-European as literally "real," it is at least a close approximation of a prehistoric reality.

The recovery of even fragments of the Proto-Indo-European language is a remarkable accomplishment, considering that it was spoken by nonliterate people many thousands of years ago and never was written down. Although the grammar and morphology of Proto-Indo-European are most important in typological studies, it is the reconstructed vocabulary, or lexicon, that holds out the most promise for archaeologists. The reconstructed lexicon is a window onto the environment, social life, and beliefs of the speakers of Proto-Indo-European.

For example, reasonably solid lexical reconstructions indicate that Proto-Indo-European contained words for otter, beaver, wolf, lynx, elk, red deer, horse, mouse, hare, and hedgehog, among wild animals; goose, crane, duck, and eagle, among birds; bee and honey; and cattle (also cow, ox, and steer), sheep (also wool and weaving), pig (also boar, sow, and piglet), and dog among the domestic animals. The horse was certainly known to the speakers of Proto-Indo-European, but the lexical evidence alone is insufficient to determine if it was domesticated. All this lexical evidence might also be attested in, and compared against, archaeological remains to reconstruct the environment, economy, and ecology of the Proto-Indo-European world.

But the proto-lexicon contains much more, including clusters of words, suggesting that the speakers of PIE inherited their rights and duties through the father's bloodline only (patrilineal descent); probably lived with the husband's family after marriage (patrilocal residence); recognized the authority of chiefs who acted as patrons and givers of hospitality for their clients; likely had formally instituted warrior bands; practiced ritual sacrifices of cattle and horses; drove wagons; recognized a male sky deity; probably avoided speaking the name of the bear for ritual reasons; and recognized two senses of the sacred ("that which is imbued with holiness" and "that which is forbidden"). Many of these practices and beliefs are simply unrecoverable through archaeology. The proto-lexicon offers the hope of recovering some of the details of daily ritual and custom that archaeological evidence alone usually fails to deliver. That is what makes the solution of the Indo-European problem important for archaeologists, and for all of us who are interested in knowing our ancestors a little better.

A New Solution for an Old Problem

Linguists have been working on cultural-lexical reconstructions of Proto-Indo-European for almost two hundred years. Archaeologists have argued about the archaeological identity of the Proto-Indo-European language community for at least a century, probably with less progress than the linguists. The problem of Indo-European origins has been intertwined with European intellectual and political history for considerably more than a century. Why hasn't a broadly acceptable union between archaeological and linguistic evidence been achieved?

Six major problems stand in the way. One is that the recent intellectual climate in Western academia has led many serious people to question the entire idea of proto-languages. The modern world has witnessed increasing cultural fusion in music (Black Ladysmith Mombasa and Paul Simon, Pavarotti and Sting), in art (Post-Modern eclecticism), in information services (News-Gossip), in the mixing of populations (international migration is at an all-time high), and in language (most of the people in the world are now bilingual or trilingual). As interest in the phenomenon of cultural convergence increased during the 1980s, thoughtful academics began to reconsider languages and cultures that had once been interpreted as individual, distinct entities. Even standard languages began to be seen as creoles, mixed tongues with multiple origins. In Indo-European studies this movement sowed doubt about the very concept of language families and the branching tree models that illustrated them, and some declared the search for any proto-language a delusion. Many ascribed the similarities between the Indo-European languages to convergence between neighboring languages that had distinct historical origins, implying that there never was a single proto-language.[13]

Much of this was creative but vague speculation. Linguists have now established that the similarities between the Indo-European languages are not the kinds of similarities produced by creolization and convergence. None of the Indo-European languages looks at all like a creole. The Indo-European languages must have replaced non–Indo-European languages rather than creolizing with them. Of course, there was inter-language borrowing, but it did not reach the extreme level of mixing and structural simplification seen in all creoles. The similarities that Sir William Jones noted among the Indo-European languages can *only* have been produced by descent from a common proto-language. On that point most linguists agree.

So we should be able to use the reconstructed Proto-Indo-European vocabulary as a source of clues about where it was spoken and when. But then the second problem arises: many archaeologists, apparently, do not believe that it is possible to reliably reconstruct any portion of the Proto-Indo-European lexicon. They do not accept the reconstructed vocabulary as real. This removes the principal reason for pursuing Indo-European origins and one of the most valuable tools in the search. In the next chapter I offer a defense of comparative linguistics, a brief explanation of how it works, and a guide to interpreting the reconstructed vocabulary.

The third problem is that archaeologists cannot agree about the antiquity of Proto-Indo-European. Some say it was spoken in 8000 BCE, others say as late as 2000 BCE, and still others regard it as an abstract idea that exists only in linguists' heads and therefore cannot be assigned to any one time. This makes it impossible, of course, to focus on a specific era. But the principal reason for this state of chronic disagreement is that most archaeologists do not pay much attention to linguistics. Some have proposed solutions that are contradicted by large bodies of linguistic evidence. By solving the second problem, regarding the question of reliability and reality, we will advance significantly toward solving problem number 3—the question of when—which occupies chapters 3 and 4.

The fourth problem is that archaeological methods are underdeveloped in precisely those areas that are most critical for Indo-European origin studies. Most archaeologists believe it is impossible to equate prehistoric language groups with archaeological artifacts, as language is not reflected in any consistent way in material culture. People who speak different languages might use similar houses or pots, and people who speak the same language can make pots or houses in different ways. But it seems to me that language and culture *are* predictably correlated under some circumstances. Where we see a *very clear* material-culture frontier—not just different pots but also different houses, graves, cemeteries, town patterns, icons, diets, and dress designs—that *persists* for centuries or millennia, it tends also to be a linguistic frontier. This does not happen everywhere. In fact, such *ethno-linguistic* frontiers seem to occur rarely. But where a robust material-culture frontier does persist for hundreds, even thousands of years, language tends to be correlated with it. This insight permits us to identify at least *some* linguistic frontiers on a map of purely archaeological cultures, which is a critical step in finding the Proto-Indo-European homeland.

Another weak aspect of contemporary archaeological theory is that archaeologists generally do not understand migration very well, and migration is an important vector of language change—certainly not the only cause but an important one. Migration was used by archaeologists before World War II as a simple explanation for any kind of change observed in prehistoric cultures: if pot type A in level one was replaced by pot type B in level two, then it was a migration of B-people that had caused the change. That simple assumption was proven to be grossly inadequate by a later generation of archaeologists who recognized the myriad *internal* catalysts of change. Shifts in artifact types were shown to be caused by changes in the size and complexity of social gatherings, shifts in economics, reorganization in the way crafts were managed, changes in the social function of crafts, innovations in technology, the introduction of new trade and exchange commodities, and so on. "Pots are not people" is a rule taught to every Western archaeology student since the 1960s. Migration disappeared entirely from the explanatory toolkit of Western archaeologists in the 1970s and 1980s. But migration is a hugely important human behavior, and you cannot understand the Indo-European problem if you ignore migration or pretend it was unimportant in the past. I have tried to use modern migration theory to understand prehistoric migrations and their probable role in language change, problems discussed in chapter 6.

Problem 5 relates to the specific homeland I defend in this book, located in the steppe grasslands of Russia and Ukraine. The recent prehistoric archaeology of the steppes has been published in obscure journals and books, in languages understood by relatively few Western archaeologists, and in a narrative form that often reminds Western archaeologists of the old "pots are people" archaeology of fifty years ago. I have tried to understand this literature for twenty-five years with limited success, but I can say that Soviet and post-Soviet archaeology is not a simple repetition of any phase of Western archaeology; it has its own unique history and guiding assumptions. In the second half of this book I present a selective and unavoidably imperfect synthesis of archaeology from the Neolithic, Copper, and Bronze Ages in the steppe zone of Russia, Ukraine, and Kazakhstan, bearing directly on the nature and identity of early speakers of Indo-European languages.

Horses gallop onstage to introduce the final, sixth problem. Scholars noticed more than a hundred years ago that the oldest well-documented Indo-European languages—Imperial Hittite, Mycenaean Greek, and the most ancient form of Sanskrit, or Old Indic—were spoken by militaristic societies that seemed to erupt into the ancient world driving chariots pulled by swift horses. Maybe Indo-European speakers invented the chariot. Maybe they were the first to domesticate horses. Could this explain the initial spread of the Indo-European languages? For about a thousand years, between 1700 and 700 BCE, chariots were the favored weapons of pharaohs and kings throughout the ancient world, from Greece to China. Large numbers of chariots, in the dozens or even hundreds, are mentioned in palace inventories of military equipment, in descriptions of battles, and in proud boasts of loot taken in warfare. After 800 BCE chariots were gradually abandoned as they became vulnerable to a new kind of warfare conducted by disciplined troops of mounted archers, the earliest cavalry. If Indo-European speakers were the first to have chariots, this could explain their early expansion; if they were the first to domesticate horses, then this could explain the central role horses played as symbols of strength and power in the rituals of the Old Indic Aryans, Greeks, Hittites, and other Indo-European speakers.

But until recently it has been difficult or impossible to determine when and where horses were domesticated. Early horse domestication left very few marks on the equine skeleton, and all we have left of ancient horses is their bones. For more than ten years I have worked on this problem with my research partner, and also my wife, Dorcas Brown, and we believe we now know where and when people began to keep herds of tamed horses. We also think that horseback riding began in the steppes long before chariots were invented, in spite of the fact that chariotry preceded cavalry in the warfare of the organized states and kingdoms of the ancient world.

LANGUAGE EXTINCTION AND THOUGHT

The people who spoke the Proto-Indo-European language lived at a critical time in a strategic place. They were positioned to benefit from innovations in transport, most important of these the beginning of horseback riding and the invention of wheeled vehicles. They were in no way superior to their neighbors; indeed, the surviving evidence suggests that their economy, domestic technology, and social organization were simpler than those of their western and southern neighbors. The expansion of their language was not a single event, nor did it have only one cause.

Nevertheless, that language did expand and diversify, and its daughters—including English—continue to expand today. Many other language families have become extinct as Indo-European languages spread. It is possible that the resultant loss of linguistic diversity has narrowed and channeled habits of perception in the modern world. For example, all Indo-European languages force the speaker to pay attention to tense and number when talking about an action: you *must* specify whether the action is past, present, or future; and you *must* specify whether the actor is singular or plural. It is impossible to use an Indo-European verb without deciding on these categories. Consequently speakers of Indo-European languages habitually frame all events in terms of when they occurred and whether they involved multiple actors. Many other language families do not *require* the speaker to address these categories when speaking of an action, so tense and number can remain unspecified.

On the other hand, other language families require that other aspects of reality be constantly used and recognized. For example, when describing an event or condition in Hopi you *must* use grammatical markers that specify whether you witnessed the event yourself, heard about it from someone else, or consider it to be an unchanging truth. Hopi speakers are forced by Hopi grammar to habitually frame all descriptions of reality in terms of the source and reliability of their information. The constant and automatic use of such categories generates habits in the perception and framing of the world that probably differ between people who use fundamentally different grammars.[14] In that sense, the spread of Indo-European grammars has perhaps reduced the diversity of human perceptual habits. It might also have caused this author, as I write this book, to frame my observations in a way that repeats the perceptual habits and categories of a small group of people who lived in the western Eurasian steppes more than five thousand years ago.

CHAPTER TWO

How to Reconstruct a Dead Language

Proto-Indo-European has been dead as a spoken language for at least forty-five hundred years. The people who spoke it were nonliterate, so there are no inscriptions. Yet, in 1868, August Schleicher was able to tell a story in reconstructed Proto-Indo-European, called "The Sheep and the Horses," or *Avis akvasas ka*. A rewrite in 1939 by Herman Hirt incorporated new interpretations of Proto-Indo-European phonology, and the title became *Owis ek'woses-k^we*. In 1979 Winfred Lehmann and Ladislav Zgusta suggested only minor new changes in their version, *Owis ekwosk^we*. While linguists debate increasingly minute details of pronunciation in exercises like these, most people are amazed that anything can be said about a language that died without written records. Amazement, of course, is a close cousin of suspicion. Might the linguists be arguing over a fantasy? In the absence of corroborative evidence from documents, how can linguists be sure about the accuracy of reconstructed Proto-Indo-European?[1]

Many archaeologists, accustomed to digging up real things, have a low opinion of those who merely reconstruct hypothetical phonemes—what is called "linguistic prehistory." There are reasons for this skepticism. Both linguists and archaeologists have made communication across the disciplines almost impossible by speaking in dense jargons that are virtually impenetrable to anyone but themselves. Neither discipline is at all simple, and both are riddled with factions on many key questions of interpretation. Healthy disagreement can resemble confusion to an outsider, and most archaeologists, including this author, are outsiders in linguistics. Historical linguistics is not taught regularly in graduate archaeology programs, so most archaeologists know very little about the subject. Sometimes we make this quite clear to linguists. Nor is archaeology taught to graduate students in linguistics. Linguists' occasional remarks about archaeology can sound simplistic and naïve to archaeologists, making some of us suspect that the entire field of historical linguistics may be riddled with simplistic and naïve assumptions.

The purpose of these first few chapters is to clear a path across the no-man's land that separates archaeology and historical linguistics. I do this with considerable uncertainty—I have no more formal training in linguistics than most archaeologists. I am fortunate that a partial way has already been charted by Jim Mallory, perhaps the only doubly qualified linguist-archaeologist in Indo-European studies. The questions surrounding Indo-European origins are, at their core, about linguistic evidence. The most basic linguistic problem is to understand how language changes with time.[2]

LANGUAGE CHANGE AND TIME

Imagine that you had a time machine. If you are like me, there would be many times and places that you would like to visit. In most of them, however, no one spoke English. If you could not afford the Six-Month-Immersion Trip to, say, ancient Egypt, you would have to limit yourself to a time and place where you could speak the language. Consider, perhaps, a trip to England. How far back in time could you go and still be understood? Say we go to London in the year 1400 CE.

As you emerge from the time machine, a good first line to speak, something reassuring and recognizable, might be the opening line of the Lord's Prayer. The first line in a conservative, old-fashioned version of Modern Standard English would be, "*Our Father, who is in heaven, blessed be your name.*" In the English of 1400, as spoken by Chaucer, you would say, "*Oure fadir that art in heuenes, halwid be thy name.*" Now turn the dial back another four hundred years to 1000 CE, and in Old English, or Anglo-Saxon, you would say, "*Fæader ure thu the eart on heofonum, si thin nama gehalgod.*" A chat with Alfred the Great would be out of the question.

Most normal spoken languages over the course of a thousand years undergo enough change that speakers at either end of the millennium, attempting a conversation, would have difficulty understanding each other. Languages like Church Latin or Old Indic (the oldest form of Sanskrit), frozen in ritual, would be your only hope for effective communication with people who lived more than a thousand years ago. Icelandic is a frequently cited example of a spoken language that has changed little in a thousand years, but it is spoken on an island isolated in the North Atlantic by people whose attitude to their old sagas and poetry has been one approaching religious reverence. Most languages undergo significantly more changes than Icelandic over far fewer than a thousand years for two reasons: first, no two people speak the same language exactly alike; and, second, most people meet a lot more people who speak differently than do the Icelanders. A language that borrows many words and phrases from another language changes more rapidly than one with a low borrowing rate. Icelandic has one of the lowest borrowing rates in the world.[3] If we are exposed to a number of different ways of speaking, our own way of speaking is likely to change more rapidly. Fortunately, however, although the speed of language change is quite variable, the structure and sequence of language change is not.

Language change is not random; it flows in the direction of accents and phrases admired and emulated by large numbers of people. Once a target accent is selected, the structure of the sound changes that moves the speaker away from his own speech to the target is governed by rules. The same rules apparently exist in all our minds, mouths, and ears. Linguists just noticed them first. If rules define how a given innovation in pronunciation affects the old speech system—if sound shifts are predictable—then we should be able to play them backward, in effect, to hear earlier language states. That is more or less how Proto-Indo-European was reconstructed.

Most surprising about sound change is its regularity, its conformation to rules no one knows consciously. In early Medieval French there probably was a time when *tsent'm* 'hundred' was heard as just a dialectical pronunciation of the Latin word *kentum* 'hundred'. The differences in sound between the two were *allophones*, or different sounds that did not create different meanings. But because of other changes in how Latin was spoken, [ts-] began to be heard as a different sound, a phoneme distinct from [k-] that could change the meaning of a word. At that point people had to decide whether *kentum* was pronounced with a [k-] or a [ts-]. When French speakers decided to use [ts-], they did so not just for the word *kentum* but in every word where Latin had the sound *k-* before a front vowel like *-e-*. And once this happened, *ts-* became confused with initial *s-*, and people had to decide again whether *tsentum* was pronounced with a [ts-] or [s-]. They chose [s-]. This sequence of shifts dropped below the level of consciousness and spread like a virus through all pre-French words with analogous sequences of sounds. Latin *cera* 'wax', pronounced [kera], became French *cire*, pronounced [seer]; and Latin *civitas* 'community', pronounced [kivitas], became French *cité*, pronounced [seetay]. Other sound changes happened, too, but they all followed the same unspoken and unconscious rules—the sound shifts were not idiosyncratic or confined to certain words; rather, they spread systematically to all similar sounds in the language. Peoples' ears were very discriminating in identifying words that fit or did not fit the analogy. In words where the Latin *k-* was followed by a *back* vowel like *-o* it remained a *k-*, as in Latin *costa*>French *côte*.

Sound changes are rule-governed probably because all humans instinctively search for order in language. This must be a hard-wired part of all human brains. We do it without committee meetings, dictionaries, or even literacy, and we are not conscious of what we are doing (unless we are linguists). Human language is defined by its rules. Rules govern sentence construction (syntax), and the relationship between the sounds of words (phonology and morphology) and their meaning. Learning these rules changes our awareness from that of an infant to a functioning member of the human tribe. Because language is central to human evolution, culture, and social identity, each member of the tribe is biologically equipped to cooperate in converting novel changes into regular parts of the language system.[4]

Historical linguistics was created as a discipline in the nineteenth century, when scholars first exposed and analyzed the rules we follow when speaking and listening. I do not pretend to know these rules adequately, and if I did I would not try to explain them all. What I hope to do is indicate, in a general way, how some of them work so that we can use the "reconstructed vocabulary" of Proto-Indo-European with some awareness of its possibilities and limitations.

We begin with phonology. Any language can be separated into several interlocking systems, each with its own set of rules. The vocabulary, or *lexicon*, composes one system; *syntax*, or word order, and sentence construction compose another; *morphology*, or word form, including much of what is called "grammar" is the third; and *phonology*, or the rules about which sounds are acceptable and meaningful, is the fourth. Each system has its own peculiar tendencies, although a change in one (say, phonology) can bring about changes in another (say, morphology).[5] We will look most closely at phonology and the lexicon, as these are the most important in understanding how the Proto-Indo-European vocabulary has been reconstructed.

Phonology: How to Reconstruct a Dead Sound

Phonology, or the study of linguistic sounds, is one of the principal tools of the historical linguist. Phonology is useful as a historical tool, because the sounds people utter tend to change over time in certain directions and not in others.

The direction of phonetic change is governed by two kinds of constraints: those that are generally applicable across most languages, and those specific to a single language or a related group of languages. General constraints are imposed by the mechanical limits of the human vocal anatomy, the need to issue sounds that can be distinguished and understood by listeners, and the tendency to simplify sound combinations that are difficult to pronounce. Constraints within languages are imposed by the limited range of sounds that are acceptable and meaningful for that language. Often these language-specific sounds are very recognizable. Comedians can make us laugh by speaking nonsense if they do it in the characteristic phonology of French or Italian, for example. Armed with a knowledge of both the *general* tendencies in the direction of phonetic change and the *specific* phonetic conventions within a given language group, a linguist can arrive at reliable conclusions about which phonetic variants are early pronunciations and which come later. This is the first step in reconstructing the phonological history of a language.

We know that French developed historically from the dialects of Latin spoken in the Roman province of Gaul (modern France) during the waning centuries of the Roman Empire around 300–400 CE. As late as the 1500s vernacular French suffered from low prestige among scholars, as it was considered nothing more than a corrupt form of Latin. Even if we knew nothing about that history, we could examine the Latin *centum* (pronounced [kentum]), and the French *cent* (pronounced [sohnt]), both meaning "hundred," and we could say that the sound of the Latin word makes it the older form, that the Modern French form could have developed from it according to known rules of sound change, and that an intermediate pronunciation, [tsohnt], probably existed before the modern form appeared—and we would be right.

Some Basic Rules of Language Change: Phonology and Analogy

Two general phonetic rules help us make these decisions. One is that initial hard consonants like *k* and hard *g* tend to change toward soft sounds like *s* and *sh* if they change at all, whereas a change from *s* to *k* would generally be unusual. Another is that a consonant pronounced as a stop in the back of the mouth (*k*) is particularly likely to shift toward the front of the mouth (*t* or *s*) in a word where it is followed by a vowel that is pronounced in the front of the mouth (*e*). Pronounce [ke-] and [se-], and note the position of your tongue. The *k* is pronounced by using the back of the tongue and both *e* and *s* are formed with the middle or the tip of the tongue, which makes it easier to pronounce the segment *se-* than the segment *ke-*. Before a front vowel like *-e* we might expect the *k-* to shift forward to [ts-] and then to [s-] but not the other way around.

This is an example of a general phonetic tendency called *assimilation*: one sound tends to assimilate to a nearby sound in the same word, simplifying the needed movements. The specific type of assimilation seen here is called *palatalization*—a back consonant (*k*) followed by a front vowel (*e*) was assimilated in French toward the front of the palate, changing the [k] to [s]. Between the Latin [k] (pronounced with the back of the tongue at the back of the palate) and the Modern French [s] (tip of the tongue at the front of the palate) there should have been an intermediate pronunciation *ts* (middle of the tongue at the middle of the palate). Such sequences permit historical linguists to reconstruct undocumented intermediate stages in the evolution of a language. Palatalization has been systematic in the development of French from Latin. It is responsible for much of the distinctive phonology of the French language.

Assimilation usually changes the quality of a sound, or sometimes removes sounds from words by slurring two sounds together. The opposite process is the *addition* of new sounds to a word. A good example of an innovation of this kind is provided by the variable pronunciations of the word *athlete* in English. Many English speakers insert [-uh] in the middle of the word, saying [ath-uh-lete], but most are not aware they are doing so. The inserted syllable always is pronounced precisely the same way, as [-uh], because it assimilates to the tongue position required to pronounce the following *-l*. Linguists could have predicted that some speakers would insert a vowel in a difficult cluster of consonants like *-thl* (a phenomenon called *epenthesis*) and that the vowel inserted in *athlete* always would be pronounced [-uh] because of the rule of assimilation.

Another kind of change is *analogical* change, which tends to affect grammar quite directly. For example, the *-s* or *-es* ending for the plural of English nouns was originally limited to one class of Old English nouns: *stān* for *stone* (nominative singular), *stānas* for *stones* (nominative plural). But when a series of sound changes (see note 5) resulted in the loss of the phonemes that had once distinguished nouns of different classes, the *-s* ending began to be reinterpreted as a *general* plural indicator and was attached to all nouns. Plurals formed with *-n* (oxen), with a zero change (sheep), and with a vowel change in the stem (women) remain as relics of Old English, but the shift to *-s* is driving out such "irregular" forms and has been doing so for eight hundred years. Similar analogical changes have affected verbs: *help/helped* has replaced Old English *help/holp* as the *-ed* ending has been reinterpreted as a general ending for the past tense, reducing the once large number of strong verbs that formed their past with a vowel change. Analogical changes can also create new words or forms by analogy with old ones. Words formed with *-able* and *-scape* exist in such great numbers in English because these endings, which were originally bound to specific words (*measurable, landscape*), were reinterpreted as suffixes that could be removed and reattached to any stem (*touchable, moonscape*).

Phonological and analogical change are the internal mechanisms through which novel forms are incorporated into a language. By examining a sequence of documents within one language lineage from several different points in the past—inscriptions in, say, classical Latin, late vulgar Latin, early Medieval French, later Medieval French, and modern French—linguists have defined virtually all the sound changes and analogical shifts in the evolution of French from Latin. Regular, systematic rules, applicable also to other cases of language change in other languages, explain most of these shifts. But how do linguists replay these shifts "backward" to discover the origins of modern languages? How can we reconstruct the sounds of a language like Proto-Indo-European, for which there are *no* documents, a language spoken before writing was invented?

"Hundred": An Example of Phonetic Reconstruction

Proto-Indo-European words were not reconstructed to create a dictionary of Proto-Indo-European vocabulary, although they are extraordinarily useful in this way. The real aim in reconstruction is to prove that a list of daughter terms are cognates, descended from the same mother term. The reconstruction of the mother term is a by-product of the comparison, the proof that every sound in every daughter word can be derived from a sound in the common parent. The first step is to gather up the suspected daughters: you must make a list of all the variants of the word you can find in the Indo-European languages (table 2.1). You have to know the rules of phonological change to do even this successfully, as some variants of the word might have changed radically in sound. Just recognizing the candidates and making up a good list can be a challenge. We will try this with the Proto-Indo-European word for "hundred." The Indo-European roots for numbers, especially 1 to 10, 100, and 1,000, have been retained in almost all the Indo-European daughters.

Our list includes Latin *centum*, Avestan *satəm*, Lithuanian *šimtas*, and Old Gothic *hunda-* (a root much like *hunda-* evolved into the English word *hundred*). Similar-looking words meaning "hundred" in other

TABLE 2.1

Indo-European Cognates for the Root "Hundred"

Branch	Language	Term	Meaning
Celtic	Welsh	cant	hundred
	Old Irish	cēt	hundred
Italic	Latin	centum	hundred
Tocharian	TochA	känt	hundred
	TochB	kante	hundred
Greek	Greek	ἑκατόν	hundred
Germanic	Old English	hund	hundred
	OldHighGerm.	hunt	hundred
	Gothic	hunda	100, 120
	OldSaxon	hunderod	(long) hundred
Baltic	Lithuanian	šimtas	hundred
	Latvian	simts	hundred
Slavic	OldChurchSlav.	sŭto	hundred
	Bulgarian	sto	hundred
Anatolian	Lycian	sñta	unit of 10 or 100
Indo-Iranian	Avestan	satəm	hundred
	OldIndic	śatám	hundred

Indo-European languages should be added, and I have already referred to the French word *cent*, but I will use only four for simplicity's sake. The four words I have chosen come from four Indo-European branches: Italic, Indo-Iranian, Baltic, and Germanic.

The question we must answer is this: Are these words phonetically transformed daughters of a single parent word? If the answer is yes, they are cognates. To prove they are cognates, we must be able to reconstruct an ancestral sequence of phonemes that could have developed into all the documented daughter sounds through known rules. We start with the first sound in the word.

The initial [k] phoneme in Latin *centum* could be explained if the parent term began with a [k] sound as well. The initial soft consonants ([s] [sh])

in Avestan *satəm* and Lithuanian *šimtas* could have developed from a Proto-Indo-European word that began with a hard consonant [k], like Latin *centum*, since hard sounds generally tend to shift toward soft sounds if they change at all. The reverse development ([s] or [sh] to [k]) would be very unlikely. Also, palatalization and sibilation (shifting to a 's' or 'sh' sound) of initial hard consonants is expected in both the Indic branch, of which Vedic Sanskrit is a member; and the Baltic branch, of which Lithuanian is a member. The general direction of sound change and the specific conventions in each branch permit us to say that the Proto-Indo-European word from which all three of these developed could have begun with 'k'.

What about *hunda*? It looks quite different but, in fact, the *h* is expected—it follows a rule that affected all initial [k] sounds in the Germanic branch. This shift involved not just *k* but also eight other consonants in Pre-Germanic.[6] The consonant shift spread throughout the prehistoric Pre-Germanic language community, giving rise to a new Proto-Germanic phonology that would be retained in all the later Germanic languages, including, ultimately, English. This consonant shift was described by and named after Jakob Grimm (the same Grimm who collected fairy tales) and so is called Grimm's Law. One of the changes described in Grimm's Law was that the archaic Indo-European sound [k] shifted in most phonetic environments to Germanic [h]. The Indo-European *k* preserved in Latin *centum* shifted to *h* in Old Gothic *hunda-*; the initial *k* seen in Latin *caput* 'head' shifted to *h* in Old English *hafud* 'head'; and so on throughout the vocabulary. (*Caput* > *hafud* shows that *p* also changed to *f*, as in *pater* > *fater*). So, although it looks very different, *hunda-* conforms: its first consonant can be derived from *k* by Grimm's Law.

The first sound in the Proto-Indo-European word for "hundred" probably was *k*. (An initial [k] sound satisfies the other Indo-European cognates for "hundred" as well.)[7] The second sound should have been a vowel, but which vowel?

The second sound was a vowel that does not exist in English. In Proto-Indo-European resonants could act as vowels, similar to the resonant *n* in the colloquial pronunciation of *fish'n'* (as in *Bob's gone fish'n'*). The second sound was a resonant, either **m* or **n*, both of which occur among the daughter terms being compared. (An asterisk is used before a reconstructed form for which there is no direct evidence.) *M* is attested in the Lithuanian cognate *šimtas*. An *m* in the Proto-Indo-European parent could account for the *m* in Lithuanian. It could have changed to *n* in Old Indic, Germanic, and other lineages by assimilating to the following *t* or *d*, as both *n* and *t* are articulated on the teeth. (Old Spanish *semda* 'path' changed to modern Spanish *senda* for the same reason.) A shift from an original *m* to an *n* before a *t* is explicable, but a shift from an original *n* to an *m* is much less likely. Therefore, the original second sound probably was ṃ. This consonant could have been lost entirely in Sanskrit *satam* by yet another assimilative tendency called total assimilation: after the *m* changed to *n*, giving **santam*, the *n* was completely assimilated to the following *t*, giving *satam*. The same process was responsible for the loss of the [k] sound in the shift from Latin *octo* to modern Italian *otto* 'eight'.

I will stop here, with an ancestral **k'ṃ -*, in my discussion of the Proto-Indo-European ancestor of *centum*. The analysis should continue through the phonemes that are attested in all the surviving cognates to reconstruct an acceptable ancestral root. By applying such rules to all the cognates, linguists have been able to reconstruct a Proto-Indo-European sequence of phonemes, **k'ṃtom*, that could have developed into all the attested phonemes in all the attested daughter forms. The Proto-Indo-European root **k'ṃtom* is the residue of a successful comparison—it is the proof that the daughter terms being compared are indeed cognates. It is also likely to be a pretty good approximation of the way this word was pronounced in at least some dialects of Proto-Indo-European.

The Limitations and Strengths of Reconstruction

The comparative method will produce the *sound* of the ancestral root and confirm a genetic relationship *only* with a group of cognates that has evolved regularly according to the rules of sound change. The result of a comparative analysis is either a demonstration of a genetic connection, if every phoneme in every cognate can be derived from a mutually acceptable parental phoneme; or no *demonstrable* connection. In many cases sounds may have been borrowed into a language from a neighboring language, and those sounds might replace the predicted shifts. The comparative method cannot force a regular reconstruction on an irregular set of sounds. Much of the Proto-Indo-European vocabulary, perhaps most of it, never will be reconstructed. Regular groups of cognates permit us to reconstruct a Proto-Indo-European root for the word *door* but not for *wall*; for *rain* but not for *river*; for *foot* but not for *leg*. Proto-Indo-European certainly had words for these things, but we cannot safely reconstruct how they sounded.

The comparative method cannot prove that two words are *not* related, but it can fail to produce proof that they *are*. For example, the Greek god Ouranos and the Indic deity Varuna had strikingly similar mythological attributes, and their names sound somewhat alike. Could Ouranos and

Varuna be reflexes of the name of some earlier Proto-Indo-European god? Possibly—but the two names cannot be derived from a common parent by the rules of sound change known to have operated in Greek and Old Indic. Similarly Latin *deus* (god) and Greek *théos* (god) look like obvious cognates, but the comparative method reveals that Latin *deus*, in fact, shares a common origin with Greek *Zéus*.[8] If Greek *théos* were to have a Latin cognate it should begin with an [f] sound (*festus* 'festive' has been suggested, but some of the other sounds in this comparison are problematic). It is still possible that *deus* and *théos* were historically related in some irregular way, but we cannot prove it.

In the end, how can we be sure that the comparative method accurately reconstructs undocumented stages in the phonological history of a language? Linguists themselves are divided on the question of the "reality" of reconstructed terms.[9] A reconstruction based on cognates from eight Indo-European branches, like *k'mtom-*, is much more reliable and probably more "true" than one based on cognates in just two branches. Cognates in at least three branches, including an ancient branch (Anatolian, Greek, Avestan Iranian, Old Indic, Latin, some aspects of Celtic) should produce a reliable reconstruction. But how reliable? One test was conceived by Robert A. Hall, who reconstructed the shared parent of the Romance languages using just the rules of sound change, and then compared his reconstruction to Latin. Making allowances for the fact that the actual parents of the Romance languages were several provincial Vulgar Latin dialects, and the Latin used for the test was the classical Latin of Cicero and Caesar, the result was reassuring. Hall was even able to reconstruct a contrast between two sets of vowels although none of the modern daughters had retained it. He was unable to identify the feature that distinguished the two vowel sets as length—Latin had long vowels and short vowels, a distinction lost in all its Romance daughters—but he was able to rebuild a system with two contrasting sets of vowels and many of the other, more obvious aspects of Latin morphology, syntax, and vocabulary. Such clever exercises aside, the best proof of the realism of reconstruction lies in several cases where linguists have suggested a reconstruction and archaeologists have subsequently found inscriptions that proved it correct.[10]

For example, the oldest recorded Germanic cognates for the word *guest* (Gothic *gasts*, Old Norse *gestr*, Old High German *gast*) are thought to be derived from a reconstructed late Proto-Indo-European *ghos-ti-* (which probably meant both "host" and "guest" and thus referred to a relationship of hospitality between strangers rather than to one of its roles) through a Proto-Germanic form reconstructed as *gastiz*. None of the known forms of the word in the later Germanic languages contained the *i* before the final consonant, but rules of sound change predicted that the *i* should theoretically have been there in Proto-Germanic. Then an archaic Germanic inscription was found on a gold horn dug from a grave in Denmark. The inscription *ek hlewagastiz holitijaz* (or *holtingaz*) *horna tawido* is translated "I, Hlewagasti of Holt (or Holting) made the horn." It contained the personal name Hlewagastiz, made up of two stems, *Hlewa-* 'fame' and *gastiz* 'guest'. Linguists were excited not because the horn was a beautiful golden artifact but because the stem contained the predicted *i*, verifying the accuracy of both the reconstructed Proto-Germanic form and its late Proto-Indo-European ancestor. Linguistic reconstruction had passed a real-world test.

Similarly linguists working on the development of the Greek language had proposed a Proto-Indo-European labiovelar $*k^w$ (pronounced [kw-]) as the ancestral phoneme that developed into Greek *t* (before a front vowel) or *p* (before a back vowel). The reconstruction of $*k^w$ was a reasonable but complex solution for the problem of how the Classical Greek consonants were related to their Proto-Indo-European ancestors. It remained entirely theoretical until the discovery and decipherment of the Mycenaean Linear B tablets, which revealed that the earliest form of Greek, Mycenaean, had the predicted k^w where later Greek had *t* or *p* before front and back vowels.[11] Examples like these confirm that the reconstructions of historical linguistics are more than just abstractions.

A reconstructed term is, of course, a phonetic idealization. Reconstructed Proto-Indo-European cannot capture the variety of dialectical pronunciations that must have existed more than perhaps one thousand years when the language was living in the mouths of people. Nevertheless, it is a remarkable victory that we can now pronounce, however stiffly, thousands of words in a language spoken by nonliterate people before 2500 BCE.

The Lexicon: How to Reconstruct Dead Meanings

Once we have reconstructed the *sound* of a word in Proto-Indo-European, how do we know what it *meant*? Some archaeologists have doubted the reliability of reconstructed Proto-Indo-European, as they felt that the original meanings of reconstructed terms could never be known confidently.[12] But we can assign reliable meanings to many reconstructed Proto-Indo-European terms. And it is in the meanings of their words that we find the best evidence for the material culture, ecological environment, social relations, and spiritual beliefs of the speakers of Proto-Indo-European. Every meaning is worth the struggle.

Three general rules guide the assignment of meaning. First, look for the most ancient meanings that can be found. If the goal is to retrieve the meaning of the original Proto–Indo–European word, modern meanings should be checked against meanings that are recorded for ancient cognates.

Second, if one meaning is consistently attached to a cognate in all language branches, like *hundred* in the example I have used, that is clearly the least problematic meaning we can assign to the original Proto-Indo-European root. It is difficult to imagine how that meaning could have become attached to all the cognates unless it were the meaning attached to the ancestral root.

Third, if the word can be broken down into roots that point to the same meaning as the one proposed, then that meaning is doubly likely. For example, Proto-Indo-European **k'ṃtom* probably was a shortened version of **dek'ṃtom*, a word that included the Proto-Indo-European root **dek'ṃ* 'ten'. The sequence of sounds in **dek'ṃ* was reconstructed independently using the cognates for the word *ten*, so the fact that the reconstructed roots for *ten* and *hundred* are linked in both meaning and sound tends to verify the reliability of both reconstructions. The root **k'ṃtom* turns out to be not just an arbitrary string of Proto-Indo-European phonemes but a meaningful compound: "(a unit) of tens." This also tells us that the speakers of Proto-Indo-European had a decimal numbering system and counted to one hundred by tens, as we do.

In most cases the meaning of a Proto-Indo-European word changed and drifted as the various speech communities using it became separated, centuries passed, and daughter languages evolved. Because the association between word and meaning is arbitrary, there is less regular directionality to change in meaning than there is in sound change (although some semantic shifts are more probable than others). Nevertheless, general meanings can be retrieved. A good example is the word for "wheel."

"Wheel": An Example of Semantic Reconstruction

The word *wheel* is the modern English descendant of a PIE root that had a sound like **kʷékʷlos* or **kʷekʷlós*. But what, exactly, did **kʷékʷlos* mean in Proto-Indo-European? The sequence of phonemes in the root **kʷékʷlos* was pieced together by comparing cognates from eight old Indo-European languages, representing five branches. Reflexes of this word survived in Old Indic and Avestan (from the Indo-Iranian branch), Old Norse and

Old English (from the Germanic branch), Greek, Phrygian, and Tocharian A and B. The meaning "wheel" is attested for the cognates in Sanskrit, Avestan, Old Norse, and Old English. The meaning of the Greek cognate had shifted to "circle" in the singular but in the plural still meant "wheels." In Tocharian and Phrygian the cognates meant "wagon" or "vehicle." What was the original meaning? (table 2.2).

Five of the eight *$k^w ék^w los$* cognates have "wheel" or "wheels" as an attested meaning, and in those languages (Phrygian, Greek, Tocharian A & B) where the meaning drifted away from "wheel(s)," it had not drifted far ("circle," "wagon," or "vehicle"). Moreover, the cognates that preserve the meaning "wheel" are found in languages that are geographically isolated from one another (Old Indic and Avestan in Iran were neighbors, but neither had any known contact with Old Norse or Old English). The meaning "wheel" is unlikely to have been borrowed into Old Norse from Old Indic, or vice versa.

Some shifts in meaning are unlikely, and others are common. It is common to name a whole ("vehicle," "wagon") after one of its most characteristic parts ("wheels"), as seems to have happened in Phrygian and Tocharian. We do the same in modern English slang when we speak of someone's car as their "wheels," or clothing as their "threads." A shift in meaning in the other direction, using a word that originally referred to the whole to refer to one of its parts (using *wagon* to refer to *wheel*), is much less probable.

The meaning of *wheel* is given additional support by the fact that it has an Indo-European etymology, like the root for *$k'ṃtom$*. It was a word created from another Indo-European root. That root was *$k^w el$–*, a verb that meant "to turn." So *$k^w ék^w los$* is not just a random string of phonemes reconstructed from the cognates for *wheel*; it meant "the thing that turns." This not only tends to confirm the meaning "wheel" rather than "circle" or "vehicle" but it also indicates that the speakers of Proto-Indo-European made up their own words for wheels. If they learned about the invention of the wheel from others they did not adopt the foreign name for it, so the social setting in which the transfer took place probably was brief, between people who remained socially distant. The alternative, that wheels were invented within the Proto-Indo-European language community, seems unlikely for archaeological and historical reasons, though it remains possible (see chapter 4).

One more rule helps to confirm the reconstructed meaning. If it fits within a semantic field consisting of other roots with closely related reconstructed meanings, we can at least be relatively confident that such a word

TABLE 2.2
Proto-Indo-European Roots for Words Referring to Parts of a Wagon

PIE Root Word	Wagon Part	Daughter Languages
*k^wek^wlos	(wheel)	*Old Norse* hvēl 'wheel'; *Old English* hweohl 'wheel'; *Middle Dutch* wiel 'wheel'; *Avestan Iranian* čaxtra- 'wheel'; *Old Indic* cakrá 'wheel, Sun disc'; *Greek* kuklos 'circle' and kukla (plural) 'wheels'; *Tocharian A* kukal 'wagon'; *Tocharian B* kokale 'wagon'
*rot-eh$_2$-	(wheel)	*Old Irish* roth 'wheel'; *Welsh* rhod 'wheel'; *Latin* rota 'wheel'; *Old High German* rad 'wheel'; *Lithuanian* rātas 'wheel'; *Latvian* rats 'wheel' and rati (plural) 'wagon'; *Albanian* rreth 'ring, hoop, carriage tire'; *Avestan Iranian* ratha 'chariot, wagon'; *Old Indic* rátha 'chariot, wagon'
*ak*s-, or	(axle)	*Latin* axis 'axle, axis'; *Old English* eax 'axle'; *Old High German* *h$_a$ek*s- ahsa 'axle'; *Old Prussian* assis 'axle'; *Lithuanian* ašís 'axle'; *Old Church Slavonic* osĭ 'axle'; *Mycenaean Greek* a-ko-so-ne 'axle'; *Old Indic* áks*a 'axle'
*ei-/*oi-, or	(thill)	*Old English* ār- 'oar'; *Russian* vojë 'shaft'; *Slovenian* oje 'shaft'; *Hittite* h$_2$ih$_3$s or hišša- 'pole, harnessing shaft'; *Greek* oisioi* 'tiller, rudderpost'; *Avestan Iranian* aēša 'pair of shafts, plow-pole'; *Old Indic* is*a 'pole, shaft'
*wéĝheti-	(ride)	*Welsh* amwain 'drive about'; *Latin* vehō 'bear, convey'; *Old Norse* vega 'bring, move'; *Old High German* wegan 'move, weigh'; *Lithuanian* vežù 'drive'; *Old Church Slavonic* vezǫ 'drive'; *Avestan Iranian* vazaiti 'transports, leads'; *Old Indic* váhati 'transports, carries, conveys'. Derivative nouns have the meaning "wagon" in *Greek*, *Old Irish*, *Welsh*, *Old High German*, and *Old Norse*.

could have existed in Proto-Indo-European. "Wheel" is part of a semantic field consisting of *words for the parts of a wagon or cart* (table 2.2). Happily, at least four other such words can be reconstructed for Proto-Indo-European. These are:

1. *$rot-eh_2$-, a second term for "wheel," with cognates in Old Indic and Avestan that meant "chariot," and cognates that meant "wheel" in Latin, Old Irish, Welsh, Old High German, and Lithuanian.
2. *aks- (or perhaps *h_2eks-) 'axle' attested by cognates that had not varied in meaning over thousands of years, and still meant "axle" in Old Indic, Greek, Latin, Old Norse, Old English, Old High German, Lithuanian, and Old Church Slavonic.
3. *h_2ih_3s- 'thill' (the harness pole) attested by cognates that meant "thill" in Hittite and Old Indic.
4. *wégheti, a verb meaning "to convey or go in a vehicle," attested by cognates carrying this meaning in Old Indic, Avestan, Latin, Old English, and Old Church Slavonic and by cognate-derived nouns ending in *-no- meaning "wagon" in Old Irish, Old English, Old High German, and Old Norse.

These four additional terms constitute a well-documented semantic field (*wheel, axle, thill,* and *wagon* or *convey in a vehicle*) that increases our confidence in reconstructing the meaning "wheel" for *$k^wék^wlos$. Of the five terms assigned to this semantic field, all but *thill* have clear Indo-European etymologies in independently reconstructed roots. The speakers of Proto-Indo-European were familiar with wheels and wagons, and used words of their own creation to talk about them.

Fine distinctions, shades of meaning, and the word associations that enriched Proto-Indo-European poetry may be forever lost, but gross meanings are recoverable for at least fifteen hundred Proto-Indo-European roots such as *dekm- 'ten', and for additional thousands of other words derived from them, such as *kṃtom- 'hundred'. Those meanings provide a window into the lives and thoughts of the speakers of Proto-Indo-European.

Syntax and Morphology: The Shape of a Dead Language

I will not try to describe in any detail the grammatical connections between the Indo-European languages. The reconstructed vocabulary is most important for our purposes. But grammar, the bedrock of language classification, provides the primary evidence for classifying languages and determining relationships between them. Grammar has two aspects: *syntax*, or the rules governing the order of words in sentences; and *morphology*, or the rules governing the forms words must take when used in particular ways.

Proto-Indo-European grammar has left its mark on all the Indo-European languages to one degree or another. In all the Indo-European language branches, nouns are declined; that is, the noun changes form depending on how it is used in a sentence. English lost most of these declinations during its evolution from Anglo-Saxon, but all the other languages in the Germanic branch retain them, and we have kept some use-dependent pronouns (*masculine*: he, his, him/*feminine*: she, hers, her). Moreover, most Indo-European nouns are declined in similar ways, with endings that are genetically cognate, and with the same formal system of cases (nominative, genitive, accusative, etc.) that intersect in the same way with the same three gender classes (masculine, feminine, neuter); and with similar formal classes, or declensions, of nouns that are declined in distinctive ways. Indo-European verbs also share similar conjugation classes (first person, second person or familiar, third person or formal, singular, plural, past tense, present tense, etc.), similar stem alterations (run-ran, give-gave), and similar endings. This particular constellation of formal categories, structures, transformations, and endings is not at all necessary or universal in human language. It is unique, as a system, and is found only in the Indo-European languages. The languages that share this grammatical system certainly are daughters of a single language from which that system was inherited.

One example shows how unlikely it would be for the Indo-European languages to share these grammatical structures by random chance. The verb *to be* has one form in the first-person singular ([I] *am*) and another in the third-person singular ([he/she/it] *is*). Our English verbs are descended from the archaic Germanic forms *im* and *ist*. The Germanic forms have exact, proven cognates in Old Indic *ásmi* and *ásti*; in Greek *eimí* and *estí*; and in Old Church Slavonic *jesmĭ* and *jestŭ*. All these words are derived from a reconstructable Proto-Indo-European pair, $*h_1e'smi$ and $*h_1e'sti$. That all these languages share the same system of verb classes (first person, second person or familiar, and third person), and that they use the same basic roots and endings to identify those classes, confirms that they are genetically related languages.

Conclusion: Raising a Language from the Dead

It will always be difficult to work with Proto-Indo-European. The version we have is uncertain in many morphological details, phonetically idealized, and fragmentary, and can be difficult to decipher. The meanings of some terms will never be fully understood, and for others only an approximate definition is possible. Yet reconstructed Proto-Indo-European captures key parts of a language that actually existed.

Some dismiss reconstructed Proto-Indo-European as nothing more than a hypothesis. But the limitations of Proto-Indo-European apply equally to the written languages of ancient Egypt and Mesopotamia, which are universally counted among the great treasures of antiquity. No curator of Assyrian records would suggest that we should discard the palace archives of Nineveh because they are incomplete, or because we cannot know the exact sound and meaning of many terms, or because we are uncertain about how the written court language related to the 'real' language spoken by the people in the street. Yet these same problems have convinced many archaeologists that the study of Proto-Indo-European is too speculative to yield any real historical value.

Reconstructed Proto-Indo-European is a long, fragmentary list of words used in daily speech by people who created no other texts. That is why it is important. The list becomes useful, however, only if we can determine where it came from. To do that we must locate the Proto-Indo-European homeland. But we cannot locate the Proto-Indo-European homeland until we first locate Proto-Indo-European in time. We have to know *when* it was spoken. Then it becomes possible to say where.

CHAPTER THREE

Language and Time 1
The Last Speakers of Proto-Indo-European

Time changes everything. Reading to my young children, I found that in mid-sentence I began to edit and replace words that suddenly looked archaic to me, in stories I had loved when I was young. The language of Robert Louis Stevenson and Jules Verne now seems surprisingly stiff and distant, and as for Shakespeare's English—we all need the glossary. What is true for modern languages was true for prehistoric languages. Over time, they changed. So what do we mean by Proto-Indo-European? If it changed over time, is it not a moving target? However we define it, for how long was Proto-Indo-European spoken? Most important, when was it spoken? How do we assign a date to a language that left no inscriptions, that died without ever being written down? It helps to divide any problem into parts, and this one can easily be divided into two: the birth date and the death date.

This chapter concentrates on the death date, the date *after which* Proto-Indo-European must have ceased to exist. But it helps to begin by considering how long a period probably preceded that. Given that the time between the birth and death dates of Proto-Indo-European could not have been infinite, precisely how long a time was it? Do languages, which are living, changing things, have life expectancies?

THE SIZE OF THE CHRONOLOGICAL WINDOW:
HOW LONG DO LANGUAGES LAST?

If we were magically able to converse with an English speaker living a thousand years ago, as proposed in the last chapter, we would not understand each other. Very few natural languages, those that are learned and spoken at home, remain sufficiently unchanged after a thousand years to be considered the "same language." How can the rate of change be measured?

Languages normally have dialects—regional accents—and, within any region, they have innovating social sectors (entertainers, soldiers, traders) and conservative sectors (the very rich, the very poor). Depending on who you are, your language might be changing very rapidly or very slowly. Unstable conditions—invasions, famines, the fall of old prestige groups and the rise of new ones—increase the rate of change. Some parts of language change earlier and faster, whereas other parts are resistant. That last observation led the linguist Morris Swadesh to develop a standard word list chosen from the most resistant vocabulary, a group of words that tend to be retained, not replaced, in most languages around the world, even after invasions and conquests. Over the long term, he hoped, the average rate of replacement in this resistant vocabulary might yield a reliable standardized measurement of the speed of language change, what Swadesh called *glottochronology*.[1]

Between 1950 and 1952 Swadesh published a hundred-word and a two-hundred-word *basic core vocabulary*, a standardized list of resistant terms. All languages, he suggested, tend to retain their own words for certain kinds of meanings, including body parts (blood, foot); lower numerals (one, two, three); some kinship terms (mother, father); basic needs (eat, sleep); basic natural features (sun, moon, rain, river); some flora and fauna (tree, domesticated animals); some pronouns (this, that, he, she); and conjunctions (and, or, if). The content of the list can be and has been modified to suit vocabularies in different languages—in fact, the preferred two-hundred-meaning list in English contains 215 words. The English core vocabulary has proven extremely resistant to change. Although English has borrowed more than 50% of its *general* vocabulary from the Romance languages, mainly from French (reflecting the conquest of Anglo-Saxon England by the French-speaking Normans) and Latin (from centuries of technical and professional vocabulary training in courts, churches, and schools), only 4% of the English *core* vocabulary is borrowed from Romance. In its core vocabulary English remains a Germanic language, true to its origins among the Anglo-Saxons who migrated from northern Europe to Britain after the fall of the Roman Empire.

Comparing core vocabularies between old and new phases in languages with long historical records (Old English/Modern English, Middle Egyptian/Coptic, Ancient Chinese/Modern Mandarin, Late Latin/Modern French, and nine other pairs), Swadesh calculated an average replacement rate of 14% per thousand years for the hundred-word list, and 19% per thousand years for the two-hundred-word list. He suggested that 19% was an acceptable average for all languages (usually rounded to 20%). To illustrate what that number means, Italian and French have distinct, unrelated words for 23% of the terms in the two-hundred-word list, and Spanish and Portuguese show a difference of 15%. As a general rule, if more than 10% of the core vocabulary is different between two dialects, they are either mutually unintelligible or approaching that state, that is, they are distinct languages or emerging languages. On average, then, with a replacement rate of 14–19% per thousand years in the core vocabulary, we should expect that most languages—including this one— would be incomprehensible to our own descendants a thousand years from now.

Swadesh hoped to use the replacement rate in the core vocabulary as a standardized clock to establish the date of splits and branches in unwritten languages. His own research involved the splits between American Indian language families in prehistoric North America, which were undatable by any other means. But the reliability of his standard replacement rate wilted under criticism. Extreme cases like Icelandic (very slow change, with a replacement rate of only 3–4% per thousand years) and English (very rapid, with a 26% replacement rate per thousand years) challenged the utility of the "average" rate.[2] The mathematics was affected if a language had multiple words for one meaning on the list. The dates given by glottochronology for many language splits contradicted known historical dates, generally by giving a date much later than it should have been. This direction in the errors suggested that real language change often was slower than Swadesh's model suggested—less than 19% per thousand years. A devastating critique of Swadesh's mathematics by Chretien, in 1962, seemed to drive a stake through the heart of glottochronology.

But in 1972 Chretien's critique was itself shown to be incorrect, and, since the 1980s, Sankoff and Embleton have introduced equations that include as critical values borrowing rates, the number of geographic borders with other languages, and a similarity index between the compared languages (because similar languages borrow in the core more easily then dissimilar languages). Multiple synonyms can each be given a fractional score. Studies incorporating these improved methods succeeded better in producing dates for splits between known languages that matched historical facts. More important, comparisons between most Indo-European languages still yielded replacement rates in the core vocabulary of about 10–20% per thousand years. Comparing the core vocabularies in ninety-five Indo-European languages, Kruskal and Black found that the most frequent date for the first splitting of Proto–Indo–European was about

3000 BCE. Although this estimate cannot be relied on absolutely, it is probably "in the ballpark" and should not be ignored.[3]

One simple point can be extracted from these debates: if the Proto-Indo-European core vocabulary changed at a rate ≥10% per millennium, or at the lower end of the expected range, Proto-Indo-European did not exist as a single language with a single grammar and vocabulary for as long as a thousand years. Proto-Indo-European grammar and vocabulary should have changed quite substantially over a thousand years. Yet the grammar of Proto-Indo-European, as reconstructed by linguists, is remarkably homogeneous both in morphology and phonology. Proto-Indo-European nouns and pronouns shared a set of cases, genders, and declensions that intersect with dozens of cognate phonological endings. Verbs had a shared system of tenses and aspects, again tagged by a shared set of phonological vowel changes (*run-ran*) and endings. This shared system of grammatical structures and phonological ways of labeling them looks like a single language. It suggests that reconstructed Proto-Indo-European probably refers to less than a thousand years of language change. It took less than a thousand years for late Vulgar Latin to evolve into seven Romance languages, and Proto-Indo-European does not contain nearly enough internal grammatical diversity to represent seven distinct grammars.

But considering that Proto-Indo-European is a fragmentary reconstruction, not an actual language, we should allow it more time to account for the gaps in our knowledge (more on this in chapter 5). Let us assign a nominal lifetime of two thousand years to the phase of language history represented by reconstructed Proto-Indo-European. In the history of English two thousand years would take us all the way back to the origins of the sound shifts that defined Proto-Germanic, and would include all the variation in all the Germanic languages ever spoken, from Hlewagasti of Holt to Puff Daddy of hip-hop fame. Proto-Indo-European does not seem to contain that much variation, so two thousand years probably is too long. But for archaeological purposes it is quite helpful to be able to say that the time period we are trying to identify is no longer than two thousand years.

What is the end date for that two-thousand-year window of time?

The Terminal Date for Proto-Indo-European: The Mother Becomes Her Daughters

The terminal date for reconstructed Proto-Indo-European—the date after which it becomes an anachronism—should be close to the date when its oldest daughters were born. Proto-Indo-European was reconstructed on the basis of systematic comparisons between all the Indo-European daughter languages. The mother tongue cannot be placed later than the daughters. Of course, it would have survived after the detachment and isolation of the oldest daughter, but as time passed, if that daughter dialect remained isolated from the Proto-Indo-European speech community, each would have developed its own peculiar innovations. The image of the mother that is retained through each of the daughters is the form the mother had *before* the detachment of that daughter branch. Each daughter, therefore, preserves a somewhat different image of the mother.

Linguists have exploited this fact and other aspects of internal variation to identify chronological phases within Proto-Indo-European. The number of phases defined by different linguists varies from three (early, middle, late) to six.[4] But if we define Proto-Indo-European as the language that was ancestral to *all* the Indo-European daughters, then it is the *oldest* reconstructable form, the *earliest* phase of Proto-Indo-European, that we are talking about. The later daughters did not evolve directly from this early kind of Proto-Indo-European but from some intermediate, evolved set of late Indo-European languages that preserved aspects of the mother tongue and passed them along.

So when did the oldest daughter separate? The answer to that question depends very much on the accidental survival of written inscriptions. And the oldest daughter preserved in written inscriptions is so peculiar that it is probably safer to rely on the image of the mother preserved within the second set of daughters. What's wrong with the oldest daughter?

THE OLDEST AND STRANGEST DAUGHTER (OR COUSIN?): ANATOLIAN

The oldest written Indo-European languages belonged to the Anatolian branch. The Anatolian branch had three early stems: Hittite, Luwian, and Palaic.[5] All three languages are extinct but once were spoken over large parts of ancient Anatolia, modern Turkey (figure 3.1). Hittite is by far the best known of the three, as it was the palace and administrative language of the Hittite Empire.

Inscriptions place Hittite speakers in Anatolia as early as 1900 BCE, but the empire was created only about 1650–1600 BCE, when Hittite warlords conquered and united several independent native Hattic kingdoms in central Anatolia around modern Kayseri. The name *Hittite* was given to them by Egyptian and Syrian scribes who failed to distinguish the Hittite kings from the Hattic kings they had conquered. The Hittites called themselves *Neshites* after the Anatolian city, Kanesh, where they

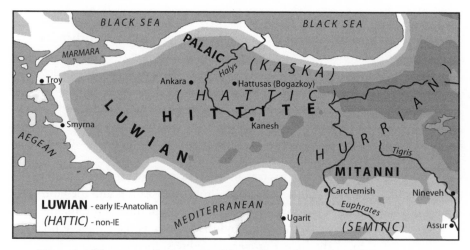

Figure 3.1 The ancient languages of Anatolia at about 1500 BCE.

rose to power. But Kanesh had earlier been a Hattic city; its name was
Hattic. Hattic-speakers also named the city that became the capital of the
Hittite Empire, Hattušas. Hattic was a non–Indo-European language,
probably linked distantly to the Caucasian languages. The Hittites borrowed Hattic words for throne, lord, king, queen, queen mother, heir apparent, priest, and a long list of palace officials and cult leaders—probably
in a historical setting where the Hattic languages were the languages of
royalty. Palaic, the second Anatolian language, also borrowed vocabulary
from Hattic. Palaic was spoken in a city called Pala probably located in
north-central Anatolia north of Ankara. Given the geography of Hattic
place-names and Hattic→ Palaic/Hittite loans, Hattic seems to have been
spoken across all of central Anatolia before Hittite or Palaic was spoken
there. The early speakers of Hittite and Palaic were intruders in a non–
Indo-European central Anatolian landscape dominated by Hattic speakers who had already founded cities, acquired literate bureaucracies, and
established kingdoms and palace cults.[6]

After Hittite speakers usurped the Hattic kingdom they enjoyed a period of prosperity enriched by Assyrian trade, and then endured defeats
that later were dimly but bitterly recalled. They remained confined to the
center of the Anatolian plateau until about 1650 BCE, when Hittite
armies became mighty enough to challenge the great powers of the Near
East and the imperial era began. The Hittites looted Babylon, took other
cities from the Assyrians, and fought the Egyptian pharaoh Ramses II to
a standstill at the greatest chariot battle of ancient times, at Kadesh, on

the banks of the Orontes River in Syria, in 1286 BCE. A Hittite monarch married an Egyptian princess. The Hittite kings also knew and negotiated with the princes who ruled Troy, probably the place referred to in the Hittite archives as *steep Wilusa (Ilios).*[7] The Hittite capital city, Hattušas, was burned in a general calamity that brought down the Hittite kings, their army, and their cities about 1180 BCE. The Hittite language then quickly disappeared; apparently only the ruling élite ever spoke it.

The third early Anatolian language, Luwian, was spoken by more people over a larger area, and it continued to be spoken after the end of the empire. During the later Hittite empire Luwian was the dominant spoken language even in the Hittite royal court. Luwian did not borrow from Hattic and so might have been spoken originally in western Anatolia, outside the Hattic core region—perhaps even in Troy, where a Luwian inscription was found on a seal in Troy level VI—the Troy of the Trojan War. On the other hand, Luwian did borrow from other, unknown non–Indo-European language(s). Hittite and Luwian texts are abundant from the empire period, 1650–1180 BCE. These are the earliest complete texts in any Indo-European language. But individual Hittite and Luwian words survive from an earlier era, before the empire began.[8]

The oldest Hittite and Luwian names and words appeared in the business records of Assyrian merchants who lived in a commercial district, or *karum*, outside the walls of Kanesh, the city celebrated by the later Hittites as the place where they first became kings. Archaeological excavations here, on the banks of the Halys River in central Anatolia, have shown that the Assyrian *karum*, a foreigners' enclave that covered more than eighty acres outside the Kanesh city walls, operated from about 1920 to 1850 BCE (level II), was burned, rebuilt, and operated again (level Ib) until about 1750 BCE, when it was burned again. After that the Assyrians abandoned the karum system in Anatolia, so the Kanesh karum is a closed archaeological deposit dated between 1920 and 1750 BCE. The Kanesh karum was the central office for a network of literate Assyrian merchants who oversaw trade between the Assyrian state and the warring kingdoms of Late Bronze Age Anatolia. The Assyrian decision to make Kanesh their distribution center greatly increased the power of its Hittite and Luwian occupants.

Most of the local names recorded by the merchants in the Kanesh karum accounts were Hittite or Luwian, beginning with the earliest records of about 1900 BCE. Many still were Hattic. But Hittite speakers seem to have controlled business with the Assyrian karum. The Assyrian merchants were so accustomed to doing business with Hittite speakers that they adopted Hittite words for *contract* and *lodging* even in their private correspondence. Palaic, the third language of the Anatolian branch, is not known from the Kanesh records. Palaic died out as a spoken language probably before 1500 BCE. It presumably was spoken in Anatolia during the karum period but not at Kanesh.

Hittite, Luwian, and Palaic had evolved already by 1900 BCE. This is a critical piece of information in any attempt to date Proto-Indo-European. All three were descended from the same root language, Proto-Anatolian. The linguist Craig Melchert described Luwian and Hittite of the empire period, ca. 1400 BCE, as sisters about as different as twentieth-century Welsh and Irish.[9] Welsh and Irish probably share a common origin of about two thousand years ago. If Luwian and Hittite separated from Proto-Anatolian two thousand years before 1400 BCE, then Proto-Anatolian should be placed at about 3400 BCE. What about *its* ancestor? When did the root of the Anatolian branch separate from the rest of Proto-Indo-European?

Dating Proto-Anatolian: The Definition of Proto- and Pre-Languages

Linguists do not use the term *proto-* in a consistent way, so I should be clear about what I mean by Proto-Anatolian. Proto-Anatolian is the language that was *immediately ancestral* to the three known daughter languages in the Anatolian branch. Proto-Anatolian can be described fairly accurately on the basis of the shared traits of Hittite, Luwian, and Palaic. But Proto-Anatolian occupies just the *later portion* of an undocumented period of linguistic change that must have occurred between it and Proto-Indo-European. The hypothetical language stage in between can be called *Pre-Anatolian*. Proto-Anatolian is a fairly concrete linguistic entity closely related to its known daughters. But Pre-Anatolian represents an *evolutionary period*. Pre-Anatolian is a phase defined by Proto-Anatolian at one end and Proto-Indo-European at the other. How can we determine when Pre-Anatolian separated from Proto-Indo-European?

The ultimate age of the Anatolian branch is based partly on objective external evidence (dated documents at Kanesh), partly on presumed rates of language change over time, and partly on internal evidence within the Anatolian languages. The Anatolian languages are quite different phonologically and grammatically from all the other known Indo-European daughter languages. They are so peculiar that many specialists think they do not really belong with the other daughters.

Many of the peculiar features of Anatolian look like archaisms, characteristics thought to have existed in an extremely early stage of Proto-

Indo-European. For example, Hittite had a kind of consonant that has become famous in Indo-European linguistics (yes, consonants can be famous): h_2, a guttural sound or *laryngeal*. In 1879 a Swiss linguist, Ferdinand de Saussure, realized that several seemingly random differences in vowel pronunciation between the Indo-European languages could be brought under one explanatory rule if he assumed that the pronunciation of these vowels had been affected by a "lost" consonant that no longer existed in any Indo-European language. He proposed that such a lost sound had existed in Proto-Indo-European. It was the first time a linguist had been so bold as to reconstruct a feature for Proto-Indo-European that no longer existed in any Indo-European language. The discovery and decipherment of Hittite forty years later proved Saussure right. In a stunning confirmation of the predictive power of comparative linguistics, the Hittite laryngeal h_2 (and traces of a slightly different laryngeal, h_3) appeared in Hittite inscriptions in just those positions Saussure had predicted for his "lost" consonant. Most Indo-Europeanists now accept that archaic Proto-Indo-European contained laryngeal sounds (probably three different ones, usually transcribed as $*h_1$, $*h_2$, $*h_3$) that were preserved clearly only in the Anatolian branch.[10] The best explanation for why Anatolian has laryngeals is that Pre-Anatolian speakers became separated from the Proto-Indo-European language community at a very early date, when a laryngeal-rich phonology was still characteristic of archaic Proto-Indo-European. But then what does *archaic* mean? What, exactly, did Pre-Anatolian separate from?

The Indo-Hittite Hypothesis

The Anatolian branch either lost or never possessed other features that were present in all other Indo-European branches. In verbs, for example, the Anatolian languages had only two tenses, a present and a past, whereas the other ancient Indo-European languages had as many as six tenses. In nouns, Anatolian had just animate and neuter; it had no feminine case. The other ancient Indo-European languages had feminine, masculine, and neuter cases. The Anatolian languages also lacked the dual, a form that was used in other early Indo-European languages for objects that were doubled like eyes or ears. (Example: Sanskrit dēvas 'one god', but dēvau 'double gods'.) Alexander Lehrman identified ten such traits that probably were innovations in Proto-Indo-European after Pre-Anatolian split away.[11]

For some Indo-Europeanists these traits suggest that the Anatolian branch did not develop from Proto-Indo-European at all but rather evolved from an older Pre-Proto-Indo-European ancestor. This ancestral language was called Indo-Hittite by William Sturtevant. According to the Indo-Hittite hypothesis, Anatolian is an Indo-European language only in the broadest sense, as it did not develop from Proto-Indo-European. But it did preserve, uniquely, features of an earlier language community from which they both evolved. I cannot solve the debate over the categorization of Anatolian here, although it is obviously true that Proto-Indo-European must have evolved from an earlier language community, and we can use *Indo-Hittite* to refer to that hypothetical earlier stage. The Proto-Indo-European language community was a chain of dialects with both geographic and chronological differences. The Anatolian branch seems to have separated from an archaic chronological stage in the evolution of Proto-Indo-European, and it probably separated from a different geographic dialect as well, but I will call it archaic Proto-Indo-European rather than Indo-Hittite.[12]

A substantial period of time is needed for the Pre-Anatolian phase. Craig Melchert and Alexander Lehrman agreed that a separation date of about 4000 BCE between Pre-Anatolian and the archaic Proto-Indo-European language community seems reasonable. The millennium or so around 4000 BCE, say 4500 to 3500 BCE, constitutes the *latest* window within which Pre-Anatolian is likely to have separated.

Unfortunately the oldest daughter of Proto-Indo-European looks so peculiar that we cannot be certain she is a daughter rather than a cousin. Pre-Anatolian could have emerged from Indo-Hittite, not from Proto-Indo-European. So we cannot confidently assign a terminal date to Proto-Indo-European based on the birth of Anatolian.

THE NEXT OLDEST INSCRIPTIONS: GREEK AND OLD INDIC

Luckily we have well-dated inscriptions in two other Indo-European languages from the same era as the Hittite empire. The first was Greek, the language of the palace-centered Bronze Age warrior kings who ruled at Mycenae, Pylos, and other strongholds in Greece beginning about 1650 BCE. The Mycenaean civilization appeared rather suddenly with the construction of the spectacular royal Shaft Graves at Mycenae, dated about 1650 BCE, about the same time as the rise of the Hittite empire in Anatolia. The Shaft Graves, with their golden death masks, swords, spears, and images of men in chariots, signified the elevation of a new Greek-speaking dynasty of unprecedented wealth whose economic power depended on long-distance sea trade. The Mycenaean kingdoms were destroyed during the same period of unrest and pillage that brought down the Hittite Empire about 1150 BCE. Mycenaean Greek, the language of palace administration as recorded in the Linear B tablets, was clearly Greek, not Proto-Greek, by 1450 BCE, the date of the oldest preserved inscriptions. The people who spoke it were the models for Nestor and Agamemnon, whose deeds, dimly remembered and elevated to epic, were celebrated centuries later by Homer in the *Iliad* and the *Odyssey*. We do not know when Greek speakers appeared in Greece, but it happened no later than 1650 BCE. As with Anatolian, there are numerous indications that Mycenaean Greek was an intrusive language in a land where non-Greek languages had been spoken before the Mycenaean age.[13] The Mycenaeans almost certainly were unaware that another Indo-European language was being used in palaces not far away.

Old Indic, the language of the *Rig Veda*, was recorded in inscriptions not long after 1500 BCE but in a puzzling place. Most Vedic specialists agree that the 1,028 hymns of the *Rig Veda* were compiled into what became the sacred form in the Punjab, in northwestern India and Pakistan, probably between about 1500 and 1300 BCE. But the deities, moral concepts, and Old Indic language of the *Rig Veda* first appeared in written documents not in India but in *northern Syria*.[14]

The Mitanni dynasty ruled over what is today northern Syria between 1500 and 1350 BCE. The Mitanni kings regularly spoke a *non*–Indo-European language, Hurrian, then the dominant local language in much of northern Syria and eastern Turkey. Like Hattic, Hurrian was a native language of the Anatolian uplands, related to the Caucasian languages. But all the Mitanni kings, first to last, took Old Indic throne names, even if they had Hurrian names before being crowned. Tus'ratta I was Old Indic *Tvesa-ratha* 'having an attacking chariot', Artatama I was *Rta-dhaaman* 'having the abode of r'ta', Artas's'umara was *Rta-smara* 'remembering r'ta', and S'attuara I was *Satvar* 'warrior'.[15] The name of the Mitanni capital city, Waššukanni, was Old Indic *vasu-khani*, literally "wealth-mine." The Mitanni were famous as charioteers, and, in the oldest surviving horse-training manual in the world, a Mitanni horse trainer named Kikkuli (a Hurrian name) used many Old Indic terms for technical details, including horse colors and numbers of laps. The Mitanni military aristocracy was composed of chariot warriors called *maryanna*, probably from an Indic term *márya* meaning "young man," employed in the *Rig Veda* to refer to the heavenly war-band assembled around Indra. Several royal Mitanni names contained the Old Indic term *r'ta*, which meant "cosmic order and truth," the central moral concept of the *Rig Veda*. The Mitanni king

Kurtiwaza explicitly named four Old Indic gods (Indra, Varuna, Mithra, and the Nāsatyas), among many native Hurrian deities, to witness his treaty with the Hittite monarch around 1380 BCE. And these were not just any Old Indic gods. Three of them—Indra, Varuna, and the Nāsatyas or Divine Twins—were the three most important deities in the *Rig Veda*. So the Mitanni texts prove not only that the Old Indic language existed by 1500 BCE but also that the central religious pantheon and moral beliefs enshrined in the *Rig Veda* existed equally early.

Why did Hurrian-speaking kings in Syria use Old Indic names, words, and religious terms in these ways? A good guess is that the Mitanni kingdom was founded by Old Indic-speaking mercenaries, perhaps charioteers, who regularly recited the kinds of hymns and prayers that were collected at about the same time far to the east by the compilers of the *Rig Veda*. Hired by a Hurrian king about 1500 BCE, they usurped his throne and founded a dynasty, a very common pattern in Near Eastern and Iranian dynastic histories. The dynasty quickly became Hurrian in almost every sense but clung to a tradition of using Old Indic royal names, some Vedic deity names, and Old Indic technical terms related to chariotry long after its founders faded into history. This is, of course, a guess, but something like it seems almost necessary to explain the distribution and usage of Old Indic by the Mitanni.

The Mitanni inscriptions establish that Old Indic was being spoken before 1500 BCE in the Near East. By 1500 BCE Proto-Indo-European had differentiated into at least Old Indic, Mycenaean Greek, and the three known daughters of Proto-Anatolian. What does this suggest about the terminal date for Proto-Indo-European?

Counting the Relatives: How Many in 1500 BCE?

To answer this question we first have to understand where Greek and Old Indic are placed among the known branches of the Indo-European family. Mycenaean Greek is the oldest recorded language in the Greek branch. It is an isolated language; it has no recorded close relatives or sister languages. It probably had unrecorded sisters, but none survived in written records. The appearance of the Shaft-Grave princes about 1650 BCE represents the latest possible arrival of Greek speakers in Greece. The Shaft-Grave princes probably already spoke an early form of Greek, not Proto-Greek, since their descendants' oldest preserved inscriptions at about 1450 BCE were in Greek. Proto-Greek might be dated *at the latest* between about 2000 and 1650 BCE. Pre-Greek, the phase that preceded

Proto-Greek, probably originated as a dialect of late Proto-Indo-European *at least* five hundred to seven hundred years before the appearance of Mycenaean Greek, and very probably earlier—*minimally* about 2400–2200 BCE. The terminal date for Proto-Indo-European can be set at about 2400–2200 BCE—it could not have been later than this—from the perspective of the Greek branch. What about Old Indic?

Unlike Mycenaean Greek, Old Indic *does* have a known sister language, Avestan Iranian, which we must take into account. Avestan is the oldest of the Iranian languages that would later be spoken by Persian emperors and Scythian nomads alike, and today are spoken in Iran and Tajikistan. Avestan Iranian was the language of the *Avesta,* the holiest text of Zorastrianism. The oldest parts of the *Avesta,* the *Gathas,* probably were composed by Zoroaster (the Greek form of the name) or by Zarathustra (the original Iranian form) himself. Zarathustra was a religious reformer who lived in eastern Iran, judging from the places he named, probably between 1200 and 1000 BCE.[16] His theology was partly a reaction against the glorification of war and blood sacrifice by the poets of the *Rig Veda.* One of the oldest Gathas was "the lament of the cow," a protest against cattle stealing from the cow's point of view. But the *Avesta* and the *Rig Veda* were closely related in both language and thought. They used the same deity names (although Old Indic gods were demonized in the *Avesta*), employed the same poetic conventions, and shared specific rituals. For example, they used a cognate term for the ritual of spreading straw for the seat of the attending god before a sacrifice (Vedic *barhis,* Avestan *baresman*); and both traditions termed a pious man "one who spread the straw." In many small details they revealed their kinship in a shared Indo-Iranian past. The two languages, Avestan Iranian and Old Indic, developed from a shared parent language, Indo-Iranian, which is not documented.

The Mitanni inscriptions establish that Old Indic had appeared as a distinct language by 1500 BCE. Common Indo-Iranian must be earlier. It probably dates back *at least* to 1700 BCE. Proto-Indo-Iranian—a dialect that had some of the innovations of Indo-Iranian but not yet all of them— has to be placed earlier still, at or before 2000 BCE. Pre-Indo-Iranian was an eastern dialect of Proto-Indo-European, and must then have existed at the *latest* around 2500–2300 BCE. As with Greek, the period from 2500 to 2300 BCE, give or take a few centuries, is the *minimal* age for the separation of Pre-Indo-Iranian from Proto-Indo-European.

So the terminal date for Proto-Indo-European—the date after which our reconstructed form of the language becomes an anachronism—can be set around 2500 BCE, more or less, from the perspective of Greek and

Old Indic. It might be extended a century or two later, but, as far as *these two languages* are concerned, a terminal date *much* later than 2500 BCE— say, as late as 2000 BCE—is impossible. And, of course, Anatolian must have separated long before 2500 BCE. By about 2500 BCE Proto-Indo-European had changed and fragmented into a variety of late dialects and daughter languages—including at least the Anatolian group, Pre-Greek and Pre-Indo-Iranian. Can other daughters be dated to the same period? How many other daughters existed by 2500 BCE?

More Help from the Other Daughters: Who's the Oldest of Them All?

In fact, some other daughters not only *can* be placed this early—they *must* be. Again, to understand why, we have to understand where Greek and Old Indic stand within the known branches of the Indo-European language family. Neither Greek nor Indo-Iranian can be placed among the very oldest Indo-European daughter branches. They are the oldest daughters to survive in inscriptions (along with Anatolian), but that is an accident of history (table 3.1). From the perspective of historical linguistics, Old Indic and Greek must be classified as *late* Indo-European daughters. Why?

Linguists distinguish older daughter branches from younger ones on the basis of shared innovations and archaisms. Older branches seem to have separated earlier because they lack innovations characteristic of the later branches, and they retain archaic features. Anatolian is a good example; it retains some phonetic traits that definitely are archaic (laryngeals) and lacks other features that probably represent innovations. Indo-Iranian, on the other hand, exhibits three innovations that identify it as a later branch.

Indo-Iranian shared one innovation with a group of languages that linguists labeled the *satəm* group: Indo-Iranian, Slavic, Baltic, Albanian, Armenian, and perhaps Phrygian. Among the *satəm* languages, Proto-Indo-European *k- before a front vowel (like *k'mtom 'hundred') was regularly shifted to š- or s- (like Avestan Iranian *satəm*). This same group of languages exhibited a second shared innovation: Proto-Indo-European *kʷ- (called a labiovelar, pronounced like the first sound in *queen*) changed to k-. The third innovation was shared between just a subgroup within the *satəm* languages: Indo-Iranian, Baltic, and Slavic. It is called the *ruki*-rule: the original sound [*-s] in Proto-Indo-European was shifted to [*-sh] after the consonants *r*, *u*, *k*, and *i*. Language branches that do not share these innovations are assumed to have split away and lost regular contact with the *satəm* and *ruki* groups before they occurred.

TABLE 3.1
The First Appearance in Written Records of the Twelve Branches of Indo–European

Language Branch	Oldest Documents or Inscriptions	Diversity at That Date	Latest Date for Proto–Language for the Branch	Grouped with
Anatolian	1920 BCE	Three closely related languages	2800–2300 BCE	No close sisters
Indo–Iranian	1450 BCE	Two very closely related languages	2000–1500 BCE	Greek, Balto–Slavic
Greek	1450 BCE	One dialect recorded, but others probably existed	2000–1500 BCE	Indo–Iranian, Armenian
Phrygian	750 BCE	Poorly documented	1200–800 BCE	Greek? Italo–Celtic?
Italic	600–400 BCE	Four languages, grouped into two quite distinct sub-branches	1600–1100 BCE	Celtic
Celtic	600–300 BCE	Three broad groups with different SVO syntax	1350–850 BCE	Italic
Germanic	0–200 CE	Low diversity; probably the innovations that defined Germanic were recent and still spreading through the Pre-Germanic speech community	500–0 BCE	Baltic/Slavic

TABLE 3.1 (*continued*)

Language Branch	Oldest Documents or Inscriptions	Diversity at That Date	Latest Date for Proto–Language for the Branch	Grouped with
Armenian	400 CE	Only one dialect documented, but Armina was a Persian province ca. 500 BCE so other dialects probably existed 400 CE	500 BCE–0 CE?	Greek, Phrygian?
Tocharian	500 CE	Two (perhaps three) quite distinct languages	500 BCE–0 CE	No close sisters
Slavic	865 CE	Only one dialect documented (OCS), but the West, South, and East Slavic branches must have existed already	0–500 CE	Baltic
Baltic	1400 CE	Three languages	0–500 CE	Slavic
Albanian	1480 CE	Two dialects	0–500 CE	Dacian–Thracian? No close sisters

The Celtic and Italic branches do not display the *satəm* innovations or the *ruki* rule; both exhibit a number of archaic features and also share a few innovations. Celtic languages, today limited to the British Isles and nearby coastal France, were spoken over much of central and western Europe, from Austria to Spain, around 600–300 BCE, when the earliest records of Celtic appeared. Italic languages were spoken in the Italian peninsula at about 600–500 BCE, but today, of course, Latin has many daughters—the Romance languages. In most comparative studies of the Indo-European languages, Italic and Celtic would be placed among the earliest branches to separate from the main trunk. The people who spoke Pre-Celtic and Pre-Italic lost contact with the eastern and northern groups of Indo-European speakers before the *satəm* and *ruki* innovations occurred. We cannot yet discuss where the boundaries of these linguistic regions were, but we can say that Pre-Italic and Pre-Celtic departed to form a western regional–chronological block, whereas the ancestors of Indo-Iranian, Baltic, Slavic, and Armenian stayed behind and shared a set of later innovations. Tocharian, the easternmost Indo-European language, spoken in the Silk Road caravan cities of the Tarim Basin in northwestern China, also lacked the *satəm* and *ruki* innovations, so it seems to have departed equally early to form an eastern branch.

Greek shared a series of linguistic features uniquely with the Indo-Iranian languages, but it did not adopt the *satəm* innovation or the *ruki* rule.[17] Pre-Greek and Pre-Indo-Iranian must have developed in neighboring regions, but the speakers of Pre-Greek departed before the *satəm* or the *ruki* innovations appeared. The shared features included morphological innovations, conventions in heroic poetry, and vocabulary. In morphology, Greek and Indo-Iranian shared two important innovations: the augment, a prefix *e-* before past tenses (although, because it is not well attested in the earliest forms of Greek and Indo-Iranian, the augment *might* have developed independently in each branch much later); and a mediopassive verb form with a suffixed *-i*. In weapon vocabulary they shared common terms for *bow* (*taksos*), *arrow* (*eis-*), *bowstring* (*jya-*), and *club* (*uágros*), or *cudgel*, the weapon specifically associated with Indra and his Greek counterpart Herakles. In ritual they shared a unique term for a specific ritual, the *hecatomb*, or sacrifice of a hundred cows; and they referred to the gods with the same shared epithet, *those who give riches*. They retained shared cognate names for at least three deities: (1) *Erinys/Saraṇ, yū*, a horse-goddess in both traditions, born of a primeval creator-god and the mother of a winged horse in Greek or of the Divine Twins in

Indo-Iranian, who are often represented as horses; (2) *Kérberos/Śárvara*, the multiheaded dog that guarded the entrance to the Otherworld; and (3) *Pan/Pūṣán*, a pastoral god that guarded the flocks, symbolically associated in both traditions with the goat. In both traditions, goat entrails were the specific funeral offering made to the hell-hound *Kérberos/Śárvara* during a funeral ceremony. In poetry, ancient Greek, like Indo-Iranian, had two kinds of verse: one with a twelve-syllable line (the Sapphic/Alcaic line) and another with an eight-syllable line. No other Indo-European poetic tradition shared both these forms. They also shared a specific poetic formula, meaning "fame everlasting," applied to heroes, found in this exact form only in the *Rig Veda* and Homer. Both Greek and Indo-Iranian used a specific verb tense, the imperfect, in poetic narratives about past events.[18]

It is unlikely that such a large bundle of common innovations, vocabulary, and poetic forms arose independently in two branches. Therefore, Pre-Greek and Pre-Indo-Iranian almost certainly were neighboring late Indo-European dialects, spoken near enough to each other so that words related to warfare and ritual, names of gods and goddesses, and poetic forms were shared. Greek did not adopt the *ruki* rule or the *satəm* shift, so we can define two strata here: the older links Pre-Greek and Pre-Indo-Iranian, and the later separates Proto-Greek from Proto-Indo-Iranian.

The Birth Order of the Daughters and the Death of the Mother

The *ruki* rule, the *centum/satəm* split, and sixty-three possible variations on seventeen other morphological and phonological traits were analyzed mathematically to generate thousands of possible branching diagrams by Don Ringe, Wendy Tarnow, and colleagues at the University of Pennsylvania.[19] The cladistic method they used was borrowed from evolutionary biology but was adapted to compare linguistic innovations rather than genetic ones. A program selected the trees that emerged most often from among *all possible* evolutionary trees. The evolutionary trees identified by this method agreed well with branching diagrams proposed on more traditional grounds. The oldest branch to split away was, without any doubt, Pre-Anatolian (figure 3.2). Pre-Tocharian probably separated next, although it also showed some later traits. The next branching event separated Pre-Celtic and Pre-Italic from the still evolving core. Germanic has some archaic traits that suggest an initial separation at about the same time as Pre-Celtic and Pre-Italic, but then later it was strongly affected by

Figure 3.2 The best branching diagram according to the Ringe–Warnow–Taylor (2002) cladistic method, with the minimal separation dates suggested in this chapter. Germanic shows a mixture of archaic and derived traits that make its place uncertain; it could have branched off at about the same time as the root of Italic and Celtic, although here it is shown branching later because it also shared many traits with Pre-Baltic and Pre-Slavic.

borrowing from Celtic, Baltic, and Slavic, so the precise time it split away is uncertain. Pre-Greek separated after Italic and Celtic, followed by Indo-Iranian. The innovations of Indo-Iranian were shared (perhaps later) with several language groups in southeastern Europe (Pre-Armenian, Pre-Albanian, partly in Pre-Phrygian) and in the forests of northeastern Europe (Pre-Baltic and Pre-Slavic). Common Indo-Iranian, we must remember, is dated *at the latest* to about 1700 BCE. The Ringe-Tarnow branching diagram puts the separations of Anatolian, Tocharian, Italic, Celtic, German, and Greek before this. Anatolian probably had split away before 3500 BCE, Italic and Celtic before 2500 BCE, Greek after 2500 BCE, and Proto-Indo-Iranian by 2000 BCE. Those are not meant to be exact dates, but they are in the right sequence, are linked to dated inscriptions in three places (Greek, Anatolian, and Old Indic), and make sense.

By 2500 BCE the language that has been reconstructed as Proto-Indo-European had evolved into something else or, more accurately, into a variety of things,—late dialects such as Pre-Greek and Pre-Indo-Iranian that continued to diverge in different ways in different places. The Indo-European languages that evolved after 2500 BCE did not develop from Proto-Indo-European but from a set of intermediate Indo-European languages that preserved and passed along aspects of the mother tongue. By 2500 BCE Proto-Indo-European was a dead language.

CHAPTER FOUR

Language and Time 2
Wool, Wheels, and Proto-Indo-European

If Proto-Indo-European was dead as a spoken language by 2500 BCE, when was it born? Is there a date *after which* Proto-Indo-European must have been spoken? This question can be answered with surprising precision. Two sets of vocabulary terms identify the date after which Proto-Indo-European must have been spoken: words related to woven wool textiles, and to wheels and wagons. Neither woven wool textiles nor wheeled vehicles existed before about 4000 BCE. It is possible that neither existed before about 3500 BCE. Yet Proto-Indo-European speakers spoke regularly about wheeled vehicles and some sort of wool textile. This vocabulary suggests that Proto-Indo-European was spoken after 4000–3500 BCE. As the Proto-Indo-European vocabulary for wheeled vehicles has already been described in chapter 2, let us begin here with the Proto-Indo-European terms for wool.

THE WOOL VOCABULARY

Woven woolen textiles are made from long wool fibers of a type that did not grow on wild sheep. Sheep with long wooly coats are genetic mutants bred just for that trait. If Proto-Indo-European contained words referring unequivocally to woven woolen textiles, then those words had to have entered Proto-Indo-European after the date when wool sheep were developed. But if we are to use the wool vocabulary as a dating tool, we need to know both the exact meaning of the reconstructed roots and the date when wool sheep first appeared. Both issues are problematic.

Proto-Indo-European contained roots that meant "sheep," "ewe," "ram," and "lamb"—a developed vocabulary that undoubtedly indicates familiarity with domesticated sheep. It also had a term that in most daughter cognates meant "wool". The root *HwlHn-* is based on cognates in almost all branches from Welsh to Indic and including Hittite, so it goes back to the archaic Proto-Indo-European era before the Anatolian branch split away. The stem is unusually long, however, suggesting to Bill Darden of the University of Chicago that it was either borrowed or derived by the addition of the *-n-* suffix from a shorter, older root. He suggested that the shorter root, and the *earliest* form, was *Hwel-* or *Hwol-* (transcribed as *Hw(e/o)l*). Its cognates in Baltic, Slavic, Greek, Germanic, and Armenian meant "felt," "roll," "beat," and "press." "Felt" seems to be the meaning that unites them, since the verbs describe operations in the manufacture of felt. Felt is made by beating or pressing wool fibers until they are pounded into a loose mat. The mat is then rolled up and pressed tightly, unrolled and wetted, then rolled and pressed again, all this repeated until the mat is tight. Wool fibers are curly, and they interlock during this pressing process. The resulting felt textile is quite warm. The winter tents of Eurasian nomads and the winter boots of Russian farmers (made to fit over regular shoes) were traditionally made from felt. If Darden is right, the most ancient Pre-Proto-Indo-European wool root, *Hw(e/o)l-*), was connected with felt. The derivative stem *HwlHn-*, the root retained in both Anatolian and classic Proto-Indo-European, meant "wool" or something made of wool, but we cannot be certain that it referred to a woven wool textile. It could have referred to the short, natural wool that grew on wild sheep or to some kind of felt textile made of short wool.[1]

Sheep (*Ovis orientalis*) were domesticated in the period from about 8000 to 7500 BCE in eastern Anatolia and western Iran as a captive source of meat, which is all they were used for during the first four thousand years of sheepherding. They were covered not with wool but with long, coarse hair called *kemp*. Wool grew on these sheep as an insulating undercoat of very short curly fibers that, in the words of textile specialist Elizabeth Barber, were "structurally unspinnable." This "wild" short wool was molted at the end of the winter. In fact, the annual shedding of short wild wool might have created the first crude (and smelly) felts, when sheep slept on their own damp sheddings. The next step would have been to intentionally pluck the wool when it loosened, just before it was shed. But *woven* wool textiles required wool *thread*.

Wool thread could only be made from unnaturally long wool fibers, as the fibers had to be long enough to cling to each other when pulled apart. A spinner of wool would pull a clump of fibers from a mass of long-fiber wool and twist them into a thread by handfeeding the strand onto a twirling weighted stick, or hand spindle (the spinning wheel was a much later invention). The spindle was suspended in the air and kept twirling with a motion of the wrist. The spindle weights are called *spindle whorls*, and they are just about the only evidence that survives of ancient thread making, although it is difficult to distinguish spindle whorls used for making woolen thread from those used for making flaxen thread, apparently the oldest kind of thread made by humans. Linen made from flax was the oldest woven textile. Woolen thread was invented only after spinners of flax and other plant fibers began to obtain the longer animal fibers that grew on mutant wool sheep. When did this genetic alteration happen? The conventional wisdom is that wool sheep appeared about 4000–3500 BCE.[2]

In southern Mesopotamia and western Iran, where the first city-based civilizations appeared, woven wool textiles were an important part of the earliest urban economies. Wool absorbed dye much better than linen did, so woolen textiles were much more colorful, and the color could be woven in with differently colored threads rather than stamped on the textile surface (apparently the oldest kind of textile decoration). But almost all the evidence for wool production appears in the Late Uruk period or later, after about 3350 BCE.[3] Because wool itself is rarely preserved, the evidence comes from animal bones. When sheep are raised for their wool, the butchering pattern should show three features: (1) sheep or goats (which differ only in a few bones) or both should make up the majority of the herded animals; (2) sheep, the wool producers, should greatly outnumber goats, the best milk producers; and (3) the sheep should have been butchered at an advanced age, after years of wool production. Susan Pollock's review of the faunal data from eight Uruk-period sites in southern Mesopotamia, northern Mesopotamia, and western Iran showed that the shift to a wool-sheep butchering pattern occurred in this heartland of cities no earlier than the Late Uruk period, after 3350 BCE (figure 4.1). Early and Middle Uruk sheep (4000–3350 BCE) did not show a wool-butchering pattern. This Mesopotamian/western Iranian date for wool sheep was confirmed at Arslantepe on the upper Euphrates in eastern Anatolia. Here, herds were dominated by cattle and goats before 3350 BCE (phase VII), but in the next phase (VIa) Late Uruk pottery appeared, and sheep suddenly rose to first place, with more than half of them living to maturity.[4]

The animal-bone evidence from the Near East suggests that wool sheep appeared after about 3400 BCE. Because sheep were not native to Europe, domesticated Near Eastern sheep were imported to Europe by the first farmers who migrated to Europe from Anatolia about 6500 BCE. But the mutation for longer wool might have appeared as an adaptation to cold winters after domesticated sheep were introduced to northern climates, so

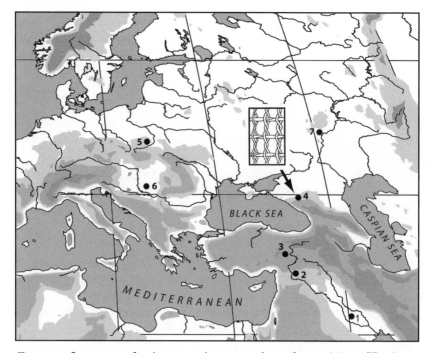

Figure 4.1 Locations of early sites with some evidence for wool sheep. The drawing is from a microscopic image of the oldest known woven wool textile published by N. Shishlina: (1) Uruk; (2) Hacinebi; (3) Arslantepe; (4) Novosvobodnaya; (5) Bronocice; (6) Kétegyháza; (7) Khvalynsk. After Shishlina 1999.

it would not be surprising if the earliest long-wool sheep were bred in Europe. At Khvalynsk, a cemetery dated about 4600–4200 BCE on the middle Volga in Russia, sheep were the principal animal sacrificed in the graves, and most of them were mature, as if being kept alive for wool or milk. But animals chosen for sacrifice might have been kept alive for a ritual reason. At Svobodnoe, a farming settlement in the North Caucasus piedmont in what is now southern Russia, dated between about 4300 and 3700 BCE, sheep were the dominant domesticated animal, and sheep outnumbered goats by 5 to 1. This is a classic wool-sheep harvesting pattern. But at other settlements of the same age in the North Caucasus this pattern is not repeated. A new large breed of sheep appeared in eastern Hungary at Kétegyháza in the Cernavoda III–Boleraz period, dated 3600–3200 BCE, which Sandor Bökönyi suggested was introduced from Anatolia and Mesopotamia; at Bronocice in southern Poland, in levels dated to the same period, sheep greatly outnumbered goats by 20 to 1. But beyond these tantalizing cases there was no broad or widespread shift to sheep keeping or to a wool-butchering pattern in Europe until after about 3300–3100 BCE, about the same time it occurred in the Near East.[5]

No actual woven woolen textiles are firmly dated before about 3000 BCE, but they were very widespread by 2800 BCE. A woven woolen textile fragment that might predate 3000 BCE was found in a grave in the North Caucasus Mountains, probably a grave of the Novosvobodnaya culture (although there is some uncertainty about the provenience). The wool fibers were dyed dark brown and beige, and then a red dye was painted on the finished fabric. The Novosvobodnaya culture is dated between 3400 and 3100 BCE, but this fabric has not been directly dated. At Shar-i Sokhta, a Bronze Age semi-urban trading center in east-central Iran, woven woolens were the only kinds of textiles recovered in levels dated 2800–2500 BCE. A woven wool fragment was found at Clairvaux-les-lacs Station III in France, dated 2900 BCE, so wool sheep and woven wool textiles were known from France to central Iran by 2900–2500 BCE.[6]

The preponderance of the evidence suggests that woven wool textiles appeared in Europe, as in the Near East, after about 3300 BCE, although wool sheep may have appeared earlier than this, about 4000 BCE, in the North Caucasus Mountains and perhaps even in the steppes. But if the root *HwlHn- referred to the short undercoat wool of "natural" sheep, it could have existed before 4000 BCE. This uncertainty in meaning weakens the reliability of the wool vocabulary for dating Proto-Indo-European. The wheeled vehicle vocabulary is different. It refers to very definite objects (wheels, axles), and the earliest wheeled vehicles are very well dated. Unlike wool textiles, wagons required an elaborate set of metal tools (chisels, axes) that preserve well, the images of wagons are easier to categorize, and the wagons themselves preserve more easily than textiles.

The Wheel Vocabulary

Proto-Indo-European contained a set of words referring to wheeled vehicles—wagons or carts or both. We can say with great confidence that wheeled vehicles were not invented until after 4000 BCE; the surviving evidence suggests a date closer to 3500 BCE. Before 4000 BCE there were no wheels or wagons to talk about.

Proto-Indo-European contained at least five terms related to wheels and wagons, as noted in chapter 2: two words for *wheel* (perhaps for different kinds of wheels), one for *axle*, one for *thill* (the pole to which the animals

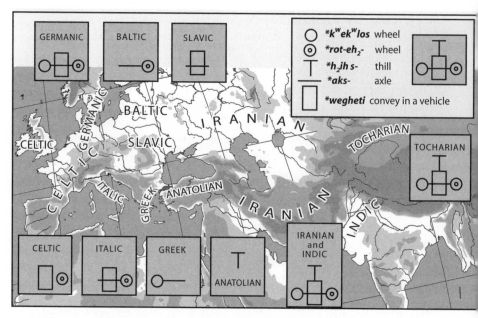

Figure 4.2 The geographic distribution of the Indo-European wheel-wagon vocabulary.

were yoked), and a verb meaning "to go or convey in a vehicle." Cognates for these terms occur in all the major branches of Indo-European, from Celtic in the west to Vedic Sanskrit and Tocharian in the east, and from Baltic in the north to Greek in the south (figure 4.2). Most of the terms have a kind of vowel structure called an o-stem that identifies a late stage in the development of Proto-Indo-European; *axle* was an older n-stem derived from a word that meant "shoulder." The o-stems are important, since they appeared only during the later end of the Proto-Indo-European period. Almost all the terms are derived from Proto-Indo-European roots, so the vocabulary for wagons and wheels was not imported from the outside but was created within the Proto-Indo-European speech community.[7]

The only branch that might *not* contain a convincing wheeled-vehicle vocabulary is Anatolian, as Bill Darden observed. Two possible Proto-Indo-European wheeled-vehicle roots are preserved in Anatolian. One (*ḫurki-* 'wheel') is thought to be descended from a Proto-Indo-European root, because the same root might have yielded Tocharian A *wärkänt* and Tocharian B *yerkwanto*, both meaning "wheel." Tocharian is an extinct Indo-European branch consisting of two (perhaps three) known languages, called A and B (and perhaps C), recorded in documents written in about 500–700 CE by Buddhist monks in the desert caravan cities of the Tarim Basin in northwestern China. But Tocharian specialist Don Ringe sees serious difficulties in deriving either Tocharian term from the same root that yielded Anatolian *hurki-*, suggesting that the Tocharian and Anatolian terms were unrelated and therefore do not require a Proto-Indo-European root.[8] The other Anatolian vehicle term (*hišša-* 'thill' or 'harness-pole') has a good Indo-European source, **ei-/*oi-* or perhaps **h₂ih₃s-*, but its original meaning might have referred to plow shafts rather than wagon shafts. So we cannot be certain that archaic Proto-Indo-European, as partially preserved in Anatolian, had a wheeled-vehicle vocabulary. But the rest of Proto-Indo-European did.

When Was the Wheel Invented?

How do we know that wheeled vehicles did not exist before 4000 BCE? First, a wheeled vehicle required not just wheels but also an axle to hold the vehicle. The wheel, axle, and vehicle together made a complicated combination of load-bearing moving parts. The earliest wagons were planed and chiseled entirely from wood, and the moving parts had to fit precisely. In a wagon with a fixed axle and revolving wheels (apparently the earliest type), the axle arms (the ends of the axle that passed through the center of the wheel) had to fit snugly, but not too snugly, in the hole through the nave, or hub. If the fit was too loose, the wheels would wobble as they turned. If it was too tight, there would be excessive drag on the revolving wheel.

Then there was the problem of the draft—the total weight, with drag, pulled by the animal team. Whereas a sledge could be pulled using traces, or flexible straps and ropes, a wagon or cart had to have a rigid draft pole, or thill, and a rigid yoke. The weight of these elements increased the overall draft. One way to reduce the draft was to reduce the diameter of the axle arms to fit a smaller hole in the wheel. A large-diameter axle was strong but created more friction between the axle arms and the revolving wheel. A smaller-diameter axle arm would cause less drag but would break easily unless the wagon was very narrow. The first wagon-wrights had to calculate the relationship between drag, axle diameter/strength, axle length/rigidity, and the width of the wagon bed. In a work vehicle meant to carry heavy loads, a short axle with small-diameter axle arms and a narrow wagon bed made good engineering sense, and, in fact, this is what the earliest wagons looked like, with a bed only about 1 m wide. Another way to reduce the draft was to reduce the number of wheels from four to two—to make a *wagon* into a *cart*. The draft of a modern two-wheeled cart is 40% less than a four-wheeled wagon *of the same weight,* and we can assume that an advantage of approximately the same magnitude applied to ancient carts. Carts were lighter and easier to pull, and on rough ground were less likely to get stuck. Large loads probably still needed wagons, but carts would have been useful for smaller loads.[9]

Archaeological and inscriptional evidence for wheeled vehicles is widespread after about 3400 BCE. One uncertain piece of evidence, a track preserved under a barrow grave at Flintbek in northern Germany, might have been made by wheels, and might be as old as 3600 BCE. But the real explosion of evidence begins about 3400 BCE. Wheeled vehicles appeared in four different media dated between about 3400 and 3000 BCE—a written sign for wagons, two-dimensional images of wagons and carts, three-dimensional models of wagons, and preserved wooden wheels and wagon parts themselves. These four independent kinds of evidence appeared across the ancient world between 3400 and 3000 BCE, about the same time as wool sheep, and clearly indicate when wheeled vehicles became widespread. The next four sections discuss the four kinds of evidence.[10]

Mesopotamian Wagons: The Oldest Written Evidence

Clay tablets with "wagon" signs impressed on them were found in the Eanna temple precinct in Uruk, one of the first cities created by humans. About thirty-nine hundred tablets were recovered from level IVa, the end of Late Uruk. In these texts, among the oldest documents in the world, a pictograph (figure 4.3.f) shows a four-wheeled wagon with some kind of canopy or superstructure. The "wagon" sign occurred just three times in thirty-nine hundred texts, whereas the sign for "sledge"—a similar kind of transport, but dragged on runners not rolled on wheels—occurred thirty-eight times. Wagons were not yet common.

The Eanna precinct tablets were inside Temple C when it burned down. Charcoal from the Temple C roof timbers yielded four radiocarbon dates averaging about 3500–3370 BCE. A radiocarbon date tells us when the dated material, in this case wood, died, not when it was burned. The wood in the center of any tree is actually dead (something few people realize); only the outer ring of bark and the sappy wood just beneath it are alive. If the timbers in Temple C were made from the center of a large tree, the wood might have died a century or two before the building was burned down, so the actual age of the Temple C tablets is later than the radiocarbon date, perhaps 3300–3100 BCE. Sledges still were far more common than wagons in the city of Uruk at that date. Ox-drawn canopied sledges might have preceded canopied wagons as a form of transport (in parades or processions? harvest rituals?) used by city officials.

A circular clay object that *might* be a model wheel, perhaps from a small ceramic model of a wagon, was found at the site of Arslantepe in eastern Turkey, in the ruins of a temple-palace from level VIa at the site, also dated 3400–3100 BCE (figure 4.3.c). Arslantepe was one of a string of native strongholds along the upper Euphrates River in eastern Anatolia that entered into close relations with faraway Uruk during the Late Uruk period. Although the kind of activities that lay behind this "Uruk expansion" northward up the Euphrates valley is not known (see chapter 12), the possible clay wheel model at Arslantepe *could* indicate that wagons were being used in eastern Anatolia during the period of Late Uruk influence.

Wagons and Carts from the Rhine to the Volga: The Oldest Pictorial Evidence

A two-dimensional image that seems to portray a four-wheeled wagon, harness pole, and yoke was incised on the surface of a decorated clay mug of the Trichterbecker (TRB) culture found at the settlement of Bronocice in southern Poland, dated about 3500–3350 BCE (figure 4.3.b). The TRB culture is recognized by its distinctive pottery shapes and tombs, which are found over a broad region in modern Poland, eastern Germany, and southern Denmark. Most TRB people were simple farmers who lived in small agricultural villages, but the Bronocice settlement was unusually large, a TRB town covering fifty-two hectares. The cup or mug with the wagon image incised on its surface was found in a rubbish pit containing animal bones, the broken sherds of five clay vessels, and flint tools. Only this cup had a wagon image. The design is unusual for TRB pottery, not an accidental combination of normal decorative motifs. The cup's date is the subject of some disagreement. A cattle bone found in the same pit yielded an average age of about 3500 BCE, whereas six of the seven other radiocarbon dates for the settlement around the pit average 150 years later, about 3350 BCE. The excavators accept an age range spanning these results, about 3500–3350 BCE. The Bronocice wagon image is the oldest well-dated image of a wheeled vehicle in the world.

Two other images could be about the same age, although they probably are somewhat later. An image of two large-horned cattle pulling what seems to be a two-wheeled cart was scratched on the wall of a Wartberg culture stone tomb at Lohne-Züschen I, Hesse, central Germany (figure 4.3.e). The

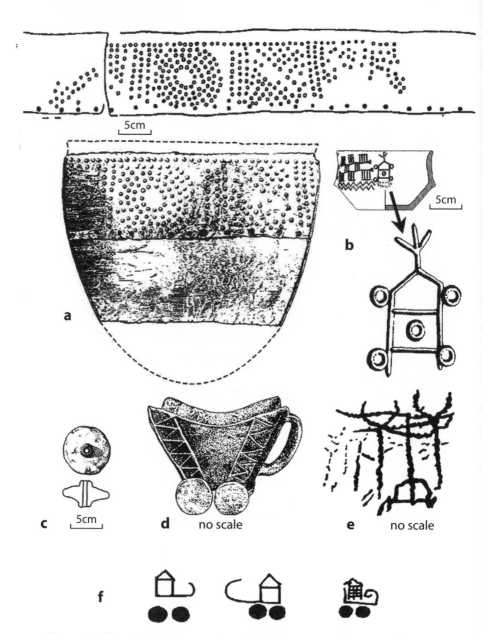

Figure 4.3 The oldest images and models of wagons and wheels: (a) bronze kettle from Evdik kurgan, lower Volga, Russia, with a design that could represent, from the left, a yoke, cart, wheel, X-braced floor, and animal head; (b) image of a four-wheeled wagon on a ceramic vessel from Bronocice, southern Poland; (c) ceramic wheel (from a clay model?) at Arslantepe, eastern Anatolia; (d) ceramic wagon model from Baden grave 177 at Budakalász,

grave was reused over a long period of time between about 3400 and 2800 BCE, so the image could have been carved any time in that span. Far away to the east, a metal cauldron from the Evdik kurgan near the mouth of the Volga River bears a repoussé image that might show a yoke, a wheel, a cart, and a draft animal; it was found in a grave with objects of the Novosvobodnaya culture, dated between 3500 and 3100 BCE (figure 4.3.a). These images of carts and wagons are distributed from central Germany through southern Poland to the Russian steppes.

Hungarian Wagons: The Oldest Clay Models

The Baden culture is recognized by its pottery and to a certain extent by its distinctive copper tools, weapons, and ornaments. It appeared in Hungary about 3500 BCE, and the styles that define it then spread into northern Serbia, western Romania, Slovakia, Moravia, and southern Poland. Baden-style polished and channeled ceramic mugs and small pots were used across southeastern Europe about 3500–3000 BCE. Similarities between Baden ceramics and those of northwestern Anatolia in the centuries before Troy I suggest one route by which wheeled vehicles could have spread between Mesopotamia and Europe. Three-dimensional ceramic models of four-wheeled wagons (figure 4.3.d) were included in sacrificial deposits associated with two graves of the Late Baden (Pécel) culture at Budakalász (Grave 177) and Szigetszentmárton in eastern Hungary, dated about 3300–3100 BCE. Paired oxen, almost certainly a team, were found sacrificed in Grave 3 at Budakalász and in other Late Baden graves in Hungary. Paired oxen also were placed in graves of the partly contemporary Globular Amphorae culture (3200–2700 BCE) in central and southern Poland. The Baden wagon models are the oldest well-dated three-dimensional models of wheeled vehicles.

Steppe and Bog Vehicles: The Oldest Actual Wagons

Remains of about 250 wagons and carts have been discovered under earthen burial mounds, or kurgans, in the steppe grasslands of Russia and Ukraine, dated about 3000–2000 BCE (figures 4.4 and 4.5). The wheels

---

Figure 4.3 (continued) Hungary; (e) cart image with two cattle incised on stone, from a tomb at Lohne-Züschen I, Hesse, central Germany; (f) earliest written symbols for a wagon, on clay tablets from Uruk IVa, southern Iraq. After (a) Shilov and Bagautdinov 1997; (b, d, e) Milisauskas 2002; (c,f) Bakker et al. 1999.

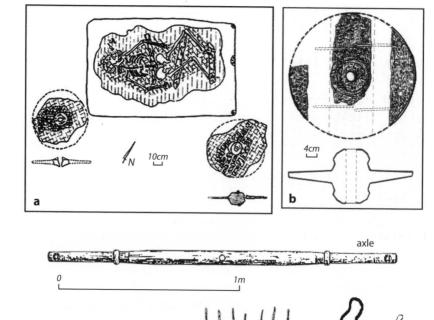

Figure 4.4 Preserved wagon parts and wheels: (a) two solid wooden wheels at the corners of grave 57, Bal'ki kurgan, Ukraine, radiocarbon dated 3330–2900 BCE; (b) Catacomb-culture tripartite wheel with dowels, probably 2600–2200 BCE; (c) preserved axle and reconstructed wagon from various preserved wheel and wagon fragments in bog deposits in northwestern Germany and Denmark dated about 3000–2800 BCE. After (a) Lyashko and Otroshchenko 1988; (b) Korpusova and Lyashko 1990; (c) Hayen 1989.

were 50–80 cm in diameter. Some were made of a single plank cut vertically from the trunk of a tree, with the grain (not like a salami). Most steppe wheels, however, were made of two or three planks cut into circular segments and then doweled together with mortice-and-tenon joints. In the center were long tapered naves (hubs), about 20–30 cm wide at the base and projecting outward about 10–20 cm on either side of the wheel. The naves were secured to the axle arms by a lynchpin that pinned the

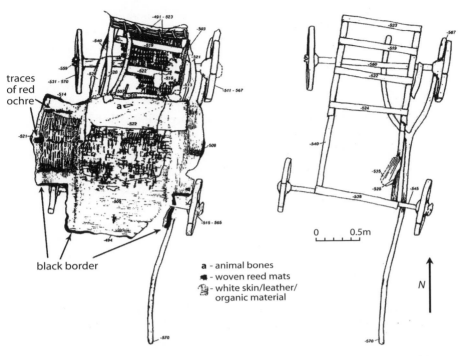

traces
of red
ochre

black border

a - animal bones
- woven reed mats
- white skin/leather/
  organic material

N

0   0.5m

Figure 4.5 The best-preserved wagon graves in the steppes are in the Kuban River region in southern Russia. This wagon was buried under Ostannii kurgan 1. Radiocarbon dated about 3300–2900 BCE, the upper part of the wagon is on the left and the lower part, on the right. After Gei 2000, figure 53.

nave to the axle, and between them they kept the wheel from wobbling. The axles had rounded axle arms for the wheel mounts and were about 2 m long. The wagons themselves were about 1 m wide and about 2 m long. The earliest radiocarbon dates on wood from steppe wagons average around 3300–2800 BCE. A wagon or cart grave at Bal'ki kurgan (grave 57) on the lower Dnieper was dated 4370±120 BP, or 3330–2880 BCE; and wood from a wagon buried in Ostanni kurgan 1 (grave 160) on the Kuban River was dated 4440 ± 40 BP, or 3320–2930 BCE. The probability distributions for both dates lie predominantly before 3000 BCE, so both vehicles probably date before 3000 BCE. But these funeral vehicles can hardly have been the very first wagons used in the steppes.

Other wooden wheels and axles have been discovered preserved in bogs or lakes in central and northern Europe. In the mountains of Switzerland and southwestern Germany wagon-wrights made the axle arms square and mortised them into a square hole in the wheel. The middle of the axle was circular and revolved under the wagon. This revolving-axle design created more drag and was less efficient than the revolving-wheel design, but it did not require carving large wooden naves and so the Alpine wheels were much easier to make. One found near Zurich in a waterlogged settlement of the Horgen culture (the Pressehaus site) was dated about 3200 BCE by associated tree-ring dates. The Pressehaus wheel tells us that separate regional European design traditions for wheel making already existed before 3200 BCE. Wooden wheels and axles also have been found in bogs in the Netherlands and Denmark, providing important evidence on the construction details of early wagons, but dated after 3000 BCE. They had fixed axles and revolving wheels, like those of the steppes and central Europe.

The Significance of the Wheel

It would be difficult to exaggerate the social and economic importance of the first wheeled transport. Before wheeled vehicles were invented, really heavy things could be moved efficiently only on water, using barges or rafts, or by organizing a large hauling group on land. Some of the heavier items that prehistoric, temperate European farmers had to haul across land all the time included harvested grain crops, hay crops, manure for fertilizer, firewood, building lumber, clay for pottery making, hides and leather, and people. In northern and western Europe, some Neolithic communities celebrated their hauling capacities by moving gigantic stones to make megalithic community tombs and stone henges; other communities hauled earth, making massive earthworks. These constructions demonstrated in a visible, permanent way the solidity and strength of the communities that made them, which depended in many ways on human hauling capacities. The importance and significance of the village community as a group transport device changed profoundly with the introduction of wagons, which passed on the burden of hauling to animals and machines, where it has remained ever since.

Although the earliest wagons were slow and clumsy, and probably required teams of specially trained oxen, they permitted single families to carry manure out to the fields and to bring firewood, supplies, crops, and people back home. This reduced the need for cooperative communal labor and made single-family farms viable. Perhaps wagons contributed to the disappearance of large nucleated villages and the dispersal of many farming populations across the European landscape after about 3500 BCE. Wagons were useful in a different way in the open grasslands of the steppes, where the economy depended more on herding than on agriculture. Here wagons made portable things that had never been portable in bulk—shelter, water, and food. Herders who had always lived in the forested river valleys and grazed their herds timidly on the edges of the steppes now could take their tents, water, and food supplies to distant pastures far from the river valleys. The wagon was a mobile home that permitted herders to follow their animals deep into the grasslands and live in the open. Again, this permitted the dispersal of communities, in this case across interior steppes that earlier had been almost useless economically. Significant wealth and power could be extracted from larger herds spread over larger pastures.

Andrew Sherratt bundled the invention of the wheel together with the invention of the plow, wool sheep, dairying, and the beginning of horse transport to explain a sweeping set of changes that occurred among European societies about 3500–3000 BCE. The Secondary Products Revolution (now often shortened to SPR), as Sherratt described it in 1981, was an economic explanation for widespread changes in settlement patterns, economy, rituals, and crafts, many of which had been ascribed by an older generation of archaeologists to Indo-European migrations. ("Secondary products" are items like wool, milk, and muscular power that can be harvested continuously from an animal without killing it, in contrast to "primary products" such as meat, blood, bone, and hides.) Much of the subject matter discussed in arguments over the SPR—the diffusion of wagons, horseback riding, and wool sheep—was also central in discussions of Indo-European expansions, but, in Sherratt's view, all of them were derived by diffusion from the civilizations of the Near East rather than from Indo-Europeans. Indo-European languages were no longer central or even necessary to the argument, to the great relief of many archaeologists. But Sherrat's proposal that all these innovations came from the Near East and entered Europe at about the same time quickly fell apart. Scratch-plows and dairying appeared in Europe long before 3500 BCE, and horse domestication was a local event in the steppes. An important fragment of the SPR survives in the conjoined diffusion of wool sheep and wagons across much of the ancient Near East and Europe between 3500 and 3000 BCE, but we do not know where either of these innovations started.[11]

The clearest proof of the wheel's impact was the speed with which wagon technology spread (figure 4.6), so rapidly, in fact, that we cannot even say where the wheel-and-axle principle was invented. Most specialists assume that the earliest wagons were produced in Mesopotamia, which was urban and therefore more sophisticated than the tribal societies of Europe; indeed, Mesopotamia had sledges that served as prototypes. But we

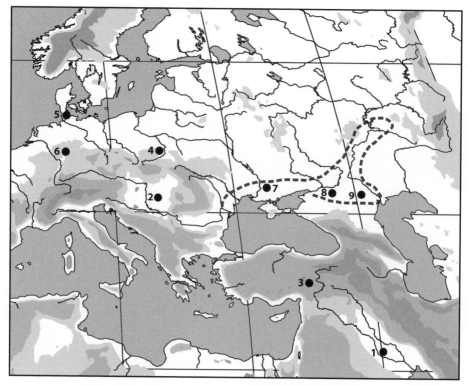

Figure 4.6 Sites with early evidence for wheels or wagons: (1) Uruk; (2) Buda-kalasz; (3) Arslantepe; (4) Bronicice; (5) Flintbek; (6) Lohne-Zuschen I; (7) Bal'ki kurgan; (8) Ostannii kurgan; (9) Evdik kurgan. Dashed line indicates the distribution of about 250 wagon graves in the Pontic-Caspian steppes.

really don't know. Another prototype existed in Europe in the form of Mesolithic and Neolithic bent-wood sleds, doweled together with fine mortice-and-tenon joints; in much of eastern Europe, in fact, right up to the twentieth century, it made sense to park your wagon or carriage in the barn for the winter and resort to sleds, far more effective than wheels in snow and ice. Bent-wood sleds were at least as useful in prehistoric Europe as in Mesopotamia, and they began to appear in northern Europe as early as the Mesolithic; thus the skills needed to make wheels and axles existed in both Europe and the Near East.[12]

Regardless of where the wheel-and-axle principle was invented, the technology spread rapidly over much of Europe and the Near East between 3400 and 3000 BCE. Proto-Indo-European speakers talked about wagons and wheels using their own words, created from Indo-European roots. Most of these words were o-stems, a relatively late development in Proto-Indo-European phonology. The wagon vocabulary shows that late Proto-Indo-European was spoken certainly after 4000 BCE, and probably after 3500 BCE. Anatolian is the only major early Indo-European branch that has a doubtful wheeled-vehicle vocabulary. As Bill Darden suggested, perhaps Pre-Anatolian split away from the archaic Proto-Indo-European dialects before wagons appeared in the Proto-Indo-European homeland. Pre-Anatolian could have been spoken before 4000 BCE. Late Proto-Indo-European, including the full wagon vocabulary, probably was spoken after 3500 BCE.

WAGONS AND THE ANATOLIAN HOMELAND HYPOTHESIS

The wagon vocabulary is a key to resolving the debate about the place and time of the Proto-Indo-European homeland. The principal alternative to a homeland in the steppes dated 4000–3500 BCE is a homeland in Anatolia and the Aegean dated 7000–6500 BCE. Colin Renfrew proposed that Indo-Hittite (Pre-Proto-Indo-European) was spoken by the first farmers in southern and western Anatolia at sites such as Çatal Höyük dated about 7000 BCE. In his scenario, a dialect of Indo-Hittite was carried to Greece with the first farming economy by pioneer farmers from Anatolia about 6700–6500 BCE. In Greece, the language of the pioneer farmers developed into Proto-Indo-European and spread through Europe and the Mediterranean Basin with the expansion of the earliest agricultural economy. By linking the dispersal of the Indo-European languages with the diffusion of the first farming economy, Renfrew achieved an appealingly elegant solution to the problem of Indo-European origins. Since 1987 he and others have shown convincingly that the migrations of pioneer farmers were one of the principal vectors for the spread of many ancient languages around the world. The "first-farming/language-dispersal" hypothesis, therefore, was embraced by many archaeologists. But it required that the first split between parental Indo-Hittite and Proto-Indo-European began about 6700–6500 BCE, when Anatolian farmers first migrated to Greece. By 3500 BCE, the earliest date for wagons in Europe, the Indo-European language family should have been bushy, multi-branched, and three thousand years old, well past the period of sharing a common vocabulary for anything.[13]

The Anatolian—origin hypothesis raises other problems as well. The first Neolithic farmers of Anatolia are thought to have migrated there from northern Syria, which, according to Renfrew's first-farming/

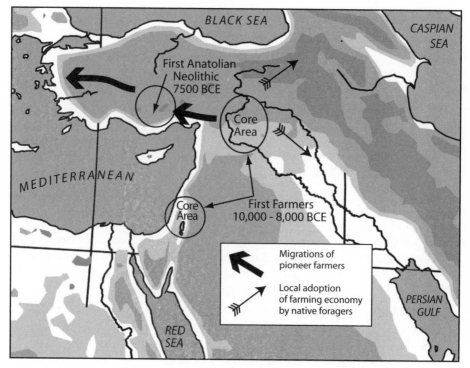

Figure 4.7 The spread of the first farming economy into Anatolia, probably by migration from the Core Area in northern Syria, about 7500 BCE. The first pioneer farmers probably spoke an Afro-Asiatic language. After Bar-Yosef 2002.

language-dispersal hypothesis, should have resulted in the spread of a north Syrian Neolithic language to Anatolia (figure 4.7). The indigenous languages of northern Syria probably belonged to the Afro-Asiatic language phylum, like Semitic and most languages of the lowland Near East. If the first Anatolian farmers spoke an Afro-Asiatic language, it was that language, not Proto-Indo-European, that should have been carried to Greece.[14] The earliest Indo-European languages documented in Anatolia—Hittite, Palaic, and Luwian—showed little diversity, and only Luwian had a significant number of speakers by 1500 BCE. All three borrowed extensively from non–Indo-European languages (Hattic, Hurrian, and perhaps others) that seem to have been older, more prestigious, and more widely spoken. The Indo-European languages of Anatolia did not have the established population base of speakers, and also lacked the kind of diversity that would be expected had they been evolving there since the Neolithic.

Phylogenetic Approaches to Dating Proto-Indo-European

Still, the Anatolian-origin hypothesis has support from new methods in phylogenetic linguistics. Cladistic methods borrowed from biology have been used for two purposes: to arrange the Indo-European languages in a chronological *order* of branching events (discussed in the previous chapter); and to estimate *dates* for the separation between any two branches, or for the root of all branches which is a much riskier proposition. Attaching time estimates to language branches using evolutionary models based on biological change is, at best, an uncertain procedure. People intentionally reshape their speech all the time but cannot intentionally reshape their genes. The way a linguistic innovation is reproduced in a speech community is quite different from the way a mutation is reproduced in a breeding population. The topography of language splits and rejoinings is much more complex and the speed of language branching far more variable. Whereas genes spread as whole units, the spread of language is always a *modular process*, and some modules (grammar and phonology) are more resistant to borrowing and spread than others (words).

Russell Gray and Quentin Atkinson attempted to work around these problems by processing a cocktail of cladistic and linguistic methods through computer programs. They suggested that pre-Anatolian detached from the rest of the Indo-European community about 6700 BCE (plus or minus twelve hundred years). Pre-Tocharian separated next (about 5900 BCE), then pre-Greek/Armenian (about 5300 BCE), and then pre–Indo-Iranian/Albanian (about 4900 BCE). Finally, a super-clade that included the ancestors of pre–Balto-Slavic and pre–Italo-Celto-Germanic separated about 4500 BCE. Archaeology shows that 6700–6500 BCE was about when the first pioneer farmers left Anatolia to colonize Greece. One could hardly ask for a closer match between archaeological and phylogentic dates.[15] But how can the presence of the wagon vocabulary in Proto-Indo-European be synchronized with a first-dispersal date of 6500 BCE?

The Slow Evolution Hypothesis

The wagon vocabulary cannot have been created *after* Proto-Indo-European was dead and the daughter languages differentiated. The wagon/wheel terms do not contain the sounds that would be expected had they been created in a later daughter language and then borrowed into the others, whereas they do contain the sounds predicted if they were inherited into the daughter branches from Proto-Indo-European. The Proto-Indo-European origin of the wagon vocabulary cannot be rejected, as it consists of at least five classic reconstructions. If they are in fact false, then the core methods of comparative linguistics—those that determine "genetic" relatedness—would be so unreliable as to be useless, and the question of Indo-European origins would be moot.

But could the wagon/wheel vocabularies have been created *independently* by the speakers of each branch from the same Proto-Indo-European roots? In the example of *$k^w ek^w los$ 'wheel', Gray suggested (in a comment on his homepage) that the semantic development from the verb *$kwel$- 'turn' to the noun *wheel* 'the turner' was so natural that it could have been repeated independently in each branch. One difficulty here is that at least four different verbs meaning "turn" or "roll" or "revolve" are reconstructed for Proto-Indo-European, which makes the repeated independent choice of *$kwel$- problematic.[16] More critical, the Proto-Indo-European pronunciations of *$kwel$- and the other wagon terms would not have survived unchanged through time. They could not have been available frozen in their Proto-Indo-European phonetic forms to speakers of nine or ten branches that originated at different times across thousands of years. We cannot assume stasis in phonetic development for the wheel vocabulary when all the rest of the vocabulary changed normally with time. But what if all the other vocabulary also changed very slowly?

This is the solution Renfrew offered (figure 4.8). For the wagon/wheel vocabulary to be brought into synchronization with the first-farming/language-dispersal hypothesis, Proto-Indo-European must have been spoken for thirty-five hundred years, requiring a very long period when Proto-Indo-European changed very little. Pre-Proto-Indo-European or Indo-Hittite was spoken in Anatolia before 6500 BCE. Archaic Proto-Indo-European evolved as the language of the pioneer farmers in Greece about 6500–6000 BCE. As their descendants migrated northward and westward, and established widely scattered Neolithic communities from Bulgaria to Hungary and Ukraine, the language they carried remained a single language, Archaic Proto-Indo-European. Their descendants paused for several centuries, and then a second wave of pioneer migration pushed across the Carpathians into the North European plain between about 5500 and 5000 BCE with the Linear Pottery farmers. These farming migrations created Renfrew's Stage 1 of Proto-Indo-European, which was spoken across most of Europe between 6500 and 5000 BCE, from the Rhine to the Dnieper and from Germany to Greece. During Renfrew's Proto-Indo-European Stage 2, between 5000 and 3000 BCE, archaic Proto-Indo-European spread into the steppes

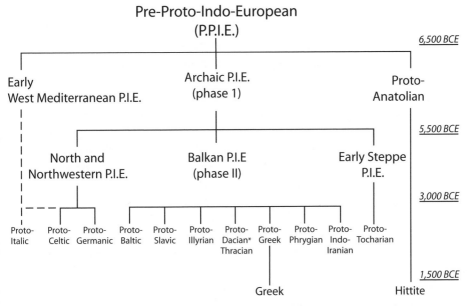

Figure 4.8 If Proto-Indo-European spread across Europe with the first farmers about 6500–5500 BCE, it must have remained almost unchanged until about 3500 BCE, when the wheeled vehicle vocabulary appeared. This diagram illustrates a division into just three dialects in three thousand years. After Renfrew 2001.

and was carried to the Volga with the adoption of herding economies. Late Proto-Indo-European dialectical features developed, including the appearance of "thematic" inflections such as o-stems, which occur in all the wagon/wheel terms. These late features were shared across the Proto-Indo-European–speaking region, which comprised two-thirds of prehistoric Europe. The wagon vocabulary appeared late in Stage 2 and was adopted from the Rhine to the Volga.[17]

It seems to me that this conception of Proto-Indo-European contains three fatal flaws. First, for Proto-Indo-European to have remained a unified dialect chain for more than thirty-five hundred years, from 6500 to 3000 BCE, would require that all its dialects changed at about the same rate and that the rate was extraordinarily slow. A *homogeneous rate of change* across most of Neolithic Europe is very unlikely, as the rate of language change is affected by a host of local factors, as Sheila Embleton showed, and these would have varied from one region to the next. And for Proto-Indo-European only to have evolved from its earlier form to its later form in thirty-five hundred years would require a pan-European condition of near stasis in the speed of language change during the Neolithic/Eneolithic, a truly unrealistic demand. In addition, Neolithic Europe evinces an almost incredible *diversity in material culture.* "This bewildering diversity," as V. Gordon Childe observed, "though embarrassing to the student and confusing on a map, is yet a significant feature in the pattern of European prehistory."[18] Long-established, undisturbed tribal languages tend to be *more* varied than tribal material cultures (see chapter 6). One would therefore expect that the linguistic diversity of Neolithic/Eneolithic Europe should have been even more bewildering than its material-culture diversity, not less so, and certainly not markedly less.

Finally, this enormous area was just too big for the survival of a single language under the conditions of tribal economics and politics, with foot travel the only means of land transport. Mallory and I discussed the likely scale of tribal language territories in Neolithic/Eneolithic Europe, and Nettles described tribal language geographies in West Africa.[19] Most tribal cultivators in West Africa spoke languages distributed over less than 10,000 km². Foragers around the world generally had much larger language territories than farmers had, and shifting farmers in poor environments had larger language territories than intensive farmers had in rich environments. Among most tribal farmers the documented size of language *families*—not languages but language families like Indo-European or Uralic—has usually been significantly less than 200,000 km². Mallory used an average of 250,000–500,000 km² for Neolithic European language families just to make room on the large end for the many uncertainties involved. Still, that resulted in twenty to forty language families for Neolithic Europe.

The actual number of language families in Europe at 3500 BCE probably was less than this, as the farming economy had been introduced into Neolithic Europe through a series of migrations that began about 6500 BCE. The dynamics of long-distance migration, particularly among pioneer farmers, *can* lead to the rapid spread of an unusually homogeneous language over an unusually large area for a few centuries (see chapter 6), but then local differentiation should have set in. In Neolithic Europe several distinct migrations flowed from different demographic recruiting pools and went to different places, where they interacted with different Mesolithic forager language groups. This should have produced incipient language differentiation among the immigrant farmers within five hundred to a thousand years, by 6000–5500 BCE. In comparison, the migrations of Bantu-speaking cattle herders across central and southern Africa occurred about two thousand years ago, and Proto-Bantu has diversified since then into more than five hundred modern Bantu languages assigned to nineteen branches, still interspersed today with enclaves belonging to non-Bantu language families. Europe in 3500 BCE, two thousand to three thousand years after the initial farming migrations, probably had at least the linguistic diversity of modern central and southern Africa— hundreds of languages that were descended from the original Neolithic farmers' speech, interspersed with pre-Neolithic language families of different types. The language of the original migrants to Greece cannot have remained a single language for three thousand years after its speakers were dispersed over many millions of square kilometers and several climate zones. Ethnographic or historic examples of such a large, stable language territory among tribal farmers simply do not exist.

That the speakers of Proto-Indo-European had wagons and a wagon vocabulary cannot be brought into agreement with a dispersal date as early as 6500 BCE. The wagon vocabulary is incompatible with the first-farming/language-dispersal hypothesis. Proto-Indo-European cannot have been spoken in Neolithic Greece and still have existed three thousand years later when wagons were invented. Proto-Indo-European therefore did not spread with the farming economy. Its first dispersal occurred much later, after 4000 BCE, in a European landscape that was already densely occupied by people who probably spoke hundreds of languages.

The Birth and Death of Proto-Indo-European

The historically known early Indo-European languages set one chronological limit on Proto-Indo-European, a *terminus ante quem*, and the reconstructed vocabulary related to wool and wheels sets another limit, a *terminus post quem*. The latest possible date for Proto-Indo-European can be set at about 2500 BCE (chapter 3). The evidence of the wool and wagon/wheel vocabularies establishes that late Proto-Indo-European was spoken after about 4000–3500 BCE, probably after 3500 BCE. If we include in our definition of Proto-Indo-European the end of the archaic Anatolian-like stage, without a securely documented wheeled-vehicle vocabulary, and the dialects spoken at the beginning of the final dispersal about 2500 BCE, the maximum window extends from about 4500 to about 2500 BCE. This two thousand-year target guides us to a well-defined archaeological era.

Within this time frame the archaeology of the Indo-European homeland is probably consistent with the following sequence, which makes sense also in terms of both traditional branching studies and cladistics. Archaic Proto-Indo-European (partly preserved only in Anatolian) probably was spoken before 4000 BCE; early Proto-Indo-European (partly preserved in Tocharian) was spoken between 4000 and 3500 BCE; and late Proto-Indo-European (the source of Italic and Celtic with the wagon/wheel vocabulary) was spoken about 3500–3000 BCE. Pre-Germanic split away from the western edge of late Proto-Indo-European dialects about 3300 BCE, and Pre-Greek split away about 2500 BCE, probably from a different set of dialects. Pre-Baltic split away from Pre-Slavic and other northwestern dialects about 2500 BCE. Pre-Indo-Iranian developed from a northeastern set of dialects between 2500 and 2200 BCE.

Now that the target is fixed in time, we can solve the old and bitter debate about *where* Proto-Indo-European was spoken.

CHAPTER FIVE

Language and Place
The Location of the Proto-Indo-European Homeland

The Indo-European homeland is like the Lost Dutchman's Mine, a legend of the American West, discovered almost everywhere but confirmed nowhere. Anyone who claims to know its *real* location is thought to be just a little odd—or worse. Indo-European homelands have been identified in India, Pakistan, the Himalayas, the Altai Mountains, Kazakhstan, Russia, Ukraine, the Balkans, Turkey, Armenia, the North Caucasus, Syria/Lebanon, Germany, Scandinavia, the North Pole, and (of course) Atlantis. Some homelands seem to have been advanced just to provide a historical precedent for nationalist or racist claims to privileges and territory. Others are enthusiastically zany. The debate, alternately dryly academic, comically absurd, and brutally political, has continued for almost two hundred years.[1]

This chapter lays out the linguistic evidence for the location of the Proto-Indo-European homeland. The evidence will take us down a well-worn path to a familiar destination: the grasslands north of the Black and Caspian Seas in what is today Ukraine and southern Russia, also known as the Pontic-Caspian steppes (figure 5.1). Certain scholars, notably Marija Gimbutas and Jim Mallory, have argued persuasively for this homeland for the last thirty years, each using criteria that differ in some significant details but reaching the same end point for many of the same reasons.[2] Recent discoveries have strengthened the Pontic-Caspian hypothesis so significantly, in my opinion, that we can reasonably go forward on the assumption that this was the homeland.

PROBLEMS WITH THE CONCEPT OF "THE HOMELAND"

At the start I should acknowledge some fundamental problems. Many of my colleagues believe that it is impossible to identify *any* homeland for Proto-Indo-European, and the following are their three most serious concerns.

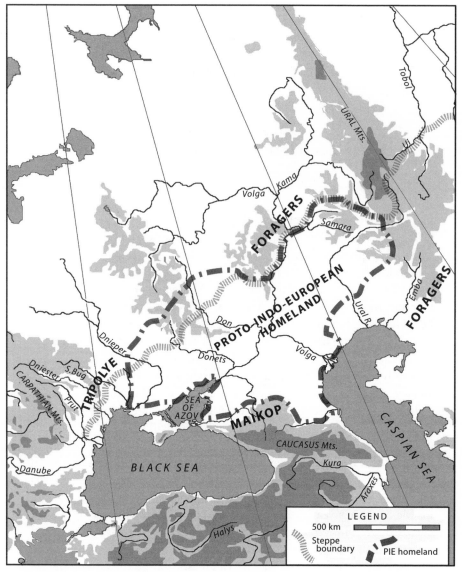

Figure 5.1 The Proto-Indo-European homeland between about 3500–3000 BCE.

Problem #1. Reconstructed Proto-Indo-European is merely a linguistic hypothesis, and hypotheses do not have homelands.

This criticism concerns the "reality" of reconstructed Proto-Indo-European, a subject on which linguists disagree. We should not imagine, some remind us, that reconstructed Proto-Indo-European was ever actually spoken anywhere. R.M.W. Dixon commented that if we cannot have "absolute certainty" about the grammatical type of a reconstructed language, it throws doubt over "every detail of the putative reconstruction."[3] But this is an extreme demand. The only field in which we can find absolute certainty is religion. In all other activities we must be content with the best (meaning both the simplest and the most data-inclusive) interpretation we can advance, given the data as they now stand. After we accept that this is true in *all* secular inquiries, the question of whether Proto-Indo-European can be thought of as "real" boils down to three sharper criticisms:

a. Reconstructed Proto-Indo-European is *fragmentary* (most of the language it represents never will be known).

b. The part that is reconstructed is *homogenized*, stripped of many of the peculiar sounds of its individual dialects, by the comparative method (although in reconstructed Proto-Indo-European some evidence of dialect survives).

c. Proto-Indo-European is not a snapshot of a moment in time but rather is "timeless": it *averages together centuries or even millennia* of development. In that sense, it is an accurate picture of no single era in language history.

These seem to be serious criticisms. But if their effect is to make Proto-Indo-European a mere fantasy, then the English language as presented in the Merriam-Webster Dictionary is a fantasy, too. My dictionary contains the English word *ombre* (a card game popular in the seventeenth and eighteenth centuries) as well as *hard disk* (a phrase that first appeared in the 1978 edition). So its vocabulary averages together at least three hundred years of the language. And its phonology, the "proper" pronuciation it describes, is quite restricted. Only one pronunciation is given for *hard disk*, and it is not the Bostonian *hard* [haahd]. The English of Merriam-Webster has never been spoken in its entirety by any one person. Nevertheless we all find it useful as a guide to real spoken English. Reconstructed Proto-Indo-European is similar, a dictionary version of a language. It is not, in itself, a real language, but it certainly *refers* to one. And we should remember that Sumerian cuneiform documents and Egyptian hieroglyphs present exactly the same problems as reconstructed Proto-Indo-European: the written scripts do not clearly indicate every sound, so their phonology is uncertain; they contain only royal or priestly dialects; and they might preserve archaic linguistic forms, like Church Latin. They are not, in themselves, real languages; they only *refer* to real languages. Reconstructed Proto-Indo-European is not so different from cuneiform Sumerian.

If Proto-Indo-European is like a dictionary, then it cannot be "timeless." A dictionary is easily dated by its most recent entries. A dictionary containing the term *hard disk* is dated after 1978 in just the way that the wagon terminology in Proto-Indo-European dates it to a time after about 4000–3500 BCE. It is more dangerous to use negative information as a dating tool, since many words that really existed in Proto-Indo-European will never be reconstructed, but it is at least interesting that Proto-Indo-European does not contain roots for items like spoke, iron, cotton, chariot, glass, or coffee—things that were invented after the evolution and dispersal of the daughter languages, or, in the metaphor we are using, after the dictionary was printed.

Of course, the dictionary of reconstructed Proto-Indo-European is much more tattered than my copy of Merriam-Webster's. Many pages have been torn out, and those that survive are obscured by the passage of time. The problem of the missing pages bothers some linguists the most. A reconstructed proto-language can seem a disappointing skeleton with a lot of bones missing and the placement of others debated between experts. The complete language the skeleton once supported certainly is a theoretical construct. So is the flesh-and-blood image of any dinosaur. Nevertheless, like the paleontologist, I am happy to have even a fragmentary skeleton. I think of Proto-Indo-European as a partial grammar and a partial set of pronunciation rules attached to the abundant fragments of a very ancient dictionary. To some linguists, that might not add up to a "real" language. But to an archaeologist it is more valuable than a roomful of potsherds.

Problem #2. The entire concept of "reconstructed Proto-Indo-European" is a fantasy: the similarities between the Indo-European languages could just as well have come about by gradual convergence over thousands of years between languages that had very different origins.

This is a more radical criticism then the first one. It proposes that the comparative method is a rigged game that automatically produces a proto-language as its outcome. The comparative method is said to ignore the linguistic changes that result from inter-language borrowing and convergence. Gradual convergence between originally diverse tongues, these scholars claim, might have produced the similarities between the Indo-European languages.[4] If this were true or even probable there would indeed be no reason to pursue a single parent of the Indo-European languages. But the Russian linguist who inspired this line of questioning, Nikolai S. Trubetzkoy, worked in the 1930s before linguists really had the tools to investigate his startling suggestion.

Since then, quite a few linguists have taken up the problem of convergence between languages. They have greatly increased our understanding of how convergence happens and what its linguistic effects are. Although they disagree strongly with one another on some subjects, all recent studies of convergence accept that the Indo-European languages owe their essential similarities to descent from a common ancestral language, and not to convergence.[5] Of course, some convergence has occurred between neighboring Indo-European languages—it is not a question of all or nothing—but specialists agree that the basic structures that define the Indo-European language family can only be explained by common descent from a mother tongue.

There are three reasons for this unanimity. First, the Indo-European languages are the most thoroughly studied languages in the world—simply put, we know a lot about them. Second, linguists know of no language where bundled similarities of the kinds seen among the Indo-European languages have come about through borrowing or convergence between languages that were originally distinct. And, finally, the features known to typify creole languages—languages that *are* the product of convergence between two or more originally distinct languages—are not seen among the Indo-European languages. Creole languages are characterized by greatly reduced noun and pronoun inflections (no case or even single/plural markings); the use of pre-verbal particles to replace verb tenses ("we bin get" for "we got"); the general absence of tense, gender, and person inflections in verbs; a severely reduced set of prepositions; and the use of repeated forms to intensify adverbs and adjectives. In each of these features Proto-Indo-European was the *opposite* of a typical creole. It is not possible to classify Proto-Indo-European as a creole by any of the standards normally applied to creole languages.[6]

Nor do the Indo-European daughter languages display the telltale signs of creoles. This means that the Indo-European vocabularies and grammars replaced competing languages rather than creolizing with them. Of course, some back-and-forth borrowing occurred—it always does in cases of language contact—but superficial borrowing and creolization are very different things. Convergence simply cannot explain the similarities between the Indo-European languages. If we discard the mother tongue, we are left with *no* explanation for the regular correspondences in sound, morphology, and meaning that define the Indo-European language family.

Problem #3. Even if there was a homeland where Proto-Indo-European was spoken, you cannot use the reconstructed vocabulary to find it because the reconstructed vocabulary is full of anachronisms that never existed in Proto-Indo-European.

This criticism, like the last one, reflects concerns about recent inter-language borrowing, focused here on just the vocabulary. Of course, many borrowed words are known to have spread through the Indo-European daughter languages long after the period of the proto-language—recent examples are *coffee* (borrowed from Arabic through Turkish) and *tobacco* (from Carib). The words for these items sound alike and have the same meanings in the different Indo-European languages, but few linguists would mistake them for ancient inherited words. Their phonetics are non–Indo-European, and their forms in the daughter branches do not represent what would be expected from inherited roots.[7] Terms like *coffee* are not a significant source of contamination.

Historical linguists do not ignore borrowing between languages. An understanding of borrowing is essential. For example, subtle inconsistencies embedded within German, Greek, Celtic, and other languages, including such fleeting sounds as the word-initial [kn-] (*knob*) can be identified as phonetically uncharacteristic of Indo-European. These fragments from extinct non–Indo-European languages are preserved only *because* they were borrowed. They can help us create maps of pre–Indo-European place-names, like the places ending with [-ssos] or [-nthos] (Corinthos, Knossos, Parnassos), borrowed into Greek and thought to show the geographic distribution of the pre-Greek language(s) of the Aegean and western Anatolia. Borrowed non–Indo-European sounds also were used to reconstruct some aspects of the long-extinct non–Indo-European languages of northern and eastern Europe. All that is left of these tongues is an occasional word or sound in the Indo-European languages that replaced them. Yet we can still identify their fragments in words borrowed thousands of years ago.[8]

Another regular use of borrowing is the study of "areal" features like *Sprachbund*s. A *Sprachbund* is a region where several different languages are spoken interchangeably in different situations, leading to their extensive borrowing of features. The most famous *Sprachbund* is in southeastern Europe, where Albanian, Bulgarian, Serbo-Croat, and Greek share many features, with Greek as the dominant element, probably because of its association with the Greek Orthodox Church. Finally, borrowing is an ever-present factor in any study of "genetic" relatedness. Whenever a linguist tries to decide whether cognate terms in two daughter languages are inherited from a common source, one alternative that must be excluded is that one language borrowed the term from the other. Many of the methods of comparative linguistics *depend* on the accurate identification of borrowed words, sounds, and morphologies.

When a root of similar sound and similar meaning shows up in widely separated Indo-European languages (including an ancient language), and phonological comparison of its forms yields a single ancestral root, that root term can be assigned with some confidence to the Proto-Indo-European vocabulary. No single reconstructed root should be used as the basis for an elaborate theory about Proto-Indo-European culture, but we do not need to work with single roots; we have clusters of terms with related meanings. At least fifteen hundred unique Proto-Indo-European roots have been reconstructed, and many of these unique roots appear in multiple reconstructed Proto-Indo-European words, so the total count of reconstructed Proto-Indo-European terms is much greater than fifteen hundred. Borrowing is a specific problem that affects specific reconstructed roots, but it does not cancel the usefulness of a reconstructed vocabulary containing thousands of terms.

The Proto-Indo-European homeland is not a racist myth or a purely theoretical fantasy. A real language lies behind reconstructed Proto-Indo-European, just as a real language lies behind any dictionary. And that language is a guide to the thoughts, concerns, and material culture of real people who lived in a definite region between about 4500 and 2500 BCE. But where was that region?

FINDING THE HOMELAND: ECOLOGY AND ENVIRONMENT

Regardless of where they ended up, most investigators of the Indo-European problem all started out the same way. The first step is to identify roots in the reconstructed Proto-Indo-European vocabulary referring to animal and plant species or technologies that existed only in certain places at particular times. The vocabulary itself should point to a homeland, at least within broad limits. For example, imagine that you were asked to identify the home of a group of people based only on the knowledge that a linguist had recorded these words in their normal daily speech:

armadillo	*sagebrush*	*cactus*
stampede	*steer*	*heifer*

calf	branding-iron	chuck-wagon
stockyard	rail-head	six-gun
saddle	lasso	horse

You could identify them fairly confidently as residents of the American southwest, probably during the late nineteenth or early twentieth centuries (*six-gun* and the absence of words for trucks, cars, and highways are the best chronological indicators). They probably were cowboys—or pretending to be. Looking closer, the combination of *armadillo*, *sagebrush*, and *cactus* would place them in west Texas, New Mexico, or Arizona.

Linguists have long tried to find animal or plant names in the reconstructed Proto-Indo-European vocabulary referring to species that lived in just one part of the world. The reconstructed Proto-Indo-European term for *salmon*, *lók*s*, was once famous as definite proof that the "Aryan" homeland lay in northern Europe. But animal and tree names seem to narrow and broaden in meaning easily. They are even reused and recycled when people move to a new environment, as English colonists used *robin* for a bird in the Americas that was a different species from the robin of England. The most specific meaning most linguists would now feel comfortable ascribing to the reconstructed term *lók*s*- is "trout-like fish." There are fish like that in the rivers across much of northern Eurasia, including the rivers flowing into the Black and Caspian Seas. The reconstructed Proto-Indo-European root for *beech* has a similar history. Because the copper beech, *Fagus silvatica*, did not grow east of Poland, the Proto-Indo-European root *bhágo- was once used to support a northern or western European homeland. But in some Indo-European languages the same root refers to other tree species (oak or elder), and in any case the common beech (*Fagus orientalis*) grows also in the Caucasus, so its original meaning is unclear. Most linguists at least agree that the fauna and flora designated by the reconstructed vocabulary are temperate-zone types (*birch*, *otter, beaver, lynx, bear, horse*), not Mediterranean (no *cypress, olive,* or *laurel*) and not tropical (no *monkey, elephant, palm,* or *papyrus*). The roots for *horse* and *bee* are most helpful.

Bee and *honey* are very strong reconstructions based on cognates in most Indo-European languages. A derivative of the term for honey, *medhu-*, was also used for an intoxicating drink, mead, that probably played a prominent role in Proto-Indo-European rituals. Honeybees were not native east of the Ural Mountains, in Siberia, because the hardwood trees (lime and oak, particularly) that wild honeybees prefer as nesting sites were rare or absent east of the Urals. If bees and honey did not exist in Siberia, the homeland could not have been there. That removes all of Siberia and much of northeastern Eurasia from contention, including the Central Asian steppes of Kazakhstan. The horse, *ek*wo-*, is solidly reconstructed and seems also to have been a potent symbol of divine power for the speakers of Proto-Indo-European. Although horses lived in small, isolated pockets throughout prehistoric Europe, the Caucasus, and Anatolia between 4500 and 2500 BCE, they were rare or absent in the Near East, Iran, and the Indian subcontinent. They were numerous and economically important only in the Eurasian steppes. The term for horse removes the Near East, Iran, and the Indian subcontinent from serious contention, and encourages us to look closely at the Eurasian steppes. This leaves temperate Europe, including the steppes west of the Urals, and the temperate parts of Anatolia and the Caucasus Mountains.[9]

Finding the Homeland: The Economic and Social Setting

The speakers of Proto-Indo-European were farmers and stockbreeders: we can reconstruct words for *bull, cow, ox, ram, ewe, lamb, pig,* and *piglet.* They had many terms for milk and dairy foods, including *sour milk, whey,* and *curds.* When they led their cattle and sheep out to the *field* they walked with a faithful *dog.* They knew how to *shear wool,* which they used to *weave* textiles (probably on a horizontal band loom). They tilled the earth (or they knew people who did) with a scratch-plow, or *ard,* which was pulled by *oxen* wearing a *yoke.* There are terms for *grain* and *chaff,* and perhaps for *furrow.* They turned their grain into flour by *grind*ing it with a hand *pestle,* and cooked their food in clay *pots* (the root is actually for *cauldron,* but that word in English has been narrowed to refer to a metal cooking vessel). They divided their possessions into two categories: movables and immovables; and the root for *movable wealth* (*peku-,* the ancestor of such English words as *pecuniary*) became the term for *herds* in general.[10] Finally, they were not averse to increasing their herds at their neighbors' expense, as we can reconstruct verbs that meant "to drive cattle," used in Celtic, Italic, and Indo-Iranian with the sense of cattle raiding or "rustling."

What was social life like? The speakers of Proto-Indo-European lived in a world of tribal politics and social groups united through kinship and marriage. They lived in households (*dómh$_a$*), containing one or more families (*génh$_1$es-*) organized into clans (*weiǩ-*), which were led by clan leaders, or chiefs (*weik-potis*). They had no word for *city*. Households appear to have been male-centered. Judging from the reconstructed kin terms, the important named kin were predominantly on the father's side, which suggests patrilocal marriages (brides moved into the husband's household). A group identity above the level of the clan was probably *tribe* (*$h_4erós$*), a root that developed into *Aryan* in the Indo-Iranian branch.[11]

The most famous definition of the basic divisions in Proto-Indo-European society was the tripartite scheme of Georges Dumézil, who suggested that there was a fundamental three-part division between the ritual specialist or priest, the warrior, and the ordinary herder/cultivator. Colors might have been associated with these three roles: white for the priest, red for the warrior, and black or blue for the herder/cultivator; and each role might have been assigned a specific type of ritual/legal death: strangulation for the priest, cutting/stabbing for the warrior, and drowning for the herder/cultivator. A variety of other legal and ritual distinctions seem to have applied to these three identities. It is unlikely that Dumézil's three divisions were groups with a limited membership. Probably they were something much less defined, like three age grades through which all males were expected to pass—perhaps herders (young), warriors (older), and lineage elders/ritual leaders (oldest), as among the Maasai in east Africa. The warrior category was regarded with considerable ambivalence, often represented in myth by a figure who alternated between a protector and a berserk murderer who killed his own father (Hercules, Indra, Thor). Poets occupied another respected social category. Spoken words, whether poems or oaths, were thought to have tremendous power. The poet's praise was a mortal's only hope for immortality.

The speakers of Proto-Indo-European were tribal farmers and stock-breeders. Societies like this lived across much of Europe, Anatolia, and the Caucasus Mountains after 6000 BCE. But regions where hunting and gathering economies persisted until after 2500 BCE are eliminated as possible homelands, because Proto-Indo-European was a dead language by 2500 BCE. The northern temperate forests of Europe and Siberia are excluded by this stockbreeders-before-2500 BCE rule, which cuts away one more piece of the map. The Kazakh steppes east of the Ural Mountains are excluded as well. In fact, this rule, combined with the exclusion of tropical regions and the presence of honeybees, makes a homeland anywhere east of the Ural Mountains unlikely.

Finding the Homeland: Uralic and Caucasian Connections

The possible homeland locations can be narrowed further by identifying the neighbors. The neighbors of the speakers of Proto-Indo-European can be identified through words and morphologies borrowed between Proto-Indo-European and other language families. It is a bit risky to discuss borrowing between reconstructed proto-languages—first, we have to reconstruct a phonological system for each of the proto-languages, then identify roots of similar form and meaning in both proto-languages, and finally see if the root in one proto-language meets all the expectations of a root borrowed from the other. If neighboring proto-languages have the same roots, reconstructed independently, and one root can be explained as a predictable outcome of borrowing from the other, then we have a strong case for borrowing. So who borrowed words from, or loaned words into, Proto-Indo-European? Which language families exhibit evidence of early contact and interchange with Proto-Indo-European?

Uralic Contacts

By far the strongest linkages can be seen with Uralic. The Uralic languages are spoken today in northern Europe and Siberia, with one southern off-shoot, Magyar, in Hungary, which was conquered by Magyar-speaking invaders in the tenth century. Uralic, like Indo-European, is a broad language family; its daughter languages are spoken across the northern forests of Eurasia from the Pacific shores of northeastern Siberia (Nganasan, spoken by tundra reindeer herders) to the Atlantic and Baltic coasts (Finnish, Estonian, Saami, Karelian, Vepsian, and Votian). Most linguists divide the family at the root into two super-branches, Finno-Ugric (the western branch) and Samoyedic (the eastern), although Salminen has argued that this binary division is based more on tradition than on solid linguistic evidence. His alternative is a "flat" division of the language family into nine branches, with Samoyedic just one of the nine.[12]

The homeland of Proto-Uralic probably was in the forest zone centered on the southern flanks of the Ural Mountains. Many argue for a homeland west of the Urals and others argue for the east side, but almost all Uralic linguists and Ural-region archaeologists would agree that Proto-Uralic was spoken somewhere in the birch-pine forests between the Oka River on the west (around modern Gorky) and the Irtysh River on the east (around modern Omsk). Today the Uralic languages spoken in this core region include, from west to east, Mordvin, Mari, Udmurt, Komi, and Mansi, of which two (Udmurt and Komi) are stems on the same branch (Permian). Some linguists have proposed homelands located farther east (the Yenisei River) or farther west (the Baltic), but the evidence for these extremes has not convinced many.[13]

The reconstructed Proto-Uralic vocabulary suggests that its speakers lived far from the sea in a forest environment. They were foragers who hunted and fished but possessed no domesticated plants or animals except the dog. This correlates well with the archaeological evidence. In the region between the Oka and the Urals, the Lyalovo culture was a center of cultural influences and interchanges among forest-zone forager cultures, with inter-cultural connections extending from the Baltic to the eastern slopes of the Urals during approximately the right period, 4500–3000 BCE.

The Uralic languages show evidence of very early contact with Indo-European languages. How that contact is interpreted is a subject of debate. There are three basic positions. First, the *Indo-Uralic* hypothesis suggests that the morphological linkages between the two families are so deep (shared pronouns), and the kinds of shared vocabulary so fundamental (words for *water* and *name*), that Proto-Indo-European and Proto-Uralic must have inherited these shared elements from some very ancient common linguistic parent—perhaps we might call it a "grandmother-tongue." The second position, the *early loan* hypothesis, argues that the forms of the shared proto-roots for terms like *name* and *water*, as reconstructed in the vocabularies of both Proto-Uralic and Proto-Indo-European, are much too similar to reflect such an ancient inheritance. Inherited roots should have undergone sound shifts in each developing family over a long period, but these roots are so similar that they can only be explained as loans from one proto-language into the other—and, in all cases, the loans went from Proto-Indo-European into Proto-Uralic.[14] The third position, the *late loan* hypothesis, is the one perhaps encountered most frequently in the general literature. It claims that there is little or no convincing evidence for borrowings even as old as the respective proto-languages; instead, the oldest well-documented loans should be assigned to contacts between Indo-Iranian and late Proto-Uralic, long after the Proto-Indo-European period. Contacts with Indo-Iranian could not be used to locate the Proto-Indo-European homeland.

At a conference dedicated to these subjects held at the University of Helsinki in 1999, not one linguist argued for a strong version of the late-loan hypothesis. Recent research on the earliest loans has reinforced the case for an early period of contact at least as early as the level of the proto-languages. This is well reflected in vocabulary loans. Koivulehto discussed at least thirteen words that are probable loans from Proto-Indo-European (PIE) into Proto-Uralic (P-U):

1. *to give* or *to sell*; P-U **mexe* from PIE **h₂mey-gʷ-* 'to change', 'exchange'
2. *to bring, lead,* or *draw*; P-U **wetä-* from PIE **wedʰ-e/o-* 'to lead', 'to marry', 'to wed'
3. *to wash*; P-U **mośke-* from PIE **mozg-eye/o-* 'to wash', 'to submerge'
4. *to fear*; P-U **pele-* from PIE **pelh₁-* 'to shake', 'cause to tremble'
5. *to plait, to spin*; P-U **puna-* from PIE **pn.H-e/o-* 'to plait', 'to spin'
6. *to walk, wander, go*; P-U **kulke-* from PIE **kʷelH-e/o-* 'it/he/she walks around', 'wanders'
7. *to drill, to bore*; P-U **pura-* from PIE **bʰr̥H-* 'to bore', 'to drill'
8. *shall, must, to have to*; P-U **kelke-* from PIE **skelH-* 'to be guilty', 'shall', 'must'
9. *long thin pole*; P-U **śalka-* from PIE **gʰalgʰo-* 'well-pole', 'gallows', 'long pole'
10. *merchandise, price*; P-U **wosa* from PIE **wosā* 'merchandise', 'to buy'
11. *water;* P-U **wete* from PIE **wed-er/en,* 'water', 'river'
12. *sinew;* P-U **sōne* from PIE **sneH(u)-* 'sinew'
13. *name;* P-U **nime-* from PIE **h₃neh₃mn-* 'name'

Another thirty-six words were borrowed from differentiated Indo-European daughter tongues into early forms of Uralic prior to the emergence of differentiated Indic and Iranian—before 1700–1500 BCE at the latest. These later words included such terms as *bread, dough, beer, to winnow,* and *piglet,* which might have been borrowed when the speakers of Uralic languages began to adopt agriculture from neighboring Indo-European–speaking farmers and herders. But the loans between the proto-languages are the important ones bearing on the location of the Proto-Indo-European homeland. And that they are so similar in form does suggest that they were loans rather than inheritances from some very ancient common ancestor.

This does not mean that there is no evidence for an older level of shared ancestry. Inherited similarities, reflected in shared pronoun forms and some noun endings, might have been retained from such a common ancestor. The pronoun and inflection forms shared by Indo-European and Uralic are the following:

Proto-Uralic		Proto-Indo-European
*te-nä	(*thou*)	*ti (?)
*te	(*you*)	*ti (clitic dative)
*me-nä	(*I*)	*mi
*tä-/to-	(*this/that*)	*te-/to-
*ke-, ku-	(*who, what*)	*kʷe/o-
*-m	(*accusative sing.*)	*-m
*-n	(*genitive plural*)	*-om

These parallels suggest that Proto-Indo-European and Proto-Uralic shared two kinds of linkages.[15] One kind, revealed in pronouns, noun endings, and shared basic vocabulary, could be ancestral: the two proto-languages shared some quite ancient common ancestor, perhaps a broadly related set of intergrading dialects spoken by hunters roaming between the Carpathians and the Urals at the end of the last Ice Age. The relationship is so remote, however, that it can barely be detected. Johanna Nichols has called this kind of very deep, apparently genetic grouping a "quasi-stock."[16] Joseph Greenberg saw Proto-Indo-European and Proto-Uralic as particularly close cousins within a broader set of such language stocks that he called "Eurasiatic."

The other link between Proto-Indo-European and Proto-Uralic seems cultural: some Proto-Indo-European words were borrowed by the speakers of Proto-Uralic. Although they seem odd words to borrow, the terms *to wash*, *price*, and *to give* or *to sell* might have been borrowed through a trade jargon used between Proto-Uralic and Proto-Indo-European speakers. These two kinds of linguistic relationship—a possible common ancestral origin and inter-language borrowings—suggest that the Proto-Indo-European homeland was situated near the homeland of Proto-Uralic, in the vicinity of the southern Ural Mountains. We also know that the speakers of Proto-Indo-European were farmers and herders whose language had disappeared by 2500 BCE. The people living east of the Urals did not adopt domesticated animals until *after* 2500 BC. Proto-Indo-European must therefore have been spoken somewhere to the *south and west of the Urals*, the only region close to the Urals where farming and herding was regularly practiced before 2500 BCE.

Caucasian Contacts and the Anatolian Homeland

Proto-Indo-European also had contact with the languages of the Caucasus Mountains, primarily those now classified as South Caucasian or Kartvelian, the family that produced modern Georgian. These connections have suggested to some that the Proto-Indo-European homeland should be placed in the Caucasus near Armenia or perhaps in nearby eastern Anatolia. The links between Proto-Indo-European and Kartvelian are said to appear in both phonetics and vocabulary, although the phonetic link is controversial. It depends on a brilliant but still problematic revision of the phonology of Proto-Indo-European proposed by the linguists T. Gamkrelidze and V. Ivanov, known as the glottalic theory.[17] The glottalic theory made Proto-Indo-European phonology sound somewhat similar to that of Kartvelian, and even to the Semitic languages (Assyrian, Hebrew, Arabic) of the ancient Near East. This opened the possibility that Proto-Indo-European, Proto-Kartvelian, and Proto-Semitic might have evolved in a region where they shared certain areal phonological features. But by itself the glottalic phonology cannot prove a homeland in the Caucasus, even if it is accepted. And the glottalic phonology still has failed to convince many Indo-European linguists.[18]

Gamkrelidze and Ivanov have also suggested that Proto-Indo-European contained terms for panther, lion, and elephant, and for southern tree species. These animals and trees could be used to exclude a northern homeland. They also compiled an impressive list of loan words which they said were borrowed from Proto-Kartvelian and the Semitic languages into Proto-Indo-European. These relationships suggested to them that Proto-Indo-European had evolved in a place where it was in close contact with both the Semitic languages and the languages of the Southern Caucasus. They suggested Armenia as the most probable Indo-European homeland. Several archaeologists, prominently Colin Renfrew and Robert Drews, have followed their general lead, borrowing some of their linguistic arguments but placing the Indo-European homeland a little farther west, in central or western Anatolia.

But the evidence for a Caucasian or Anatolian homeland is weak. Many of the terms suggested as loans from Semitic into Proto-Indo-European have been rejected by other linguists. The few Semitic-to-Proto-Indo-European loan words that are widely accepted, words for items like silver and bull, might be words that were carried along trade and migration routes far from the Semites' Near Eastern homeland. Johanna Nichols has shown from the phonology of the loans that the Proto-Indo-European/Proto-Kartvelian/Proto-Semitic contacts were indirect—all the loan words passed through unknown intermediaries between the known three. One intermediary is required by chronology, as Proto-Kartvelian is generally thought to have existed after Proto-Indo-European and Proto-Semitic.[19]

The Semitic and Caucasian vocabulary that was borrowed into Proto-Indo-European through Kartvelian therefore contains roots that belonged to some *Pre-Kartvelian* or *Proto-Kartvelian* language in the Caucasus. This language had relations, through unrecorded intermediaries, with Proto-Indo-European on one side and Proto-Semitic on the other. That is not a particularly close lexical relationship. If Proto-Kartvelian was spoken on the south side of the North Caucasus Mountain range, as seems likely, it might have been spoken by people associated with the Early Transcaucasian Culture (also known as the Kura-Araxes culture), dated about 3500–2200 BCE. They could have had indirect relations with the speakers of Proto-Indo-European through the Maikop culture of the North Caucasus region. Many experts agree that Proto-Indo-European shared some features with a language ancestral to Kartvelian but not necessarily through a direct face-to-face link. Relations with the speakers of Proto-Uralic were closer.

So who were the neighbors? Proto-Indo-European exhibits strong links with Proto-Uralic and weaker links with a language ancestral to Proto-Kartvelian. The speakers of Proto-Indo-European lived somewhere between the Caucasus and Ural Mountains but had deeper linguistic relationships with the people who lived around the Urals.

The Location of the Proto-Indo-European Homeland

The speakers of Proto-Indo-European were tribal farmers who cultivated grain, herded cattle and sheep, collected honey from honeybees, drove wagons, made wool or felt textiles, plowed fields at least occasionally or knew people who did, sacrificed sheep, cattle, and horses to a troublesome array of sky gods, and fully expected the gods to reciprocate the favor. These traits guide us to a specific kind of material culture—one with wagons, domesticated sheep and cattle, cultivated grains, and sacrificial deposits with the bones of sheep, cattle, and horses. We should also look for a specific kind of ideology. In the reciprocal exchange of gifts and favors between their patrons, the gods, and human clients, humans offered a portion of their herds through sacrifice, accompanied by well-crafted verses of praise; and the gods in return provided protection from disease and misfortune, and the blessings of power and prosperity. Patron-client reciprocity of this kind is common among chiefdoms, societies with institutionalized differences in prestige and power, where some clans or lineages claim a right of patronage over others, usually on grounds of holiness or historical priority in a given territory.

Knowing that we are looking for a society with a specific list of material culture items and institutionalized power distinctions is a great help in locating the Proto-Indo-European homeland. We can exclude all regions where hunter-gatherer economies survived up to 2500 BCE. That eliminates the northern forest zone of Eurasia and the Kazakh steppes east of the Ural Mountains. The absence of honeybees east of the Urals eliminates any part of Siberia. The temperate-zone flora and fauna in the reconstructed vocabulary, and the absence of shared roots for Mediterranean or tropical flora and fauna, eliminate the tropics, the Mediterranean, and the Near East. Proto-Indo-European exhibits some very ancient links with the Uralic languages, overlaid by more recent lexical borrowings into Proto-Uralic from Proto-Indo-European; and it exhibits less clear linkages to some Pre- or Proto-Kartvelian language of the Caucasus region. All these requirements would be met by a Proto-Indo-European homeland placed west of the Ural Mountains, between the Urals and the Caucasus, in the steppes of eastern Ukraine and Russia. The internal coherence of reconstructed Proto-Indo-European—the absence of evidence for radical internal variation in grammar and phonology—indicates that the period of language history it reflects was less than two thousand years, probably less than one thousand. The heart of the Proto-Indo-European period probably fell between 4000 and 3000 BCE, with an early phase that might go back to 4500 BCE and a late phase that ended by 2500 BCE.

What does archaeology tell us about the steppe region between the Caucasus and the Urals, north of the Black and Caspian Seas—the Pontic-Caspian region—during this period? First, archaeology reveals a set of cultures that fits all the requirements of the reconstructed vocabulary: they sacrificed domesticated horses, cattle, and sheep, cultivated grain at least occasionally, drove wagons, and expressed institutionalized status distinctions in their funeral rituals. They occupied a part of the world—the steppes—where the sky is by far the most striking and magnificent part of

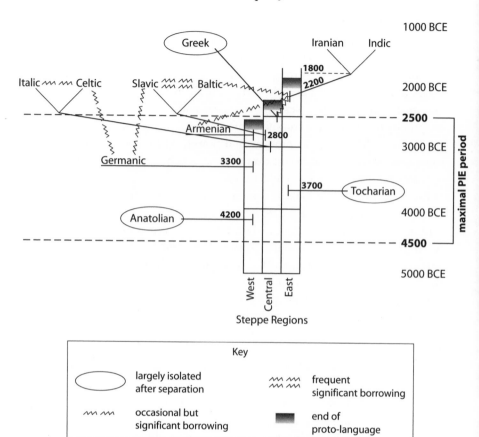

Figure 5.2 A diagram of the sequence and approximate dates of splits in early Indo-European as proposed in this book, with the maximal window for Proto-Indo-European indicated by the dashed lines. The dates of splits are determined by archaeological events described in chapters 11 (Anatolian) through 16 (Iranian and Indic).

the landscape, a fitting environment for people who believed that all their most important deities lived in the sky. Archaeological evidence for migrations from this region into neighboring regions, both to the west and to the east, is well established. The sequence and direction of these movements matches the sequence and direction suggested by Indo-European linguistics and geography (figure 5.2). The first identifiable migration out of the Pontic-Caspian steppes was a movement toward the west about 4200–3900 BCE that could represent the detachment of the Pre-Anatolian branch, at a time before wheeled vehicles were introduced to the steppes (see chapter

4). This was followed by a movement toward the east (about 3700–3300 BCE) that could represent the detachment of the Tocharian branch. The next visible migration out of the steppes flowed toward the west. Its earliest phase might have separated the Pre-Germanic branch, and its later, more visible phase detached the Pre-Italic and Pre-Celtic dialects. This was followed by movements to the north and east that probably established the Baltic-Slavic and Indo-Iranian tongues. The remarkable match between the archaeologically documented pattern of movements out of the steppes and that expected from linguistics is fascinating, but it has absorbed, for too long, most of the attention and debate that is directed at the archaeology of Indo-European origins. Archaeology also adds substantially to our cultural and economic understanding of the speakers of Proto-Indo-European. Once the homeland has been located with linguistic evidence, the archaeology of that region provides a wholly new kind of information, a new window onto the lives of the people who spoke Proto-Indo-European and the process by which it became established and began to spread.

Before we step into the archaeology, however, we should pause and think for a moment about the gap we are stepping across, the void between linguistics and archaeology, a chasm most Western archaeologists feel cannot be crossed. Many would say that language and material culture are completely unrelated, or are related in such changeable and complicated ways that it is impossible to use material culture to identify language groups or boundaries. If that is true, then even if we *can* identify the place and time of the Indo-European homeland using the reconstructed vocabulary, the link to archaeology is impossible. We cannot expect any correlation with material culture. But is such pessimism warranted? Is there *no* predictable, regular link between language and material culture?

CHAPTER SIX

The Archaeology of Language

A language homeland implies a bounded space of some kind. How can we define those boundaries? Can ancient linguistic frontiers be identified through archaeology?

Let us first define our terms. It would be helpful if anthropologists used the same vocabulary used in geography. According to geographers, the word *border* is neutral—it has no special or restricted meaning. A *frontier* is a specific kind of border—a transitional zone with some depth, porous to cross-border movement, and very possibly dynamic and moving. A frontier can be cultural, like the Western frontier of European settlement in North America, or ecological. An *ecotone* is an ecological frontier. Some ecotones are very subtle and small-scale—there are dozens of tiny ecotones in any suburban yard—and others are very large-scale, like the border between steppe and forest running east-west across central Eurasia. Finally, a sharply defined border that limits movement in some way is a *boundary*; for example, the political borders of modern nations are boundaries. But nation-like political and linguistic boundaries were unknown in the Pontic-Caspian region between 4500 and 2500 BCE. The cultures we are interested in were tribal societies.[1]

Archaeologists' interpretations of premodern tribal borders have changed in the last forty years. Most pre-state tribal borders are now thought to have been porous and dynamic—frontiers, not boundaries. More important, most are thought to have been ephemeral. The tribes Europeans encountered in their colonial ventures in Africa, South Asia, the Pacific, and the Americas were at first assumed to have existed for a long time. They often claimed antiquity for themselves. But many tribes are now believed to have been transient political communities of the historical moment. Like the Ojibwa, some might have crystallized only after contact with European agents who wanted to deal with bounded groups to facilitate the negotiation of territorial treaties. And the same critical attitude toward bounded tribal territories is applied to European history. Ancient European tribal identities—Celt, Scythian, Cimbri, Teuton, and Pict—are now frequently seen as convenient names for chameleon-like political alliances that had no true ethnic identity, or as brief ethnic phenomena that were *unable* to persist for any length of time, or even as entirely imaginary later inventions.[2]

Pre-state language borders are thought to have been equally fluid, characterized by intergrading local dialects rather than sharp boundaries. Where language and material culture styles (house type, town type, economy, dress, etc.) did coincide geographically to create a tribal ethnolinguistic frontier, we should expect it to have been short-lived. Language and material culture can change at different speeds for different reasons, and so are thought to grow apart easily. Historians and sociologists from Eric Hobsbawm to Anthony Giddens have proposed that there were no really distinct and stable ethnolinguistic borders in Europe until the late eighteenth century, when the French Revolution ushered in the era of nation-states. In this view of the past only the state is accorded both the need and the power to warp ethnolinguistic identity into a stable and persistent phenomenon, like the state itself. So how can we hope to identify ephemeral language frontiers in 3500 BCE? Did they even exist long enough to be visible archaeologically?[3]

Unfortunately this problem is compounded by the shortcomings of archaeological methods. Most archaeologists would agree that we do not really know how to recognize tribal ethnolinguistic frontiers, even if they *were* stable. Pottery styles were often assumed by pre–World War II archaeologists to be an indicator of social identity. But we now know that no simple connection exists between pottery types and ethnicity; as noted in chapter 1, every modern archaeology student knows that "pots are not people." The same problem applies to other kinds of material culture. Arrow-point types did seem to correlate with language families among the San hunter-gatherers of South Africa; however, among the Contact-period Native Americans in the northeastern U.S., the "Madison"-type arrow point was used by both Iroquoian and Algonkian speakers—its distribution had no connection to language. Almost any object could have been used to signal linguistic identity, or not. Archaeologists have therefore rejected the possibility that language and material culture are correlated in any predictable or recognizable way.[4]

But it seems that language and material culture are related in at least two ways. One is that tribal languages are generally more numerous in any long-settled region than tribal material cultures. Silver and Miller noticed, in 1997, that most tribal regions had more languages than material cultures. The Washo and Shoshone in the Great Basin had very different languages, of distinct language families, but similar material cultures; the Pueblo Indians had more languages than material cultures; the California Indians had more languages than stylistic groups; and the Indians of the central Amazon are well known for their amazing linguistic variety and broadly similar material cultures. A Chicago Field Museum study of language and material culture in northern New Guineau, the most detailed of its type, confirmed that regions defined by material culture were crisscrossed with numerous materially invisible language borders.[5] But the opposite pattern seems to be rare: a homogeneous tribal language is rarely separated into two very distinct bundles of material culture. This regularity seems discouraging, as it guarantees that many prehistoric language borders must be archaeologically invisible, but it does help to decide such questions as whether one language could have covered all the varied material culture groups of Copper Age Europe (probably not; see chapter 4).

The second regularity is more important: language is correlated with material culture at very long-lasting, distinct material-culture borders.

Persistent Frontiers

Persistent cultural frontiers have been ignored, because, I believe, they were dismissed on theoretical grounds.[6] They are not supposed to be there, since pre-state tribal borders are interpreted today as ephemeral and unstable. But archaeologists have documented a number of remarkably long-lasting, prehistoric, material-culture frontiers in settings that must have been tribal. A robust, persistent frontier separated Iroquoian and Algonkian speakers along the Hudson Valley, who displayed different styles of smoking pipes, subtle variations in ceramics, quite divergent house and settlement types, diverse economies, and very different languages for at least three centuries prior to European contact. Similarly the Linear Pottery/Lengyel farmers created a robust material-culture frontier between themselves and the indigenous foragers in northern Neolithic Europe, a moving border that persisted for at least a thousand years; the Criş/Tripolye cultures were utterly different from the Dnieper-Donets culture on a moving frontier between the Dniester and Dnieper Rivers in Ukraine for twenty-five hundred years during the Neolithic and Eneolithic; and the Jastorf and Halstatt cultures maintained distinct identities for centuries on either side of the lower Rhine in the Iron Age.[7] In each of these cases cultural norms changed; house designs, decorative aesthetics, and religious rituals were not frozen in a single form on either side. It was the *persistent opposition of bundles of customs* that defined the frontier rather than any one artifact type.

Persistent frontiers need not be stable geographically—they can move, as the Romano-Celt/Anglo-Saxon material-culture frontier moved across Britain between 400 and 700 CE, or the Linear Pottery/forager frontier moved across northern Europe between 5400 and 5000 BCE. Some material-culture frontiers, described in the next chapters, survived for millennia, in a pre-state social world governed just by tribal politics—no border guards, no national press. Particularly clear examples defined the edges of the Pontic-Caspian steppes on the west (Tripolye/Dnieper), on the north (Russian forest forager/steppe herder), and on the east (Volga-Ural steppe herder/Kazakh steppe forager). These were the borders of the region that probably was the homeland of Proto-Indo-European. If ancient ethnicities were ephemeral and the borders between them short-lived, how do we understand premodern tribal material-culture frontiers that persisted for thousands of years? And can language be connected to them?

I think the answer is yes. Language is strongly associated with persistent material-culture frontiers that are defined by bundles of opposed customs, what I will call *robust* frontiers.[8] The migrations and frontier formation processes that followed the collapse of the Roman Empire in western Europe provide the best setting to examine this association, because documents and place-names establish the linguistic identity of the migrants, the locations of newly formed frontiers, and their persistence over many centuries in political contexts where centralized state governments were weak or nonexistent. For example, the cultural frontier between the Welsh (Celtic branch) and the English (Germanic branch) has persisted since the Anglo-Saxon conquest of Romano-Celtic Britain during the sixth century. Additional conquests by Norman-English feudal barons after 1277 pushed the frontier back to the *landsker*, a named and overtly recognized ethnolinguistic frontier between Celtic Welsh-speaking and Germanic English-speaking populations that persisted to the present day. They spoke different languages (Welsh/English), built different kinds of churches (Celtic/Norman English), managed agriculture differently and with different tools, used diverse systems of land measurement, employed dissimilar standards of justice, and maintained a wide variety of distinctions in dress, food, and custom. For many centuries men rarely married across this border, maintaining a genetic difference between modern Welsh and English men (but not women) in traits located on the male Y chromosome.

Other post-Roman ethnolinguistic frontiers followed the same pattern. After the fall of Rome German speakers moved into the northern cantons of Switzerland, and the Gallic kingdom of Burgundy occupied what had been Gallo-Roman western Switzerland. The frontier between them still separates ecologically similar regions within a single modern state that differ in language (German-French), religion (Protestant-Catholic), architecture, the size and organization of landholdings, and the nature of the agricultural economy. Another post-Roman migration created the Breton/French frontier across the base of the peninsula of Brittany, after Romano-Celts migrated to Brittany from western Britain around 400–600 CE, fleeing the Anglo-Saxons. For more than fifteen hundred years the Celtic-speaking Bretons have remained distinct from their French-speaking neighbors in rituals, dress, music, and cuisine. Finally, migrations around 900–1000 CE brought German speakers into what is now northeastern Italy, where the persistent frontier between Germans and Romance speakers inside Italy was studied by Eric Wolf and John Cole in the 1960s. Although in this case both cultures were Catholic Christians, after a thousand years they still maintained different languages, house types, settlement organizations, land tenure and inheritance systems, attitudes toward authority and cooperation, and quite unfavorable stereotypes of each other. In all these cases documents and inscriptions show that the ethnolinguistic oppositions were not recent or invented but deeply historical and persistent.[9]

These examples suggest that most persistent, robust material-culture frontiers were ethnolinguistic. Robust, persistent, material-culture frontiers are not found everywhere, so only exceptional language frontiers can be identified. But that, of course, is better than nothing.

Population Movement across Persistent Frontiers

Unlike the men of Wales and England, most people moved back and forth across persistent frontiers easily. A most interesting fact about stable ethnolinguistic frontiers is that they were not necessarily biological; they persisted for an extraordinarily long time despite people regularly moving across them. As Warren DeBoer described in his study of native pottery styles in the western Amazon basin, "ethnic boundaries in the Ucayali basin are highly permeable with respect to bodies, but almost inviolable with respect to style."[10] The back-and-forth movement of *people* is indeed the principal focus of most contemporary borderland studies. The persistence of the *borders* themselves has remained understudied, probably because modern nation-states insist that all borders are permanent and inviolable, and many nation-states, in an attempt to naturalize their borders, have tried to argue that they have persisted from ancient times. Anthropologists and historians alike dismiss this as a fiction; the borders I have discussed frequently persist *within* modern nation-states rather than corresponding to their modern boundaries. But I think we have failed to recognize that we have internalized the modern nation-state's basic premise by insisting that ethnic borders must be inviolable boundaries or they did not really exist.

If people move across an ethno-linguistic frontier freely, then the frontier is often described in anthropology as, in some sense, a fiction. Is this just because it was not a boundary *like that of a modern nation?* Eric Wolf used this very argument to assert that the North American Iroquois did not exist as a distinct tribe during the Colonial period; he called them a multiethnic trading company. Why? Because their communities were full of captured and adopted non-Iroquois. But if biology is independent of language and culture, then the simple movement of Delaware and Nanticoke *bodies* into Iroquoian towns should not imply a dilution of Iroquoian *culture*. What matters is how the immigrants acted. Iroquoian adoptees were required to behave as Iroquois or they might be killed. The Iroquoian cultural identity remained distinct, and it was long established and persistent. The idea that European nation-states created the Iroquois "nation" in their own European image is particularly ironic in view of the fact that the five nations or tribes of the pre-European Northern Iroquois can be traced back archaeologically *in their traditional five tribal territories* to 1300 CE, more than 250 years before European contact. An Iroquois might argue that the borders of the original five nations of the Northern Iroquois were demonstrably older than those of many European nation-states at the end of the sixteenth century.[11]

Language frontiers in Europe are not generally strongly correlated with genetic frontiers; people mated across them. But persistent ethnolinguistic frontiers probably did originate in places where relatively *few* people moved between neighboring mating and migration networks. Dialect borders usually are correlated with borders between socioeconomic "functional zones," as linguists call a region marked by a strong network of intra-migration and socioeconomic interdependence. (Cities usually are divided into several distinct socioeconomic-linguistic functional zones.) Labov, for example, showed that dialect borders in central Pennsylvania correlated with reduced cross-border traffic flow densities at the borders of functional zones. In some places, like the Welsh/English border, the cross-border flow of people was low enough to appear genetically as a contrast in gene pools, but at other persistent frontiers there was enough cross-border movement to blur genetic differences. What, then, maintained the frontier itself, the persistent sense of difference?[12]

Persistent, robust premodern ethnolinguistic frontiers seem to have survived for long periods under one or both of two conditions: at *large-scale ecotones* (forest/steppe, desert/savannah, mountain/river bottom, mountain/ coast) and at places where long-distance migrants stopped migrating and formed a *cultural frontier* (England/Wales, Britanny/France, German Swiss/ French Swiss). Persistent identity depended partly on the continuous confrontation with Others that was inherent in these kinds of borders, as Frederik Barth observed, but it also relied on a home culture behind the border, a font of imagined tradition that could continuously feed those contrasts, as Eric Wolf recognized in Italy.[13] Let us briefly examine how these factors worked together to create and maintain persistent frontiers. We begin with borders created by long-distance migration.

MIGRATION AS A CAUSE OF PERSISTENT MATERIAL-CULTURE FRONTIERS

During the 1970s and 1980s the very idea of folk migrations was avoided by Western archaeologists. Folk migrations seemed to represent the boiled-down essence of the discredited idea that ethnicity, language, and material culture were packaged into neatly bounded societies that careened across the landscape like self-contained billiard balls, in a famously dismissive simile. Internal causes of social change—shifts in production and the means of production, in climate, in economy, in access to wealth and prestige, in political structure, and in spiritual beliefs—all got a good long look by archaeologists during these decades. While archaeologists were ignoring migration, modern demographers became very good at picking apart the various causes, recruiting patterns, flow dynamics, and targets of modern migration streams. Migration models moved far beyond the billiard ball analogy. The acceptance of modern migration models in the archaeology of the U.S. Southwest and in Iroquoian archaeology in the Northeast during the 1990s added new texture to the interpretation of Anasazi/Pueblo and Iroquoian societies, but in most other parts of the world the archaeological database was simply not detailed enough to test the very specific behavioral predictions of modern migration theories.[14] History, on the other hand, contains a very detailed record of the past, and among modern historians migration is accepted as a cause of persistent cultural frontiers.

The colonization of North America by English speakers is one prominent example of a well-studied, historical connection between migration
and ethnolinguistic frontier formation. Decades of historical research have
shown, surprisingly, that while the borders separating Europeans and
Native Americans were important, those that separated different British
cultures were just as significant. Eastern North America was colonized by
four distinct migration streams that originated in four different parts of
the British Isles. When they touched down in eastern North America,
they created four clearly bounded ethnolinguistic regions between about
1620 and 1750. The Yankee dialect was spoken in New England. The
same region also had a distinctive form of domestic architecture—the
salt-box clapboard house—as well as its own barn and church architecture, a distinctive town type (houses clustered around a common grazing
green), a peculiar cuisine (often baked, like Boston baked beans), distinct
fashions in clothing, a famous style of gravestones, and a fiercely legalistic
approach to politics and power. The geographic boundaries of the New
England folk-culture region, drawn by folklorists on the basis of these
traits, and the Yankee dialect region, drawn by linguists, coincide almost
exactly. The Yankee dialect was a variant of the dialect of East Anglia, the
region from which most of the early Pilgrim migrants came; and New England folk culture was a simplified version of East Anglian folk culture.
The other three regions also exhibited strongly correlated dialects and folk
cultures, as defined by houses, barn types, fence types, the frequency of
towns and their organization, food preferences, clothing styles, and religion. One was the mid-Atlantic region (Pennsylvania Quakers from the
English Midlands), the third was the Virginia coast (Royalist Anglican
tobacco planters from southern England, largely Somerset and Wessex),
and the last was the interior Appalachians (borderlanders from the Scotch-
Irish borders). Both dialect and folk culture are traceable in each case to a
particular region in the British Isles from which the first effective European settlers came.[15]

The four ethnolinguistic regions of Colonial eastern North America
were created by four separate migration streams that imported people with
distinctive ethnolinguistic identities into four different regions where simplified versions of their original linguistic and material differences were
established, elaborated, and persisted for centuries (table 6.1). In some
ways, including modern presidential voting patterns, the remnants of these
four regions survive even today. But can modern migration patterns be applied to the past, or do modern migrations have purely modern causes?

TABLE 6.1

Migration Streams to Colonial North America

Colonial Region	Source	Religion
New England	East Anglia/Kent	Puritan
Mid–Atlantic	English Midlandss/ Southern Germany	Quaker/German Protestant
Tidewater Virginia–Carolina	Somerset/Wessex	Anglican
Southern Appalachian	Scots–Irish borderlands	Calvinist/Celtic church

The Causes of Migration

Many archaeologists think that modern migrations are fueled principally by overpopulation and the peculiar boundaries of modern nation-states, neither of which affected the prehistoric world, making modern migration studies largely irrelevant to prehistoric societies.[16] But migrations have many causes besides overpopulation within state borders. People do not migrate, even in today's crowded world, simply because there are too many at home. Crowding would be called a "push" factor by modern demographers, a negative condition at home. But there are other kinds of "push" factors—war, disease, crop failure, climate change, institutionalized raiding for loot, high bride-prices, the laws of primogeniture, religious intolerance, banishment, humiliation, or simple annoyance with the neighbors. Many causes of today's migrations and those in the past were social, not demographic. In ancient Rome, feudal Europe, and many parts of modern Africa, *inheritance rules* favored older siblings, condemning the younger ones to find their own lands or clients, a strong motive for them to migrate.[17] Pushes could be even more subtle. The persistent outward migrations and conquests of the pre-Colonial East African Nuer were caused, according to Raymond Kelley, not by overpopulation within Nuerland but rather by a cultural system of *bride-price regulations* that made it very expensive for young Nuer men to obtain a socially desirable bride. A bride-price was a payment made by the groom to the bride's family to compensate for the loss of her labor. Escalation in bride-prices encouraged Nuer men to raid their non-Nuer neighbors for cattle (and pastures to support them) that could be used to pay the elevated bride-price for a high-status marriage. Tribal status rivalries supported by high brideprices in an arid, low-productivity environment led to out-migration and the rapid territorial expansion of the Nuer.[18] Grassland migrations among tribal pastoralists can be "pushed" by many things other than absolute resource shortages.

Regardless of how "pushes" are defined, *no* migration can be adequately explained by "pushes" alone. Every migration is affected as well by "pull" factors (the alleged attractions of the destination, regardless of whether they are true), by communication networks that bring information to potential migrants, and by transport costs. Changes in any of these factors will raise or lower the threshold at which migration becomes an attractive option. Migrants weigh these dynamics, for far from being an instinctive response to overcrowding, migration is often a *conscious social strategy* meant to improve the migrant's position in competition for status and riches. If possible, migrants recruit clients and followers among the people at home, convincing them also to migrate, as Julius Caesar described the recruitment speeches of the chiefs of the Helvetii prior to their migration from Switzerland into Gaul. Recruitment in the homeland by potential and already departed migrants has been a continuous pattern in the expansion and reproduction of West African clans and lineages, as Igor Kopytoff noted. There is every reason to believe that similar social calculations have inspired migrations since humans evolved.

Effects: The Archaeological Identification of Ancient Migrations

Large, sustained migrations, particularly those that moved a long distance from one cultural setting into a very different one, or *folk migrations*, can be identified archaeologically. Emile Haury knew most of what to look for already in his excavations in Arizona in the 1950s: (1) the sudden appearance of a new material culture that has no local antecedents or prototypes; (2) a simultaneous shift in skeletal types (biology); (3) a neighboring territory where the intrusive culture evolved earlier; and (4) (a sign not recognized by Haury) the introduction of new *ways* of making things, new technological styles, which we now know are more "fundamental" (like the core vocabulary in linguistics) than decorative styles.

Smaller-scale migrations by specialists, mercenaries, skilled craft workers, and so on, are more difficult to identify. This is partly because archaeologists have generally stopped with the four simple criteria just described and neglected to analyze the internal workings even of folk migrations. To really understand why and how folk migrations occurred, and to have any hope of identifying small-scale migrations, archaeologists have to study the internal structure of long-distance migration streams, both large and small.

The organization of migrating groups depends on the identity and social connections of the scouts (who select the target destination); the social organization of information sharing (which determines who gets access to the scouts' information); transportation technology (cheaper and more effective transport makes migration easier); the targeting of destinations (whether they are many or few); the identity of the first effective settlers (also called the "charter group"); return migration (most migrations have a counterflow going back home); and changes in the goals and identities of migrants who join the stream later. If we look for all these factors we can better understand why and how migrations happened. Sustained migrations, particularly by pioneers looking to settle in new homes, can create very long-lasting, persistent ethnolinguistic frontiers.

The Simplification of Dialect and Culture among Long-distance Migrants

Access to the scouts' information defines the pool of potential migrants. Studies have found that the first 10% of new migrants into a region is an accurate predictor of the social makeup of the population that will follow them. This restriction on information at the source produces two common behaviors: leapfrogging and chain migration. In leapfrogging, migrants go only to those places about which they have heard good things, skipping over other possible destinations, sometimes moving long distances in one leap. In chain migration, migrants follow kin and co-residents to familiar places with social support, not to the objectively "best" place. They jump to places where they can rely on people they know, from point to targeted point. Recruitment usually is relatively restricted, and this is clearly audible in their speech.

Colonist speech generally is more homogeneous than the language of the homeland they left behind. Dialectical differences were fewer among Colonial-era English speakers in North America than they were in the British Isles. The Spanish dialects of Colonial South America were more homogeneous than the dialects of Southern Spain, the home region of most of the original colonists. Linguistic simplification has three causes. One is chain migration, where colonists tend to recruit family and friends from the same places and social groups that the colonists came from. Simplification also is a normal linguistic outcome of mixing between dialects in a contact situation at the destination.[19] Finally, simplification is encouraged among long-distance migrants by the social influence of the charter group.

The first group to establish a viable social system in a new place is called the *charter group*, or the first effective settlers.[20] They generally get the best land. They might claim rights to perform the highest-status rituals, as among the Maya of Central America or the Pueblo Indians of the American Southwest. In some cases, for example, Puritan New England, their councils choose who is permitted to join them. Among Hispanic migrants in the U.S. Southwest, charter groups were called *apex families* because of their structural position in local prestige hierarchies. Many later migrants were indebted to or dependent on the charter group, whose dialect and material culture provided the cultural capital for a new group identity. Charter groups leave an inordinate cultural imprint on later generations, as the latter copy the charter group's behavior, at least publicly. This explains why the English language, English house forms, and English settlement types were retained in nineteenth-century Ohio, although the overwhelming majority of later immigrants was German. The charter group, already established when the Germans arrived, was English. It also explains why East Anglian English traits, typical of the earliest Puritan immigrants, continued to typify New England dialectical speech and domestic architecture long after the majority of later immigrants arrived from other parts of England or Ireland. As a font of tradition and success in a new land, the charter group exercised a kind of historical cultural hegemony over later generations. Their genes, however, could easily be swamped by later migrants, which is why it is often futile to pursue a genetic fingerprint associated with a particular language.

The combination of chain migration, which restricted the pool of potential migrants at home, and the influence of the charter group, which encouraged conformity at the destination, produced a leveling of differences among many colonists. Simplification (fewer variants than in the home region) and leveling (the tendency toward a standardized form) affected both dialect *and* material culture. In material culture, domestic architecture and settlement organization—the external form and construction of the house and the layout of the settlement—particularly tended toward standardization, as these were the most visible signals of identity in any social landscape.[21] Those who wished to declare their membership in the mainstream culture adopted its external domestic forms, whereas those who retained their old house and barn styles (as did some Germans in Ohio) became political, as well as architectural and linguistic, minorities. Linguistic and cultural homogeneity among long-distance migrants facilitated stereotyping by Others, and strengthened the illusion of shared interests and origins among the migrants.

ECOLOGICAL FRONTIERS: DIFFERENT WAYS OF MAKING A LIVING

Franz Boas, the father of American anthropology, found that the borders of American Indian tribes rarely correlated with geographic borders. Boas decided to study the diffusion of cultural ideas and customs *across* borders. But a certain amount of agreement between ecology and culture is not at all surprising, particularly among people who were farmers and animal herders, which Boas's North American tribes generally were not. The length of the frost-free growing season, precipitation, soil fertility, and topography affect many aspects of daily life and custom among farmers: herding systems, crop cultivation, house types, the size and arrangement of settlements, favorite foods, sacred foods, the size of food surpluses, and the timing and richness of public feasts. At large-scale ecotones these basic differences in economic organization, diet, and social life can blossom into oppositional ethnic identities, which sometimes are complementary and mutually supportive, sometimes are hostile, and often are both. Frederick Barth, after working among the societies of Iran and Afghanistan, was among the first anthropologists to argue that ethnic identity was continuously created, even invented, at frontiers, rather than residing in the genes or being passively inherited from the ancestors. Oppositional politics crystallize who we are *not*, even if we are uncertain who we *are*, and therefore play a large role in the definition of ethnic identities. Ecotones were places where contrasting identities were likely to be reproduced and maintained for long periods because of structural differences in how politics and economics were played.[22]

Ecotones coincide with ethnolinguistic frontiers at many places. In France the Mediterranean provinces of the South and the Atlantic provinces of the North have been divided by an ethnolinguistic border for at least eight hundred years; the earliest written reference to it dates to 1284. The flat, tiled roofs of the South sheltered people who spoke the *langue d'oc*, whereas the steeply pitched roofs of the North were home to people who spoke the *langue d'oil*. They had different cropping systems, and different legal systems as well until they were forced to conform to a national legal standard. In Kenya the Nilotic-speaking pastoralist Maasai maintained a purely cattle-herding economy (or at least that was their ideal) in the dry plains and plateaus, whereas Bantu-speaking farmers occupied moister environments on the forested slopes of the mountains or in low wetlands. Probably the most famous anthropological example of this type was described by Sir Edmund Leach in his classic *Political Systems of High-*

land Burma. The upland Kachin forest farmers, who lived in the hills of Burma (Myanmar), were distinct linguistically, and also in many aspects of ritual and material culture, from the Thai-speaking Shan paddy farmers who occupied the rich bottomlands in the river valleys. Some Kachin leaders adopted Shan identities on certain occasions, moving back and forth between the two systems. But the broader distinction between the two cultures, Kachin and Shan, persisted, a distinction rooted in different ecologies, for example, the contrasting reliability and predictability of crop surpluses, the resulting different potentials for surplus wealth, and the dissimilar social organizations required for upland forest and lowland paddy farming. Cultural frontiers rooted in ecological differences could survive for a long time, even with people regularly moving across them.[23]

Language Distributions and Ecotones

Why do some language frontiers follow ecological borders? Does language just ride on the coattails of economy? Or is there an independent relationship between ecology and the way people speak? The linguists Daniel Nettle at Oxford University and Jane Hill at the University of Arizona proposed, in 1996 (independently, or at least without citing each other), that the geography of language reflects an underlying ecology of social relationships.[24]

Social ties require a lot of effort to establish and maintain, especially across long distances, and people are unlikely to expend all that energy unless they think they *need* to. People who are self-sufficient and fairly sure of their economic future tend to maintain *strong* social ties with a small number of people, usually people very much like themselves. Jane Hill calls this a *localist* strategy. Their own language, the one they grew up with, gets them everything they need, and so they tend to speak only that language— and often only one dialect of that language. (Most college-educated North Americans fit nicely in this category.) Secure people like this tend to live in places with productive natural ecologies or at least secure access to pockets of high productivity. Nettles showed that the average size of language groups in West Africa is inversely correlated with agricultural productivity: the richer and more productive the farmland, the smaller the language territory. This is one reason why a single pan-European Proto-Indo-European language during the Neolithic is so improbable.

But people who are moderately uncertain of their economic future, who live in less-productive territories and have to rely on multiple sources of income (like the Kachin in Burma or most middle-class families with two income earners), maintain numerous *weak* ties with a wider variety of people. They often learn two or more languages or dialects, because they need a wider network to feel secure. They pick up new linguistic habits very rapidly; they are innovators. In Jane Hill's study of the Papago Indians in Arizona, she found that communities living in rich, productive environments adopted a "localist" strategy in both their language and social relations. They spoke just one homogeneous, small-territory Papago dialect. But communities living in more arid environments knew many different dialects, and combined them in a variety of nonstandard ways. They adopted a "distributed" strategy, one that distributed alliances of various kinds, linguistic and economic, across a varied social and ecological terrain. She proposed that arid, uncertain environments were natural "spread zones," where new languages and dialects would spread quickly between communities that relied on diverse social ties and readily picked up new dialects from an assortment of people. The Eurasian steppes had earlier been described by the linguist Johanna Nichols as the prototypical linguistic spread zone; Hill explained why. Thus the association between language and ecological frontiers is not a case of language passively following culture; instead, there are independent socio-linguistic reasons why language frontiers tend to break along ecological frontiers.[25]

Summary: Ecotones and Persistent Ethnolinguistic Frontiers

Language frontiers did not universally coincide with ecological frontiers or natural geographic barriers, even in the tribal world, because migration and all the other forms of language expansion prevented that. But the heterogeneity of languages—the number of languages per $1,000 \, km^2$—certainly was affected by ecology. Where an ecological frontier separated a predictable and productive environment from one that was unpredictable and unproductive, societies could not be organized the same way on both sides. Localized languages and small language territories were found among settled farmers in ecologically productive territories. More variable languages, fuzzier dialect boundaries, and larger language territories appeared among mobile hunter-gatherers and pastoralists occupying territories where farming was difficult or impossible. In the Eurasian steppes the ecological frontier between the steppe (unproductive, unpredictable, occupied principally by hunters or herders) and the neighboring agricultural lands (extremely productive and reliable, occupied by rich farmers) was a linguistic frontier through recorded history. Its persistence was one of the guiding factors in the history of China at one end of the steppes and of eastern Europe at the other.[26]

Small-scale Migrations, Elite Recruitment, and Language Shift

Persistent ecological and migration-related frontiers surrounded the Proto-Indo-European homeland in the Pontic-Caspian steppes. But the spread of the Indo-European languages *beyond* that homeland probably did not happen principally through chain-type folk migrations. A folk movement is not required to establish a new language in a strange land. Language change flows in the direction of accents that are admired and emulated by large numbers of people. Ritual and political elites often introduce and popularize new ways of speaking. Small elite groups can encourage widespread language shift toward their language, even in tribal contexts, in places where they succeed at introducing a new religion or political ideology or both while taking control of key territories and trade commodities. An ethnohistorical study of such a case in Africa among the Acholi illustrates how the introduction of a new ideology and control over trade can result in language spread even where the initial migrants were few in number.[27]

The Acholi are an ethnolinguistic group in northern Uganda and southern Sudan. They speak Luo, a Western Nilotic language. In about 1675, when Luo-speaking chiefs first migrated into northern Uganda from the south, the overwhelming majority of people living in the area spoke Central Sudanic or Eastern Nilotic languages—Luo was very much a minority language. But the Luo chiefs imported symbols and regalia of royalty (drums, stools) that they had adopted from Bantu kingdoms to the south. They also imported a new ideology of chiefly religious power, accompanied by demands for tribute service. Between about 1675 and 1725 thirteen new chiefdoms were formed, none larger than five villages. In these islands of chiefly authority the Luo-speaking chiefs recruited clients from among the lineage elders of the egalitarian local populations, offering them positions of prestige in the new hierarchy. Their numbers grew through marriage alliances with the locals, displays of wealth and generosity, assistance for local families in difficulty, threats of violence, and, most important, control over the inter-regional trade in iron prestige objects used to pay bride-prices. The Luo language spread slowly through recruitment.[28] Then an external stress, a severe drought beginning in 1790–1800, affected the region. One ecologically favored Lou chiefdom—an old one, founded by one of the first Luo charter groups—rose to paramount status as its wealth was maintained through the crisis. The Luo language then spread rapidly. When European traders arrived from Egypt in the 1850s they designated the local people by the name of this widely spoken language, which they called *Shooli*, which became *Achooli*. The paramount chiefs acquired so much wealth through trade with the Europeans that they quickly became an aristocracy. By 1872 the British recorded a single Luo-speaking tribe called the Acholi, an inter-regional ethnic identity that had not existed two hundred years earlier.

Indo-European languages probably spread in a similar way among the tribal societies of prehistoric Europe. Out-migrating Indo-European chiefs probably carried with them an ideology of political clientage like that of the Acholi chiefs, becoming patrons of their new clients among the local population; and they introduced a new ritual system in which they, in imitation of the gods, provided the animals for public sacrifices and feasts, and were in turn rewarded with the recitation of praise poetry—all solidly reconstructed for Proto-Indo-European culture, and all effective public recruiting activities. Later Proto-Indo-European migrations also introduced a new, mobile kind of pastoral economy made possible by the combination of ox-drawn wagons and horseback riding. Expansion beyond a few islands of authority might have waited until the new chiefdoms successfully responded to external stresses, climatic or political. Then the original chiefly core became the foundation for the development of a new regional ethnic identity. Renfrew has called this mode of language shift *elite dominance* but *elite recruitment* is probably a better term. The Normans conquered England and the Celtic Galatians conquered central Anatolia, but both failed to establish their languages among the local populations they dominated. Immigrant elite languages are adopted only where an elite status system is not only dominant but is also open to recruitment and alliance. For people to change to a new language, the shift must provide a key to integration within the new system, and those who join the system must see an opportunity to rise within it.[29]

A good example of how an open social system can encourage recruitment and language shift, cited long ago by Mallory, was described by Frederik Barth in eastern Afghanistan. Among the Pathans (today usually called Pashtun) on the Kandahar plateau, status depended on agricultural surpluses that came from circumscribed river-bottom fields. Pathan landowners competed for power in local councils (*jirga*) where no man admitted to being subservient and all appeals were phrased as requests among equals. The Baluch, a neighboring ethnic group, lived in the arid mountains and were, of necessity, pastoral herders. Although poor, the Baluch had an openly hierarchical political system, unlike the Pathan. The Pathan had more weapons than the Baluch, more people, more wealth, and generally more power and status. Yet, at the Baluch-Pathan frontier, many dispossessed Pathans crossed over to a new life as clients of Baluchi chiefs. Because Pathan status was tied to land ownership, Pathans who had lost their land in feuds were doomed to menial and peripheral lives. But Baluchi status was linked to herds, which could grow rapidly if the herder was lucky; and to political alliances, not to land. All Baluchi chiefs were the clients of more powerful chiefs, up to the office of *sardar*, the highest Baluchi authority, who himself owed allegiance to the khan of Kalat. Among the Baluch there was no shame in being the client of a powerful chief, and the possibilities for rapid economic and political improvement were great. So, in a situation of chronic low-level warfare at the Pathan-Baluch frontier, former agricultural refugees tended to flow toward the pastoral Baluch, and the Baluchi language thus gained new speakers. Chronic tribal warfare might generally favor pastoral over sedentary economies as herds can be defended by moving them, whereas agricultural fields are an immobile target.

Migration and the Indo-European Languages

Folk migrations by pioneer farmers brought the first herding-and-farming economies to the edge of the Pontic-Caspian steppes about 5800 BCE. In the forest-steppe ecological zone northwest of the Black Sea the incoming pioneer farmers established a cultural frontier between themselves and the native foragers. This frontier was robust, defined by bundles of cultural and economic differences, and it persisted for about twenty-five hundred years. If I am right about persistent frontiers and language, it was a linguistic frontier; if the other arguments in the preceding chapters are correct, the incoming pioneers spoke a non–Indo-European language, and the foragers spoke a Pre-Proto-Indo-European language. Selected aspects of the new farming economy (a little cattle herding, a little grain cultivation) were adopted by the foragers who lived on the frontier, but away from the frontier the local foragers kept hunting and fishing for many centuries. At the frontier both societies could reach back to very different sources of tradition in the lower Danube valley or in the steppes, providing a continuously renewed source of contrast and opposition.

Eventually, around 5200–5000 BCE, the new herding economy was adopted by a few key forager groups on the Dnieper River, and it then diffused very rapidly across most of the Pontic-Caspian steppes as far east as the Volga and Ural rivers. This was a revolutionary event that transformed not just the economy but also the rituals and politics of steppe societies.

A new set of dialects and languages probably spread across the Pontic-Caspian steppes with the new economic and ritual-political system. These dialects were the ancestors of Proto-Indo-European.

With a clearer idea of how language and material culture are connected, and with specific models indicating how migrations work and how they might be connected with language shifts, we can now begin to examine the archaeology of Indo-European origins.

PART TWO

The Opening of the Eurasian Steppes

CHAPTER SEVEN

How to Reconstruct a Dead Culture

The archaeology of Indo-European origins usually is described in terms that seem arcane to most people, and that even archaeologists define differently. So I offer a short explanation of how I approach the archaeological evidence. To begin at the beginning, surprisingly enough, we must start out in Denmark.

In 1807 the kingdom of Denmark was unsure of its prospects for survival. Defeated by Britain, threatened by Sweden, and soon to be abandoned by Norway, it looked to its glorious past to reassure its citizens of their greatness. Plans for a National Museum of Antiquities, the first of its type in Europe, were developed and promoted. The Royal Cabinet of Antiquities quickly acquired vast collections of artifacts that had been plowed or dug from the ground under a newly expanded agricultural policy. Amateur collectors among the country gentry, and quarrymen or ditch diggers among the common folk, brought in glimmering hoards of bronze and boxes of flint tools and bones.

In 1816, with dusty specimens piling up in the back room of the Royal Library, the Royal Commission for the Preservation of Danish Antiquities selected Christian J. Thomsen, a twenty-seven-year-old without a university degree but known for his practicality and industry, to decide how to arrange this overwhelming trove of strange and unknown objects in some kind of order for its first display. After a year of cataloguing and thinking, Thomsen elected to put the artifacts in three great halls. One would be for the stone artifacts, which seemed to come from graves or sediments belonging to a Stone Age, lacking any metals at all; one for the bronze axes, trumpets, and spears of the Bronze Age, which seemed to come from sites that lacked iron; and the last for the iron tools and weapons, made during an Iron Age that continued into the era of the earliest written references to Scandinavian history. The exhibit opened in 1819 and was a triumphant success. It inspired an animated discussion among

European intellectuals about whether these three ages truly existed in this chronological order, how old they were, and whether a science of archaeology, like the new science of historical linguistics, was possible. Jens Worsaae, originally an assistant to Thomsen, proved, through careful excavation, that the Three Ages indeed existed as distinct prehistoric eras, with some qualifications. But to do this he had to dig much more carefully than the ditch diggers, borrowing stratigraphic methods from geology. Thus professional field archaeology was born to solve a problem, not to acquire things.[1]

It was no longer possible, after Thomsen's exhibit, for an educated person to regard the prehistoric past as a single undifferentiated era into which mammoth bones and iron swords could be thrown together. Forever after time was to be divided, a peculiarly satisfying task for mortals, who now had a way to triumph over their most implacable foe. Once chronology was discovered, tinkering with it quickly became addictive. Even today chronological arguments dominate archaeological discussions in Russia and Ukraine. Indeed, a chief problem preventing Western archaeologists from really understanding steppe archaeology is that Thomsen's Three Ages are defined differently in the steppes than in western Europe. The Bronze Age seems like a simple concept, but if it began at different times in places very close to each other, it can be complicated to apply.

The Bronze Age can be said to begin when bronze tools and ornaments began to appear regularly in excavated graves and settlements. But what is bronze? It is an alloy, and the oldest bronze was an alloy of copper and arsenic. Arsenic, recognized by most of us simply as a poison, is in fact a naturally occurring whitish mineral typically in the form of arsenopyrite, which is frequently associated with copper ores in quartzitic copper deposits, and is probably how the alloy was discovered. In nature, arsenic rarely comprises more than about 1% of a copper ore, and usually much less than that. Ancient metalsmiths discovered that, if the arsenic content was boosted to about 2–8% of the mixture, the finished metal was lighter in color than pure copper, harder when cool, and, when molton, less viscous and easier to cast. A bronze alloy even lighter in color, harder, and more workable was copper and about 2–8% tin, but tin was rare in the ancient Old World, so tin-bronzes only appeared later, after tin deposits were discovered. The Bronze Age, therefore, marks that moment when metalsmiths regularly began to mix molten minerals to make alloys that were superior to naturally occurring copper. From that perspective, it immediately becomes clear that the Bronze Age would have started in different places at different times.

The Three Ages in the Pontic-Caspian Steppes

The oldest Bronze Age in Europe began about 3700–3500 BCE, when smiths started to make arsenical bronze in the North Caucasus Mountains, the natural frontier between the Near East and the Pontic-Caspian steppes. Arsenical bronzes, and the Bronze Age they signaled, appeared centuries later in the steppes and eastern Europe including the lower Danube valley, beginning about 3300–3200 BCE; and the beginning of the Bronze Age in central and western Europe was delayed a thousand years after that, starting only about 2400–2200 BCE. Yet, an archaeologist trained in western Europe may commonly ask why a Caucasian culture dated 3700 BCE is called a Bronze Age culture, when this would be the Stone Age (or Neolithic) in Britain or France. The answer is that bronze metallurgy appeared first in eastern Europe and then spread to the west, where it was adopted only after a surprisingly long delay. The Bronze Age began in the Pontic-Caspian steppes, the probable Indo-European homeland, much earlier than in Denmark.

The age preceding the Bronze Age in the steppes is called the Eneolithic; Christian Thomsen did not recognize that period in Denmark. The Eneolithic was a Copper Age, when metal tools and ornaments were used widely but were made of unalloyed copper. This was the first age of metal, and it lasted a long time in southeastern Europe, where European copper metallurgy was invented. The Eneolithic did not appear in northern or western Europe, which skipped directly from the Neolithic to the Bronze Age. Experts in southeastern Europe disagree on how to divide the Eneolithic internally; the chronological boundaries of the Early, Middle, and Late Eneolithic are set at different times by different archaeologists in different regions. I have tried to follow what I see as an emerging interregional consensus among Russian and Ukrainian archaeologists, and between them and the archaeologists of eastern Poland, Bulgaria, Romania, Hungary, and the former Yugoslavia.[2]

Before the Eneolithic was the Neolithic, the later end of Thomsen's Stone Age. Eventually the Stone Age was divided into the Old, Middle, and New Stone Ages, or the Paleolithic, Mesolithic, and Neolithic. In Soviet archaeology and in current Slavic or post-Soviet terminology the word *Neolithic* is applied to prehistoric societies that made pottery but had not yet discovered how to make metal. The invention of ceramics defined the beginning of the Neolithic. Pottery, of course, was an important discovery. Fire-resistant clay pots made it possible to cook stews and soups all day over a low fire, breaking down complex starches and proteins so that they were easier to digest for people with delicate stomachs—babies and elders. Soups that simmered in clay pots helped infants survive and kept old people alive longer. Pottery also is a convenient "type fossil" for archaeologists, easily recognized in archaeological sites. But Western archaeologists defined the Neolithic differently. In Western archaeology, societies can only be called *Neolithic* if they had economies based on food production—herding or farming or both. Hunters and gatherers who had pottery are called *Mesolithic*. It is oddly ironic that capitalist archaeologists made the mode of production central to their definition of the Neolithic, and Marxist archaeologists ignored it. I'm not sure what this might say about archaeologists and their politics, but here I must use the Eastern European definition of the Neolithic—which includes both foragers and early farmers who made pottery but used no metal tools or ornaments—because this is what *Neolithic* means in Russian and Ukrainian archaeology.

DATING AND THE RADIOCARBON REVOLUTION

Radiocarbon dating created a revolution in prehistoric archaeology. From Christian Thomsen's museum exhibit until the mid-twentieth century archaeologists had no clear idea how old their artifacts were, even if they knew how to place them in a sequence of types. The only way even to guess their age was to attempt to relate dagger or ornament styles in Europe to similar styles of known age in the Near East, where inscriptions provided dates going back to 3000 BCE. These long-distance stylistic comparisons, risky at best, were useless for dating artifacts older than the earliest Near Eastern inscriptions. Then, in 1949, Willard Libby demonstrated that the absolute age (literally the number of years since death) of any organic material (wood, bone, straw, shell, skin, hair, etc.) could be determined by counting its ^{14}C content, and thus radiocarbon dating was born. A radiocarbon date reveals when the dated sample died. Of course, the sample had to have been alive at some point, which disqualified Libby's discovery for dating rocks or minerals, but archaeologists often found charred wood from ancient fireplaces or discarded animal bones in places where humans had lived. Libby was awarded a Nobel Prize, and Europe acquired its own prehistory independent of the civilizations of the Near East. Some important events such as the invention of copper metallurgy were shown to have happened so early in Europe that influence from the Near East was almost ruled out.[3]

Chronological schemes based on radiocarbon dates have struggled through several significant changes in methods since 1949 (see the appendix in this volume). The most significant changes were the introduction of a new method (Accelerator Mass Spectrometry, or AMS) for counting how much ^{14}C remained in a sample, which made all dates much more accurate; and the realization that all radiocarbon dates, regardless of counting method, had to be corrected using calibration tables, which revealed large errors in old, uncalibrated dates. These periodic changes in methods and results slowed the scientific reception of radiocarbon dates in the former Soviet Union. Many Soviet archaeologists resisted radiocarbon dating, partly because it sometimes contradicted their theories and chronologies; partly because the first radiocarbon dates were later proved wrong by changes in methods, making it possible that all radiocarbon dates might soon be proved wrong by a newer refinement; and partly because the dates themselves, even when corrected and calibrated, sometimes made no sense—the rate of error in radiocarbon dating in Soviet times seemed high.

A new problem affecting radiocarbon dates in the steppes is that old carbon in solution in river water is absorbed by fish and then enters the bones of people who eat a lot of fish. Many steppe archaeological sites are cemeteries, and many radiocarbon dates in steppe archaeology are from human bones. Analysis of ^{15}N isotopes in human bone can tell us how much fish a person ate. Measurements of ^{15}N in skeletons from early steppe cemeteries show that fish was very important in the diet of most steppe societies, including cattle herders, often accounting for about 50% of the food consumed. Radiocarbon dates measured on the bones of these humans might come out too old, contaminated by old carbon in the fish they ate. This is a newly realized problem, one still without a solution widely agreed on. The errors should be in the range of 100–500 radiocarbon years too old, meaning that the person actually died 100–500 years *after* the date given by the count of ^{14}C. I note in the text places where old carbon contamination might be a problem making the dates measured on human bones too old, and, in the appendix, I explain my own interim approach to fixing the problem.[4]

Attitudes toward radiocarbon dating in the CIS have changed since 1991. The major universities and institutes have thrown themselves into new radiocarbon dating programs. The field collection of samples for dating has become more careful and more widespread, laboratories continuously improve their methods, and the error rate has fallen. It is difficult now to keep up with the flow of new radiocarbon dates. They have overthrown many old ideas and chronologies, including my own. Some of the chronological relationships outlined in my 1985 Ph.D. dissertation have now been proved wrong, and entire cultures I barely knew about in 1985 have become central to any understanding of steppe archaeology.[5]

But to understand people we need to know more than just *when* they lived; we also need to know something about their economy and culture. And in the specific case of the people of the Pontic-Caspian region, some of the most important questions are about *how* they lived—whether they were wandering nomads or lived in one place all year, whether they had chiefs or lived in egalitarian groups without formal full-time leaders, and how they went about getting their daily bread, if indeed they ate bread at all. But to talk about these matters I first need to introduce some additional methods archaeologists use.

WHAT DID THEY EAT?

One of the most salient signals of cultural identity is food. Long after immigrants give up their native clothing styles and languages, they retain and even celebrate their traditional food. How the members of a society get food is, of course, a central organizing fact of life for all humans. The supermarkets we use so casually today are microcosms of modern Western life: they would not exist without a highly specialized, capital-financed, market-based economic structure; a consumer-oriented culture of profligate consumption (Do we really need fifteen kinds of mushrooms?); interstate highways; suburbs; private automobiles; and dispersed nuclear families lacking a grandma at home who could wash, chop, process, and prepare meat and produce. Long ago, before all these modern conveniences appeared, getting food determined how people spent much of their day, every day: what time they woke in the morning, where they went to work, what skills and knowledge they needed there, whether they could live in independent family homes or needed the much larger communal labor resources of a village, how long they were away from home, what kind of ecological resources they needed, what cooking and food-preparation skills they had to know, and even what foods they offered to the gods. In a world dominated by the rhythms and values of raising crops and caring for animals, clans with productive fields or large herds of cattle were the envy of everyone. Wealth and the political power it conveyed were equated with cultivated land and pasture.

To understand ancient agricultural and herding economies, archaeologists have to collect the animal bones from ancient garbage dumps with the same care they devote to broken pottery, and they must also make special efforts to recover carbonized plant remains. Luckily ancient people often buried their food trash in dumps or pits, restricting it to one place where archaeologists can find it more easily. Although cow bones and charred seeds cannot easily be displayed in the national museum, archaeology is not about collecting pretty things but about solving problems, so in the following pages much attention is devoted to animal bones and charred seeds.

Archaeologists count animal bones in two principal ways. Many bones in garbage dumps had been broken into such small pieces for cooking that they cannot be assigned to a specific animal species. Those that are big enough or distinctive enough to assign to a definite species constitute the NISP, or the "number of identified specimens," where *identified* means assignable to a species. Thus, the NISP count, which describes the number of bones found for each species, is the first way to count bones: three hundred cattle, one hundred sheep, five horse. The second counting method is to calculate the MNI, or the "minimum number of individuals" those bones represent. If the five horse bones were each from a different animal, they would represent five horses, whereas the hundred sheep bones might all be from a single skeleton. The MNI is used to convert bones into minimum meat weights—how much beef, for example, would be represented, minimally, by a certain number of cattle bones. Meat weight, comprised of fat and muscle, in most adult mammals averages about half the live body weight, so by identifying the minimum number, age, and species of animals butchered at the site, the minimum meat weight, with some qualifications, can be estimated.

Seeds, like wheat and barley, were often parched by charring them lightly over a fire to help preserve them for storage. Although many charred seeds are accidentally lost in this process, without charring they would soon rot into dust. The seeds preserved in archaeological sites have been charred just enough to carbonize the seed hull. Seeds tell us which plant foods were eaten, and can reveal the nature of the area's gardens, fields, forests, groves, and vineyards. The recovery of charred seeds from excavated sediments requires a flotation tank and a pump to force water through the tank. Excavated dirt is dumped into the tank and the moving water helps the seeds to float to the surface. They are then collected in screens as the water flows out the top of the tank through an exit spout. In the laboratory the species of plants are identified and counted, and domesticated varieties of wheat, barley, millet, and oats are distinguished from wild plant seeds. Flotation was rarely used in Western archaeology before the late 1970s and was almost never used in Soviet archaeology. Soviet paleobotanical experts relied on chance finds of seeds charred in burned pots or on seed impressions preserved in the damp clay of a pot before it had been fired. These lucky finds occur rarely. A true understanding of the importance of plant foods in the steppes will come only after flotation methods are widely used in excavations.

ARCHAEOLOGICAL CULTURES AND LIVING CULTURES

The story that follows is populated rarely by individuals and more often by cultures, which, although created and reproduced by people, act quite differently than people do. Because "living cultures" contain so many subgroups and variants, anthropologists have difficulty describing them in the abstract, leading many anthropologists to discard the concept of a "unitary culture" entirely. However, when cultural identities are contrasted with other bordering cultures, they are much easier to describe. .

Frederik Barth's investigations of border identities in Afghanistan suggested that the reproduction and perhaps even the invention of cultural identities often was generated by the continuous confrontation with Others inherent in border situations. Today many anthropologists find this a productive way to understand cultural identities, that is, as responses to particular historical situations rather than as long-term phenomena, as noted in the previous chapter. But cultural identities also carry emotional and historical weight in the hearts of those who believe in them, and the source of this shared emotional attachment is more complicated. It must be derived from a shared set of customs and historical experiences, a font of tradition that, even if largely imagined or invented, provides the fuel that feeds border confrontations. If that font of tradition is given a geographic location or a homeland it is often away from the border, dispersed, for example, across shrines, burial grounds, coronation sites, battlefields, and landscape features like mountains and forests, all thought to be imbued with culture-specific spiritual forces.[6]

Archaeological cultures are defined on the basis of potsherds, grave types, architecture, and other material remains, so the relationship between archaeological cultures and living cultures might seem tenuous. When Christian Thomsen and Jens Worsaae first began to divide artifacts into types, they were trying to arrange them in a chronological sequence; they soon realized, however, that a lot of regional variation also cut across the chronological types. Archaeological cultures are meant to capture and define that regional variation. An archaeological culture is a recurring set of artifact types that co-occur in a particular region during a set time period.

In practice, pottery types are often used as the key identifiers of archaeological cultures, as they are easy to find and recognize even in small excavations, whereas the recognition of distinct house types, for example, requires much larger exposures. But archaeological cultures should never be defined on the basis of pottery alone. What makes an archaeological culture interesting, and meaningful, is the co-occurrence of many similar customs, crafts, and dwelling styles across a region, including, in addition to ceramics, grave types, house types, settlement types (the arrangement of houses in the typical settlement), tool types, and ritual symbols (figurines, shrines, and deities.) Archaeologists worry about individual types changing through time and shifting their areas of distribution, and we *should* worry about these things, but we should not let problems with defining individual tree species and ranges convince us that the forest is not there. Archaeological cultures (like forests) are particularly recognizable and definable at their borders, whereas regional variation in the back country, away from the borders, might often present a more confusing picture. It is at robust borders, defined by bundles of material-culture contrasts, where archaeological cultures and living cultures or societies might actually correspond. As I argued in the previous chapter, robust borders that persist for centuries probably were not just archaeological or cultural but also linguistic.

Within archaeological cultures a few traits, archaeologists have learned, are particularly important as keys to cultural identity. Most Western archaeologists accept that technological style, or the way an object is made, is a more fundamental indicator of craft tradition than the way it is decorated, its decorative style. The technology of production is more culture-bound and resistant to change, rather like the core vocabulary in linguistics. So clay tempering materials and firing methods usually are better indicators of a potter's cultural origin than the decorative styles the potter produced, and the same probably was true for metallurgy, weaving, and other crafts.[7]

One important alternative to archaeological cultures is the archaeological *horizon*. A *horizon*, more like a popular fashion than a culture, can be defined by a single artifact type or cluster of artifact types that spreads suddenly over a very wide geographic area. In the modern world the blue jeans and T-shirt complex is a horizon style, superimposed on diverse populations and cultures around the planet but still representing an important diffusion of cultural influence, particularly youth culture, from an area of origin in the United States. It is important, as it tells us something about the place the United States occupied in world youth culture at the moment of initial diffusion (the 1960s and 1970s), but it is not a migration or cultural replacement. Similarly the Beaker horizon in Late Neolithic Europe is defined primarily by a widespread style of decorated drinking cups (beakers) and in many places by a few weapon types (copper daggers, polished stone wrist-guards) that diffused with a new fashion in social drinking. In most places these styles were superimposed on preexisting archaeological cultures. A horizon is different from an archaeological culture because it is less robust—it is defined on the basis of just a few traits—and is often superimposed on local archaeological cultures. Horizons were highly significant in the prehistoric Eurasian steppes.

THE BIG QUESTIONS AHEAD

We will proceed on the assumption that Proto-Indo-European probably was spoken in the steppes north of the Black and Caspian Seas, the Pontic-Caspian steppes, broadly between 4500 and 2500 BCE. But we have to start somewhat earlier to understand the evolution of Indo-European-speaking societies. The speakers of Proto-Indo-European were a cattle-keeping people. Where did the cattle come from? Both cattle and sheep were introduced from outside, probably from the Danube valley (although we also have to consider the possibility of a diffusion route through the Caucasus Mountains). The Neolithic pioneers who imported domesticated cattle and sheep into the Danube valley probably spoke non–Indo-European languages ultimately derived from western Anatolia. Their arrival in the eastern Carpathians, northwest of the Black Sea, around 5800 BCE, created a cultural frontier between the native foragers and the immigrant farmers that persisted for more than two thousand years.

The arrival of the first pioneer farmers and the creation of this cultural frontier is described in chapter 8. A recurring theme will be the development of the relationship between the farming cultures of the Danube valley and the steppe cultures north of the Black Sea. Marija Gimbutas called the Danubian farming cultures "Old Europe." The agricultural towns of Old Europe were the most technologically advanced and aesthetically sophisticated in all of Europe between about 6000 and 4000 BCE.

Chapter 9 describes the diffusion of the earliest cattle-and-sheep-herding economy across the Pontic-Caspian steppes after about 5200–5000 BCE. This event laid the foundation for the kinds of power politics and rituals that defined early Proto-Indo-European culture. Cattle herding was not just a new way to get food; it also supported a new division of society between high-status and ordinary people, a social hierarchy that had not existed when daily sustenance was based on fishing and hunting. Cattle and the cleavage of society into distinct statuses appeared together. Right away, cattle, sheep—and horses—were offered together in sacrifices at the funerals of a select group of people, who also carried unusual weapons and ornamented their bodies in unique and ostentatious ways. They were the new leaders of a new kind of steppe society.

Chapter 10 describes the discovery of horseback riding—a subject of intense controversy—by these archaic steppe herding societies, probably before 4200 BCE. The intrusion into Old Europe of steppe herders, probably mounted on horses, who either caused or took advantage of the collapse of Old Europe, is the topic of chapter 11. Their spread into the lower Danube valley about 4200–4000 BCE likely represented the initial expansion of archaic Proto-Indo-European speakers into southeastern Europe, speaking dialects that were ancestral to the later Anatolian languages.

Chapter 12 considers the influence of the earliest Mesopotamian urban civilizations on steppe societies—and vice versa—at a very early age, about 3700–3100 BCE. The chiefs who lived in the North Caucasus Mountains overlooking the steppes grew incredibly rich from long-distance trade with the southern civilizations. The earliest wheeled vehicles, the first wagons, probably rolled into the steppes through these mountains.

The societies that probably spoke classic Proto-Indo-European—the herders of the Yamnaya horizon—are introduced in chapter 13. They were the first people in the Eurasian steppes to create a herding economy that required regular seasonal movements to new pastures throughout the year. Wagons pulled by cattle allowed them to carry tents, water, and food into the deep steppes, far from the river valleys, and horseback riding enabled them to scout rapidly and over long distances and to herd on a large scale, necessities in such an economy. Herds were spread out across the enormous grasslands between the river valleys, making those grasslands useful, which led to larger herds and the accumulation of greater wealth.

Chapters 14 through 16 describe the initial expansions of societies speaking Proto-Indo-European dialects, to the east, the west, and finally to the south, to Iran and the Indian subcontinent. I do not attempt to follow what happened after the initial migrations of these groups; my effort is just to understand the development and the first dispersal of speakers of Proto-Indo-European and, along the way, to investigate the influence of technological innovations in transportation—horseback riding, wheeled vehicles, and chariots—in the opening of the Eurasian steppes.

CHAPTER EIGHT

First Farmers and Herders
The Pontic-Caspian Neolithic

At the beginning of time there were two brothers, twins, one named Man (*Manu*, in Proto-Indo-European) and the other Twin (*Yemo*). They traveled through the cosmos accompanied by a great cow. Eventually Man and Twin decided to create the world we now inhabit. To do this, Man had to sacrifice Twin (or, in some versions, the cow). From the parts of this sacrificed body, with the help of the sky gods (*Sky Father, Storm God of War, Divine Twins*), Man made the wind, the sun, the moon, the sea, earth, fire, and finally all the various kinds of people. Man became the first priest, the creator of the ritual of sacrifice that was the root of world order.

After the world was made, the sky-gods gave cattle to "Third man" (*Trito*). But the cattle were treacherously stolen by a three-headed, six-eyed serpent (*Ng^whi*, the Proto-Indo-European root for *negation*). Third man entreated the storm god to help get the cattle back. Together they went to the cave (or mountain) of the monster, killed it (or the storm god killed it alone), and freed the cattle. *Trito* became the first warrior. He recovered the wealth of the people, and his gift of cattle to the priests insured that the sky gods received their share in the rising smoke of sacrificial fires. This insured that the cycle of giving between gods and humans continued.[1]

These two myths were fundamental to the Proto-Indo-European system of religious belief. *Manu* and *Yemo* are reflected in creation myths preserved in many Indo-European branches, where *Yemo* appears as Indic *Yama*, Avestan *Yima*, Norse *Ymir*, and perhaps Roman *Remus* (from *iemus*, the archaic Italic form of *yemo*, meaning "twin"); and Man appears as Old Indic *Manu* or Germanic *Mannus*, paired with his twin to create the world. The deeds of *Trito* have been analyzed at length by Bruce Lincoln, who found the same basic story of the hero who recovered primordial lost cattle from a three-headed monster in Indic, Iranian, Hittite, Norse, Roman and Greek myths. The myth of Man and Twin established the importance of the sacrifice and the priest who regulated it. The myth of the "Third one" defined the role of the warrior, who obtained animals for the people and the gods. Many other themes are also reflected in these two stories: the Indo-European fascination with binary doublings combined with triplets, two's and three's, which reappeared again and again, even in the metric structure of Indo-European poetry; the theme of pairs who represented magical and legal power (Twin and Man, Varuna-Mitra, Odin-Tyr); and the partition of society and the cosmos between three great functions or roles: the priest (in both his magical and legal aspects), the warrior (the Third Man), and the herder/cultivator (the cow or cattle).[2]

For the speakers of Proto-Indo-European, domesticated cattle were basic symbols of the generosity of the gods and the productivity of the earth. Humans were created from a piece of the primordial cow. The ritual duties that defined "proper" behavior revolved around the value, both moral and economic, of cattle. Proto-Indo-European mythology was, at its core, the worldview of a male-centered, cattle-raising people—not necessarily cattle nomads but certainly people who held sons and cattle in the highest esteem. Why were cattle (and sons) so important?

Domesticated Animals and Pontic-Caspian Ecology

Until about 5200–5000 BCE most of the people who lived in the steppes north of the Black and Caspian Seas possessed no domesticated animals at all. They depended instead on gathering nuts and wild plants, fishing, and hunting wild animals; in other words, they were foragers. But the environment they were able to exploit profitably was only a small fraction of the total steppe environment. The archaeological remains of their camps are found almost entirely in river valleys. Riverine gallery forests provided shelter, shade, firewood, building materials, deer, aurochs (European wild cattle), and wild boar. Fish supplied an important part of the diet. Wider river valleys like the Dnieper or Don had substantial gallery forests, kilometers wide; smaller rivers had only scattered groves. The wide grassy plateaus between the river valleys, the great majority of the steppe environment, were forbidding places occupied only by wild equids and saiga antelope. The foragers were able to hunt the wild equids, including horses. The wild horses of the steppes were stout-legged, barrel-chested, stiff-maned animals that probably looked very much like modern Przewalski horses, the only truly wild horses left in the world.[3] The most efficient hunting method would have been to ambush horse bands in a ravine, and the easiest opportunity would have been when they came into the river valleys to drink or to find shelter. In the steppe regions, where wild horses were most numerous, wild equid hunting was common. Often it supplied most of the foragers' terrestrial meat diet.

The Pontic-Caspian steppes are at the western end of a continuous steppe belt which rolls east all the way to Mongolia. It is possible, if one is so inclined, to walk, 5,000 km from the Danube delta across the center of the Eurasian continent to Mongolia without ever leaving the steppes. But a person on foot in the Eurasian steppes feels very small. Every footfall raises the scent of crushed sage, and a puff of tiny white grasshoppers skips ahead of your boot. Although the flowers that grow among the fescue and feathergrass (*Festuca* and *Stipa*) make a wonderful boiled tea, the grass is inedible, and outside the forested river valleys there is not much else to eat. The summer temperature frequently rises to 110–120°F (43–49°C), although it is a dry heat and usually there is a breeze, so it is surprisingly tolerable. Winter, however, kills quickly. The howling, snowy winds drive temperatures below –35°F (–37°C). The bitter cold of steppe winters (think North Dakota) is the most serious limiting factor for humans and animals, more restricting even than water, since there are shallow lakes in most parts of the Eurasian steppes.

The dominant mammal of the interior steppes at the time our account begins was the wild horse, *Equus caballus*. In the moister, lusher western steppes of Ukraine, north of the Black Sea (the North Pontic steppes), there was another, smaller equid that ranged into the lower Danube valley and down to central Anatolia, *Equus hydruntinus*, the last one hunted to extinction between 4000 and 3000 BCE. In the drier, more arid steppes of the Caspian Depression was a third ass-like, long-eared equid, the onager, *Equus hemionus*, now endangered in the wild. Onagers then lived in Mesopotamia, Anatolia, Iran, and in the Caspian Depression. Pontic-Caspian foragers hunted all three.

The Caspian Depression was itself a sign of another important aspect of the Pontic-Caspian environment: its instability. The Black and Caspian Seas were not placid and unchanging. Between about 14,000 and 12,000 BCE the warming climate that ended the last Ice Age melted the northern glaciers and the permafrost, releasing their combined meltwater in a torrential surge that flowed south into the Caspian basin. The late Ice-Age Caspian ballooned into a vast interior sea designated the Khvalynian Sea. For two thousand years the northern shoreline stood near Saratov on the middle Volga and Orenburg on the Ural River, restricting east-west movement south of the Ural Mountains. The Khvalynian Sea separated the already noticeably different late-glacial forager cultures that prospered east and west of the Ural Mountains.[4] Around 11,000–9,000 BCE the water finally rose high enough to overflow catastrophically through a southwestern outlet, the Manych Depression north of the North Caucasus Mountains, and a violent flood poured into the Black Sea, which was then well below the world ocean level. The Black Sea basin filled up until it overflowed, also through a southwestern outlet, the narrow Bosporus valley, and finally poured into the Aegean. By 8000 BCE the Black Sea, now about the size of California and seven thousand feet deep, was in equilibrium with the Aegean and the world ocean. The Caspian had fallen back into its own basin and remained isolated thereafter. The Black Sea became the Pontus Euxeinos of the Greeks, from which we derive the term *Pontic* for the Black Sea region in general. The North Caspian Depression, once the bottom of the northern end of the Khvalynian Sea, was left an enormous flat plain of salty clays, incongruous beds of sea shells, and sands, dotted with brackish lakes and covered with dry steppes that graded into red sand deserts (the Ryn Peski) just north of the Caspian Sea. Herds of saiga antelopes, onagers, and horses were hunted across these saline plains by small bands of post-glacial Mesolithic and Neolithic hunters. But, by the time the sea receded, they had become very different culturally and probably linguistically on the eastern and western sides of the Ural-Caspian frontier. When domesticated cattle were accepted by societies west of the Urals, they were rejected by those east of the Urals, who remained foragers for thousands of years.[5]

Domesticated cattle and sheep started a revolutionary change in how humans exploited the Pontic-Caspian steppe environment. Because cattle and sheep were cultured, like humans, they were part of everyday work and worry in a way never approached by wild animals. Humans identified with their cattle and sheep, wrote poetry about them, and used them as a currency in marriage gifts, debt payments, and the calculation of social status. And they were grass processors. They converted plains of grass, useless and even hostile to humans, into wool, felt, clothing, tents, milk, yogurt, cheese, meat, marrow, and bone—the foundation of both life and wealth. Cattle and sheep herds can grow rapidly with a little luck. Vulnerable to bad weather and theft, they can also decline rapidly. Herding was a volatile, boom-bust economy, and required a flexible, opportunistic social organization.

Because cattle and sheep are easily stolen, unlike grain crops, cattle-raising people tend to have problems with thieves, leading to conflict and warfare. Under these circumstances brothers tend to stay close together. In Africa, among Bantu-speaking tribes, the spread of cattle raising seems to have led to the loss of matrilineal social organizations and the spread of male-centered patrilineal kinship systems.[6] Stockbreeding also created entirely new kinds of political power and prestige by making possible elaborate public sacrifices and gifts of animals. The connection between animals, brothers, and power was the foundation on which new forms of male-centered ritual and politics developed among Indo-European-speaking societies. That is why the cow (and brothers) occupied such a central place in Indo-European myths relating to how the world began.

So where did the cattle come from? When did the people living in the Pontic-Caspian steppes begin to keep and care for herds of dappled cows?

The First Farmer-Forager Frontier in the Pontic-Caspian Region

The first cattle herders in the Pontic-Caspian region arrived about 5800–5700 BCE from the Danube valley, and they probably spoke languages unrelated to Proto-Indo-European. They were the leading edge of a broad movement of farming people that began around 6200 BCE when pioneers from Greece and Macedonia plunged north into the temperate forests of the Balkans and the Carpathian Basin (figure 8.1). Domesticated sheep and cattle had been imported from Anatolia to Greece by their ancestors centuries before, and now were herded northward into forested southeastern Europe. Genetic research has shown that the cattle did interbreed with the native European aurochs, the huge wild cattle of Europe, but only the male calves (traced on the Y chromosome) of aurochs were kept, perhaps because they could improve the herd's size or resistance to disease without affecting milk yields. The cows, probably already kept for their milk, all were descended from mothers that had come from Anatolia (traced through MtDNA). Wild aurochs cows probably were relatively poor milk producers and might have been temperamentally difficult to milk, so Neolithic European farmers made sure that all their cows were born of long-domesticated mothers, but they did not mind a little crossbreeding with native wild bulls to obtain larger domestic bulls.[7]

Comparative studies of chain migration among recent and historical pioneer farmers suggest that, in the beginning, the farming-and-herding groups that first moved into temperate southeastern Europe probably spoke similar dialects and recognized one another as cultural cousins. The thin native population of foragers was certainly seen as culturally and linguistically Other, regardless of how the two cultures interacted.[8] After an initial rapid burst of exploration (sites at Anzabegovo, Karanovo

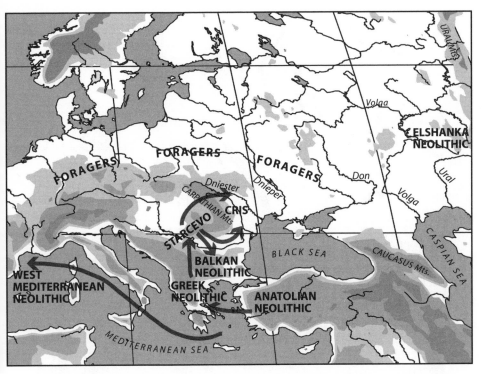

Figure 8.1 The migrations of pioneer farmers into Greece and across Europe between 6500 and 5500 BCE, including the colonization of the eastern Carpathian piedmont by the Criş culture.

I, Gura Baciului, Cîrçea) pioneer groups became established in the Middle Danube plains north of Belgrade, where the type site of Starčevo and other similar Neolithic settlements are located. This central Danubian lowland produced two streams of migrants that leapfrogged in one direction down the Danube, into Romania and Bulgaria, and in the other up the Mureş and Körös Rivers into Transylvania. Both migration streams created similar pottery and tool types, assigned today to the Criş culture (figure 8.2).[9]

First Farmers in the Pontic Region: The Criş Culture

The names Criş in Romania and Körös in eastern Hungary are two variants of the same river name and the same prehistoric culture. The northern Criş people moved up the Hungarian rivers into the mountains of Transylvania and then pushed over the top of the Carpathian ridges into

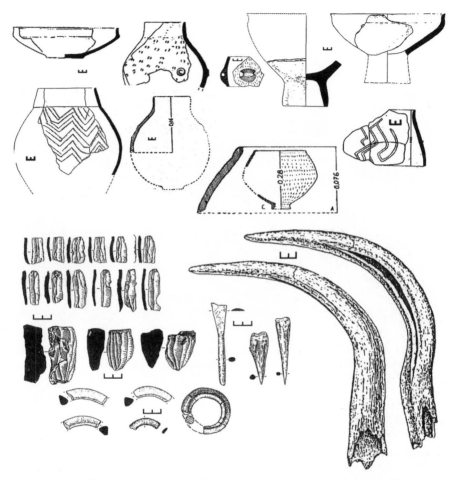

Figure 8.2 Criş-culture ceramic shapes and decorative motifs (*top half*), flint blades and cores (*left*), antler and bone tools (*right*), and ceramic rings (*bottom*) dated 5700–5300 BCE. After Dergachev 1999; and Ursulescu 1984.

an ecologically rich and productive piedmont region east of the Carpathians. They herded their cattle and sheep down the eastern slopes into the upper valleys of the Seret and Prut rivers about 5800–5700 BCE. (Criş radiocarbon dates are unaffected by reservoir effects because they were not measured on human bone; see table 8.1.) The other migration stream in the lower Danube valley moved into the same eastern Carpathian piedmont from the south. These two groups created a northern and a southern variant of the East Carpathian Criş culture, which survived from about 5800 to about 5300 BCE. Criş farms in the East Carpathian piedmont

TABLE 8.1
Radiocarbon Dates for the Late Mesolithic and Early Neolithic of the Pontic–Caspian Region.

Lab Number	BP Date	Sample	Calibrated Date
1. Criş Culture Farming Settlements			
Trestiana (Romania), phase III of the Criş culture			
GrN–17003	6665±45	Charcoal	5640–5530 BCE
Cârcea–Viaduct (Romania), phase IV of the Criş culture			
Bln–1981	6540±60	?	5610–5390 BCE
Bln–1982	6530±60	?	5610–5380 BCE
Bln–1983	6395±60	?	5470–5310 BCE
2. Linear Pottery (LBK) Farming Settlements			
Tirpeşti, Siret River, (Romania)			
Bln–800	6170±100	?	5260–4960 BCE
Bln–801	6245±100	?	5320–5060 BCE
3. Bug–Dniester Mesolithic–Neolithic Settlements			
Soroki II, level 1 early Bug–Dniester, Dniester valley			
Bln–586	6825±150	?	5870–5560 BCE
Soroki II, level 2 pre-ceramic Bug–Dniester, Dniester valley			
Bln–587	7420±80	?	6400–6210 BCE
Savran settlement, late Bug–Dniester, Dniester valley			
Ki–6654	6985±60	?	5980–5790 BCE
Bazkov Ostrov settlement, with early ceramics, South Bug valley			
Ki–6651	7235±60	?	6210–6010 BCE
Ki–6696	7215±55	?	6200–6000 BCE
Ki–6652	7160±55	?	6160–5920 BCE
Sokolets II settlement, with early ceramics, South Bug valley			
Ki–6697	7470±60	?	6400–6250 BCE
Ki–6698	7405±55	?	6390–6210 BCE
4. Early Neolithic Elshanka-type Settlements, Middle Volga Region			
Chekalino 4, Sok River, Samara oblast			
Le–4781	8990±100	shell	8290–7960 BCE

TABLE 8.1 (*continued*)

Lab Number	BP Date	Sample	Calibrated Date
GrN–7085	8680±120	shell	7940–7580 BCE
Le–4783	8050±120	shell	7300–6700 BCE
Le–4782	8000±120	shell	7080–6690 BCE
GrN–7086	7950±130	shell	7050–6680 BCE
Le–4784	7940±140	shell	7050–6680 BCE

Chekalino 6, Sok River, Samara oblast

Le–4883	7940±140	shell	7050–6650 BCE

Ivanovka, upper Samara River, Orenburg oblast

Le–2343	8020±90	bone	7080–6770 BCE

5. Steppe Early Neolithic Settlements

Matveev Kurgan I, very primitive ceramics, Azov steppes

GrN–7199	7505±210	charcoal	6570–6080 BCE
Le–1217	7180±70	charcoal	6160–5920 BCE

Matveev Kurgan II, same material culture, Azov steppes

Le–882	5400±200	charcoal	4450–3980 BCE

Varfolomievka, Layer 3 (bottom ceramic layer), North Caspian steppes

GIN–6546	6980±200	charcoal	6030–5660 BCE

Kair–Shak III, North Caspian steppes

GIN–5905	6950±190	?	6000–5660 BCE
GIN 5927	6720±80	?	5720–5550 BCE

Rakushechni Yar, lower Don shell midden, layers 14–15

Ki–6479	6925±110	?	5970–5710 BCE
Ki–6478	6930±100	?	5970–5610 BCE
Ki–6480	7040±100	?	6010–5800 BCE

Surskii Island, Dnieper Rapids forager settlement

Ki–6688	6980±65	?	5980–5780 BCE
Ki–6989	7125±60	?	6160–5910 BCE
Ki–6690	7195±55	?	6160–5990 BCE
Ki–6691	7245±60	?	6210–6020 BCE

were the source of the first domesticated cattle in the North Pontic region. The Criş pioneers moved eastward through the forest-steppe zone in the piedmont northwest of the Black Sea, where rainfall agriculture was possible, avoiding the lowland steppes on the coast and the lower courses of the rivers that ran through them into the sea.

Archaeologists have identified at least thirty Criş settlement sites in the East Carpathian piedmont, a region of forests interspersed with natural meadows cut by deep, twisting river valleys (figure 8.3). Most Criş farming hamlets were built on the second terraces of rivers, overlooking the floodplain; some were located on steep-sided promontories above the floodplain (Suceava); and a few farms were located on the high forested ridges between the rivers (Sakarovka I). Houses were one room, built with timber posts and beams, plaster-on-wattle walls, and probably reed-thatched roofs. Larger homes, sometimes oval in outline, were built over dug-out floors and contained a kitchen with a domed clay oven; lighter, smaller structures were built on the surface with an open fire in the center. Most villages consisted of just a few families living in perhaps three to ten smoky thatched pit-dwellings, surrounded by agricultural fields, gardens, plum orchards, and pastures for the animals. No Criş cemeteries are known. We do not know what they did with their dead. We do know, however, that they still prized and wore white shell bracelets made from imported *Spondylus*, an Aegean species that was first made into bracelets by the original pioneers in Early Neolithic Greece.[10]

Criş families cultivated barley, millet, peas, and four varieties of wheat (emmer, einkorn, spelt, and bread wheats). Wheat and peas were not native to southeastern Europe; they were exotics, domesticated in the Near East, carried into Greece by sea-borne immigrant farmers, and propagated through Europe from Greece. Residues inside pots suggest that grains were often eaten in the form of a soup thickened with flour. Charred fragments of Neolithic bread from Germany and Switzerland suggest that wheat flour was also made into a batter that was fried or baked, or the grains were moistened and pressed into small whole-grain baked loaves. Criş harvesting sickles used a curved red deer antler inset with flint blades 5–10 cm long, angled so that their corners formed teeth. Their working corners show "sickle gloss" from cutting grain. The same type of sickle and flint blade is found in all the Early Neolithic farming settlements of the Danube-Balkans-Carpathians. Most of the meat in the East Carpathian Criş diet was from cattle and pigs, with red deer a close third, followed by sheep—a distribution of species reflecting their largely forested environment. Their small-breed cows and pigs were slightly different from the

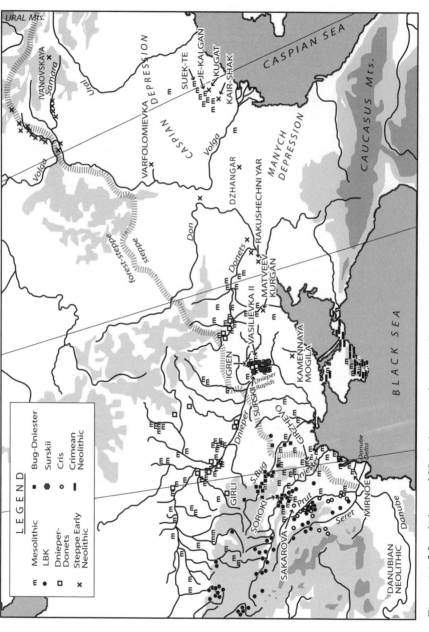

Figure 8.3 Mesolithic and Neolithic sites from the Carpathian Mountains to the Ural River.

local wild aurochs or wild boar but not markedly so. The sheep, however, were exotic newcomers, an invasive species like wheat and peas, brought into the steep Carpathian valleys by strange people whose voices made a new kind of sound.[11]

Criş ceramic vessels were hand-made by the coiling method, and included plain pots for cooking and storage, and a variety of fine wares with polished reddish-brown surfaces—tureens, bowls, and cups on pedestals (figure 8.2). Decorative designs were incised with a stick on the clay surface before firing or were impressed with a fingernail. Very rarely they were painted in broad brown stripes. The shapes and designs made by Criş settlers in the East Carpathians were characteristic of periods III and IV of the Criş culture; older sites of stages I and II are found only in eastern Hungary, the Danube valley, and Transylvania.

Criş farmers never penetrated east of the Prut-Dniester watershed. In the Dniester valley they came face-to-face with a dense population of local foragers, known today as the Bug-Dniester culture, named after the two river valleys (Dniester and South Bug) where most of their sites are found. The Bug-Dniester culture was the filter through which farming and stock-breeding economies were introduced to Pontic-Caspian societies farther east (figure 8.3).

The Criş people were different from their Bug-Dniester neighbors in many ways: Criş flint tool kits featured large blades and few scrapers, whereas the foragers used microlithic blades and many scrapers; most Criş villages were on the better-drained soils of the second terrace, convenient for farming, and most foragers lived on the floodplain, convenient for fishing; whereas Criş woodworkers used polished stone axes, the foragers used chipped flint axes; Criş pottery was distinct both in the way it was made and its style of decoration; and Criş farmers raised and ate various exotic foods, including mutton, which has a distinctive taste. Four forged cylindrical copper beads were found at the Criş site of Selishte, dated 5800–5600 BCE (6830±100 BP).[12] They show an early awareness of the metallic minerals in the mountains of Transylvania (copper, silver, gold) and the Balkans (copper), something the foragers of southeastern Europe had never noticed.

Some archaeologists have speculated that the East Carpathian Criş culture could have been an acculturated population of local foragers who had adopted a farming economy, rather than immigrant pioneers.[13] This is unlikely given the numerous similarities between the material culture and economy of Criş sites in the Danube valley and the East Carpathians, and the sharp differences between the East Carpathian Criş culture and the local foragers. But it really is of no consequence—no one seriously believes that the East Carpathian Criş people were *genetically* "pure" anyway. The important point is that the people who lived in Criş villages in the East Carpathians were *culturally* Criş in almost all the material signs of their identity, and given how they got there, almost certainly in nonmaterial signs like language as well. The Criş *culture* came, without any doubt, from the Danube valley.

The Language of the Criş Culture

If the Starcevo–Criş–Karanovo migrants were at all similar to pioneer farmers in North America, Brazil, southeast Asia, and other parts of the world, it is very likely that they retained the language spoken in their parent villages in northern Greece. Forager languages were more apt to decline in the face of agricultural immigration. Farmers had a higher birth rate; their settlements were larger, and were occupied permanently. They produced food surpluses that were easier to store over the winter. Owning and feeding "cultured" animals has always been seen as an utterly different ethos from hunting wild ones, as Ian Hodder emphasized. The material and ritual culture and economy of the immigrant farmers were imposed on the landscapes of Greece and southeastern Europe and persisted there, whereas the external signs of forager identity disappeared. The language of the foragers might have had substrate effects on that of the farmers, but it is difficult to imagine a plausible scenario under which it could have competed with the farmers' language.[14]

What languages were spoken by Starčevo, Criş, and Karanovo I pioneers? The parent language for all of them was spoken in the Thessalian plain of Greece, where the first Neolithic settlements were founded about 6700–6500 BCE probably by seafarers who island-hopped from western Anatolia in open boats. Katherine Perlés has convincingly demonstrated that the material culture and economy of the first farmers in Greece was transplanted from the Near East or Anatolia. An origin somewhere in western Anatolia is suggested by similarities in pottery, flint tools, ornaments, female figurines, pintadera stamps, lip labrets, and other traits. The migrants leapfrogged to the Thessalian plain, the richest agricultural land in Greece, almost certainly on the basis of information from scouts (probably Aegean fishermen) who told their relatives in Anatolia about the destination. The population of farmers in Thessaly grew rapidly. At least 120 Early Neolithic settlements stood on the Thessalian plain by 6200–6000 BCE, when pioneers began to move north into the temperate forests of southeastern Europe. The Neolithic villages of Thessaly provided the original breeds of domesticated sheep, cattle, wheat, and barley, as well as red-on-white pottery, female-centered domestic rituals, bracelets and beads made of Aegean *Spondylus* shell, flint tool types, and other traditions that were carried into the Balkans. The language of Neolithic Thessaly probably was a dialect of a language spoken in western Anatolia about 6500 BCE. Simplification and leveling should have occurred among the first colonist dialects in Thessaly, so the 120 villages occupied five hundred years later spoke a language that had passed through a bottleneck and probably was just beginning to separate again into strongly differentiated dialects.[15]

The tongue spoken by the first Criş farmers in the East Carpathian foothills about 5800–5600 BCE was removed from the parent tongue spoken by the first settlers in Thessaly by less than a thousand years—the same interval that separates Modern American English from Anglo-Saxon. That was long enough for several new Old European Neolithic languages to have emerged from the Thessalian parent, but they would have belonged to a single language family. That language family was not Indo-European. It came from the wrong place (Anatolia and Greece) at the wrong time (before 6500 BCE). Curiously a fragment of that lost language might be preserved in the Proto-Indo-European term for bull, *tawro-s*, which many linguists think was borrowed from an Afro-Asiatic term. The Afro-Asiatic super-family generated both Egyptian and Semitic in the Near East, and one of its early languages might have been spoken in Anatolia by the earliest farmers. Perhaps the Criş people spoke a language of Afro-Asiatic type, and as they drove their cattle into the East Carpathian valleys they called them something like *tawr-*.[16]

FARMER MEETS FORAGER: THE BUG-DNIESTER CULTURE

The first indigenous North Pontic people to adopt Criş cattle breeding and perhaps also the Criş word for bull were the people of the Bug-Dniester culture, introduced a few pages ago. They occupied the frontier where the expansion of the Criş farmers came to a halt, apparently blocked by the Bug-Dniester culture itself. The initial contact between farmers and foragers must have been a fascinating event. The Criş immigrants brought herds of cultured animals that wandered up the hillsides among the deer. They introduced sheep, plum orchards, and hot wheat-cakes. Their families lived in the same place all year, year after year; they cut down the trees to make houses and orchards and gardens; and they spoke a foreign language.

The foragers' language might have been part of the broad language family from which Proto-Indo-European later emerged, although, since the ultimate fate of the Bug-Dniester culture was extinction and assimilation, their dialect probably died with their culture.[17]

The Bug-Dniester culture grew out of Mesolithic forager cultures that dwelt in the region since the end of the last Ice Age. Eleven Late Mesolithic technological-typological groups have been defined by differences in flint tool kits just in Ukraine; other Late Mesolithic flint tool-based groups have been identified in the Russian steppes east of the Don River, in the North Caspian Depression, and in coastal Romania. Mesolithic camps have been found in the lower Danube valley and the coastal steppes northwest of the Black Sea, not far from the Criş settlement area. In the Dobruja, the peninsula of rocky hills skirted by the Danube delta at its mouth, eighteen to twenty Mesolithic surface sites were found just in one small area northwest of Tulcea on the southern terraces of the Danube River. Late Mesolithic groups also occupied the northern side of the estuary. Mirnoe is the best-studied site here. The Late Mesolithic hunters at Mirnoe hunted wild aurochs (83% of bones), wild horse (14%), and the extinct *Equus hydruntinus* (1.1%). Farther up the coast, away from the Danube delta, the steppes were drier, and at Late Mesolithic Girzhevo, on the lower Dniester, 62% of the bones were of wild horses, with fewer aurochs and *Equus hydruntinus*. There is no archaeological trace of contact between these coastal steppe foragers and the Criş farmers who were advancing into the upland forest-steppe.[18]

The story is different in the forest-steppe. At least twenty-five Bug-Dniester sites have been excavated in the forest-steppe zone in the middle and upper parts of the South Bug and Dniester River valleys, in the transitional ecological zone where rainfall was sufficient for the growth of forests but there were still open meadows and some pockets of steppe. This environment was favored by the Criş immigrants. In it the native foragers had for generations hunted red deer, roe deer, and wild boar, and caught riverine fish (especially the huge river catfish, *Siluris glanis*). Early Bug-Dniester flint tools showed similarities both to coastal steppe groups (Grebenikov and Kukrekskaya types of tool kits) and northern forest groups (Donets types).

Pottery and the Beginning of the Neolithic

The Bug-Dniester culture was a Neolithic culture; Bug-Dniester people knew how to make fired clay pottery vessels. The first pottery in the Pontic-Caspian region, and the beginning of the Early Neolithic, is associated with the Elshanka culture in the Samara region in the middle Volga River valley. It is dated by radiocarbon (on shell) about 7000–6500 BCE, which makes it, surprisingly, the oldest pottery in all of Europe. The pots were made of a clay-rich mud collected from the bottoms of stagnant ponds. They were formed by the coiling method and were baked in open fires at 450–600°C (figure 8.4).[19] From this northeastern source ceramic technology diffused south and westward. It was adopted widely by most foraging and fishing bands across the Pontic-Caspian region about 6200–6000 BCE, before any clear contact with southern farmers. Early Neolithic pottery tempered with vegetal material and crushed shells appeared at Surskii Island in the Dnieper Rapids in levels dated about 6200–5800 BCE. In the lower Don River valley a crude vegetal-tempered pottery decorated with incised geometric motifs appeared at Rakushechni Yar and other sites such as Samsonovka in levels dated 6000–5600 BCE.[20] Similar designs and vessel shapes, but made with a shell-tempered clay fabric, appeared on the lower Volga, at Kair Shak III dated about 5700–5600 BCE (6720±80 BP). Older pottery was made in the North Caspian at Kugat, where a different kind of pottery was stratified beneath Kair Shak-type pottery, possibly the same age as the pottery at Surskii Island. Primitive, experimental ceramic fragments appeared about 6200 BCE also at Matveev Kurgan in the steppes north of the Sea of Azov. The oldest pottery south of the middle Volga appeared at the Dnieper Rapids (Surskii), on the lower Don (Rakushechni Yar), and on the lower Volga (Kair Shak III, Kugat) at about the same time, around 6200–6000 BCE (figure 8.4).

The earliest pottery in the South Bug valley was excavated by Danilenko at Bas'kov Ostrov and Sokolets II, dated by five radiocarbon dates about 6200–6000 BCE, about the same age as Surskii on the Dnieper.[21] In the Dniester River valley, just west of the South Bug, at Soroki II, archaeologists excavated two stratified Late Mesolithic occupations (levels 2 and 3) dated by radiocarbon to about 6500–6200 BCE. They contained no pottery. Pottery making was adopted by the early Bug-Dniester culture about 6200 BCE, probably the same general time it appeared in the Dnieper valley and the Caspian Depression.

Farmer-Forager Exchanges in the Dniester Valley

After about 5800–5700 BCE, when Criş farmers moved into the East Carpathian foothills from the west, the Dniester valley became a frontier between two very different ways of life. At Soroki II the uppermost

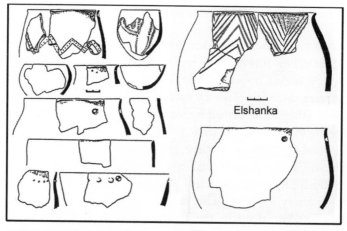

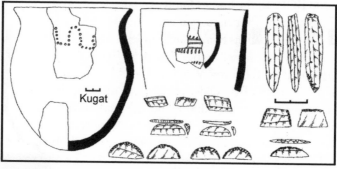

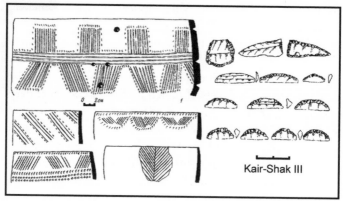

Figure 8.4 Top: Early Neolithic ceramics of Elshanka type on the middle Volga (7000–6500 BCE); *middle*: ceramics and flint tools from Kugat (perhaps 6000 BCE), North Caspian; *bottom*: ceramics and flint tools from Kair-Shak III (5700–5600 BCE) North Caspian. After (*top*) Mamonov 1995; and (*middle and bottom*) Barynkin and Kozin 1998.

occupation level (1) was left by Bug-Dniester people who clearly had made contact with the incoming Criş farmers, dated by good radiocarbon dates at about 5700–5500 BCE. Some of the ceramic vessels in level 1 were obvious copies of Criş vessels—round-bodied, narrow-mouthed jars on a ring base and bowls with carinated sides. But they were made locally, using clay tempered with sand and plant fibers. The rest of the pottery in level 1 looked more like indigenous bag-shaped South Bug ceramics (figure 8.5). Continuity in the flint tools between level 1 and the older levels 2 and 3 suggests that it was the same basic culture, and all three levels are traditionally assigned to the Bug-Dniester culture.

The Bug-Dniester people who lived at Soroki II in the level 1 camp copied more than just Criş pottery. Botanists found seed impressions in the clay vessels of three kinds of wheat. Level 1 also yielded a few bones from small domesticated cattle and pigs. This was the beginning of a significant shift—the adoption of an imported food-production economy by the native foragers. It is perhaps noteworthy that the exotic ceramic types copied by Soroki II potters were small Criş pedestaled jars and bowls, probably used to serve drink and food rather than to store or cook it. Perhaps Criş foods were served to visiting foragers in jars and bowls like these inside Criş houses, inspiring some Bug-Dniester families to re-create both the new foods and the vessels in which they were served. But the original decorative motifs on Bug-Dniester pottery, the shapes of the largest pots, the vegetal and occasional shell temper in the clay, and the low-temperature firing indicate that early Bug-Dniester potters knew their own techniques, clays, and tempering formulas. The largest pots they made (for cooking? storage?) were shaped like narrow-mouthed baskets, unlike any shape made by Criş potters.

Three kinds of wheat impressions appeared in the clay of early Bug-Dniester pots at two sites in the Dniester valley: Soroki II/level 1 and Soroki III. Both sites had impressions of emmer, einkorn, and spelt.[22] Was the grain actually grown locally? Both sites had a variety of wheats, with impressions of chaff and spikelets, parts removed during threshing. The presence of threshing debris suggests that at least some grain was grown and threshed locally. The foragers of the Dniester valley seem to have cultivated at least small plots of grain very soon after their initial contact with Criş farmers. What about the cattle?

In three Early Bug-Dniester Neolithic sites in the Dniester valley occupied about 5800–5500 BCE, domesticated cattle and swine averaged 24% of the 329 bones recovered from garbage pits, if each bone is counted for the NISP; or 20% of the animals, if the bones are converted into a

Figure 8.5 Pottery types of the Bug-Dniester culture. The four vessels in the top row appear to have been copied after Criş types seen in Figure 8.2. After Markevich 1974; and Dergachev 1999.

minimum number of individuals, or MNI. Red deer and roe deer remained more important than domesticated animals in the meat diet. Middle Bug-Dniester sites (Samchin phase), dated about 5600–5400 BCE, contained more domesticated pigs and cattle: at Soroki I/level 1a, a Middle-phase site, cattle and swine made up 49% of the 213 bones recovered (32% MNI). By the Late (Savran) phase, about 5400–5000 BCE, domesticated pigs and cattle totaled 55% of the animal bones (36% MNI) in two sites.[23] In contrast, the Bug-Dniester settlement sites in the South

Bug valley, farther away from the source of the domesticated animals, never showed more than 10% domesticated animal bones. But even in the South Bug valley a few domesticated cattle and pigs appeared at Bas'kov Ostrov and Mit'kov Ostrov very soon after the Criş farmers entered the Eastern Carpathian foothills. The "availability" phase, in Zvelebil's three-phase description of farmer-forager interactions, was very brief.[24] Why? What was so attractive about Criş foods and even the pottery vessels in which they were served?

There are three possibilities: intermarriage, population pressure, and status competition. Intermarriage is an often-repeated but not very convincing explanation for incremental changes in material culture. In this case, imported Criş-culture wives would be the vehicle through which Criş-culture pottery styles and foods should have appeared in Bug-Dniester settlements. But Warren DeBoer has shown that wives who marry into a foreign tribe among tribal societies often feel so exposed and insecure that they become hyper-correct imitators of their new cultural mores rather than a source of innovation. And the technology of Bug-Dniester ceramics, the method of manufacture, was local. Technological styles are often better indicators of ethnic origin than decorative styles. So, although there may have been intermarriage, it is not a persuasive explanation for the innovations in pottery or economy on the Dniester frontier.[25]

Was it population pressure? Were the pre-Neolithic Bug-Dniester foragers running out of good hunting and fishing grounds, and looking for ways to increase the amount of food that could be harvested within their hunting territories? Probably not. The forest-steppe was an ideal hunting territory, with maximal amounts of the forest-edge environment preferred by deer. The abundant tree pollen in Criş-period soils indicates that the Criş pioneers had little impact on the forest around them, so their arrival did not greatly reduce deer populations. A major component of the Bug-Dniester diet was riverine fish, some of which supplied as much meat as a small adult pig, and there is no evidence that fish stocks were falling. Cattle and pigs might have been acquired by cautious foragers as a hedge against a bad year, but the immediate motive probably was not hunger.

The third possibility is that the foragers were impressed by the abundance of food available for feasting and seasonal festivals among Criş farmers. Perhaps some Bug-Dniester locals were invited to such festivals by the Criş farmers in an attempt to encourage peaceful coexistence. Socially ambitious foragers might have begun to cultivate gardens and raise cattle to sponsor feasts among their own people, even making serving bowls and cups like those used in Criş villages—a political explanation, and one that also explains why Criş pots were copied. Unfortunately neither culture had cemeteries, and so we cannot examine graves to look for evidence of a growing social hierarchy. Status objects seem to have been few, with the possible exception of food itself. Probably both economic insurance and social status played roles in the slow but steady adoption of food production in the Dniester valley.

The importance of herding and cultivation in the Bug-Dniester diet grew very gradually. In Criş settlements domesticated animals contributed 70–80% of the bones in kitchen middens. In Bug-Dniester settlements domesticated animals exceeded hunted wild game only in the latest phase, and only in the Dniester valley, immediately adjacent to Criş settlements. Bug-Dniester people never ate mutton—not one single sheep bone has been found in a Bug-Dniester site. Early Bug-Dniester bakers did not use Criş-style saddle querns to grind their grain; instead, they initially used small, rhomboidal stone mortars of a local style, switching to Criş-style saddle querns only in the middle Bug-Dniester phase. They preferred their own chipped flint axe types to the smaller polished stone Criş axes. Their pottery was quite distinctive. And their historical trajectory led directly back to the local Mesolithic populations, unlike the Criş culture.

Even after 5500–5200 BCE, when a new farming culture, the Linear Pottery culture, moved into the East Carpathian piedmont from southern Poland and replaced the Criş culture, the Dniester valley frontier survived. No Linear Pottery sites are known east of the Dniester valley.[26] The Dniester was a cultural frontier, not a natural one. It persisted despite the passage of people and trade goods across it, and through significant cultural changes on each side. Persistent cultural frontiers, particularly at the edges of ancient migration streams, usually are ethnic and linguistic frontiers. The Bug-Dniester people may well have spoken a language belonging to the language family that produced Pre-Proto-Indo-European, while their Criş neighbors spoke a language distantly related to those of Neolithic Greece and Anatolia.

BEYOND THE FRONTIER: PONTIC-CASPIAN FORAGERS BEFORE CATTLE ARRIVED

The North Pontic societies east of the Dniester frontier continued to live as they always had, by hunting, gathering wild plants, and fishing until about 5200 BCE. Domesticated cattle and hot wheatcakes might have seemed irresistibly attractive to the foragers who were in direct contact with the farmers who presented and legitimized them, but, away from that active frontier, North Pontic forager-fishers were in no rush to become animal tenders. Domesticated animals can only be raised by people who are committed morally and ethically to watching their families go hungry rather than letting them eat the breeding stock. Seed grain and breeding stock must be saved, not eaten, or there will be no crop and no calves the next year. Foragers generally value immediate sharing and generosity over miserly saving for the future, so the shift to keeping breeding stock was a moral as well as an economic one. It probably offended the old morals. It is not surprising that it was resisted, or that when it did begin it was surrounded by new rituals and a new kind of leadership, or that the new leaders threw big feasts and shared food when the deferred investment paid off. These new rituals and leadership roles were the foundation of Indo-European religion and society.[27]

The most heavily populated part of the Pontic-Caspian steppes was the place where the shift to cattle keeping happened next after the Bug-Dniester region. This was around the Dnieper Rapids. The Dnieper Rapids started at modern Dnepropetrovsk, where the Dnieper River began to cut down to the coastal lowlands through a shelf of granite bedrock, dropping 50 m in elevation over 66 km. The Rapids contained ten major cascades, and in early historical accounts each one had its own name, guardian spirits, and folklore. Fish migrating upstream, like the sudak (*Lucioperca*), could be taken in vast quantities at the Rapids, and the swift water between the cascades was home to wels (*Silurus glanis*), a type of catfish that grows to 16 feet. The bones of both types of fish are found in Mesolithic and Neolithic camps near the Rapids. At the southern end of the Rapids there was a ford near Kichkas where the wide Dnieper could be crossed relatively easily on foot, a strategic place in a world without bridges.

The Rapids and many of the archaeological sites associated with them were inundated by dams and reservoirs built between 1927 and 1958. Among the many sites discovered in connection with reservoir construction was Igren 8 on the east bank of the Dnieper. Here the deepest level F contained Late Mesolithic Kukrekskaya flint tools; levels E and E1 above contained Surskii Early Neolithic pottery (radiocarbon dated 6200–5800 BCE); and stratum D1 above that contained Middle Neolithic Dnieper-Donets I pottery tempered with plant fibers and decorated with incised chevrons and small comb stamps (probably about 5800–5200 BCE but not directly dated by radiocarbon). The animal bones in the Dnieper-Donets I garbage were from red deer and fish. The shift to cattle keeping had not yet begun. Dnieper-Donets I was contemporary with the Bug-Dniester culture.[28]

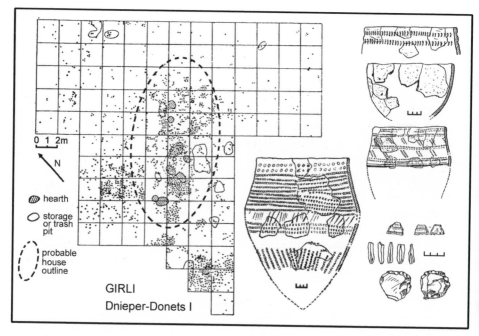

Figure 8.6 Dnieper-Donets I camp at Girli, Ukraine, probably about 5600–5200 BCE. After Neprina 1970 , Figures 3, 4, and 8.

Campsites of foragers who made Dnieper-Donets I (DDI) pottery have been excavated on the southern borders of the Pripet Marshes in the northwest and in the middle Donets valley in the east, or over much of the forest-steppe and northern steppe zone of Ukraine. At Girli (figure 8.6) on the upper Teterev River near Zhitomir, west of Kiev, a DDI settlement contained eight hearths arranged in a northeast-southwest line of four pairs, each pair about 2–3 m apart, perhaps representing a shelter some 14 m long for four families. Around the hearths were thirty-six hundred flint tools including microlithic blades, and sherds of point-based pots decorated with comb-stamped and pricked impressions. The food economy depended on hunting and gathering. Girli was located on a trail between the Dnieper and South Bug rivers, and the pottery was similar in shape and decoration to some Bug-Dniester ceramics of the middle or Samchin phase. But DDI sites did not contain domesticated animals or plants, or even polished stone axes like those of the Criş and late Bug-Dniester cultures; DDI axes were still chipped from large pieces of flint.[29]

Forager Cemeteries around the Dnieper Rapids

Across most of Ukraine and European Russia post-glacial foragers did not create cemeteries. The Bug-Dniester culture was typical: they buried their dead by ones and twos, often using an old campsite, perhaps the one where the death occurred. Graveside rituals took place but not in places set aside just for them. Cemeteries were different: they were formal plots of ground reserved just for funerals, funeral monuments, and public remembrance of the dead. Cemeteries were visible statements connecting a piece of land with the ancestors. During reservoir construction around the Dnieper Rapids archaeologists found eight Mesolithic and forager Neolithic cemeteries, among them Vasilievka I (twenty-four graves), Vasilievka II (thirty-two graves), Vasilievka III (forty-five graves), Vasilievka V (thirty-seven graves), Marievka (fifteen graves), and Volos'ke (nineteen graves). No comparable cluster of forager cemeteries exists anywhere else in the Pontic-Caspian region.

Several different forager populations seem to have competed with one another around the Dnieper Rapids at the end of the Ice Age. Already by about 8000 BCE, as soon as the glaciers melted, at least three skull-and-face types, a narrow-faced gracile type (Volos'ke), a broad-faced medium-weight type (Vasilievka I), and a broad-faced robust type (Vasilievka III) occupied different cemeteries and were buried in different poses (contracted and extended). Two of the nineteen individuals buried at Volos'ke and two (perhaps three) of the forty-five at Vasilevka III were wounded by weapons tipped with Kukrekskaya-type microlithic blades. The Vasilievka III skeletal type and burial posture ultimately spread over the whole Rapids during the Late Mesolithic, 7000–6200 BCE. Two cemeteries that were assumed to be Early Neolithic (Vasilievka II and Marievka) because of the style of the grave now are dated by radiocarbon to 6500–6000 BCE, or the Late Mesolithic.

Only one of the Dnieper Rapids cemeteries, Vasilievka V, is dated to the Middle Neolithic DDI period by radiocarbon dates (5700–5300 BCE). At Vasilevka V thirty-seven skeletons were buried in supine positions (on their backs) with their hands near the pelvis, with their heads to the northeast. Some were buried singly in individual pits, and others apparently were layered in reused graves. Sixteen graves in the center of the cemetery seem to represent two or three superimposed layers of burials, the first hint of a collective burial ritual that would be elaborated greatly in the following centuries. Eighteen graves out of thirty-seven were sprinkled with red ochre, again a hint of things to come. The grave gifts at Vasilievka V, however, were very simple, limited to microlithic flint blades and flint scrapers. These were the last people on the Dnieper Rapids who clung to the old morality and rejected cattle keeping.[30]

Foragers on the Lower Volga and Lower Don

Different styles of pottery were made among the Early Neolithic foragers who lived even farther east, a longer distance away from the forager/farmer frontier on the Dniester. Forager camps on the lower Volga River dated between 6000 and 5300 BCE contained flat-based open bowls made of clay tempered with crushed shell and vegetal material, and were decorated by stabbing rows of impressions with a triangular-ended stick or drawing incised diamond and lozenge shapes. These decorative techniques were different from the comb-stamps used to decorate DDI pottery in the Dnieper valley. Flint tool kits on the Volga contained many geometric microliths, 60–70% of the tools, like the flint tools of the earlier Late Mesolithic foragers. Important Early Neolithic sites included Varfolomievka level 3 (radiocarbon dated about 5900–5700 BCE) and Kair-Shak III (also dated about 5900–5700 BCE) in the lower Volga region; and the lower levels at Rakushechni Yar, a dune on the lower Don (dated 6000–5600 BCE).[31] At Kair Shak III, located in an environment that was then semi-desert, the economy was based almost entirely on hunting onagers (*Equus hemionus*). The animal bones at Varfolomievka, located in a small river valley in the dry steppe, have not been reported separately by level, so it is impossible to say what the level 3 Early Neolithic economy was, but half of all the animal bones at Varfolomievka were of horses (*Equus caballus*), with some bones of aurochs (*Bos primigenius*). Fish scales (unidentified) were found on the floors of the dwellings. At Rakushechni Yar, then surrounded by broad lower-Don valley gallery forests, hunters pursued red deer, wild horses, and wild pigs. As I noted in several endnotes in this chapter, some archaeologists have claimed that the herding of cattle and sheep began earlier in the lower Don-Azov steppes, but this is unlikely. Before 5200 BCE the forager-farmer frontier remained confined to the Dniester valley.[32]

THE GODS GIVE CATTLE

The Criş colonization of the Eastern Carpathians about 5800 BCE created a robust and persistent cultural frontier in the forest-steppe zone at the Dniester valley. Although the Bug-Dniester culture quickly acquired at least some domesticated cereals, pigs, and cattle, it retained an economy based primarily on hunting and gathering, and remained culturally and economically distinct in most ways. Beyond it, both in the forest-steppe zone and the steppe river valleys to the east, no other indigenous societies seem to have adopted cereal cultivation or domesticated animals until after about 5200 BCE.

In the Dniester valley, native North Pontic cultures had direct, face-to-face contact with farmers who spoke a different language, had a different religion, and introduced an array of invasive new plants and animals as if they were something wonderful. The foragers on the frontier itself rapidly accepted some cultivated plants and animals but rejected others, particularly sheep. Hunting and fishing continued to supply most of the diet. They did not display obvious signs of a shift to new rituals or social structures. Cattle keeping and wheat cultivation seem to have been pursued part-time, and were employed as an insurance policy against bad years and perhaps as a way of keeping up with the neighbors, not as a replacement of the foraging economy and morality. For centuries even this halfway shift to partial food production was limited to the Dniester valley, which became a narrow and well-defined frontier. But after 5200 BCE a new threshold in population density and social organization seems to have been crossed among European Neolithic farmers. Villages in the East Carpathian piedmont adopted new customs from the larger towns in the lower Danube valley, and a new, more complex culture appeared, the Cucuteni-Tripolye culture. Cucuteni-Tripolye villages spread eastward. The Dniester frontier was breached, and large western farming communities pushed into the Dniester and South Bug valleys. The Bug-Dniester culture, the original frontier society, disappeared into the wave of Cucuteni-Tripolye immigrants.

But away to the east, around the Dnieper Rapids, the bones of domesticated cattle, pigs, and, remarkably, even sheep began to appear regularly in garbage dumps. The Dnieper Rapids was a strategic territory, and the clans that controlled it already had more elaborate rituals than clans elsewhere in the steppes. When they accepted cattle keeping it had rapid economic and social consequences across the steppe zone.

CHAPTER NINE

Cows, Copper, and Chiefs

The Proto-Indo-European vocabulary contained a compound word (*weik-potis*) that referred to a village chief, an individual who held power within a residential group; another root (*reǵ-*) referred to another kind of powerful officer. This second root was later used for *king* in Italic (*rēx*), Celtic (*rīx*), and Old Indic (*raj-*), but it might originally have referred to an official more like a priest, literally a "*regulator*" (from the same root) or "one who makes things *right*" (again the same root), possibly connected with drawing "*correct*" (same root) boundaries. The speakers of Proto-Indo-European had institutionalized offices of power and social ranks, and presumably showed deference to the people who held them, and these powerful people, in return, sponsored feasts at which food and gifts were distributed.[1] When did a hierarchy of social power first appear in the Pontic-Caspian region? How was it expressed? And who were these powerful people?

Chiefs first appeared in the archaeological record of the Pontic-Caspian steppes when domesticated cattle, sheep, and goats first became widespread, after about 5200–5000 BCE.[2] An interesting aspect of the spread of animal keeping in the steppes was the concurrent rapid rise of chiefs who wore multiple belts and strings of polished shell beads, bone beads, beaver-tooth and horse-tooth beads, boars tusk pendants, boars-tusk caps, boars-tusk plates sewed to their clothing, pendants of crystal and porphyry, polished stone bracelets, and gleaming copper rings. Their ornaments must have clacked and rustled when they walked. Older chiefs carried maces with polished stone mace-heads. Their funerals were accompanied by the sacrifice of sheep, goats, cattle, and horses, with most of the meat and bones distributed to the celebrants so only a few symbolic lower leg pieces and an occasional skull, perhaps attached to a hide, remained in the grave. No such ostentatious leaders had existed in the old hunting and gathering bands of the Neolithic. What made their sudden rise even more intriguing is that the nitrogen levels in their bones suggest that more than 50% of their meat diet continued to come from fish. In the Volga region the bones of horses, the preferred wild prey of the earlier hunters, still outnumbered cattle and sheep in kitchen trash. The domesticated cattle and sheep that played such a large ritual role were eaten only infrequently, particularly in the east.

What seems at first to be the spread of a new food economy on second look appears to be deeply interwined in new rituals, new values associated with them, and new institutions of social power. People who did not accept the new animal currency, who remained foragers, did not even use formal cemeteries, much less sponsor such aggrandizing public funeral feasts. Their dead still were buried simply, in plain clothing, in their old camping places. The cultural gap widened between those who tended domesticated animals, including foreign sheep and goats, and those who hunted native wild animals.

The northern frontier of the new economy coincided with the ecological divide between the forests in the north and the steppes in the south. The northern hunters and fishers refused to be shackled to domesticated animals for another two thousand years. Even in the intervening zone of forest-steppe the percentage of domesticated animal bones declined and the importance of hunted game increased. In contrast, the eastern frontier of the new economy did not coincide with an ecotone but instead ran along the Ural River, which drained the southern flanks of the Ural Mountains and flowed south through the Caspian Depression into the Caspian Sea. East of the Ural River, in the steppes of northern Kazakhstan, steppe foragers of the Atbasar type continued to live by hunting wild horses, deer, and aurochs. They lived in camps sheltered by grassy bluffs on low river terraces or on the marshy margins of lakes in the steppes. Their rejection of the new western economy possibly was rooted in ethnic and linguistic differences that had sharpened during the millennia between 14,000 and 9,000 BCE, when the Khvalynian Sea had divided the societies of the Kazakh and the Russian steppes. Regardless of its cause, the Ural valley became a persistent frontier dividing western steppe societies that accepted domesticated animals from eastern steppe societies that rejected them.

Copper ornaments were among the gifts and baubles traded eastward across the steppes from the Danube valley to the Volga-Ural region with the first domesticated animals. The regular, widespread appearance of copper in the Pontic-Caspian steppes signals the beginning of the Eneolithic. The copper was Balkan in origin and probably was obtained with the animals through the same trade networks. From this time forward Pontic-Caspian steppe cultures were drawn into increasingly complicated social, political, and economic relations with the cultures of the Balkans and the lower Danube valley. The gulf between them, however, only intensified. By 4400–4200 BCE, when the Old European cultures were at their peak of economic productivity, population size, and stability, their frontier with the Pontic-Caspian herding cultures was the most pronounced cultural divide in prehistoric Europe, an even starker contrast than that between the northern forest hunters and the steppe herders. The Neolithic and Eneolithic cultures of the Balkans, Carpathians, and middle and lower Danube valley had more productive farming economies in an age when that really mattered, their towns and houses were much more substantial, and their craft techniques, decorative aesthetics, and metallurgy were more sophisticated than those of the steppes. The Early Eneolithic herding cultures of the steppes certainly were aware of the richly ornamented and colorfully decorated people of Old Europe, but steppe societies developed in a different direction.[3]

THE EARLY COPPER AGE IN OLD EUROPE

There is an overall rhythm to the Eneolithic over most of southeastern Europe: a rise to a new level of social and technological complexity, its flourishing, and its subsequent disintegration into smaller-scale, more mobile, and technologically simpler communities at the opening of the Bronze Age. But it began, developed, and ended differently in different places. Its beginning is set at about 5200–5000 BCE in Bulgaria, which was in many ways the heart and center of Old Europe. Pontic-Caspian steppe societies were pulled into the Old European copper-trade network at least as early as 4600 BCE, more than six hundred years before copper was regularly used in Germany, Austria, or Poland.[4]

The scattered farming hamlets of Bulgaria and southern Romania, about 5200–5000 BCE, blossomed into increasingly large and solidly built agricultural villages of large multiroomed timber and mud-plaster houses, often two-storied, set in cleared and cultivated landscapes surrounded by herds of cattle, pigs, and sheep. Cattle pulled ards, primitive scratch-plows, across the fields.[5] In the Balkans and the fertile plains of the lower Danube valley, villages were rebuilt on the same spot generation after generation, creating stratified tells that grew to heights of 30–50 feet, lifting the village above its surrounding fields. Marija Gimbutas has made Old Europe famous for the ubiquity and variety of its goddesses. Household cults symbolized by broad-hipped female figurines were practiced everywhere. Marks incised on figurines and pots suggest the appearance of a notation system.[6] Fragments of colored plaster suggest that house walls were painted with the same swirling, curvilinear designs that appeared on decorated pottery. Potters invented kilns that reached temperatures of 800–1100°C. They used a low-oxygen reducing atmosphere to create a black ceramic surface that was painted with graphite to make silver designs; or a bellows-aided high-oxygen atmosphere to create a red or orange surface, intricately painted in white ribbons bordered with black and red.

Pottery kilns led to metallurgy. Copper was extracted from stone by mixing powdered green-blue azurite or malachite minerals (possibly used for pigments) with powdered charcoal and baking the mixture in a bellows-aided kiln, perhaps accidentally at first. At 800°C the copper separated from the powdered ore in tiny shining beads. It could then be tapped out, reheated, forged, welded, annealed, and hammered into a wide variety of tools (hooks, awls, blades) and ornaments (beads, rings, and other pendants). Ornaments of gold (probably mined in Transylvania and coastal Thrace) began to circulate in the same trade networks. The early phase of copper working began before 5000 BCE.

Balkan smiths, about 4800–4600 BCE, learned to fashion molds that withstood the heat of molten copper, and began to make cast copper tools and weapons, a complicated process requiring a temperature of 1,083°C to liquefy copper metal. Molten copper must be stirred, skimmed, and poured correctly or it cools into a brittle object full of imperfections. Well-made cast copper tools were used and exchanged across southeastern Europe by about 4600–4500 BCE in eastern Hungary with the Tiszapolgar culture; in Serbia with the Vinča D culture; in Bulgaria at Varna and in the Karanovo VI tell settlements; in Romania with the Gumelnitsa culture; and in Moldova and eastern Romania with the Cucuteni-Tripolye culture. Metallurgy was a new and different kind of craft. It was obvious to anyone that pots were made of clay, but even after being told that a shiny copper ring was made from a green-stained rock, it was difficult to see how. The magical aspect of copperworking set metalworkers apart, and the demand for copper objects increased trade. Prospecting, mining, and long-distance trade for ore and finished products introduced a new era in inter-regional politics and interdependence that quickly reached deep into the steppes as far as the Volga.[7]

Kilns and smelters for pottery and copper consumed the forests, as did two-storied timber houses and the bristling palisade walls that protected many Old European settlements, particularly in northeastern Bulgaria. At Durankulak and Sabla Ezerec in northeastern Bulgaria and at Tîrpeşti in Romania, pollen cores taken near settlements show significant reductions in local forest cover.[8] The earth's climate reached its post-glacial thermal maximum, the Atlantic period, about 6000–4000 BCE, and was at its warmest during the late Atlantic (paleoclimatic zone A3), beginning about 5200 BCE. Riverine forests in the steppe river valleys contracted because of increased warmth and dryness, and grasslands expanded. In the forest-steppe uplands majestic forests of elm, oak, and lime trees spread from the Carpathians to the Urals by 5000 BCE. Wild honeybees, which preferred lime and oak trees for nests, spread with them.[9]

THE CUCUTENI-TRIPOLYE CULTURE

The Cucuteni-Tripolye culture occupied the frontier between Old Europe and the Pontic-Caspian cultures. More than twenty-seven hundred Cucuteni-Tripolye sites have now been discovered and examined with small excavations, and a few have been entirely excavated (figure 9.1). The Cucuteni-Tripolye culture first appeared around 5200–5000 BCE and survived a thousand years longer than any other part of the Old European world. Tripolye people were still creating large houses and villages, advanced pottery and metals, and female figurines as late as 3000 BCE. They were the sophisticated western neighbors of the steppe people who probably spoke Proto-Indo-European.

Cucuteni-Tripolye is named after two archaeological sites: Cucuteni, discovered in eastern Romania in 1909, and Tripolye, discovered in central Ukraine in 1899. Romanian archaeologists use the name Cucuteni and Ukrainians use Tripolye, each with its own system of internal chronological divisions, so we must use cumbersome labels like Pre-Cucuteni III/Tripolye A to refer to a single prehistoric culture. There is a Borges-like dreaminess to the Cucuteni pottery sequence: one phase (Cucuteni C) is not a phase at all but rather a type of pottery probably made outside the Cucuteni-Tripolye culture; another phase (Cucuteni A1) was defined before it was found, and never was found; still another (Cucteni A5) was created in 1963 as a challenge for future scholars, and is now largely forgotten; and the whole sequence was first defined on the assumption, later proved wrong, that the Cucuteni A phase was the oldest, so later archaeologists had to invent the Pre-Cucuteni phases I, II, and III, one of which (Pre-Cucuteni I) might not exist. The positive side of this obsession with pottery types and phases is that the pottery is known and studied in minute detail.[10]

The Cucuteni-Tripolye culture is defined most clearly by its decorated pottery, female figurines, and houses. They first appeared about 5200–5000 BCE in the East Carpathian piedmont. The late Linear Pottery

Figure 9.1 Early Eneolithic sites in the Pontic-Caspian region.

people of the East Carpathians acquired these new traditions from the late Boian-Giuleşti and late Hamangia cultures of the lower Danube valley. They adopted Boian and Hamangia design motifs in pottery, Boian-style female figurines, and some aspects of Boian house architecture (a clay floor fired before the walls were raised, called a *ploshchadka* floor in Russian). They acquired objects made of Balkan copper and Dobrujan flint, again from the Danube valley. The borrowed customs were core aspects of any tribal farming culture—domestic pottery production, domestic architecture, and domestic female-centered rituals—and so it seems likely that at least some Boian people migrated up into the steep, thickly forested valleys at the peakline of the East Carpathians. Their appearance defined the beginning of the Cucuteni-Tripolye culture—phases Pre-Cucuteni I (?) and II (about 5200–4900 BCE).

The first places that showed the new styles were clustered near high Carpathian passes, and perhaps attracted migrants partly because they controlled passage through the mountains. From these high Carpathian valleys the new styles and domestic rituals spread quickly northeastward to Pre-Cucuteni II settlements located as far east as the Dniester valley. As the culture developed (during pre-Cucteni III/Tripolye A) it was carried across the Dniester, erasing a cultural frontier that had existed for six hundred to eight hundred years, and into the South Bug River valley in Ukraine. Bug-Dniester sites disappeared. Tripolye A villages occupied the South Bug valley from about 4900–4800 BCE to about 4300–4200 BCE.

The Cucuteni-Tripolye culture made a visible mark on the forest-steppe environment, reducing the forest and creating pastures and cultivated fields over wider areas. At Floreşti, on a tributary of the Seret River, the remains of a late Linear Pottery homestead, radiocarbon dated about 5200–5100 BCE, consisted of a single house with associated garbage pits, set in a clearing in an oak-elm forest—tree pollen was 43% of all pollen. Stratified above it was a late Pre-Cucuteni III village, dated about 4300 BCE, with at least ten houses set in a much more open landscape—tree pollen was only 23%.[11]

Very few Bug-Dniester traits can be detected in early Cucuteni-Tripolye artifacts. The late Bug-Dniester culture was absorbed or driven away, removing the buffer culture that had mediated interchanges on the frontier.[12] The frontier shifted eastward to the uplands between the Southern Bug and Dnieper rivers. This soon became the most clearly defined, high-contrast cultural frontier in all of Europe.

The Early Cucuteni-Tripolye Village at Bernashevka

A good example of an early Cucuteni-Tripolye farming village on that moving frontier is the site of Bernashevka, wholly excavated by V. G. Zbenovich between 1972 and 1975.[13] On a terrace overlooking the Dniester River floodplain six houses were built in a circle around one large structure (figure 9.2). The central building, 12 by 8 m, had a foundation of horizontal wooden beams, or sleeper beams, probably with vertical wall posts morticed into them. The walls were wattle-and-daub, the roof thatched, and the floor made of smooth fired clay 8–17 cm thick on a subfloor of timber beams (a *ploshchadka*). The door had a flat stone threshold, and inside was the only domed clay oven in the settlement—perhaps a

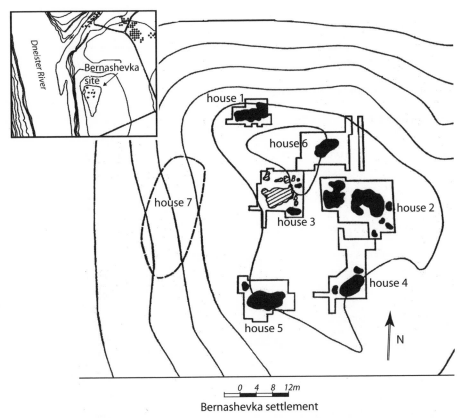

Figure 9.2 Bernashevka settlement on the Dniester River. After Zbenovich 1980, figure 3.

central bakery and work building for the village. The houses ranged from 30 m² to 150 m² in floor area. The population of the village probably was forty to sixty people. Two radiocarbon dates (5500–5300 BCE) seem two hundred years too old (table 9.1), perhaps because the dated wood fragments were from burned heartwood that had died centuries before the village was occupied.

No cemetery was found at Bernashevka or at any other Cucuteni-Tripolye village. Like the Criș people, the Cucuteni-Tripolye people did not ordinarily bury their dead. Parts of human skeletons are occasionally found in ritual deposits beneath house floors, human teeth were used occasionally as beads, and at Drăgușeni (Cucuteni A4, about 4300–4000 BCE) loose human bones were found in the litter between houses. Perhaps

TABLE 9.1
Early Eneolithic Radiocarbon Dates

Lab Number	BP Date	Sample	Calibrated Date
1. Pre-Cucuteni II Settlements			
Bernashevka			
Ki-6670	6440±60	?	5490–5300 BCE
Ki-6681	6510±55	?	5620–5360 BCE
Okopi			
Ki-6671	6330±65	?	5470–5210 BCE
2. Tripolye A Settlements			
Sabatinovka 2			
Ki-6680	6075±60	?	5060–4850 BCE
Ki-6737	6100±55	?	5210–4850 BCE
Luka Vrublevetskaya			
Ki-6684	5905±60	?	4850–4710 BCE
Ki-6685	5845±50	?	4780–4610 BCE
Grenovka			
Ki-6683	5860±45	?	4790–4620 BCE
Ki-6682	5800±50	?	4720–4550 BCE

3. Dnieper-Donets II Cemeteries (average ^{15}N = 11.8, average offset 228±30 too old)

Lab Number	BP Date	Sample	Calibrated Date
Osipovka cemetery		*Skeleton #*	
OxA6168	7675±70	skeleton 20, bone (invalid?)*	6590–6440 BCE
Ki 517	6075±125	skeleton 53	5210–4800 BCE
Ki 519	5940±420	skeleton 53	5350–4350 BCE
Nikol'skoe cemetery		*Grave Pit, Skeleton #*	
OxA 5029	6300±80	E, skeleton 125	5370–5080 BCE
OxA 6155	6225±75	Z, skeleton 94	5300–5060 BCE
Ki 6603	6160±70	E, skeleton 125	5230–4990 BCE
OxA 5052	6145±70	Z, skeleton 137	5210–4950 BCE
Ki 523	5640±400	skeleton ?	4950–4000 BCE
Ki 3125	5560±30	Z, bone	4460–4350 BCE

TABLE 9.1 (*continued*)

Lab Number	BP Date	Sample	Calibrated Date
Ki 3575	5560±30	B, skeleton 1	4460–4350 BCE
Ki 3283	5460±40	E, skeleton 125 (invalid?)	4450–4355 BCE
Ki 5159	5340±50	Z, skeleton 105 (invalid?)	4250–4040 BCE
Ki 3158	5230±40	Z, bone (invalid?)	4220–3970 BCE
Ki 3284	5200±30	E, skeleton 115 (invalid?)	4040–3970 BCE
Ki 3410	5200±30	D, skeleton 79a (invalid?)	4040–3970 BCE

Yasinovatka cemetery

OxA 6163	6465±60	skeleton 5	5480–5360 BCE
OxA 6165	6370±70	skeleton 19	5470–5290 BCE
Ki-6788	6310±85	skeleton 19	5470–5080 BCE
OxA 6164	6360±60	skeleton 45	5470–5290 BCE
Ki-6791	6305±80	skeleton 45	5370–5080 BCE
Ki-6789	6295±70	skeleton 21	5370–5080 BCE
OxA 5057	6260±180	skeleton 36	5470–4990 BCE
Ki-1171	5800±70	skeleton 36	4770–4550 BCE
OxA 6167	6255±55	skeleton 18	5310–5080 BCE
Ki-3032	5900±90	skeleton 18	4910–4620 BCE
Ki-6790	5860±75	skeleton 39	4840–4610 BCE
Ki-3160	5730±40	skeleton 15	4670–4490 BCE

Dereivka 1 cemetery

OxA 6159	6200±60	skeleton 42	5260–5050 BCE
OxA 6162	6175±60	skeleton 33	5260–5000 BCE
Ki-6728	6145±55	skeleton 11	5210–4960 BCE

4. Rakushechni Yar Settlement, Lower Don River

Bln 704	6070±100	level 8, charcoal	5210–4900 BCE
Ki-955	5790±100	level 5, shell	4790–4530 BCE
Ki-3545	5150±70	level 4, ?	4040–3800 BCE
Bln 1177	4360±100	level 3, ?	3310–2880 BCE

5. Khvalynsk Cemetery (average ^{15}N = 14.8, average offset 408±52 too old)

AA12571	6200±85	cemetery II, grave 30	5250–5050 BCE
AA12572	5985±85	cemetery II, grave 18	5040–4780 BCE
OxA 4310	6040±80	cemetery II, ?	5040–4800 BCE

Tᴀʙʟᴇ 9.1 (*continued*)

Lab Number	BP Date	Sample	Calibrated Date
OxA 4314	6015±85	cemetery II, grave 18	5060–4790 BCE
OxA 4313	5920±80	cemetery II, grave 34	4940–4720 BCE
OxA 4312	5830±80	cemetery II, grave 24	4840–4580 BCE
OxA 4311	5790±80	cemetery II, grave 10	4780–4570 BCE
UPI119	5903±72	cemetery I, grave 4	4900–4720 BCE
UPI120	5808±79	cemetery I, grave 26	4790–4580 BCE
UPI132	6085±193	cemetery I, grave 13	5242–4780 BCE

6. Lower Volga Cultures

Varfolomievka settlement, North Caspian
Lu2642	6400±230	level 2B, unknown material	5570–5070 BCE
Lu2620	6090±160	level 2B, "	5220–4840 BCE
Ki-3589	5430±60	level 2A, "	4350–4170 BCE
Ki-3595	5390±60	level 2A, "	4340–4050 BCE

Kombak-Te, Khvalynsk hunting camp in the North Caspian
GIN 6226	6000±150	?	5210–4710 BCE

Kara-Khuduk, Khvalynsk hunting camp in the North Caspian
UPI 431	5110±45	?	3800–3970 BCE

*"Invalid" means the date was contradicted by stratigraphy or by another date.

bodies were exposed and permitted to return to the birds somewhere near the village. As Gimbutas noted, some Tripolye female figurines seem to be wearing bird masks.

Half the pottery at Bernashevka was coarse ware: thick-walled, relatively crude vessels tempered with sand, quartz, and grog (crushed ceramic sherds) decorated with rows of stabbed impressions or shallow channels impressed with a spatula in swirling patterns (figure 9.3). Some of these were perforated strainers, perhaps used for making cheese or yogurt. Another 30% were thin-walled, fine-tempered jugs, lidded bowls, and ladles. The last 20% were very fine, thin-walled, quite beautiful lidded jugs and bowls (probably for individual servings of food), ladles (for serving), and hollow-pedestaled "fruit-stands" (perhaps for food presentation), elaborately decorated over the entire surface with stamped, incised, and channeled motifs, some enhanced with white paint against the orange clay.

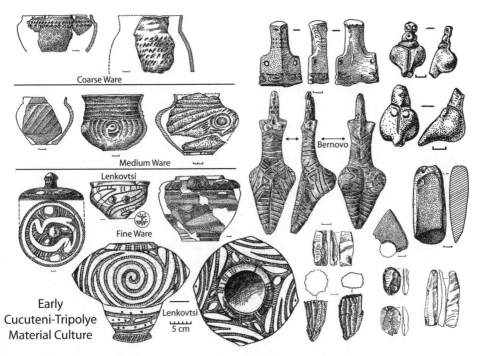

Coarse Ware

Medium Ware

Bernovo

Lenkovtsi

Fine Ware

Early
Cucuteni-Tripolye
Material Culture

Lenkovtsi

5 cm

Figure 9.3 Artifacts of the Pre-Cucuteni II/III-Tripolye A period from the sites of Bernashevka (most), Bernovo (labeled), and Lenkovtsi (labeled). After Zbenovich 1980, figures 55, 57, 61, 69, 71, 75, 79; and Zbenovich 1989, figure 65, 74.

Lidded bowls and jugs imply that food was served in individual containers at some distance from the hearth where it was cooked, and their careful decoration implies that the presentation of food involved an element of social theater, an unveiling.

Every house at Bernashevka contained fragmented ceramic female figurines with joined legs, exaggerated hips and buttocks, and schematic rod-like heads, about 10 cm long (figure 9.3). Simple incisions indicated the pubis and a girdle or waistband. Figurines were found at various places on the house floors; there was no obvious domestic shrine or altar. The number of figurines per house ranged from one to twenty-one, but four houses had nine or more. Almost two thousand similar figurines have been found in other Pre-Cucuteni II-III/Tripolye A sites, occasionally arranged in groups seated in chairs. At the Tripolye A site of Luka-Vrublevetskaya on the Dniester, they were made of clay tempered with a mixture of wheat, barley, and millet grains—all the grains cultivated in the village—and with finely ground flour. These, at least, seem to have symbolized the generative fertility of cultivated grain. But they were only one aspect of domestic cults. Under every house at Bernashevka was the skull of a domesticated cow or bull. One house also had wild animal symbols: the skull of a wild aurochs and the antlers of a red deer. Preconstruction foundation deposits of cattle horns and skulls, and occasionally of human skulls, are found in many Tripolye A villages. Bovine and female spirit powers were central to domestic household cults.

The Bernashevka farmers cultivated emmer and spelt wheats, with some barley and millet. Fields were prepared with mattocks made of antler (nineteen examples were found) and polished slate (twenty examples); some of these might have been attached to ards, which were primitive plows. The grain was harvested with flint blades of the Karanovo type (figure 9.3).

The animal bones from Bernashevka are the largest sample from any early Cucuteni-Tripolye site: 12,657 identifiable bones from a minimum of 804 animals. About 50% of the bones (60% of the individuals) were from wild animals, principally red deer (*Cervus elaphus*) and wild pig. Roe deer (*Capreolus capreolus*) and the wild aurochs (*Bos primigenius*) were hunted occasionally. Many early Cucuteni-Tripolye sites have about 50% wild animal bones. Like Bernashevka, most were frontier settlements established in places not previously cleared or farmed. In contrast, at the long-settled locale of Tirpeşti the Pre-Cucuteni III settlement produced 95% domesticated animal bones. And even in frontier settlements like Bernashevka, about 50% of all animal bones were from cattle, sheep/goat, and pigs. Cattle and pigs were more important in heavily forested areas like Bernashevka, where cattle constituted 75% of the domesticated animal bones, whereas sheep and goats were more important in villages closer to the steppe border.

Pre-Cucuteni II Bernashevka was abandoned before copper tools and ornaments became common enough to lose casually; no copper artifacts were left in the settlement. But only a few centuries later small copper artifacts became common. At Tripolye A Luka-Vrublevetskaya, probably occupied about 4800–4600 BCE, 12 copper objects (awls, fishhooks, a bead, a ring) were found among seven houses in piles of discarded shellfish, animal bones, and broken crockery. At Karbuna, near the steppe boundary, probably occupied about 4500–4400 BCE, a spectacular hoard of 444 copper objects was buried in a fine late Tripolye A pot closed with a Tripolye A bowl (figure 9.4). The hoard contained two cast copper hammer-axes 13–14 cm long, hundreds of copper beads, and dozens of flat

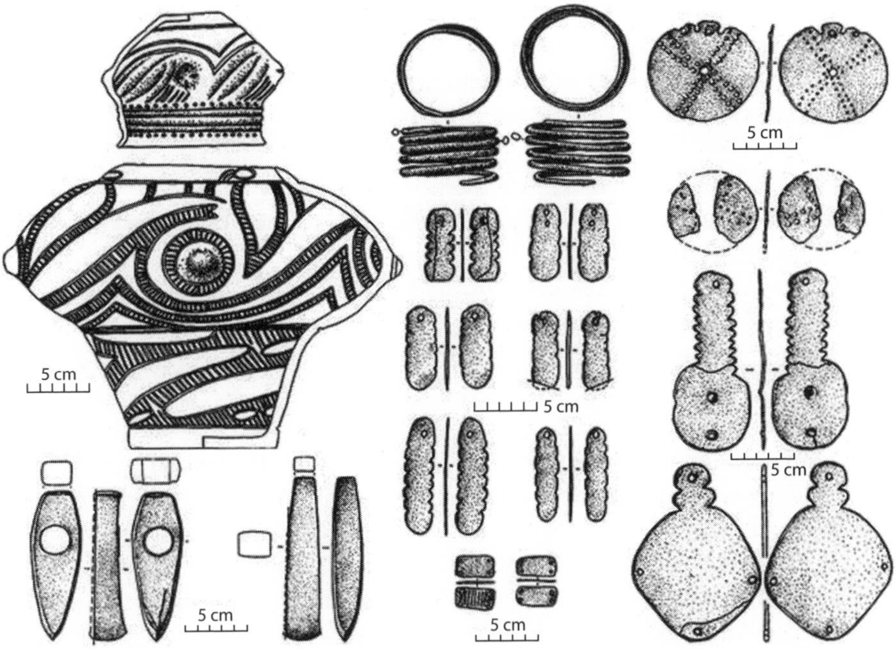

Figure 9.4 Part of the Karbuna hoard with the Tripolye A pot and bowl-lid in which it was found. All illustrated objects except the pot and lid are copper, and all are the same scale. After Dergachev 1998.

"idols," or wide-bottomed pendants made of flat sheet copper; two hammer-axes of marble and slate with drilled shaft-holes for the handle; 127 drilled beads made of red deer teeth; 1 drilled human tooth; and 254 beads, plaques, or bracelets made of *Spondylus* shell, an Aegean shell used for ornaments continuously from the first Greek Neolithic through the Old European Eneolithic. The Karbuna copper came from Balkan ores, and the Aegean shell was traded from the same direction, probably through the tell towns of the lower Danube valley. By about 4500 BCE social prestige had become closely linked to the accumulation of exotic commodities, including copper.[14]

As Cucuteni-Tripolye farmers moved eastward out of the East Carpathian piedmont they began to enter a more open, gently rolling, drier landscape. East of the Dniester River annual precipitation declined and the forests thinned. The already-old cultural frontier moved to the Southern Bug river valley. The Tripolye A town of Mogil'noe IV, among the first established in the South Bug valley, had more than a hundred buildings and covered 15–20 hectares, with a population of perhaps between four hundred and seven hundred. East of the Southern Bug, in the Dnieper valley, were people of a very different cultural tradition: the Dnieper-Donets II culture.

THE DNIEPER-DONETS II CULTURE

Dimitri Telegin defined the Dnieper-Donets II culture based on a series of excavated cemeteries and settlement sites in the Dnieper valley, in the steppes north of the Sea of Azov, and in the Donets valley. Dnieper-Donets II societies created large, elaborate cemeteries, made no female figurines, had open fires rather than kilns or ovens in their homes, lived in bark-covered huts rather than in large houses with fired clay floors, had no towns, cultivated little or no grain, and their pottery was very different in appearance and technology from Tripolye ceramics. The trajectory of the Cucuteni-Tripolye culture led back to the Neolithic societies of Old Europe, and that of Dnieper-Donets II led to the local Mesolithic foragers. They were fundamentally different people and almost certainly spoke different languages. But around 5200 BCE, the foragers living around the Dnieper Rapids began to keep cattle and sheep.

The bands of fishers and hunters whose cemeteries had overlooked the Rapids since the Early Mesolithic might have been feeling the pinch of growing populations. Living by the rich resources of the Rapids they might have become relatively sedentary, and women, when they live a settled life, generally have more children. They controlled a well-known, strategic area in a productive territory. Their decision to adopt cattle and sheep herding could have opened the way for many others in the Pontic-Caspian steppes. In the following two or three centuries domesticated cattle, sheep, and goats were walked and traded from the Dnieper valley eastward to the Volga-Ural steppes, where they had arrived by about 4700–4600 BCE. The evidence for any cereal cultivation east of the Dnieper before about 4200 BCE is thin to absent, so the initial innovation seems to have involved animals and animal herding.

Dating the Shift to Herding

The traditional Neolithic/Eneolithic chronology of the Dnieper valley is based on several sites near the Dnieper Rapids; the important ones are Igren 8, Pokhili, and Vovchok, where a repeated stratigraphic sequence was found. At the bottom were Surskii-type Neolithic pots and microlithic flint tools associated with the bones of hunted wild animals, principally red deer, wild pigs, and fish. These assemblages defined the Early Neolithic (dated about 6200–5700 BCE). Above them were Dnieper-Donets phase I occupations with comb-impressed and vegetal-tempered pottery, still associated with wild fauna; they defined the Middle Neolithic (probably about 5700–5400 BCE, contemporary with the Bug-Dniester culture). Stratified above these deposits were layers with Dnieper-Donets II pottery, sand-tempered with "pricked" or comb-stamped designs, and large flint blade tools, associated with the bones of domesticated cattle and sheep. These DDII assemblages represented the beginning of the Early Eneolithic and the beginning of herding economies east of the Dnieper River.[15]

Unlike the dates from DDI and Surskii, most DDII radiocarbon dates were measured on human bone from cemeteries. The average level of ^{15}N in DDII human bones from the Dnieper valley is 11.8%, suggesting a meat diet of about 50% fish. Correcting the radiocarbon dates for this level of ^{15}N, I obtained an age range of 5200–5000 BCE for the oldest DDII graves at the Yasinovatka and Dereivka cemeteries near the Dnieper Rapids. This is probably about when the DDII culture began. Imported pots of the late Tripolye A^2 Borisovka type have been found in DDII settlements at Grini, Piliava, and Stril'cha Skelia in the Dnieper valley, and sherds from three Tripolye A pots were found at the DDII Nikol'skoe cemetery. Tripolye A^2 is dated about 4500–4200 BCE by good dates (not on human bone) in the Tripolye heartland, and late DDII radiocarbon dates (when corrected for ^{15}N) agree with this range. The DDII period began about 5200–5000 BCE and lasted until about 4400–4200 BCE. Contact with Tripolye A people seems to have intensified after about 4500 BCE.[16]

The Evidence for Stockbreeding and Grain Cultivation

Four Dnieper-Donets II settlement sites in the Dnieper valley have been studied by zoologists—Surskii, Sredni Stog 1, and Sobachki in the steppe zone near the Rapids; and Buz'ki in the moister forest-steppe to the north (table 9.2). Domesticated cattle, sheep/goat, and pig accounted for 30–75% of the animal bones in these settlements. Sheep/goat contributed more than 50% of the bones at Sredni Stog 1 and 26% at Sobachki. Sheep finally were accepted into the meat diet in the steppes. Perhaps they were already being plucked for felt making; the vocabulary for wool might have first appeared among Pre-Proto-Indo-European speakers at about this time. Wild horses were the most important game (?) animal at Sredni Stog 1 and Sobachki, whereas red deer, roe deer, wild pig, and beaver were hunted in the more forested parts of the river at Buz'ki and Surskii 2–4.

TABLE 9.2

Dnieper–Donets II Animal Bones from Settlements

Mammal Bones	Sobachki	Sredni Stog 1	Buz'ki
		(Bones / MNI)*	
Cattle	56 / 5	23 / 2	42 / 3
Sheep/goat	54 / 8	35 / 4	3 / 1
Pig	10 / 3	1 / 1	4 / 1
Dog	9 / 3	12 / 1	8 / 2
Horse	48 / 4	8 / 1	—
Onager	1 / 1	—	—
Aurochs	2 / 1	—	—
Red deer	16 / 3	12 / 1	16 / 3
Roe deer	—	—	28 / 4
Wild pig	3 / 1	—	27 / 4
Beaver	—	—	34 / 5
Other mammal	8 / 4	—	7 / 4
Domestic	129 bones / 62%	74 bones / 78%	57 bones / 31%
Wild	78 bones / 38%	20 bones / 22%	126 bones / 69%

*MNI=minimum number of individuals

Fishing net weights and hooks suggest that fish remained important. This is confirmed by levels of ^{15}N in the bones of people who lived on the Dnieper Rapids, which indicate a meat diet containing more than 50% fish. Domesticated cattle, pig, and sheep bones occurred in all DDII settlements and in several cemeteries, and constituted more than half the bones at two settlement sites (Sredni Stog I and Sobachki) in the steppe zone. Domesticated animals seem indeed to have been an important addition to the diet around the Dnieper Rapids.[17]

Flint blades with sickle gloss attest to the harvesting of cereals at DDII settlements. But they could have been wild seed plants like *Chenopodium* or *Amaranthus*. If cultivated cereals were harvested there was very little evidence found. Two impressions of barley (*Hordeum vulgare*) were recovered on a potsherd from a DDII settlement site at Vita Litovskaya, near Kiev, west of the Dnieper. In the forests northwest of Kiev, near the Pripet marshes, there were sites with pottery that somewhat resembled DDII pottery but there were no elaborate cemeteries or other traits of the DDII culture. Some of these settlements (Krushniki, Novosilki, Obolon') had pottery with a few seed impressions of wheat (*T. monococcum* and *T. dicoccum*) and millet (*Panicum sativum*). These sites probably should be dated before 4500 BCE, since Lengyel-related cultures replaced them in Volhynia and the Polish borderlands after about that date. Some forest-zone farming seems to have been practiced in the southern Pripet forests west of the Dnieper. But in steppe-zone DDII cemeteries east of the Dnieper, Malcolm Lillie recorded almost no dental caries, suggesting that the DDII people ate a low-carbohydrate diet similar to that of the Mesolithic. No cultivated cereal imprints have been found east of the Dnieper River in pots dated before about 4000 BCE.[18]

Pottery and Settlement Types

Pottery was more abundant in DDII living sites than it had been in DDI, and appeared for the first time in cemeteries (figure 9.5). The growing importance of pottery perhaps implies a more sedentary lifestyle, but shelters were still lightly built and settlements left only faint footprints. A typical DDII settlement on the Dnieper River was Buz'ki. It consisted of five hearths and two large heaps of discarded shellfish and animal bones. No structures were detected, although some kind of shelter probably did exist.[19] Pots here and in other DDII sites were made in larger sizes (30–40 cm in diameter) with flat bottoms (pots seen in DDI sites had mainly pointed or rounded bottoms) and an applied collar around the rim. Decoration usually covered the entire outside of the vessel, made by pricking the surface with a stick, stamping designs with a small comb-stamp, or incising thin lines in horizontal-linear and zig-zag motifs—quite different from the spirals and swirls of Tripolye A potters. The application of a "collar" to thicken the rim was a popular innovation, widely adopted across the Pontic-Caspian steppes about 4800 BCE.

Polished (not chipped) stone axes now became common tools, perhaps for felling forests, and long unifacial flint blades (5–15 cm long) also became increasingly common, perhaps as a standardized part of a trade or gift package, since they appeared in graves and in small hoards in settlements.

Dnieper-Donets II Funeral Rituals

DDII funerals were quite different from those of the Mesolithic or Neolithic. The dead usually were exposed, their bones were collected, and they were finally buried in layers in communal pits. Some individuals were

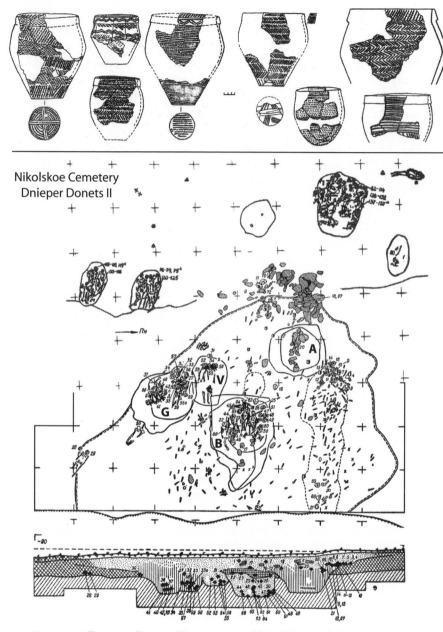

Figure 9.5 Dnieper-Donets II cemetery at Nikol'skoe with funerary ceramics. Pits A,B,G, and V were in an area deeply stained with red ochre. The other five burial pits were on a slightly higher elevation. Broken pots and animal bones were found near the cluster of rocks in the center. After Telegin 1991, figures 10, 20; and Telegin 1968, figure 27.

buried in the flesh, without exposure. This communal pit type of cemetery, with several treatments of the body in one pit, spread to other steppe regions. The thirty known DDII communal cemeteries were concentrated around the Dnieper Rapids but occurred also in other parts of the Dnieper valley and in the steppes north of the Sea of Azov. The largest cemeteries were three times larger than those of any earlier era, with 173 bodies at Dereivka, 137 at Nikol'skoe, 130 at Vovigny II, 124 at Mariupol, 68 at Yasinovatka, 50 at Vilnyanka, and so on. Pits contained up to four layers of burials, some whole and in an extended supine position, others consisting of only skulls. Cemeteries contained up to nine communal burial pits. Traces of burned structures, perhaps charnel houses built to expose dead bodies, were detected near the pits at Mariupol and Nikol'skoe. At some cemeteries, including Nikol'skoe (figure 9.5), loose human bones were widely scattered around the burial pits.

At Nikol'skoe and Dereivka some layers in the pits contained only skulls, without mandibles, indicating that some bodies were cleaned to the bone long before final burial. Other individuals were buried in the flesh, but the pose suggests that they were tightly wrapped in some kind of shroud. The first and last graves in the Nikol'skoe pits were whole skeletons. The standard burial posture for a body buried in the flesh was extended and supine, with the hands by the sides. Red ochre was densely strewn over the entire ritual area, inside and outside the grave pits, and pots and animal bones were broken and discarded near the graves.[20]

The funerals at DDII cemeteries were complex events that had several phases. Some bodies were exposed, and sometimes just their skulls were buried. In other cases whole bodies were buried. Both variants were placed together in the same multilayered pits, strewn with powdered red ochre. The remains of graveside feasts—cattle and horse bones—were thrown in the red-stained soil at Nikol'skoe, and cattle bones were found in grave 38, pit A, at Vilnyanka.[21] At Nikol'skoe almost three thousand sherds of pottery, including three Tripolye A cups, were found among the animal bones and red ochre deposited over the graves.

Power and Politics

The people of the DDII culture looked different than people of earlier periods in two significant respects: the profusion of new decorations for the human body and the clear inequality in their distribution. The old fisher-gatherers of the Dnieper Rapids were buried wearing, at most, a few beads of deer or fish teeth. But in DDII cemeteries a few individuals were

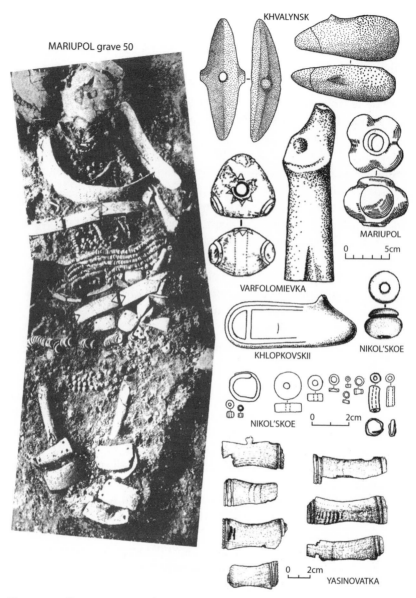

Figure 9.6 Ornaments and symbols of power in the Early Eneolithic, from Dnieper-Donets II graves, Khvalynsk, and Varfolomievka. The photo of grave 50 at Mariupol, skull at the top, is adapted from Gimbutas 1956, plate 8. The beads from Nikol'skoe include two copper beads and a copper ring on the left, and a gold ring on the lower right. The other beads are polished and drilled stone. The maces from Mariupol and Nikol'skoe, and beads from Nikol'skoe are after Telegin 1991, figures 29, 38; and Telegin and Potekhina 1987, figure 39.

buried with thousands of shell beads, copper and gold ornaments, imported crystal and porphyry ornaments, polished stone maces, bird-bone tubes, and ornamental plaques made of boar's tusk (figure 9.6). Boar's-tusk plaques were restricted to very few individuals. The tusks were cut into rectangular flat pieces (not an easy thing to do), polished smooth, and pierced or incised for attachment to clothing. They may have been meant to emulate Tripolye A copper and *Spondylus*-shell plaques, but DDII chiefs found their own symbols of power in the tusks of wild boars.

At the Mariupol cemetery 310 (70%) of the 429 boar's-tusk plaques accompanied just 10 (8%) of the 124 individuals. The richest individual (gr. 8) was buried wearing forty boars-tusk plaques sewn to his thighs and shirt, and numerous belts made of hundreds of shell and mother-of-pearl beads. He also had a polished porphyry four-knobbed mace head (figure 9.6), a bull figurine carved from bone, and seven bird-bone tubes. At Yasinovatka, only one of sixty-eight graves had boars-tusk plaques: an adult male wore nine plaques in grave 45. At Nikol'skoe, a pair of adults (gr. 25 and 26) was laid atop a grave pit (B) equipped with a single boar's-tusk plaque, a polished serpentine mace head, four copper beads, a copper wire ring, a gold ring, polished slate and jet beads, several flint tools, and an imported Tripolye A pot. The copper contained trace elements that identify it as Balkan in origin. Surprisingly few children were buried at Mariupol (11 of 124 individuals), suggesting that a selection was made—not all children who died were buried here. But one was among the richest of all the graves: he or she (sex is indeterminate in immature skeletons) wore forty-one boar's-tusk plaques, as well as a cap armored with eleven whole boar's tusks, and was profusely ornamented with strings of shell and bone beads. The selection of only a few children, including some who were very richly ornamented, implies the inheritance of status and wealth. Power was becoming institutionalized in families that publicly advertised their elevated status at funerals.

The valuables that signaled status were copper, shell, and imported stone beads and ornaments; boars-tusk plaques; polished stone maceheads; and bird-bone tubes (function unknown). Status also might have been expressed through the treatment of the body after death (exposed, burial of the skull/not exposed, burial of the whole body); and by the

---

Figure 9.6 (continued) The Varfolomievka mace (or pestle?) is after Yudin 1988, figure 2; Khvalynsk maces are after Agapov, Vasliev, and Pestrikova 1990, figure 24. Boars-tusk plaques, at the bottom, are after Telegin 1991, figure 38.

public sacrifice of domesticated animals, particularly cattle. Similar markers of status were adopted across the Pontic-Caspian steppes, from the Dnieper to the Volga. Boars-tusk plaques with exactly the same flower-like projection on the upper edge (figure 9.6, top plaque from Yasinovatka) were found at Yasinovatka in the Dnieper valley and in a grave at S'yezzhe in the Samara valley, 400 km to the east. Ornaments made of Balkan copper were traded across the Dnieper and appeared on the Volga. Polished stone mace-heads had different forms in the Dnieper valley (Nikol'skoe), the middle Volga (Khvalynsk), and the North Caspian region (Varfolomievka), but a mace is a weapon, and its wide adoption as a symbol of status suggests a change in the politics of power.

THE KHVALYNSK CULTURE ON THE VOLGA

The initial spread of stockbreeding in the Pontic-Caspian steppes was notable for the various responses it provoked. The DDII culture, where the shift began, incorporated domesticated animals not just as a ritual currency but also as an important part of the daily diet. Other people reacted in quite different ways, but they were all clearly interacting, perhaps even competing, with one another. A key regional variant was the Khvalynsk culture.

A prehistoric cemetery was discovered at Khvalynsk in 1977 on the west bank of the middle Volga. Threatened by the water impounded behind a Volga dam, it was excavated by teams led by Igor Vasiliev of Samara (figure 9.7). Its location has since been completely destroyed by bank erosion. Sites of the Khvalynsk type are now known from the Samara region southward along the banks of the Volga into the Caspian Depression and the Ryn Peski desert in the south. The characteristic pottery included open bowls and bag-like, round-bottomed pots, thick-walled and shell-tempered, with very distinctive sharply everted thick "collars" around the rims. They were densely embellished with bands of pricked and comb-stamped decoration that often covered the entire exterior surface. Early Khvalynsk, well documented at the Khvalynsk cemetery, began around 4700–4600 BCE in the middle Volga region (after adjusting the dates downward for the ^{15}N content of the humnan bones on which the dates were measured). Late Khvalynsk on the lower Volga is dated 3900–3800 BCE at the site of Kara-Khuduk but probably survived even longer than this on the lower Volga.[22]

The first excavation at the Khvalynsk cemetery, in 1977–79 (excavation I), uncovered 158 graves; the second excavation in 1980–85 (excavation II) recovered, I have been told, 43 additional graves.[23] Only Khvalynsk I has

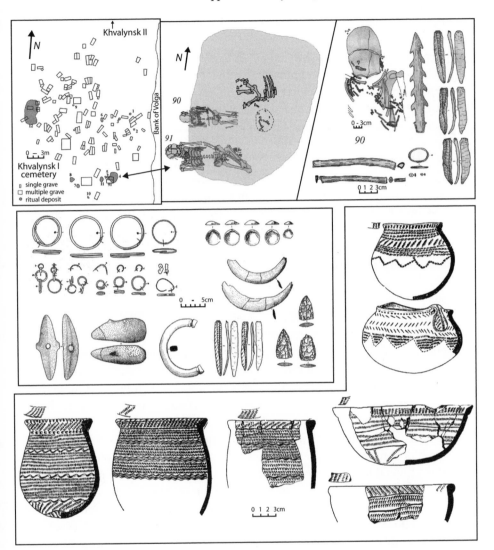

Figure 9.7 Khvalynsk cemetery and grave gifts. Grave 90 contained copper beads and rings, a harpoon, flint blades, and a bird-bone tube. Both graves (90 and 91) were partly covered by Sacrificial Deposit 4 with the bones from a horse, a sheep, and a cow.

Center: grave goods from the Khvalynsk cemetery—copper rings and bracelets, polished stone mace heads, polished stone bracelet, Cardium shell ornaments, boars tusk chest ornaments, flint blades, and bifacial projectile points.

Bottom: shell-tempered pottery from the Khvalynsk cemetery. After Agapov, Vasiliev, and Pestrikova 1990; and Ryndina 1998, Figure 31.

been published, so all statistics here are based on the first 158 graves (figure 9.7). Khvalynsk was by far the largest excavated Khvalynsk-type cemetery; most others had fewer than 10 graves. At Khvalynsk most of the deceased were layered in group pits, somewhat like DDII graves, but the groups were much smaller, containing only two to six individuals (perhaps families) buried on top of one another. One-third of the graves were single graves, a move away from the communal DDII custom. Only mature males, aged thirty to fifty, were exposed and disarticulated prior to burial, probably an expression of enhanced male status, associated with the introduction of herding economies elsewhere in the world.[24] Few children were buried in the cemetery (13 of 158), but those who were included some of the most profusely ornamented individuals, again possibly indicating that status was inherited. The standard burial posture was on the back with the knees raised, a distinctive pose. Most had their heads to the north and east, a consistent orientation that was absent in DDII cemeteries. Both the peculiar posture and the standard orientation later became widespread in steppe funeral customs.

Khvalynsk had many more animal sacrifices than any DDII cemetery: 52 (or 70) sheep/goat, 23 cattle, and 11 horses, to accompany the burials of 158 humans. (The published reports are inconsistent on the number of sheep/goat.) The head-and-hoof form of sacrifice appeared for the first time: at least 17 sheep/goat and 9 cattle were slaughtered and only the skull and lower leg bones were buried, probably still attached to the animal's hide. In later steppe funerals the custom of hanging a hide containing the head and hooves over the grave or burying it in the grave was very common. The head and hide symbolized a gift to the gods, and the flesh was doled out to guests at the funeral feast. Parts of domesticated animals were offered in all phases of the funerals at Khvalynsk: on the grave floor, in the grave fill, at the edge of the grave, and in twelve special sacrificial deposits stained with red ochre, found above the graves (figure 9.7). The distribution of animal sacrifices was unequal: 22 graves of 158 (14 percent) had animal sacrifices in the grave or above it, and enough animals were sacrificed to supply about half of the graves were they distributed equally. Only 4 graves (100, 127, 139, and 55–57) contained multiple species (cattle and sheep, sheep and horse, etc.) and all four of those also were covered by ochre-stained ritual deposits above the grave, with additional sacrifices. About one in five people had sacrificed domestic animals, and one in forty had multiple domestic animals.

The role of the horse in the Khvalynsk sacrifices is intriguing. The only animals sacrificed at Khvalysnk I were domesticated sheep/goat, domesticated cattle, and horses. Horse leg parts occurred by themselves, without other animal bones, in eight graves. They were included with a sheep/goat head-and-hoof offering in grave 127, and were included with sheep/goat and cattle remains in sacrificial deposit 4 (figure 9.7). It is not possible to measure the bones—they were discarded long ago—but horses certainly were treated symbolically like domesticated animals at Khvalynsk: they were grouped with cattle and sheep/goat in human funeral rituals that excluded obviously wild animals. Carved images of horses were found at other cemeteries dated to this same period (see below). Horses certainly had a new ritual and symbolic importance at Khvalynsk. If they were domesticated, they would represent the oldest domesticated horses.[25]

There is much more copper at Khvalynsk than is known from the entire DDII culture, and the copper objects there are truly remarkable (figure 9.7). Unfortunately most of it, an astonishing 286 objects, came from the 43 (?) graves of the Khvalynsk II excavation, still unpublished though analyses of some of the objects have been published by Natalya Ryndina. The Khvalynsk I excavation yielded 34 copper objects found in 11 of the 158 published graves. The copper from excavations I and II showed the same trace elements and technology, the former characteristic of Balkan copper. Ryndina's study of 30 objects revealed three technological groups: 14 objects made at 300–500°C, 11 made at 600–800°C, and 5 made at 900–1,000°C. The quality of welding and forging was uniformly low in the first two groups, indicating local manufacture, but was strongly influenced by the methods of the Tripolye A culture. The third group, which included two thin rings and three massive spiral rings, was technically identical to Old European status objects from the cemeteries of Varna and Durankulak in Bulgaria. These objects were made in Old Europe and were traded in finished form to the Volga. In the 158 graves of Khvalynsk I, adult males had the most copper objects, but the number of graves with *some* copper was about equal between the sexes, five adult male graves and four adult female graves. An adolescent (gr. 90 in figure 9.7) and a child were also buried with copper rings and beads.[26]

Polished stone mace-heads and polished serpentine and steatite stone bracelets appeared with copper as status symbols. Two polished stone maces occurred in one adult male grave (gr. 108) and one in another (gr. 57) at Khvalynsk. Grave 108 also contained a polished steatite bracelet. Similar bracelets and mace-heads were found in other Khvalynsk-culture cemeteries on the Volga, for example, at Krivoluchie (Samara oblast) and Khlopkovskii (Saratov oblast). Some mace heads were given "ears" that made them seem vaguely zoomorphic, and some observers have seen horse heads in them. A clearly zoomorphic polished stone mace head appeared at Varfolomievka, part of a different culture group on the lower Volga. Maces, copper, and elaborate decoration of the body appeared with domesticated animals, not before.[27]

Khvalynsk settlements have been found at Gundurovka and Lebyazhinka I on the Sok River, north of the Samara. But the Khvalynsk artifacts and pottery are mixed with artifacts of other cultures and ages, making it difficult to isolate features or animal bones that can be ascribed to the Khvalynsk period alone. We do know from the bones of the Khvalynsk people themselves that they ate a lot of fish; with an average ^{15}N measurement of 14.8%, fish probably represented 70% of their meat diet. Pure Khvalynsk camps have been found on the lower Volga in the Ryn Peski desert, but these were specialized hunters' camps where onagers and saiga antelope were the quarry, comprising 80–90 percent of the animal bones. Even here, at Kara Khuduk I, we find a few sheep/goat and cattle bones (6–9 %), perhaps provisions carried by Khvalynsk hunters.

In garbage dumps found at sites of other steppe cultures of the same period east of the Don (see below), horse bones usually made up more than half the bones found, and the percentage of cattle and sheep was usually under 40%. In the east, cattle and sheep were more important in ritual sacrifices than in the diet, as if they were initially regarded as a kind of ritual currency used for occasional (seasonal?) sanctified meals and funeral feasts. They certainly were associated with new rituals at funerals, and probably with other new religious beliefs and myths as well. The set of cults that spread with the first domesticated animals was at the root of the Proto-Indo-European conception of the universe as described at the beginning of chapter 8.

Nalchik and North Caucasian Cultures

Many archaeologists have wondered if domesticated cattle and sheep might have entered the steppes through the Eneolithic farmers of the Caucasus as well as from Old Europe.[28] Farming cultures had spread from the Near East into the southern Caucasus Mountains (Shulaveri, Arukhlo, and Shengavit) by 5800–5600 BCE. But these earliest farming communities in the Caucasus were not widespread; they remained concentrated in a few river-bottom locations in the upper Kura and Araxes River valleys. No bridging sites linked them to the distant European steppes, more than 500 km to the north and west. The permanently glaciated North Caucasus Mountains, the highest and most impassable mountain range in Europe, stood between them and the steppes. The bread wheats (*Triticum aestivum*) preferred in the

Caucasus were less tolerant of drought conditions than the hulled wheats (emmer, einkorn) preferred by Criş, Linear Pottery, and Bug-Dniester cultivators. The botanist Zoya Yanushevich observed that the cultivated cereals that appeared in Bug-Dniester sites and later in the Pontic-Caspian steppe river valleys were a Balkan/Danubian crop suite, not a Caucasian crop suite.[29] Nor is there an obvious stylistic connection between the pottery or artifacts of the earliest Caucasian farmers at Shulaveri and those of the earliest herders in the steppes off to the north. If I had to guess at the linguistic identity of the first Eneolithic farmers at Shulaveri, I would link them with the ancestors of the Kartvelian language family.

The Northwest Caucasian languages, however, are quite unlike Kartvelian. Northwest Caucasian seems to be an isolate, a survival of some unique language stock native to the northern slopes of the North Caucasus Mountains. In the western part of the North Caucasian piedmont, overlooking the steppes, the few documented Eneolithic communities had stone tools and pottery somewhat like those of their northern steppe neighbors; these communities were southern participants in the steppe world, not northern extensions of Shulaveri-type Caucasian farmers. I would guess they spoke languages ancestral to Northwest Caucasian, but only a few early sites are published. The most important is the cemetery at Nalchik.

Near Nalchik, in the center of the North Caucasus piedmont, was a cemetery containing 147 graves with contracted skeletons lying on their sides in red ochre–stained pits in groups of two or three under stone cairns. Females lay in a contracted pose on the left side and males on their right.[30] A few copper ornaments, beads made of deer and cattle teeth, and polished stone bracelets (like those found in grave 108 at Khvalynsk and at Krivoluchie) accompanied them. One grave yielded a date on human bone of 5000–4800 BCE (possibly too old by a hundred to five hundred years, if the dated sample was contaminated by old carbon in fish). Five graves in the same region at Staronizhesteblievsk were provided with boars-tusk plaques of the DDII Mariupol type, animal-tooth beads, and flint blades that seem at home in the Early Eneolithic.[31] An undated cave occupation in the Kuban valley at Kamennomost Cave, level 2, which could be of the same date, has yielded sheep/goat and cattle bones stratified beneath a later level with Maikop-culture materials. Carved stone bracelets and ornamental stones from the Caucasus—black jet, rock crystal, and porphyry—were traded into Khvalynsk and Dnieper-Donets II sites, perhaps from people like those at Nal'chik and Kamennomost Cave 2. The Nalchik-era sites clearly represent a community that had at least a few domesticated cattle and sheep/goats, and was in contact with Khvalynsk.

They probably got their domesticated animals from the Dnieper, as the Khvalynsk people did.

The Lower Don and North Caspian Steppes

In the steppes between Nalchik and Khvalynsk many more sites, of different kinds, are dated to this period. Rakushechni Yar on the lower Don, near the Sea of Azov, is a deeply stratified settlement site with a cluster of six graves at the edge of the settlement area. The lowest cultural levels, with shell-tempered pottery lightly decorated with incised linear motifs and impressions made with a triangular-ended stick, probably dated about 5200–4800 BCE, contained the bones of sheep/goat and cattle. But in the interior steppes, away from the major river valleys, equid hunting was still the focus of the economy. In the North Caspian Depression the forager camp of Dzhangar, also dated 5200 BCE (on animal bone) and with pottery similar to Rakushechni Yar, yielded only the bones of wild horses and onagers.[32]

On the eastern side of the lower Volga, sites such as Varfolomievka were interspersed with Khvalynsk hunters' camps such as Kara Khuduk I.[33] The settlement at Varfolomievka is stratified and well dated by radiocarbon, and clearly shows the transition from foraging to herding in the North Caspian Depression. Varfolomievka was first occupied around 5800–5600 BCE by pottery-making foragers who hunted onagers and horses (level 3). The site was reoccupied twice more (levels 2B and 2A). In level 2B, dated about 5200–4800 BCE, people constructed three pit-houses. They used copper (one copper awl and some amorphous lumps of copper were found) and kept domesticated sheep/goats, though "almost half" the animal bones at Varfolomievka were of horses. Bone plaques were carved in the shape of horses, and horse metacarpals were incised with geometric decorations. Three polished stone mace-head fragments were found here. One was carved into an animal head at one end, perhaps a horse (figure 9.6). Four graves were dug rather casually into abandoned house depressions at Varfolomievka, like the similar group of graves at the edge of Rakushechni Yar. Hundreds of beads made of drilled and polished horse teeth were deposited in ochre-stained sacrificial deposits near the human graves. There were also a few deer teeth, several kinds of shell beads, and whole boars' tusk ornaments.

These sites in the southern steppes, from the lower Don to the lower Volga, are dated 5200–4600 BCE and exhibit the bones of sheep/goat and occasionally cattle, small objects of copper, and casual disposal of the dead. Small settlements provide most of the data, unlike the cemetery-based archaeological record for Khvalynsk. Pots were shell-tempered and decorated with designs incised or pricked with a triangular-ended stick. Motifs included diamond-like lozenges and, rarely, incised meanders filled with pricked ornament. Most rims were simple but some were thickened on the inside. A. Yudin has grouped these sites together under the name of the Orlovka culture, after the settlement of Orlovka, excavated in 1974, on the Volga. Nalchik seems to have existed at the southern fringe of this network.[34]

THE FOREST FRONTIER: THE SAMARA CULTURE

One other culture interacted with northern Khvalynsk in the middle Volga region, along the forest-steppe boundary (see figure 9.1). The Samara Neolithic culture, distinguished by its own variety of "collared" pots covered with pricked, incised, and rocker-stamped motifs, developed at the northern edge of the steppe zone along the Samara River. The pottery, tempered with sand and crushed plants, was similar to that made on the middle Don River. Dwellings at Gundurovka near Samara had dug-out floors, 20 m by 8 m, with multiple hearths and storage pits in the floors (this settlement also contained Khvalynsk pottery). Domesticated sheep/goat (13% of 3,602 bones) and cattle (21%) were identified at Ivanovskaya on the upper Samara River, although 66% of the bones were of horses. The settlement of Vilovatoe on the Samara River yielded 552 identifiable bones, of which 28.3% were horse, 19.4% were sheep/goat, and 6.3% were cattle, in addition to beaver (31.8%) and red deer (12.9%). The Samara culture showed some forest-culture traits: it had large polished stone adzes like those of forest foragers to the north.

Samara people created formal cemeteries (figure 9.8). The cemetery at S'yezzhe (see-YOZH-yay) contained nine burials in an extended position on their backs, different from the Khvalynsk position and more like that of DDII. Above the graves at the level of the original ground surface was a ritual deposit of red ochre, broken pottery, shell beads, a bone harpoon, and the skulls and lower leg bones (astragali and phalanges) of two horses—funeral-feast deposits like the above-grave deposits at Khvalynsk. S'yezzhe had the oldest horse head-and-hoof deposit in the steppes. Near the horse head-and-hoof deposit, but outside the area of ochre-stained soil, were two figurines of horses carved on flat pieces of bone, similar to others found at Varfolomievka, and one bone figurine of a bull. The S'yezzhe people wore boar's-tusk plaques like those of the Dnieper-Donets II culture, one of which was shaped exactly like one found at the DDII cemetery of Yasinovatka in the Dnieper valley.[35]

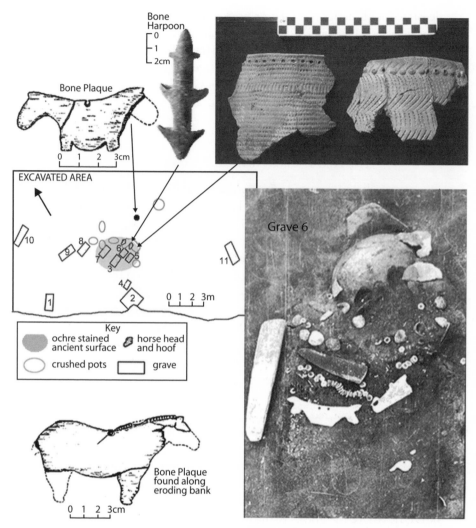

Figure 9.8 S'yezzhe cemetery, Samara oblast. Graves 1–9 were a cemetery of the Samara culture, Early Eneolithic. Graves 10 and 11 were later. After Vasiliev and Matveeva 1979.

Cows, Social Power, and the Emergence of Tribes

It is impossible to say how much the people buried at Khvalynsk really *knew* of the societies of Old Europe, but they certainly were connected by a trade network of impressive reach. Cemeteries across the Pontic-Caspian steppes (DDII, Khvalynsk, S'yezzhe, Nalchik) became larger or appeared for the first time, suggesting the growth of larger, more stable communities.

Cattle and sheep were important in the diet at some DDII settlements on the Dnieper River, but farther east they seem initially to have been more important in funeral rituals than in the daily diet, which was still dominated by horse meat. In the east, domesticated cattle and sheep seem to have served as a kind of currency in a new set of rituals and religious beliefs.

Participation in long-distance trade, gift exchange, and a new set of cults requiring public sacrifices and feasting became the foundation for a new kind of social power. Stockbreeding is by nature a volatile economy. Herders who lose animals always borrow from those who still have them. The social obligations associated with these loans are institutionalized among the world's pastoralists as the basis for a fluid system of status distinctions. Those who loaned animals acquired power over those who borrowed them, and those who sponsored feasts obligated their guests. Early Proto-Indo-European included a vocabulary about verbal contracts bound by oaths (*$h_1óitos$-), used in later religious rituals to specify the obligations between the weak (humans) and the strong (gods). Reflexes of this root were preserved in Celtic, Germanic, Greek, and Tocharian. The model of political relations it references probably began in the Eneolithic. Only a few Eneolithic steppe people wore the elaborate costumes of tusks, plaques, beads, and rings or carried the stone maces that symbolized power, but children were included in this exceptional group, suggesting that the rich animal loaners at least tried to see that their children inherited their status. Status competition between regional leaders, *weik-potis* or *reĝ-* in later Proto-Indo-European, resulted in a surprisingly widespread set of shared status symbols. As leaders acquired followers, political networks emerged around them—and this was the basis for tribes.

Societies that did not accept the new herding economy became increasingly different from those that did. The people of the northern forest zone remained foragers, as did those who lived in the steppes east of the Ural Mountains. These frontiers probably were linguistic as well as economic, given their persistence and clarity. The Pre-Proto-Indo-European language family probably expanded with the new economy during the Early Eneolithic in the western steppes. Its sister-to-sister linguistic links may well have facilitated the spread of stockbreeding and the beliefs that went with it.

One notable aspect of the Pontic-Caspian Early Eneolithic is the importance of horses, in both diet and funeral symbolism. Horse meat was a major part of the meat diet. Images of horses were carved on bone plaques at Varfolomievka and S'yezzhe. At Khvalynsk, horses were included with cattle and sheep in funeral rituals that excluded obviously wild animals. But, zoologically, we cannot say whether they looked very different from wild horses—the bones no longer exist. The domestication of the horse, an enormously important event in human history, is not at all well understood. Recently, however, a new kind of evidence has been obtained straight from the horse's mouth.

CHAPTER TEN

The Domestication of the Horse
and the Origins of Riding
The Tale of the Teeth

The importance of the horse in human history is matched only by the difficulties
inherent in its study; there is hardly an incident in the story which is not the
subject of controversy, often of a violent nature.
—Grahame Clark, 1941

In the summer of 1985 I went with my wife Dorcas Brown, a fellow archaeologist, to the Veterinary School at the University of Pennsylvania to
ask a veterinary surgeon a few questions. Do bits create pathologies on
horse teeth? If they do, then shouldn't we be able to see the signs of
bitting—scratches or small patches of wear—on ancient horse teeth?
Wouldn't that be a good way to identify early bitted horses? Could he point
us toward the medical literature on the dental pathologies associated with
horse bits? He replied that there really was no literature on the subject. A
properly bitted horse wearing a well-adjusted bridle, he said, really *can't*
take the bit in its teeth very easily, so contact between the bit and the teeth
would have been too infrequent to show up with any regularity. Nice idea,
but it wouldn't work. We decided to get a second opinion.

At the Veterinary School's New Bolton Center for large mammals, outside Philadelphia, the trainers, who worked every day with horses, responded very differently. Horses chewed their bits all the time, they said.
Some rolled the bit around in their mouths like candy. You could hear it
clacking against their teeth. Of course, it was a vice—properly trained and
harnessed horses were not supposed to do it, but they did. And we should
talk to Hilary Clayton, formerly at New Bolton, who had gone to a university job somewhere in Canada. She had been studying the mechanics of
bits in horses' mouths.

We located Hilary Clayton at the University of Saskatchewan and found that she had made X-ray fluoroscopic videos of horses chewing bits (figure 10.1). She bitted horses and manipulated the reins from a standing position behind. An X-ray fluoroscope mounted beside the horses' heads took pictures of what was happening inside their mouths. No one had done this before. She sent us two articles co-authored with colleagues in Canada.[1] Their images showed just how horses manipulated a bit inside their mouths and precisely where it sat between their teeth. A well-positioned bit is supposed to sit on the tongue and gums in the space between the front and back teeth, called the "bars" of the mouth. When the rider pulls the reins, the bit presses the tongue and the gums into the lower jaw, squeezing the sensitive gum tissue between the bit and the underlying bone. That hurts. The horse will dip its head toward a one-sided pull (a turn) or lower its

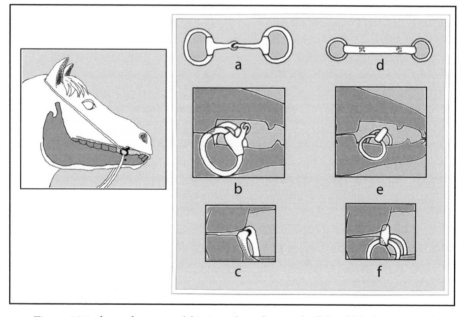

Figure 10.1 A modern metal bit in a horse's mouth. Mandible bone tinted gray. (a) jointed snaffle bit; (b) X-ray of jointed snaffle sitting on the tongue in proper position; (c) X-ray of snaffle being grasped in the teeth; (d) bar bit showing chewing wear; (e) X-ray of bar bit sitting on the tongue in proper position; (f) X-ray of bar bit being grasped in the teeth. After Clayton and Lee 1984; and Clayton 1985.

chin into a two-sided pull (a brake) to avoid the bit's pressure on its tongue and gums.

Clayton's X-rays showed how horses use their tongues to elevate the bit and then retract it, pushing it back into the grip of their premolars, where it can no longer cause pressure on soft tissue no matter how hard the rider pulls on the reins. The soft corners of the mouth are positioned in front of the molars, so in order to get a bit into its teeth the horse has to force it back against the corners of its mouth. These stretched tissues act like a spring. If the bit is not held *very* firmly between the tips of the teeth it will pop forward again onto the bars of the mouth. It seemed likely to us that this repeated back-and-forth movement over the tips of the front premolars should affect the lower teeth more than the uppers just because of gravity—the bit sat on the lower jaw. The wear from bit chewing should be concentrated on one small part of two teeth (the lower second premolars, or P_2s), unlike the wear from chewing anything else. Clayton's X-rays made it possible, for the first time, to say positively that a specific part of a single tooth was the place to look for bit wear. We found several published photographs of archaeological horse P_2s with wear facets or bevels on precisely that spot. Two well-known archaeological zoologists, Juliet Clutton-Brock in London and Antonio Azzaroli in Rome, had described this kind of wear as "possibly" made by a bit. Other zoologists thought it was impossible for horses to get a bit that far back into their mouth with any frequency, like our first veterinary surgeon. No one knew for sure. But they had not seen Clayton's X-rays.[2]

Encouraged and excited, we visited the anthropology department at the Smithsonian Museum of Natural History in Washington, and asked Melinda Zeder, then a staff archaeozoologist, if we could study some never-bitted ancient wild horse teeth—a control sample—and if she could offer us some technical advice about how to proceed. We were not trained as zoologists, and we did not know much about horse teeth. Zeder and a colleague who knew a lot about dental microwear, Kate Gordon, sat us down in the staff cafeteria. How would we distinguish bit wear from tooth irregularities caused by malocclusion? Or from dietary wear, created by normal chewing on food? Would the wear caused by a bit survive very long, or would it be worn away by dietary wear? How long would that take? How fast do horse teeth grow? Aren't they the kind of teeth that grow out of the jaw and are worn away at the crown until they become little stubs? Would that change bit wear facets with increasing age? What about rope or leather bits—probably the oldest kind? Do they cause wear? What kind? Is the action of the bit different when a horse is ridden from when it pulls a chariot? And what, exactly, causes wear—if it exists? Is it the rider pulling the bit into the *front* of the tooth, or is it the horse chewing on the bit, which would cause wear on the *occlusal* (chewing) surface of the tooth? Or is it both? And if we did find wear under the microscope, how would we describe it so that the difference between a tooth with and without wear could be quantified?

Mindy Zeder took us through her collections. We made our first molds of ancient equid P_2s, from the Bronze Age city of Malyan in Iran, dated about 2000 BCE. They had wear facets on their mesial corners; later we would be able to say that the facets were created by a hard bit of bone or metal. But we didn't know that yet, and, as turned out, there really was not a large collection of never-bitted wild horse teeth at the Smithsonian. We had to find our own, and we left thinking that we could do it if we took one problem at a time. Twenty years later we still feel that way.[3]

WHERE WERE HORSES FIRST DOMESTICATED?

Bit wear is important, because other kinds of evidence have proven uncertain guides to early horse domestication. Genetic evidence, which we might hope would solve the problem, does not help much. Modern horses are genetically schizophrenic, like cattle (chapter 8) but with the genders reversed. The *female* bloodline of modern domesticated horses shows extreme diversity. Traits inherited through the mitochondrial DNA, which passes unchanged from mother to daughter, show that this part of the bloodline is so diverse that *at least* seventy-seven ancestral mares, grouped into seventeen phylogenetic branches, are required to account for the genetic variety in modern populations around the globe. Wild mares must have been taken into domesticated horse herds in many different places at different times. Meanwhile, the *male* aspect of modern horse DNA, which is passed unchanged on the Y chromosome from sire to colt, shows remarkable homogeneity. It is possible that just a single wild stallion was domesticated. So horse keepers apparently have felt free to capture and breed a variety of wild mares, but, according to these data, they universally rejected wild males and even the male progeny of any wild stallions that mated with domesticated mares. Modern horses are descended from very few original wild males, and many, varied wild females.[4]

Why the Difference?

Wildlife biologists have observed the behavior of feral horse bands in several places around the world, notably at Askania Nova, Ukraine, on the barrier islands of Maryland and Virginia (the horses described in the childrens' classic *Misty of Chincoteague*), and in northwestern Nevada. The standard feral horse band consists of a stallion with a harem of two to seven mares and their immature offspring. Adolescents leave the band at about two years of age. Stallion-and-harem bands occupy a home range, and stallions fight one another, fiercely, for control of mares and territory. After the young males are expelled they form loose associations called "bachelor bands," which lurk at the edges of the home range of an established stallion. Most bachelors are unable to challenge mature stallions or keep mares successfully until they are more than five years old. Within established bands, the mares are arranged in a social hierarchy led by the lead mare, who chooses where the band will go during most of the day and leads it in flight if there is a threat, while the stallion guards the flanks or the rear. Mares are therefore instinctively disposed to accept the dominance of others, whether dominant mares, stallions—or humans. Stallions are headstrong and violent, and are instinctively disposed to challenge authority by biting and kicking. A relatively docile and controllable mare could be found at the bottom of the pecking order in many wild horse bands, but a relatively docile and controllable stallion was an unusual individual—and one that had little hope of reproducing in the wild. Horse domestication might have depended on a lucky coincidence: the appearance of a relatively manageable and docile male in a place where humans could use him as the breeder of a domesticated bloodline. From the horse's perspective, humans were the only way he could get a girl. From the human perspective, he was the only sire they wanted.

Where Did He Live? And When?

Animal domestication, like marriage, is the culmination of a long prior relationship. People would not invest the time and energy to attempt to care for an animal they were unfamiliar with. The first people to think seriously about the benefits of keeping, feeding, and raising tame horses must have been familiar with wild horses. They must have lived in a place where humans spent a lot of time hunting wild horses and learning their behavior. The part of the world where this was possible contracted significantly about ten thousand to fourteen thousand years ago, when the Ice Age steppe—a favorable environment for horses—was replaced by dense forest over much of the Northern Hemisphere. The horses of North America became extinct as the climate shifted, for reasons still poorly understood. In Europe and Asia large herds of wild horses survived only in the

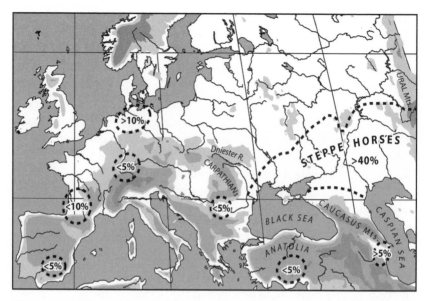

Figure 10.2 Map of the distribution of wild horses (*Equus caballus*) in the mid-Holocene, about 5000 BCE. The numbers show the approximate frequencies of horse bones in human kitchen garbage in each region, derived from charts in Benecke 1994 and from various Russian sources.

steppes in the center of the Eurasian continent, leaving smaller populations isolated in pockets of naturally open pasture (marsh-grass meadows, alpine meadows, arid *mesetas*) in Europe, central Anatolia (modern Turkey), and the Caucasus Mountains. Horses disappeared from Iran, lowland Mesopotamia, and the Fertile Crescent, leaving these warm regions to other equids (onagers and asses) (figure 10.2).

In western and central Europe, central Anatolia, and the Caucasus the isolated pockets of horses that survived into the Holocene never became important in the human food quest—there just weren't enough of them. In Anatolia, for example, a few wild horses probably were hunted occasionally by the Neolithic occupants of Catal Hüyök, Pinarbaşi, and other farming villages in the central plateau region between about 7400 and 6200 BCE. But most of the equids hunted at these sites were *Equus hydruntinus* (now extinct) or *Equus hemionus* (onagers), both ass-like equids smaller than horses. Only a few bones are large enough to qualify as possible horses. Horses were not present in Neolithic sites in western Anatolia, or in Greece or Bulgaria, or in the

Mesolithic and Early Neolithic of Austria, Hungary, or southern Poland. In western and northern Europe, Mesolithic foragers hunted horses occasionally. But horse bones accounted for more than 5% of the animals in only a few post-Glacial sites in the coastal plain of Germany/Poland and in the uplands of southern France. In the Eurasian steppes, on the other hand, wild horses and related wild equids (onagers, *E. hydruntinus*) were the most common wild grazing animals. In early Holocene steppe archaeological sites (Mesolithic and early Neolithic) wild horses regularly account for more than 40% of the animal bones, and probably more than 40% of the meat diet because horses are so big and meaty. For this reason alone we should look first to the Eurasian steppes for the earliest episode of domestication, the one that probably gave us our modern male bloodline.[5]

Early and middle Holocene archaeological sites in the Pontic-Caspian steppes contain the bones of three species of equids. In the Caspian Depression, at Mesolithic sites such as Burovaya 53, Je-Kalgan, and Istai IV, garbage dumps dated before 5500 BCE contain almost exclusively the bones of horses and onagers (see site map, figure 8.3). The onager, *Equus hemionus*, also called a "hemione" or "half-ass," was a fleet-footed, long-eared animal smaller than a horse and larger than an ass. The natural range of the onager extended from the Caspian steppes across Central Asia and Iran and into the Near East. A second equid, *Equus hydruntinus*, was hunted in the slightly moister North Pontic steppes in Ukraine, where its bones occur in small percentages in Mesolithic and Early Neolithic components at Girzhevo and Matveev Kurgan, dated to the late seventh millennium BCE. This small, gracile animal, which then lived from the Black Sea steppes westward into Bulgaria and Romania and south into Anatolia, became extinct before 3000 BCE. The true horse, *Equus caballus*, ranged across both the Caspian Depression and the Black Sea steppes, and it survived in both environments long after both *E. hemionus* and *E. hydruntinus* were hunted out. Horse bones contributed more than 50% of the identified animal bones at Late Mesolithic Girzhevo in the Dniester steppes and Meso/Neolithic Matveev Kurgan and Kammenaya Mogila in the Azov steppes; also at Neo/Eneolithic Varfolomievka and Dzhangar in the Caspian Depression, Ivanovskaya on the Samara River, and Mullino in the southern foothills of the Ural Mountains. The long history of human dependence on wild equids in the steppes created a familiarity with their habits that would later make the domestication of the horse possible.[6]

Why Were Horses Domesticated?

The earliest evidence for possible horse domestication in the Pontic-Caspian steppes appeared after 4800 BCE, long after sheep, goats, pigs, and cattle were domesticated in other parts of the world. What was the incentive to tame wild horses if people already had cattle and sheep? Was it for transportation? Almost certainly not. Horses were large, powerful, aggressive animals, more inclined to flee or fight than to carry a human. Riding probably developed only after horses were already familiar as domesticated animals that could be controlled. The initial incentive probably was the desire for a cheap source of winter meat.

Horses are easier to feed through the winter than cattle or sheep, as cattle and sheep push snow aside with their noses and horses use their hard hooves. Sheep can graze on winter grass through soft snow, but if the snow becomes crusted with ice than their noses will get raw and bloody, and they will stand and starve in a field where there is ample winter forage just beneath their feet. Cattle do not forage through even soft snow if they cannot see the grass, so a snow deep enough to hide the winter grass will kill range cattle if they are not given fodder. Neither cattle nor sheep will break the ice on frozen water to drink. Horses have the instinct to break through ice and crusted snow with their hooves, not their noses, even in deep snows where the grass cannot be seen. They paw frozen snow away and feed themselves and so do not need water or fodder. In 1245 the Franciscan John of Plano Carpini journeyed to Mongolia to meet Güyük Khan (the successor to Genghis) and observed the steppe horses of the Tartars, as he called them, digging for grass from under the snow, "since the Tartars have neither straw nor hay nor fodder." During the historic blizzard of 1886 in the North American Plains hundreds of thousands of cattle were lost on the open range. Those that survived followed herds of mustangs and grazed in the areas they opened up.[7] Horses are supremely well adapted to the cold grasslands where they evolved. People who lived in cold grasslands with domesticated cattle and sheep would soon have seen the advantage in keeping horses for meat, just because the horses did not need fodder or water. A shift to colder climatic conditions or even a particularly cold series of winters could have made cattle herders think seriously about domesticating horses. Just such a shift to colder winters occurred between about 4200 and 3800 BCE (see chapter 11).

Cattle herders would have been particularly well suited to manage horses because cattle and horse bands both follow the lead of a dominant female. Cowherds already knew they needed only to control the lead cow to control the whole herd, and would easily have transferred that knowledge to controlling lead mares. Males presented a similar management problem in both species, and they had the same iconic status as symbols of virility and strength. When people who depended on equid-hunting began to keep domesticated cattle, someone would soon have noticed these similarities and applied cattle-management techniques to wild horses. And that would quickly have produced the earliest domesticated horses.

This earliest phase of horse keeping, when horses were primarily a recalcitrant but convenient source of winter meat, may have begun as early as 4800 BCE in the Pontic-Caspian steppes. This was when, at Khvalynsk and S'yezzhe in the middle Volga region, and Nikol'skoe on the Dnieper Rapids, horse heads and/or lower legs were first joined with the heads and/or lower legs of cattle and sheep in human funeral rituals; and when bone carvings of horses appeared with carvings of cattle in a few sites like S'yezzhe and Varfolomievka. Certainly horses were linked symbolically with humans and the cultured world of domesticated animals by 4800 BCE. Horse keeping would have added yet another element to the burst of economic, ritual, decorative, and political innovations that swept across the western steppes with the initial spread of stockbreeding about 5200– 4800 BCE.

WHAT IS A DOMESTICATED HORSE?

We decided to investigate bit wear on horse teeth, because it is difficult to distinguish the bones of early domesticated horses from those of their wild cousins. The Russian zoologist V. Bibikova tried to define a domesticated skull type in 1967, but her small sample of horse skulls did not define a reliable type for most zoologists.

The bones of wild animals usually are distinguished from those of domesticated animals by two quantifiable measurements: measurements of variability in size, and counts of the ages and sexes of butchered animals. Other criteria include finding animals far outside their natural range and detecting domestication-related pathologies, of which bit wear is an example. Crib biting, a stall-chewing vice of bored horses, might cause another domestication-related pathology on the incisor teeth of horses kept in stalls, but it has not been studied systematically. Marsha Levine of the McDonald Institute at Cambridge University has examined riding-related pathologies in vertebrae, but vertebrae are difficult to study. They break and rot easily, their frequency is low in most archaeological samples, and only eight caudal thoracic vertebrae (T11–18) are known to exhibit pathologies from riding. Discussions of horse domestication still tend to focus on the first two methods.[8]

The Size-Variability Method

The size-variability method depends on two assumptions: (1) domesticated populations, because they are protected, should contain a wider variety of sizes and statures that survive to adulthood, or *more variability*; and (2) the average size of the domesticated population as a whole should decline, because penning, control of movement, and a restricted diet should *reduce average stature*. Measurements of leg bones (principally the width of the condyle and shaft) are used to look for these patterns. This method seems to work quite well with the leg bones of cattle and sheep: an increase in variability and reduction in average size does apparently identify domesticated cattle and sheep.

But the underlying assumptions are not known to apply to the earliest domesticated horses. American Indians controlled their horses not in a corral but with a "hobble" (a short rope tied between the two front legs, permitting a walk but not a run). The principal advantage of early horse keeping—its low cost in labor—could be realized only if horses were permitted to forage for themselves. Pens and corrals would defeat this purpose. Domesticated horses living and grazing in the same environment with their wild cousins probably would not show a reduction in size, and might not show an increase in variability. These changes could be expected if and when horses were restricted to shelters and fed fodder over the winter, like cattle and sheep were, or when they were separated into different herds that were managed and trained differently, for example, for riding, chariot teams, or meat and milk production.

During the earliest phase of horse domestication, when horses were free-ranging and kept for their meat, any size reductions caused by human control probably would have been obscured by natural variations in size between different regional wild populations. The scattered wild horses living in central and western Europe were smaller than the horses that lived in the steppes. In figure 10.3, the three bars on the left of the graph represent wild horses from Ice Age and Early Neolithic Germany. They were quite small. Bars 4 and 5 represent wild horses from forest-steppe and steppe-edge regions, which were significantly bigger. The horses from Dereivka, in the central steppes of Ukraine, were bigger still; 75% stood between 133 and 137 cm at the withers, or between 13 and 14 hands. The

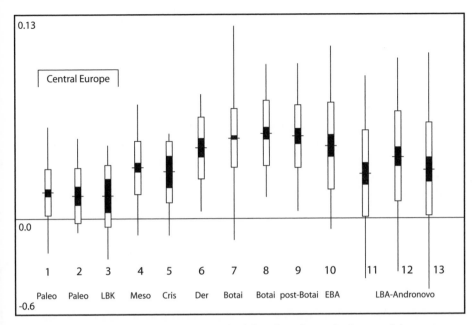

Figure 10.3 The size-variability method for identifying the bones of domesticated horses. The box-and-whisker graphs show the thickness of the leg bones for thirteen archaeological horse populations, with the oldest sites (Paleolithic) on the left and the youngest (Late Bronze Age) on the right. The whiskers, showing the extreme measurements, are most affected by sample size and so are unreliable indicators of population variability. The white boxes, showing two standard deviations from the mean, are reliable indicators of variability, and it is these that are usually compared. The increase in this measurement of variability in bar 10 is taken as evidence for the beginning of horse domestication. After Benecke and von den Dreisch 2003, figures 6.7 and 6.8 combined.

horses of Botai in northern Kazakhstan were even bigger, often over 14 hands. West-east movements of horse populations could cause changes in their average sizes, without any human interference. This leaves an increase in variability as the only indicator of domestication during the earliest phase. And variability is very sensitive to sample size—the larger the sample of bones, the better the chance of finding very small and very large individuals—so changes in variability alone are difficult to separate from sample-size effects.

The domestication of the horse is dated about 2500 BCE by the size-variability method. The earliest site that shows both a significant decrease in average size and an increase in variability is the Bell Beaker settlement of Csepel-Háros in Hungary, represented by bar 10 in figure 10.3, and dated about 2500 BCE. Subsequently many sites in Europe and the steppes show a similar pattern. The absence of these statistical indicators at Dereivka in Ukraine, dated about 4200–3700 BCE (see chapter 11), and at Botai-culture sites in northern Kazakhstan, dated about 3700–3000 BCE, are widely accepted as evidence that horses were not domesticated before about 2500 BCE. But marked regional size differences among early wild horses, the sensitivity of variability measurements to sample size effects, and the basic question of the applicability of these methods to the earliest domesticated horses are three reasons to look at other kinds of evidence. The appearance of significant new variability in horse herds after 2500 BCE could reflect the later development of specialized breeds and functions, not the earliest domestication.[9]

Age-at-Death Statistics

The second quantifiable method is the study of the ages and sexes of butchered animals. The animals selected for slaughter from a domesticated herd should be different ages and sexes from those obtained by hunting. Herders would probably cull young males as soon as they reached adult meat weight, at about two to three years of age. A site occupied by horse herders might contain very few obviously male horses, since the eruption of the canine teeth in males, the principal marker of gender in horse bones, happens at about age four or five, after the age when the males should have been slaughtered for food. Females should have been kept alive as breeders, up to ten years old or more. In contrast, hunters prey on the most predictable elements of a wild herd, so they would concentrate their efforts on the standard wild horse social group, the stallion-with-harem bands, which move along well-worn paths and trails within a defined territory. Regular hunting of stallion-with-harem bands would yield a small number of prime stallions (six to nine years old) and a large number of breeding-age females (three to ten years old) and their immature young.[10]

But many other hunting and culling patterns are possible, and might be superimposed on one another in a long-used settlement site. Also, only a few bones in a horse's body indicate sex—a mature male (more than five years old) has canine teeth whereas females usually do not, and the pelvis of a mature female is distinctive. Horse jaws with the canines still embedded are not often preserved, so data on gender are spotty. Age is estimated based on molar teeth, which preserve well, so the sample for age estimation usually is bigger. But assigning a precise age to a loose horse molar, not found in the jaw, is difficult, and teeth are often found loose in archaeological sites. We had to invent a way to narrow down the very broad range of ages that could be assigned to each tooth. Further, teeth are part of the head, and heads may receive special treatment. If the goal of the analysis is to determine which horses were culled for food, heads are not necessarily the most direct indicators of the human diet. If the occupants of the site kept and used the heads of prime-age stallions for rituals, the teeth found in the site would reflect that, and not culling for food.[11]

Marsha Levine studied age and sex data at Dereivka in Ukraine (4200–3700 BCE) and Botai in northern Kazakhstan (3700–3000 BCE), two critical sites for the study of horse domestication in the steppes. She concluded that the horses at both sites were wild. At Dereivka the majority of the teeth were from animals whose ages clustered between five and seven years old, and fourteen of the sixteen mandibles were from mature males.[12] This suggested that most of the horse heads at Dereivka came from prime-age stallions, not the butchering pattern expected for a managed population. But, in fact, it is an odd pattern for a hunted population as well. Why would hunters kill only prime stallions? Levine suggested that the Dereivka hunters had *stalked* wild horse bands, drawing the attention of the stallions, which were killed when they advanced to protect their harems. But stalking in the open steppe is probably the least productive way for a pedestrian hunter to attack a wild horse band, as stallions are more likely to alarm their band and run away than to approach a predator. Pedestrian hunters should have used ambush methods, shooting at short range on a habitually used horse trail. Moreover, the odd stallion-centered slaughter pattern of Dereivka closely matches the slaughter pattern at the Roman military cemetery at Kestren, the Netherlands (figure 10.4), where the horses certainly were domesticated. At Botai, in contrast, the age–and–sex profile matched what would be expected if whole wild herds were slaughtered en masse, with no selection for age or sex. The two profiles were dissimilar, yet Levine concluded that horses were wild at both places. Age and sex profiles are open to many different interpretations.

If it is difficult to distinguish wild from domesticated horses, it is doubly problematic to distinguish the bones of a mount from those of a horse merely eaten for dinner. Riding leaves few traces on horse bones. But a bit leaves marks on the teeth, and teeth usually survive very well. Bits are used only to guide horses from behind, to drive or to ride. They are not used if the horse is pulled from the front, as a packhorse is, as this would just pull the bit out of the mouth. Thus bit wear on the teeth indicates

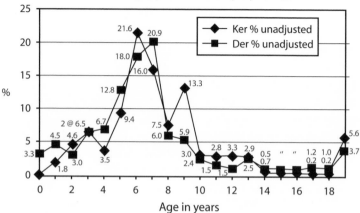

Figure 10.4 The age-at-death method for identifying the bones of domesticated horses. This graph compares the age-at-death statistics for Late Eneolithic horses from Dereivka, Ukraine, to domesticated horses from the Roman site of Kesteren, Netherlands. The two graphs are strikingly similar, but one is interpreted as a "wild" profile and the other is "domesticated." After Levine 1999, figure 2.21.

riding or driving. The *absence* of bit wear means nothing, since other forms of control (nosebands, hackamores) might leave no evidence. But its *presence* is an unmistakable sign of riding or driving. That is why we pursued it. Bit wear could be the smoking gun in the long argument over the origins of horseback riding and, by extension, in debates over the domestication of the horse.

BIT WEAR AND HORSEBACK RIDING

After Brown and I left the Smithsonian in 1985 we spent several years gathering a collection of horse lower second premolars (P_2s), the teeth most affected by bit chewing. Eventually we collected 139 P_2s from 72 modern horses. Forty were domesticated horses processed through veterinary autopsy labs at the University of Pennsylvania and Cornell University. All had been bitted with modern metal bits. We obtained information on their age, sex, and usage—hunting, leisure, driving, racing, or draft—and for some horses we even knew how often they had been bitted, and with what kind of bit. Thirteen additional horses came from the Horse

Training and Behavior program at the State University of New York at Cobleskill. Some had never been bitted. We made casts of their teeth in their mouths, much as a dentist makes an impression to fit a crown—we think that we were the first people to do this to a living horse. A few feral horses, never bitted, were obtained from the Atlantic barrier island of Assateague, MD. Their bleached bones and teeth were found by Ron Keiper of Penn State, who regularly followed and studied the Assateague horses and generously gave us what he had found. Sixteen Nevada mustangs, killed in 1988 by ranchers, supplied most of our never-bitted P_2s. I read about the event, made several telephone calls, and was able to get their mandibles from the Bureau of Land Management after the kill sites were documented. Many years later, in a separate study, Christian George at the University of Florida applied our methods to 113 more never-bitted P_2s from a minimum of 58 fossil equids 1.5 million years old. These animals, of the species *Equus "leidyi,"* were excavated from a Pleistocene deposit near Leisey, Florida. George's Leisey equids (the same size, diet, and dentition as modern horses) had never seen a human, much less a bit.[13]

We studied high-resolution casts or replicas of all the P_2s under a Scanning Electron Microscope (SEM). The SEM revealed that the vice of bit chewing was amazingly widely practiced (figure 10.5). More than 90% of the bitted horses showed some wear on their P_2s from chewing the bit, often just on one side. Their bits also showed wear from being chewed. Riding creates the same wear as driving, because it is not the rider or driver who creates bit wear—it is the horse grasping and releasing the bit between its teeth. A metal bit or even a bone bit creates distinctive microscopic abrasions on the occlusal enamel of the tooth, usually confined to the first or metaconid cusp, but extending back to the second cusp in many cases. These abrasions (type "a" wear, in our terminology) are easily identified under a microscope. All bits, whether hard (metal or bone) or soft (rope or leather) also create a second kind of wear: a wear facet or bevel on the front (mesial) corner of the tooth. The facet is caused both by direct pressure (particularly with a hard bit of bone or metal), which weakens and cracks the enamel when the bit is squeezed repeatedly between the teeth; and by the bit slipping back and forth over the front or mesial corner of the P_2. Metal bits create both kinds of wear: abrasions on the occlusal enamel and wear facets on the mesial corner of the tooth. But rope bits probably were the earliest kind. Can a rope bit alone create visible wear on the enamel of horse teeth?

With a grant from the National Science Foundation and the cooperation of the State University of New York (SUNY) at Cobleskill we acquired

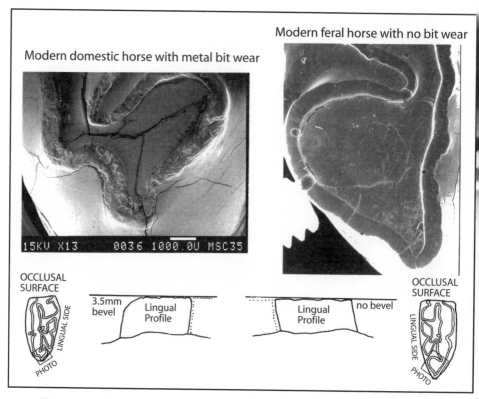

Figure 10.5 Bit wear and no wear on the lower second premolars (P$_2$s) of modern horses.

Left: a Scanning Electron Micrograph (SEM) taken at 13x of "a-wear" abrasions on the first cusp of a domesticated horse that was bitted with a metal bit. The profile shows a 3.5 mm bevel or facet on the same cusp.

Right: An SEM taken at 15x of the smooth surface of the first cusp of a feral horse from Nevada, never bitted. The profile shows a 90° angle with no bevel.

four horses that had never been bitted. They were kept and ridden at SUNY Cobleskill, which has a Horse Training and Behavior Program and a thirty-five-horse stable. They ate only hay and pasture, no soft feeds, to mimic the natural dental wear of free-range horses. Each horse was ridden with a different organic bit—leather, horsehair rope, hemp rope, or bone—for 150 hours, or 600 hours of riding for all four horses. The horse with the horsehair rope bit was bitted by tying the rope around its lower jaw in the classic "war bridle" of the Plains Indians, yet

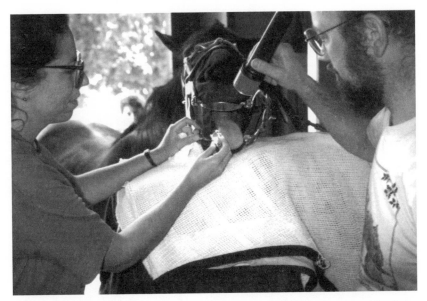

Figure 10.6 Brown and Anthony removing a high-resolution mold of the P_2 of a horse bitted with an organic bit at State University of New York, Cobleskill, in 1992.

it was still able to loosen the loop with its tongue and chew the rope. The other horses' bits were kept in place by antler cheek-pieces made with flint tools. At four intervals each horse was anaesthetized by a bemused veterinarian, and we propped open its mouth, brushed its teeth, dried them, pulled its tongue to the side, and made molds of its P_2s (figure 10.6). We tracked the progress of bit wear over time, and noted the differences between the wear made by the bone bit (hard) and the leather and rope bits (soft).[14]

The riding experiment demonstrated that soft bits do create bit wear. The actual cause of wear might have been microscopic grit trapped in and under the bit, since all the soft bits were made of materials softer than enamel. After 150 hours of riding, bits made of leather and rope wore away about 1 mm of enamel on the first cusp of the P_2 (figure 10.7). The mean bevel measurement for the three horses with rope or leather bits at the end of the experiment was more than 2 standard deviations greater than the pre-experiment mean.[15] The rope and leather mouthpieces stood up well to chewing, although the horse with the hemp rope bit chewed through it several times. The horses bitted with soft bits showed the same

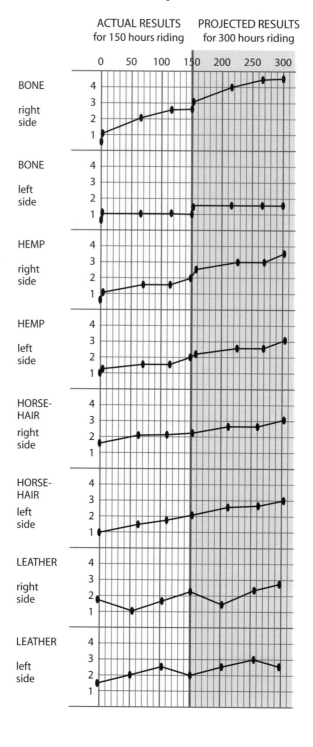

TABLE 10.1

Bevel Measurements on the P$_2$s of Bitted and Never–Bitted Mature (>3yr) Horses

	Never–Bitted, Feral and Domestic (16 horses / 31 teeth)	*Pleistocene Leisey equids* (44 h. / 74t.)	*Domestic Bitted* (39 h. / 73 t.)	*Domestic Bitted Daily* (13 h. / 24t.)
Median	0.5 mm	1.1 mm	2.5 mm	4.0 mm
Mean	0.79 mm	1.1 mm	3.11 mm	3.6 mm
Standard Deviation	0.63 mm	0.71 mm	1.93 mm	1.61 mm
Range	0–2 mm	0–2.9 mm	0–10 mm	1–7 mm

wear facet on the same part of the P$_2$ as horses bitted with metal and bone bits, but the surface of the facet was microscopically smooth and polished, not abraded. Hard bits, including our experimental bone bit, create distinctive "a" wear on the occlusal enamel of the facet, but soft bits do not. Soft bit wear is best identified by measuring the depth of the wear facet or bevel on the P$_2$, not by looking for abrasions on its surface.

Table 10.1 shows bevel measurements for modern horses that never were bitted (left column); Pleistocene North American equids that never were bitted (center left column); domestic horses that were bitted, including some that were bitted infrequently (center right column); and a smaller sub-group of domestic horses that were bitted at least five times a week up to the day we made molds of their teeth (right column). Measurements of the depth of the wear facet easily distinguished the 73 teeth of bitted horses from the 105 teeth of never-bitted horses. The never-bitted/bitted means are different at better than the .001 level of significance. The never-bitted/daily-bitted means are more than 4 standard deviations apart. Bevel measurements segregate mature bitted from mature never-bitted horses, *as populations.*[16]

We set a bevel measurement of 3.0 mm as the minimum threshold for recognizing bit wear on archaeological horse teeth (figure 10.8). More than half of our occasionally bitted teeth did not exhibit a bevel measuring as much as 3 mm . But all horses in our sample with a bevel of 3 mm or more

Figure 10.7 Graph showing the increase in bevel measurements in millimeters caused by organic bits over 150 hours of riding, with projections of measurements if riding had continued for 300 hours.

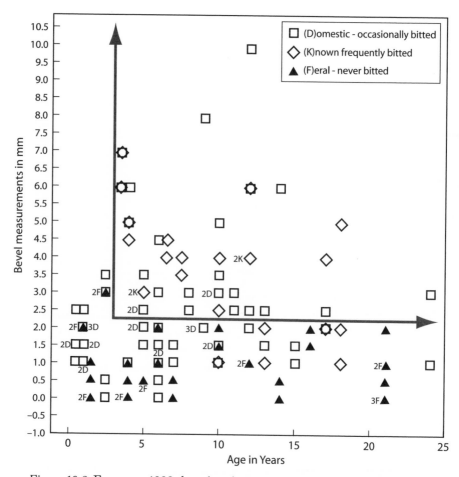

Figure 10.8 From our 1998 data: bevel measurements of never bitted, occasionally bitted, and frequently bitted horse teeth plotted against age. All domesticated horses had precisely known ages; all feral horses were aged by examining entire mandibles with intact incisor teeth. The line excludes feral horses and horses aged ≤3 yr. and includes only bitted horses. After Brown and Anthony 1998.

had been bitted. So the last question was, how adequate was our sample? Could a 3 mm wear facet occur naturally on a wild horse P_2, caused by malocclusion? Criticisms of bit wear have centered on this problem.[17]

Very young horses with newly erupted permanent premolars do display natural dips and rises on their teeth. New permanent premolars are uneven because they have not yet been worn flat by occlusion with the opposing tooth. We had to exclude the teeth of horses two to three years old for that reason. But among the 105 measurable P_2s from *mature* equids that had never been bitted, Pleistocene to modern, we found that a "natural" bevel measurement of more than 2.0 mm is unusual (less than 3% of teeth), and a bevel of 2.5 mm is exceedingly rare (less than 1%). Only one of the 105 never-bitted teeth had a bevel measurement greater than 2.5 mm—a single tooth from the Leisey equids with a mesial bevel of 2.9 mm (the next-nearest bevel was 2.34 mm). In contrast, bevels of 2.5 mm and more occurred in 58% of the teeth of mature horses that were bitted.[18]

A bevel of 3 mm or more on the P_2 of a mature horse is evidence for either an exceedingly rare malocclusion or a very common effect of bitting. If even one mature horse from an archaeological site shows a bevel ≥3 mm bit wear is suggested, but is not a closed case. If multiple mature horses from a single site show mesial bevel measurements of 3 mm or more, they probably were bitted. I should stress that our method depends on the accurate measurement of a very small feature—a bevel or facet just a few millimeters deep. According to our measurements on 178 P_2 teeth of mature equids the difference between a 2 mm and a 3 mm bevel is extremely important. In any discussion of bit wear, precise measurements are required and young animals must be eliminated. But until someone finds a population of mature wild horses that displays many P_2 teeth with bevels ≥3 mm , bit wear as we have defined it indicates that a horse has been ridden or driven.[19]

INDO-EUROPEAN MIGRATIONS AND BIT WEAR AT DEREIVKA

Many archaeologists and historians in the first half of the twentieth century thought that horses were first domesticated by Indo-European–speaking peoples, often specifically characterized as Aryans, who also were credited with inventing the horse-drawn chariot. This fascination with the Aryans, or *Ariomania*, to use Peter Raulwing's term, dominated the study of horseback riding and chariots before World War II.[20]

In 1964 Dimitri Telegin discovered the head-and-hoof bones of a seven- to eight-year-old stallion buried together with the remains of two dogs at Dereivka in Ukraine, apparently a cultic deposit of some kind (see figure 11.9). The Dereivka settlement contained three excavated structures of the Sredni Stog culture and the bones of a great many horses, 63% of the bones found. Ten radiocarbon dates placed the Sredni Stog settlement about 4200–3700 BCE, after the Dnieper-Donets II and Early Khvalynsk era. V. I. Bibikova, the chief paleozoologist at the Kiev

Institute of Archaeology, declared the stallion a domesticated horse in 1967. The respected Hungarian zoologist and head of the Hungarian Institute of Archaeology, Sandor Bökönyi, agreed, noting the great variabity in the leg dimensions of the Dereivka horses. The German zoologist G. Nobis also agreed. During the late 1960s and 1970s horse domestication at Dereivka was widely accepted.[21]

For Marija Gimbutas of UCLA, the domesticated horses at Dereivka were part of the evidence which proved that horse-riding, Indo-European–speaking "Kurgan-culture" pastoralists had migrated in several waves out of the steppes between 4200 and 3200 BCE, destroying the world of egalitarian peace and beauty that she imagined for the Eneolithic cultures of Old Europe. But the idea of Indo-European migrations sweeping westward out of the steppes was not accepted by most Western archaeologists, who were increasingly suspicious of any migration-based explanation for culture change. During the 1980s Gimbutas's scenario of massive "Kurgan-culture" invasions into eastern and central Europe was largely discredited, notably by the German archaeologist A. Häusler. Jim Mallory's 1989 masterful review of Indo-European archaeology retained Gimbutas's steppe homeland and her three waves as periods of increased movement in and around the steppes, but he was much less optimistic about linking specific archaeological cultures with specific migrations by specific Indo-European branches. Others, myself included, criticized both Gimbutas's archaeology and Bibikova's interpretation of the Dereivka horses. In 1990 Marsha Levine seemed to nail the coffin shut on the horse-riding, Kurgan-culture invasion hypothesis when she declared the horse age and sex ratios at Dereivka to be consistent with a wild, hunted population.[22]

Brown and I visited the Institute of Zoology in Kiev in 1989, the year after Levine, learning of her trip only after we arrived. With the cheerful help of Natalya Belan, a senior zoologist, we made molds of dozens of horse P_2s from many archaeological sites in Ukraine. We examined one P_2 from Early Eneolithic Varfolomievka in the Caspian Depression (no wear), one from the Tripolye A settlement of Luka Vrublevetskaya (no wear), several from Mesolithic and Paleolithic sites in Ukraine (no wear), many from Scythian and Roman-era graves (a lot of bit wear, some of it extreme), and those of the cult stallion and four other horse P_2s from Dereivka. As soon as we saw the Dereivka cult stallion we knew it had bit wear. Its P_2s had bevels of 3.5 mm and 4 mm , and the enamel on the first cusp was deeply abraded. Given its stratigraphic position at the base of a Late Eneolithic cultural level almost 1 m deep, dated by ten radiocarbon dates to 4200–3700 BCE, the cult stallion should have been about two thousand years older than the previously known oldest evidence for horseback riding. Only four other P$_2$s still survived in the Dereivka collection: two deciduous teeth from horses less than 2.5 years old (not measurable), and two others from adult horses but with no bit wear. So our case rested on a single horse. But it was very clear wear—surprisingly similar to modern metal bit wear. In 1991 we published articles in *Scientific American* and in the British journal *Antiquity* announcing the discovery of bit wear at Dereivka. Levine's conclusion that the Dereivka horses were wild had been published just the year before. Briefly we were too elated to worry about the argument that would follow.[23]

It began when A. Häusler challenged us at a conference in Berlin in 1992. He did not think the Dereivka stallion was Eneolithic or cultic; he deemed it a Medieval garbage deposit, denying there was evidence for a horse cult anywhere in the steppes during the Eneolithic. That the wear looked like metal bit wear was part of the problem, since a metal bit was improbable in the Eneolithic. Häusler's target was bigger than bit wear or even horse domestication: he had dedicated much of his career to refuting Gimbutas's "Kurgan-culture" migrations and the entire notion of a steppe Indo-European homeland.[24] The horses at Dereivka were just a small piece in a larger controversy. But criticisms like his forced us to obtain a direct date on the skull itself.

Telegin first sent us a bone sample from the same excavation square and level as the stallion. It yielded a date between 90 BCE and 70 BCE (OxA 6577), our first indication of a problem. He obtained another anomalous radiocarbon date, ca. 3000 BCE, on a piece of bone that, like our first sample, seems not to have been from the stallion itself (Ki 5488). Finally, he sent us one of the bit-worn P$_2$s from the cult stallion. The Oxford radiocarbon laboratory obtained a date of 410–200 BCE from this tooth (OxA 7185). Simultaneously the Kiev radiocarbon laboratory obtained a date of 790–520 BCE on a piece of bone from the skull (Ki 6962). Together these two samples suggest a date between 800 and 200 BCE.

The stallion-and-dog deposit at Dereivka was of the Scythian era. No wonder it had metal bit wear—so did many other Scythian horse teeth. It had been placed in a pit dug into the Eneolithic settlement between 800 and 200 BCE. The archaeologists who excavated this part of the site in 1964 did not see the intrusive pit. In 2000, nine years after our initial publication in *Antiquity*, we published another *Antiquity* article retracting the early date for bit wear at Dereivka. We were disappointed, but by then Dereivka was no longer the only prehistoric site in the steppes with bit wear.[25]

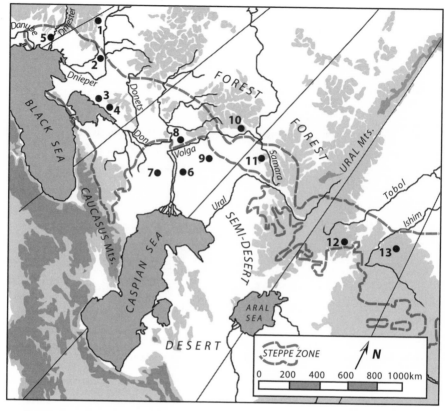

Figure 10.9 Horse-related sites of Eneolithic or older age in the western and central Eurasian steppes. The steppe ecological zone is enclosed in dashed lines.
(1) Moliukhor Bugor; (2) Dereivka; (3) Mariupol; (4) Matveev Kurgan; (5) Girzhevo; (6) Kair Shak; (7) Dzhangar; (8) Orlovka; (9) Varfolomievka; (10) Khvalynsk; (11) S'yezzhe; (12) Tersek; (13) Botai

Botai and Eneolithic Horseback Riding

The oldest horse P_2s showing wear facets of 3 mm and more are from the Botai and Tersek cultures of northern Kazakhstan (figure 10.9). Excavated through the 1980s by Victor Zaibert, Botai was a settlement of specialized hunters who rode horses to hunt horses, a peculiar kind of economy that existed only between 3700 and 3000 BCE, and only in the steppes of northern Kazakhstan. Sites of the Botai type, east of the Ishim River, and of the related Tersek type, west of the Ishim, contain

Figure 10.10 A concentration of horse bones in an excavated house pit at the Botai settlement in north-central Kazakhstan, dated about 3700–3000 BCE. Archaeozoologist Lubomir Peske takes measurements during an international conference held in Kazakhstan in 1995 "Early Horsekeepers of the Eurasian Steppe 4500–1500 BC." Photo by Asko Parpola.

65–99.9%/horse bones. Botai had more than 150 house-pits (figure 10.10) and 300,000 animal bones, 99.9% of them horse. A partial list of the other species represented at Botai (primarily by isolated teeth and phalanges) includes a very large bovid, probably bison but perhaps aurochs, as well as elk, red deer, roe deer, boar, bear, beaver, saiga antelope, and gazelle. Horses, not the easiest prey for people on foot, were overwhelmingly preferred over these animals.[26]

We visited Zaibert's lab in Petropavlovsk, Kazakhstan, in 1992, again unaware that Marsha Levine had arrived the year before. Among the forty-two P_2s we examined from Botai, nineteen were acceptable for study (many had heavily damaged surfaces, and others were from horses younger than three years old). Five of these nineteen teeth, representing at least three different horses, had significant bevel measurements: two 3 mm, one 3.5 mm, one 4 mm, and one 6 mm . Wear facets on undamaged portions of the Botai P_2s were polished smooth, the same kind of polish created by "soft" bits in our experiment. The five teeth were found in different places across the settlement—they did not come from a single intrusive pit. The

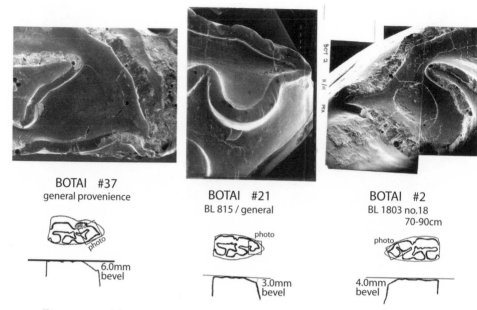

Figure 10.11 Three horse P$_2$s with bit wear from the Botai settlement. The photos show extensive postmortem damage to the occlusal surfaces. The undamaged middle tooth showed smooth enamel surfaces but had a significant wear facet, like a horse ridden with a "soft" bit of rope or leather.

proportion of P$_2$s exhibiting bit wear at Botai was 12% of the entire sample of P$_2$s provided, or 26% of the nineteen measurable P$_2$s. Either number was just too high to explain by appealing to a rare natural malocclusion (figure 10.11). We also examined the horse P$_2$s from a Tersek site, Kozhai 1, dated to the same period, 3700–3000 BCE. At Kozhai 1 horses accounted for 66.1% of seventy thousand identified animal bones (others were saiga antelope at 21.8%, onager at 9.4%, and bison, perhaps including some very large domesticated cattle, at 2.1%). We found a 3 mm wear facet on two P$_2$s of the twelve we examined from Kozhai 1. Most of the P$_2$s at Botai and Kozhai 1 did not exhibit bit wear, but a small percentage (12–26%) did, consistent with the interpretation that the Botai-Tersek people were mounted horse hunters.[27]

Botai attracted the attention of everyone interested in early horse domestication. Two field excavations by Western archaeologists (Marsha Levine and Sandra Olsen) have occurred at Botai or Botai-culture sites. The original excavator, Victor Zaibert, the Kazakh zoologist L.A.

Makarova, and the American archaeozoologist Sandra Olsen of the Carnegie Museum of Natural History in Pittsburgh all concluded that at least some of the Botai horses were domesticated. In opposition, the archaeozoologists N. M. Ermolova, Marsha Levine, and the German team Norbert Benecke and Angela von den Dreisch concluded that all the Botai horses were wild.[28] Levine found some pathologies in the Botai vertebrae but attributed them to age. Benecke and von den Dreisch showed that the Botai horses exhibited a narrow range of variability in size, like Paleolithic wild populations. The ages and sexes of the Botai horses were typical of a wild population, with a 1:1 ratio between the sexes, including all age groups, even colts and pregnant mares with gestating fetuses. Everyone agrees that whole herds of wild horses were killed by the Botai people, using herd-driving hunting techniques that had never been used before in the Kazakh steppes, certainly not on this scale. Were the hunters riding or on foot? Native American hunters on foot drove bison herds over cliffs before the introduction of horses to the Americas by Europeans, so herd driving was possible without riding.

Sandra Olsen of the Carnegie Museum concluded that at least some Botai horses were used for transport, because whole horse carcasses were butchered regularly over the course of several centuries *in the settlement* at Botai.[29] How would pedestrian hunters drag eight-hundred-pound carcasses to the settlement, not just once or twice but as a regular practice that continued for centuries? Pedestrian hunters who used herd-driving hunting methods in the European Paleolithic at Solutré (where Olsen had worked earlier) and in the North American Plains butchered large animals where they died at the kill site. But the Botai settlement is located on the open, south-facing slope of a broad ridge top in a steppe environment—wild horses could not have been trapped in the settlement. Either some horses were tamed and could be led into the settlement or horses were used to drag whole carcasses of killed animals into the settlement, perhaps on sleds. Olsen's interpretation was supported by soil analysis from a house pit at Botai (Olsen's excavation 32) that revealed a distinct layer of soil filled with horse dung. This "must have been the result of redeposition of material from stabling layers," according to the soil scientists who examined it.[30] This dung-rich soil was removed from a horse stable or corral. The stabling of horses at Botai obviously suggests domestication.

One more argument for horseback riding is that the slaughter of wild populations with a 1:1 sex ratio could only be achieved by sweeping up both stallion-with-harem bands *and* bachelor bands, and these two kinds of social groups normally live far apart in the wild. If stallion-with-harem bands were driven into traps, the female:male ratio would be more than 2:1. The only way to capture both bachelor bands and harem bands in herd drives is to actively search and sweep up all the wild horses in a very large region. That would be impossible on foot.

Finally, the beginning of horseback riding provides a good explanation for the economic and cultural changes that appeared with the Botai-Tersek cultures. Before 3700 BCE foragers in the northern Kazkah steppes lived in small groups at temporary lakeside camps such as Vinogradovka XIV in Kokchetav district and Tel'manskie in Tselinograd district. Their remains are assigned to the Atbasar Neolithic.[31] They hunted horses but also a variety of other game: short-horned bison, saiga antelope, gazelle, and red deer. The details of their foraging economy are unclear, as their camp sites were small and ephemeral and have yielded relatively few animal bones. Around 3700–3500 BCE they shifted to specialized horse hunting, started to use herd-driving hunting methods, and began to aggregate in large settlements—a new hunting strategy and a new settlement pattern. The number of animal bones deposited at each settlement rose to tens or even hundreds of thousands. Their stone tools changed from microlithic tool kits to large bifacial blades. They began to make large polished stone weights with central perforations, probably for manufacturing multi-stranded rawhide ropes (weights are hung from each strand as the strands are twisted together). Rawhide thong manufacture was one of the principal activities Olsen identified at Botai based on bone tool microwear. For the first time the foragers of the northern Kazakh steppes demonstrated the ability to drive and trap whole herds of horses and transport their carcasses into new, large communal settlements. No explanation other than the adoption of horseback riding has been offered for these changes.

The case for horse management and riding at Botai and Kozhai 1 is based on the presence of bit wear on seven Botai-Tersek horse P_2s from two different sites, carcass transport and butchering practices, the discovery of horse-dung–filled stable soils, a 1:1 sex ratio, and changes in economy and settlement pattern consistent with the beginning of riding. The case against riding is based on the low variability in leg thickness and the absence of riding-related pathologies in a small sample of horse vertebrae, possibly from wild hunted horses, which probably made up 75–90% of the horse bones at Botai. We are reasonably certain that horses were bitted and ridden in northern Kazakhstan beginning about 3700–3500 BCE.

THE ORIGIN OF HORSEBACK RIDING

Horseback riding probably did not begin in northern Kazakhstan. The Botai-Tersek people were mounted foragers. A few domesticated cattle (?) bones might be found in some Tersek sites, but there were none in Botai sites, farther east; and neither had sheep.[32] It is likely that Botai-Tersek people acquired the idea of domesticated animal management from their western neighbors, who had been managing domesticated cattle and sheep, and probably horses, for a thousand years before 3700–3500 BCE.

The evidence for riding at Botai is not isolated. Perhaps the most interesting parallel from beyond the steppes is a case of severe wear on a mesial horse P_2 with a bevel much deeper than 3 mm , on a five-year-old stallion jaw excavated from Late Chalcolithic levels at Mokhrablur in Armenia, dated 4000–3500 BCE. This looks like another case of early bit wear perhaps even older than Botai, but we have not examined it for confirmation.[33] Also, after about 3500 BCE horses began to appear in greater numbers or appeared regularly for the first time outside the Pontic-Caspian steppes. Between 3500 and 3000 BCE horses began to show up regularly in settlements of the Maikop and Early Transcaucasian Culture (ETC) in the Caucasus, and also for the first time in the lower and middle Danube valley in settlements of the Cernavoda III and Baden-Boleraz cultures as at Cernavoda and Kétegyháza. Around 3000 BCE horse bones rose to about 10–20% of the bones in Bernberg sites in central Germany and to more than 20% of the bones at the Cham site of Galgenberg in Bavaria. The Galgenburg horses included a native small type and a larger type probably imported from the steppes. This general increase in the importance of horses from Kazakhstan to the Caucasus, the Danube valley, and Germany after 3500 BCE suggests a significant change in the relationship between humans and horses. Botai and Tersek show what that change was: people had started to ride.[34]

Over the long term it would have been very difficult to manage horse herds without riding them. Anywhere that we see a sustained, long-term dependence on domesticated horses, riding is implied for herd management alone. Riding began in the Pontic-Caspian steppes before 3700 BCE, or before the Botai-Tersek culture appeared in the Kazakh steppes. It may well have started before 4200 BCE. It spread outside the Pontic-Caspian steppes between 3700 and 3000 BCE, as shown by increases in horse bones in southeastern Europe, central Europe, the Caucasus, and northern Kazakhstan.

The Economic and Military Effects
of Horseback Riding

A person on foot can herd about two hundred sheep with a good herding dog. On horseback, with the same dog, that single person can herd about five hundred.[35] Riding greatly increased the efficiency and therefore the scale and productivity of herding in the Eurasian grasslands. More cattle and sheep could be owned and controlled by riders than by pedestrian herders, which permitted a greater accumulation of animal wealth. Larger herds, of course, required larger pastures, and the desire for larger pastures would have caused a general renegotiation of tribal frontiers, a series of boundary conflicts. Victory in tribal warfare depended largely on forging alliances and mobilizing larger forces than your enemy, and so intensified warfare stimulated efforts to build alliances through feasts and the redistribution of wealth. Gifts were effective both in building alliances before conflicts and in sealing agreements after them. An increase in boundary conflicts would thus have encouraged more long-distance trade to acquire prestigious goods, as well as elaborate feasts and public ceremonies to forge alliances. This early phase of conflict, caused partly by herding on horseback, might be visible archaeologically in the horizon of polished stone mace-heads and body decorations (copper, gold, boars-tusk, and shell ornaments) that spread across the western steppes with the earliest herding economies about 5000–4200 BCE.[36]

Horses were valuable and easily stolen, and riding increased the efficiency of stealing cattle. When American Indians in the North American Plains first began to ride, chronic horse-stealing raids soured relationships even between tribes that had been friendly. Riding also was an excellent way to retreat quickly; often the most dangerous part of tribal raiding on foot was the running retreat after a raid. Eneolithic war parties might have left their horses under guard and attacked on foot, as many American Indians did in the early decades of horse warfare in the Plains. But even if horses were used for nothing more than transportation to and from the raid, the rapidity and reach of mounted raiders would have changed raiding tactics, status-seeking behaviors, alliance-building, displays of wealth, and settlement patterns. Thus riding cannot be cleanly separated from warfare.[37]

Many experts have suggested that horses were not ridden in warfare until after about 1500–1000 BCE, but they failed to differentiate between *mounted raiding*, which probably is very old, and *cavalry*, which was invented in the Iron Age after about 1000 BCE.[38] Eneolithic tribal herders probably rode horses in inter-clan raids before 4000 BCE, but they were not like the Huns sweeping out of the steppes on armies of shaggy horses. What is intriguing about the Huns and their more ancient cousins, the Scythians, was that they formed armies. During the Iron Age the Scythians, essentially tribal in most other aspects of their political organization, became organized in their military operations like the formal armies of urban states. That required a change in ideology—how a warrior thought about himself, his role, and his responsibilities—as well as in the technology of mounted warfare—how weapons were used from horseback. Probably the change in weapons came first.

Mounted archery probably was not yet very effective before the Iron Age, for three reasons. The bows reconstructed from their traces in steppe Bronze Age graves were more than 1 m long and up to 1.5 m, or almost five feet, in length, which would clearly have made them clumsy to use from horseback; the arrowheads were chipped from flint or made from bone in widely varying sizes and weights, implying a nonstandardized, individualized array of arrow lengths and weights; and, finally, the bases of most arrowheads were made to fit into a hollow or split shaft, which weakened the arrow or required a separate hollow foreshaft for the attachment of the point. The more powerful the bow, and the higher the impact on striking a target, the more likely the arrow was to split, if the shaft had already been split to secure the point. Stemmed and triangular flint points, common before the Iron Age, were made to be inserted into a separate foreshaft with a hollow socket made of reed or wood (for stemmed points), or were set into a split shaft (for triangular points). The long bows, irregular arrow sizes, and less-than-optimal attachments between points and arrows together reduced the military effectiveness of early mounted archery. Before the Iron Age mounted raiders could harass tribal war bands, disrupt harvests in farming villages, or steal cattle, but that is not the same as defeating a disciplined army. Tribal raiding by small groups of riders in eastern Europe did not pose a threat to walled cities in Mesopotamia, and so was ignored by the kings and generals of the Near East and the eastern Mediterranean.[39]

The invention of the short, recurved, compound bow (the "cupid" bow) around 1000 BCE made it possible for riders to carry a powerful bow short enough to swing over the horse's rear. For the first time arrows could be fired behind the rider with penetrating power. This maneuver, later known as the "Parthian shot," was immortalized as the iconic image of the steppe archer. Cast bronze socketed arrowheads of standard weights and sizes also appeared in the Early Iron Age. A socketed arrowhead did not require a split-shaft mount, so arrows with socketed arrowheads did not split despite the power of the bow; they also did not need a separate fore-shaft, and so arrows could be simpler and more streamlined. Reusable moulds were invented so that smiths could produce hundreds of socketed arrowheads of standard weight and size. Archers now had a much wider field of fire—to the rear, the front, and the left—and could carry dozens of standardized arrows. An army of mounted archers could now fill the sky with arrows that struck with killing power.[40]

But organizing an army of mounted archers was not a simple matter. The technical advances in bows, arrows, and casting were meaningless without a matching change in mentality, in the identity of the fighter, from a heroic single warrior to a nameless soldier. An ideological model of fighting appropriate for a state had to be grafted onto the mentality of tribal horseback riders. Pre-Iron-Age warfare in the Eurasian steppes, from what we can glean from sources like the *Iliad* and the *Rig Veda*, prob-ably emphasized personal glory and heroism. Tribal warfare generally was conducted by forces that never drilled as a unit, often could choose to ig-nore their leaders, and valued personal bravery above following orders.[41] In contrast, the tactics and ideology of state warfare depended on large disciplined units of anonymous soldiers who obeyed a general. These tac-tics, and the soldier mentality that went with them, were not applied to riders before 1000 BCE, partly because the short bows and standardized arrows that would make mounted archery truly threatening had not yet been invented. As mounted archers gained in firepower, someone on the edge of the civilized world began to organize them into armies. That seems to have occurred about 1000–900 BCE. Cavalry soon swept chari-otry from the battlefield, and a new era in warfare began. But it would be grossly inappropriate to apply that later model of mounted warfare to the Eneolithic.

Riding began in the region identified as the Proto-Indo-European homeland. To understand how riding affected the spread of Indo-European languages we have to pick up the thread of the archaeological narrative that ended in chapter 9.

CHAPTER ELEVEN

The End of Old Europe and the Rise of the Steppe

By 4300–4200 BCE Old Europe was at its peak. The Varna cemetery in eastern Bulgaria had the most ostentatious funerals in the world, richer than anything of the same age in the Near East. Among the 281 graves at Varna, 61 (22%) contained more than three thousand golden objects together weighing 6 kg (13.2 lb). Two thousand of these were found in just four graves (1, 4, 36, and 43). Grave 43, an adult male, had golden beads, armrings, and rings totaling 1,516 grams (3.37 lb), including a copper axe-adze with a gold-sheathed handle.[1] Golden ornaments have also been found in tell settlements in the lower Danube valley, at Gumelniţa, Vidra, and at Hotnitsa (a 310-gm cache of golden ornaments). A few men in these communities played prominent social roles as chiefs or clan leaders, symbolized by the public display of shining gold ornaments and cast copper weapons.

Thousands of settlements with broadly similar ceramics, houses, and female figurines were occupied between about 4500 and 4100 BCE in eastern Bulgaria (Varna), the upland plains of Balkan Thrace (KaranovoVI), the upper part of the Lower Danube valley in western Bulgaria and Romania (Krivodol-Sălcuţa), and the broad riverine plains of the lower Danube valley (Gumelniţa) (figure 11.1). Beautifully painted ceramic vessels, some almost 1 m tall and fired at temperatures of over 800°C, lined the walls of their two-storied houses. Conventions in ceramic design and ritual were shared over large regions. The crafts of metallurgy, ceramics, and even flint working became so refined that they must have required master craft specialists who were patronized and supported by chiefs. In spite of this, power was not obviously centralized in any one village. Perhaps, as John Chapman observed, it was a time when the restricted resources (gold, copper, *Spondylus* shell) were not critical, and the critical resources (land, timber, labor, marriage partners) were not seriously restricted. This could have prevented any one region or town from dominating others.[2]

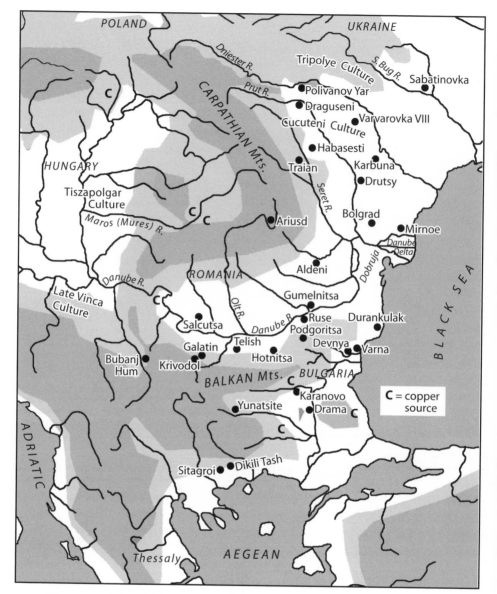

Figure 11.1 Map of Old Europe at 4500–4000 BCE.

Towns in the high plains atop the Balkans and in the fertile lower Danube valley formed high tells. Settlements fixed in one place for so long imply fixed agricultural fields and a rigid system of land tenure around each tell. The settlement on level VI at Karanovo in the Balkans was the type site for the period. About fifty houses crowded together in orderly rows inside a protective wooden palisade wall atop a massive 12-m (40-ft) tell. Many tells were surrounded by substantial towns. At Bereket, not far from Karanovo, the central part of the tell was 250 m in diameter and had cultural deposits 17.5 m (57 ft) thick, but even 300–600 m away from this central eminence the occupation deposits were 1–3 m thick. Surveys at Podgoritsa in northeastern Bulgaria also found substantial off-tell settlement.[3]

Around 4200–4100 BCE the climate began to shift, an event called the Piora Oscillation in studies of Swiss alpine glaciers. Solar insolation decreased, glaciers advanced in the Alps (which gave this episode its name), and winters became much colder.[4] Variations in temperature in the northern hemisphere are recorded in the annual growth rings in oaks preserved in bogs in Germany and in annual ice layers in the GISP2 glacial ice core from Greenland. According to these sources, extremely cold years happened first in 4120 and 4040 BCE. They were harbingers of a 140-year-long, bitterly cold period lasting from 3960 to 3821 BCE, with temperatures colder than at any time in the previous two thousand years. Investigations led by Douglass Bailey in the lower Danube valley showed that floods occurred more frequently and erosion degraded the riverine floodplains where crops were grown. Agriculture in the lower Danube valley shifted to more cold-tolerant rye in some settlements.[5] Quickly these and perhaps other stresses accumulated to create an enormous crisis.

Between about 4200 and 3900 BCE more than six hundred tell settlements of the Gumelniţa, Karanovo VI, and Varna cultures were burned and abandoned in the lower Danube valley and eastern Bulgaria. Some of their residents dispersed temporarily into smaller villages like the Gumelniţa B1 hamlet of Jilava, southwest of Bucharest, with just five to six houses and a single-level cultural deposit. But Jilava was burned, apparently suddenly, leaving behind whole pots and many other artifacts.[6] People scattered and became much more mobile, depending for their food on herds of sheep and cattle rather than fixed fields of grain. The forests did not regenerate; in fact, pollen cores show that the countryside became even more open and deforested.[7] Relatively mild climatic conditions returned after 3760 BCE according to the German oaks, but by then the cultures of the lower Danube valley and the Balkans had changed dramatically. The cultures that appeared after about 3800 BCE did not regularly use female figurines in domestic rituals, no longer wore copper spiral bracelets or *Spondylus*-shell ornaments, made relatively plain pottery in a limited number of shapes, did not live on tells, and depended more on stockbreeding. Metallurgy, mining, and ceramic technology declined sharply in both volume and technical skill, and ceramics and metal objects changed markedly in style. The copper mines in the Balkans abruptly ceased production; copper-using cultures in central Europe and the Carpathians switched to Transylvanian and Hungarian ores about 4000 BCE, at the beginning of the Bodrogkeresztur culture in Hungary (see ore sources in figure 11.1). Oddly this was when metallurgy really began in western Hungary and nearby in Austria and central Europe.[8] Metal objects now were made using new arsenical bronze alloys, and were of new types, including new weapons, daggers being the most important. "We are faced with the complete replacement of a culture," the foremost expert on Eneolithic metallurgy E. N. Chernykh said. It was "a catastrophe of colossal scope . . . a complete cultural caesura," according to the Bulgarian archaeologist H. Todorova.[9]

The end of Old Europe truncated a tradition that began with the Starcevo-Criş pioneers in 6200 BCE. Exactly what happened to Old Europe is the subject of a long, vigorous debate. Graves of the Suvorovo type, ascribed to immigrants from the steppes, appeared in the lower Danube valley just before the destruction of the tells. Settlements of the Cernavoda I type appeared just after. They regularly contain horse bones and ceramics exhibiting a mixture of steppe technology and indigenous Danubian shapes, and are ascribed to a mixed population of steppe immigrants and people from the tells. The number of abandoned sites and the rapid termination of many long-standing traditions in crafts, domestic rituals, decorative customs, body ornaments, housing styles, living arrangements, and economy suggest not a gradual evolution but an abrupt and probably violent end. At Hotnitsa on the Danube in north-central Bulgaria the burned houses of the final Eneolithic occupation contained human skeletons, interpreted as massacred inhabitants. The final Eneolithic destruction level at Yunatsite on the Balkan upland plain contained forty-six human skeletons. It looks like the tell towns of Old Europe fell to warfare, and, somehow, immigrants from the steppes were involved. But the primary causes of the crisis could have included climate change and related agricultural failures, or soil erosion and environmental degradation accumulated from centuries of intensive farming, or internecine warfare over declining timber and copper resources, or a combination of all these.[10]

The crisis did not immediately affect all of southeastern Europe. The most widespread settlement abandonments occurred in the lower Danube valley

(Gumelniṭa, northeastern Bulgaria, and the Bolgrad group), in eastern Bulgaria (Varna and related cultures), and in the mountain valleys of the Balkans (Karanovo VI), east of the Yantra River in Bulgaria and the Olt in Romania. This was where tell settlements, and the stable field systems they imply, were most common. In the Balkans, a well-cultivated, densely populated landscape occupied since the earliest Neolithic, no permanent settlements can be dated between 3800 and 3300 BCE. People probably still lived there, but herds of sheep grazed on the abandoned tells.

The traditions of Old Europe survived longer in western Bulgaria and western Romania (Krivodol-Sălcuṭa IV–Bubanj Hum Ib). Here the settlement system had always been somewhat more flexible and less rooted; the sites of western Bulgaria usually did not form high tells. Old European ceramic types, house types, and figurine types were abandoned gradually during Sălcuṭa IV, 4000–3500 BCE. Settlements that were occupied during the crisis, places like Telish-Redutite III and Galatin, moved to high, steep-sided promontories, but they retained mud-brick architecture, two-story houses, and cult and temple buildings.[11] Many caves in the region were newly occupied, and since herders often use upland caves for shelter, this might suggest an increase in upland-lowland seasonal migrations by herders. The Krivodol–Salcutsa–Bubanj Hum Ib people reoriented their external trade and exchange connections to the north and west, where their influence can be seen on the Lasinja-Balaton culture in western Hungary.

The Old European traditions of the Cucuteni-Tripolye culture also survived and, in fact, seemed curiously reinvigorated. After 4000 BCE, in its Tripolye B2 phase, the Tripolye culture expanded eastward toward the Dnieper valley, creating ever larger agricultural towns, although none was rebuilt in one place long enough to form a tell. Domestic cults still used female figurines, and potters still made brightly painted fine lidded pots and storage jars 1 m high. Painted fine ceramics were mass-produced in the largest towns (Varvarovka VIII), and flint tools were mass-produced at flint-mining villages like Polivanov Yar on the Dniester.[12] Cucuteni AB/Tripolye B2 settlements such as Veseli Kut (150 ha) contained hundreds of houses and apparently were preeminent places in a new settlement hierarchy. The Cucuteni-Tripolye culture forged new relationships with the copper-using cultures of eastern Hungary (Borogkeresztur) in the west and with the tribes of the steppes in the east.

The languages spoken by those steppe tribes, around 4000 BCE, probably included archaic Proto-Indo-European dialects of the kind partly preserved later in Anatolian. The steppe people who spoke in that way probably already rode horses. Were the Suvorovo sites in the lower Danube valley created by Indo-European invaders on horseback? Did they play a role in the destruction of the tell settlements of the lower Danube valley, as Gimbutas suggested? Or did they just slip into an opening created by climate change and agricultural failures? In either case, why did the Cucuteni-Tripolye culture survive and even prosper? To address these questions we first have to examine the Cucuteni-Tripolye culture and its relations with steppe cultures.

WARFARE AND ALLIANCE: THE CUCUTENI-TRIPOLYE CULTURE AND THE STEPPES

The crisis in the lower Danube valley corresponded to late Cucuteni A3/Tripolye B1, around 4300–4000 BCE. Tripolye B1 was marked by a steep increase in the construction of fortifications—ditches and earthen banks—to protect settlements (figure 11.2). Fortifications might have appeared just about when the climate began to deteriorate and the collapse of Old Europe occurred, but Cucuteni-Tripolye fortifications then *decreased* during the coldest years of the Piora Oscillation, during Tripolye B2, 4000–3700 BCE. If climate change destabilized Old Europe and caused the initial construction of Cucuteni-Tripolye fortifications, the first phase of change was sufficient by itself to tip the system into crisis. Probably there was more to it than just climate.

Only 10% of Tripolye B1 settlements were fortified even in the worst of times. But those that *were* fortified required substantial labor, implying a serious, chronic threat. Fortified Cucteni-Tripolye villages usually were built at the end of a steep-sided promontory, protected by a ditch dug across the promontory neck. The ditches were 2–5 m wide and 1.5–3 m deep, made by removing 500–1,500 m³ of earth. They were relocated and deepened as settlements grew in size, as at Traian and Habaşeşti I. In a database of 2,017 Cucuteni/Tripolye settlements compiled by the Moldovan archaeologist V. Dergachev, half of *all* fortified Cucuteni/Tripolye sites are dated just to the Tripolye B1 period. About 60% of all the flint projectile points from all the Cucuteni/Tripolye culture also belonged just to the Tripolye B1 period. There was no corresponding increase in hunting during Tripolye B1 (no increase in wild animal bones in settlements), and so the high frequency of projectile points was not connected with hunting. Probably it was associated with increased warfare.

The number of Cucuteni-Tripolye settlements increased from about 35 settlements per century during Tripolye A to about 340 (!) during Tripolye

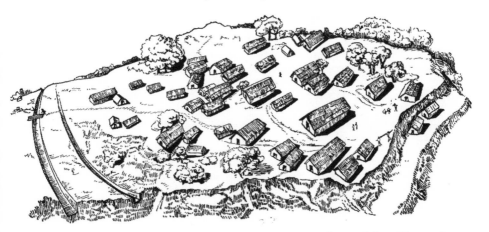

Figure 11.2 Habaşesti I, a fortified Tripolye B1 village. After Chernysh 1982.

B1, a tenfold rise in the number of settlements without a significant expansion of the area settled (figure 11.3b).[13] Part of this increase in settlement density during Tripolye B1 might be ascribed to refugees fleeing from the towns of the Gumelniţa culture. At least one Tripolye B1 settlement in the Prut drainage, Drutsy 1, appears to have been attacked. More than one hundred flint points (made of local Carpathian flint) were found around the walls of the three excavated houses as if they had been peppered with arrows.[14] Compared to its past and its future, the Tripolye B1 period was a time of sharply increased conflict in the Eastern Carpathians.

Contact with Steppe Cultures during Tripolye B: Cucuteni C Ware

Simultaneously with the increase in fortifications and weapons, Tripolye B1 towns showed widespread evidence of contact with steppe cultures. A new pottery type, Cucuteni C ware,[15] shell-tempered and similar to steppe pottery, appeared in Tripolye B1 settlements of the South Bug valley (Sabatinovka I) and in Romania (Draguşeni and Fedeleşeni, where Cucuteni C ware amounted to 10% of the ceramics). Cucuteni C ware is usually thought to indicate contact with and influence from steppe pottery traditions (figure 11.4).[16] Cucuteni C ware might have been used in ordinary homes with standard Cucuteni-Tripolye fine wares as a new kind of coarse or kitchen pottery, but it did not replace traditional coarse kitchen wares tempered with grog (ground-up ceramic sherds). Some Cucuteni C pots look very much like steppe pottery, whereas others had shell-temper,

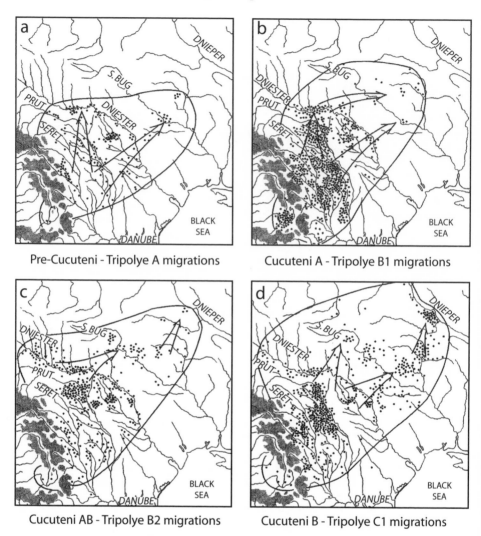

Pre-Cucuteni - Tripolye A migrations

Cucuteni A - Tripolye B1 migrations

Cucuteni AB - Tripolye B2 migrations

Cucuteni B - Tripolye C1 migrations

Figure 11.3. Tripolye B1-B2 migrations. After Dergachev 2002, figure 6.2.

gray-to-brown surface color and some typical steppe decorative techniques (like "caterpillar" impressions, made with a cord-wrapped, curved pressing tool) but were made in typical Cucuteni-Tripolye shapes with other decorative elements typical of Cucuteni-Tripolye wares.

The origin of Cucuteni C ware is disputed. There were good utilitarian reasons for Tripolye potters to adopt shell-tempering. Shell-temper in the clay can increase resistance to heat shock, and shell-tempered pots

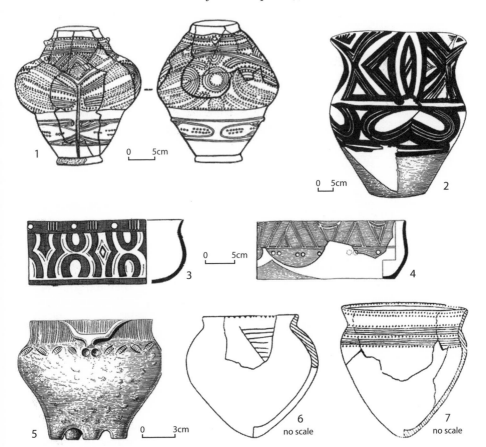

Figure 11.4 Cucuteni C (bottom row) and standard Cucuteni B wares (top two rows): (1) fine ware, Novye Ruseshti I$_{1a}$ (Tripolye B1); (2) fine ware, Geleshti (Tripolye B2); (3–4) fine ware, Frumushika I (Tripolye B1); (5) Cucuteni C ware, Frumushika II (Tripolye B2); (6–7) Cucuteni C ware, Berezovskaya GES. After Danilenko and Shmagli 1972, Figure 7; Chernysh 1982, Figure LXV.

can harden at lower firing temperatures, which could save fuel.[17] Changes in the organization of pottery making could also have encouraged the spread of Cucuteni C wares. Ceramic production was beginning to be taken over by specialized ceramic-making towns during Tripolye B1 and B2, although local household production also continued in most places. Rows of reusable two-chambered kilns appeared at the edges of a few settlements, with 11 kilns at Ariuşd in southeastern Transylvania. If fine painted wares were beginning to be produced in villages that specialized in making pottery and the coarse wares remained locally produced, the change in coarse wares could have reflected the changing organization of production.

On the other hand, these particular coarse wares obviously resembled the pottery of steppe tribes. Many Cucuteni C pots look like they were made by Sredni Stog potters. This suggests familiarity with steppe cultures and even the presence of steppe people in some Tripolye B villages, perhaps as hired herders or during seasonal trade fairs. Although it is unlikely that *all* Cucuteni C pottery was made by steppe potters—there is just too much of it—the appearance of Cucuteni C ware suggests intensified interactions with steppe communities.

Steppe Symbols of Power: Polished Stone Maces

Polished stone maces were another steppe artifact type that appeared in Tripolye B1 villages. A mace, unlike an axe, cannot really be used for anything except cracking heads. It was a new weapon type and symbol of power in Old Europe, but maces had appeared across the steppes centuries earlier in DDII, Khvalynsk, and Varfolomievka contexts. There were two kinds—zoomorphic and eared types—and both had steppe prototypes that were older (figure 11.5; also see figure 9.6). Mace heads carved and polished in the shape of horse heads were found in two Cucuteni A3/A4-Tripolye B1 settlements, Fitioneşti and Fedeleşeni, both of which also had significant amounts of Cucuteni C ware. The eared type appeared at the Cucuteni-Tripolye settlements of Obarşeni and Berezovskaya GES, also with Cucuteni C ware that at Berezovskaya looked like it was imported from steppe communities. Were steppe people present in these Tripolye B1 towns? It seems likely. The integration of steppe pottery and symbols of power into Cucuteni-Tripolye material culture suggests some kind of social integration, but the maintenance of differences in economy, house form, fine pottery, metallurgy, mortuary rituals, and domestic rituals indicates that it was limited to a narrow social sector.[18]

Other Signs of Contact

Most settlements of the Tripolye B period, even large ones, continued to dispose of their dead in unknown ways. But inhumation graves appeared in or at the edge of a few Tripolye B1 settlement sites. A grave in the settle-

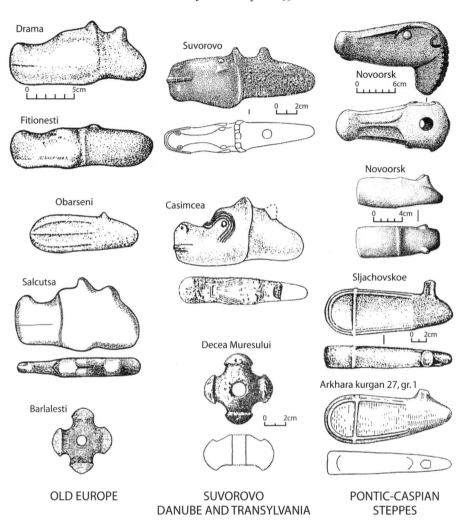

Drama

Suvorovo

Novoorsk

Fitionesti

Novoorsk

Obarseni

Casimcea

Salcutsa

Sljachovskoe

Decea Muresului

Barlalesti

Arkhara kurgan 27, gr. 1

OLD EUROPE

SUVOROVO
DANUBE AND TRANSYLVANIA

PONTIC-CASPIAN
STEPPES

Figure 11.5 Eared and horse-head maces of Old Europe, the Suvorovo migrants, and the Pontic-Caspian steppes. Stone mace heads appeared first and were more common in the steppes. After Telegin et al. 2001; Dergachev 1999; Gheorgiu 1994; Kuzmina 2003.

ment of Nezvisko contained a man with a low skull and broad, thick-boned face like those of steppe people—a type of skull-and-face configuration called "Proto-Europoid" by Eastern European physical anthropologists. Tripolye, Varna, and Gumelniţa people generally had taller heads, narrower faces, and more gracile facial bones, a configuration called "Mediterranean."[19]

Another indicator of movement across the steppe border was the little settlement near Mirnoe in the steppes north of the Danube delta. This is the only known classic-period Tripolye settlement in the coastal steppe lowlands. It had just a few pits and the remains of a light structure containing sherds of Tripolye B1 and Cucuteni C pots, a few bones of cattle and sheep, and more than a hundred grape seeds, identified as wild grapes. Mirnoe seems to have been a temporary Tripolye B1 camp in the steppes, perhaps for grape pickers.[20] Some people, though not many, were moving across the cultural-ecological frontier in both directions.

During Tripolye B2, around 4000–3700 BCE, there was a significant migration out of the Prut-Seret forest-steppe uplands, the most densely settled part of the Tripolye B1 landscape, eastward into the South Bug and Dnieper valleys (figure 11.3c). Settlement density in the Prut-Seret region declined by half.[21] Tripolye, the type site first explored in 1901, was an eastern frontier village of the Tripolye B2 period, situated on a high terrace overlooking the broad, fertile valley of the Dnieper River. The population consolidated into fewer, larger settlements (only about 180 settlements per century during Tripolye B2). The number of fortified settlements decreased sharply.

These signs of demographic expansion and reduced conflict appeared after the tell settlements of the Danube valley were burned and abandoned. It appears that any external threat from the steppes, if there was one, turned away from Cucuteni-Tripolye towns. Why?

Steppe Riders at the Frontiers of Old Europe

Frontiers can be envisioned as peaceful trade zones where valuables are exchanged for the mutual benefit of both sides, with economic need preventing overt hostilities, or as places where distrust is magnified by cultural misunderstandings, negative stereotypes, and the absence of bridging institutions. The frontier between agricultural Europe and the steppes has been seen as a border between two ways of life, farming and herding, that were implacably opposed. Plundering nomads like the Huns and Mongols are old archetypes of savagery. But this is a misleading stereotype, and one derived from a specialized form of militarized pastoral nomadism that did not exist before about 800 BCE. As we saw in the previous chapter, Bronze Age riders in the steppes used bows that were too long for effective mounted archery. Their arrows were of varied weights and sizes. And Bronze Age war bands were not organized like armies. The Hunnic invasion analogy is anachronistic, yet that does not mean that mounted raiding never occurred in the Eneolithic.[22]

There is persuasive evidence that steppe people rode horses to hunt horses in Kazakhstan by about 3700–3500 BCE. Almost certainly they were not the first to ride. Given the symbolic linkage between horses, cattle, and sheep in Pontic-Caspian steppe funerals as early as the Khvalynsk period, horseback riding might have begun in a limited way before 4500 BCE. But western steppe people began to *act* like they were riding only about 4300–4000 BCE, when a pattern consistent with long-distance raiding began, seen most clearly in the Suvorovo-Novodanilovka horizon described at the end of this chapter. Once people began to ride, there was nothing to prevent them from riding into tribal conflicts—not the supposed shortcomings of rope and leather bits (an organic bit worked perfectly well, as our students showed in the organic-bit riding experiment, and as the American Indian "war bridle" demonstrated on the battlefield); not the size of Eneolithic steppe horses (most were about the size of Roman cavalry horses, big enough); and certainly not the use of the wrong "seat" (an argument that early riders sat on the rump of the horse, perhaps for millennia, before they discovered the more natural forward seat—based entirely on Near Eastern images of riders probably made by artists who were unfamiliar with horses).[23]

Although I *do* see evidence for mounted raiding in the Eneolithic, I do *not* believe that any Eneolithic army of pitiless nomads ever lined up on the horizon mounted on shaggy ponies, waiting for the command of their bloodthirsty general. Eneolithic warfare was tribal warfare, so there were no armies, just the young men of this clan fighting the young men of that clan. And early Indo-European warfare seems from the earliest myths and poetic traditions to have been conducted principally to gain glory—*imperishable fame*, a poetic phrase shared between Pre-Greek and Pre-Indo-Iranian. If we are going to indict steppe raiders in the destruction of Old Europe, we first have to accept that they did not fight like later cavalry. Eneolithic warfare probably was a strictly seasonal activity conducted by groups organized more like modern neighborhood gangs than modern armies. They would have been able to disrupt harvests and frighten a sedentary population, but they were not nomads. Steppe Eneolithic settlements like Dereivka cannot be interpreted as pastoral nomadic camps. After nomadic cavalry is removed from the picture, how do we understand social and political relations across the steppe/Old European frontier?

A mutualist interpretation of steppe/farming-zone relations is one alternative. Conflict is not denied, but it is downplayed, and mutually beneficial trade and exchange are emphasized.[24] Mutualism might well explain the relationship between the Cucuteni-Tripolye and Sredni Stog cultures during the Tripolye B period. Among historically known pastoralists in close contact with farming populations there has been a tendency for wealthy herd owners to form alliances with farmers to acquire land as insurance against the loss of their more volatile wealth in herds. In modern economies, where land is a market commodity, the accumulation of property could lead the wealthiest herders to move permanently into towns. In a pre-state tribal world this was not possible because agricultural land was not for sale, but the strategy of securing durable alliances and assets in agricultural communities as insurance against future herd losses could still work. Steppe herders might have taken over the management of some Tripolye herds in exchange for metal goods, linen textiles, or grain; or steppe clans might have attended regular trading fairs at agricultural towns. Annual trading fairs between mounted hunters and river-valley corn farmers were a regular feature of life in the northern Plains of the U.S.[25] Alliances and trade agreements sealed by marriages could account for the increased steppe involvement in Tripolye communities during Tripolye B1, about 4400–4000 BCE. The institutions that normalized these cross-cultural relations probably included gift partnerships. In archaic Proto-Indo-European as partly preserved in Hittite, the verb root that in all other Indo-European languages meant "give" (*dō-) meant "take" and another root (*pai*) meant "give." From this give-and-take equivalence and a series of other linguistic clues Emile Benveniste concluded that, during the archaic phase of Proto-Indo-European, "exchange appears as a round of gifts rather than a genuine commercial operation."[26]

On the other hand, mutualism cannot explain everything, and the end of the Varna-Karanovo VI–Gumelniţa culture is one of those events it does not explain. Lawrence Keeley sparked a heated debate among archaeologists by insisting that warfare was common, deadly, and endemic among prehistoric tribal societies. Tribal frontiers might be creative places, as Frederik Barth realized, but they often witnessed pretty nasty behavior. Tribal borders commonly were venues for insults: the Sioux called the Bannock the "Filthy-Lodge People"; the Eskimo called the Ingalik "Nit-heads"; the Hopi called the Navaho "Bastards"; the Algonkian called the Mohawk "Maneaters"; the Shuar called the Huarani "Savages"; and the simple but eloquent "Enemies" is a very common meaning for names given by neighboring tribes. Because tribal frontiers displayed things people needed just beyond the limits of their own society, the temptation to take them by force was strong. It was doubly strong when those things had legs, like cattle.[27]

Cattle raiding was encouraged by Indo-European beliefs and rituals. The myth of Trito, the warrior, rationalized cattle theft as the recovery of cattle that the gods had *intended* for the people who sacrificed properly. Proto-Indo-European initiation rituals included a requirement that boys initiated into manhood *had* to go out and become like a band of dogs or wolves—to raid their enemies.[28] Proto-Indo-European also had a word for bride-price, **üedmo-*.[29] Cattle, sheep, and probably horses would have been used to pay bride-prices, since they generally are valued higher than other currencies for bride-price payments in pastoral societies without formal money.[30] Already in the preceding centuries domesticated animals had become the proper gifts for gods at funerals (e.g., at Khvalynsk). A relatively small elite already competed across very large regions, adopting the same symbols of status—maces with polished stone heads, boar's tusk plaques, copper rings and pendants, shell disc beads, and bird-bone tubes. When bride-prices escalated as one aspect of this competition, the result would be increased cattle raiding by unmarried men. Combined with the justification provided by the Trito myth and the institution of male-initiation-group raiding, rising bride-prices calculated in animals would have made cross-border raiding almost inevitable.

If they were on foot, Eneolithic steppe cattle raiders might have attacked one another or attacked neighboring Tripolye settlements. But, if they were mounted, they could pick a distant target that did not threaten valued gift partnerships. Raiding parties of a dozen riders could move fifty to seventy-five head of cattle or horses fairly quickly over hundreds of kilometers.[31] Thieving raids would have led to deaths, and then to more serious killing and revenge raids. A cycle of warfare evolving from thieving to revenge raids probably contributed to the collapse of the tell towns of the Danube valley.

What kinds of societies lived on the steppe side of the frontier? Is there good archaeological evidence that they were indeed deeply engaged with Old Europe and the Cucuteni-Tripolye culture in quite different ways?

THE SREDNI STOG CULTURE: HORSES AND RITUALS FROM THE EAST

The Sredni Stog culture is the best-defined Late Eneolithic archaeological culture in steppe Ukraine. Sredni Stog, or "middle stack," was the name of a small haystack-shaped island in the Dnieper at the southern end of the Dnieper Rapids, the central one of three. All were inundated by a dam, but before that happened, archaeologists found and excavated a site there in 1927. It contained a stratified sequence of settlements with Early

Eneolithic (DDII) pottery in level I and Late Eneolithic pottery in level II.[32] Sredni Stog II became the type site for this Late Eneolithic kind of pottery. Sredni Stog–style pottery was found stratified above older DDII settlements at several other sites, including Strilcha Skelya and Aleksandriya. Dimitri Telegin, who had earlier defined the Dnieper-Donets culture, in 1973 first pulled together and mapped all the sites with Sredni Stog material culture, about 150 in all (figure 11.6). He found Sredni Stog sites across the Ukrainian steppes from the Ingul valley, west of the Dnieper, on the west to the lower Don on the east.

The Sredni Stog culture became the archaeological foundation for the Indo-European steppe pastoralists of Marija Gimbutas. The horse bones from the Sredni Stog settlement of Dereivka, excavated by Telegin, played a central role in the ensuing debates between pro-Kurgan-culture and anti-Kurgan-culture archaeologists. I described in the last chapter how Gimbutas's interpretation of the horses of Dereivka was challenged by Levine. Simultaneously Yuri Rassamakin challenged Telegin's concept of the Sredni Stog culture.[33]

Rassamakin separated Telegin's Sredni Stog culture into at least three separate cultures, reordered and redated some of the resulting pieces, and refocused the central cause of social and political change away from the development of horse riding and agro-pastoralism in the steppes (Telegin's themes) to the integration of steppe societies into the cultural sphere of Old Europe, which was Rassamakin's new mutualist theme. But Rassamakin assigned well-dated sites like Dereivka and Khvalynsk to periods inconsistent with their radiocarbon dates.[34] Telegin's groupings seem to me to be better documented and explained, so I retain the Sredni Stog culture as a framework for ordering Eneolithic sites in Ukraine, while disagreeing with Telegin in some details.

This was the critical era when innovative early Proto-Indo-European dialects began to spread across the steppes. The principal causes of change in the steppes included both the internal maturation of new economic systems and new social networks (Telegin's theme) and the inauguration of new interactions with Old Europe (Rassamakin's theme).

The Origins and Development of the Sredni Stog Culture

We should not imagine that Sredni Stog, or any other archaeological culture, appeared or disappeared everywhere at the same time. Telegin defined four broad phases (Ia, Ib, IIa, IIb) in its evolution, but a phase might last longer in some regions than others. In his scheme, the settlements at

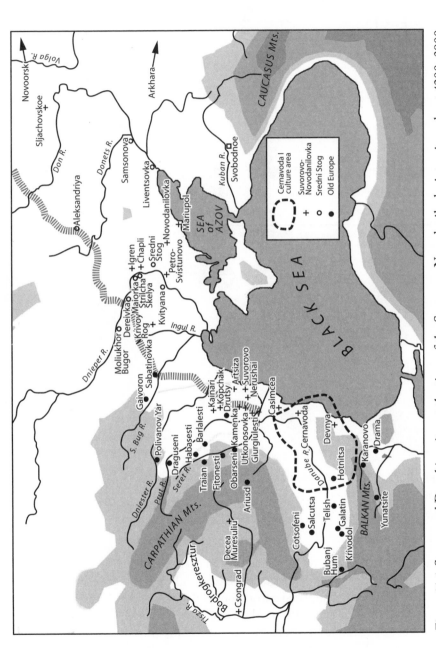

Figure 11.6 Steppe and Danubian sites at the time of the Suvorovo-Novodanilovka intrusion, about 4200–3900 BCE.

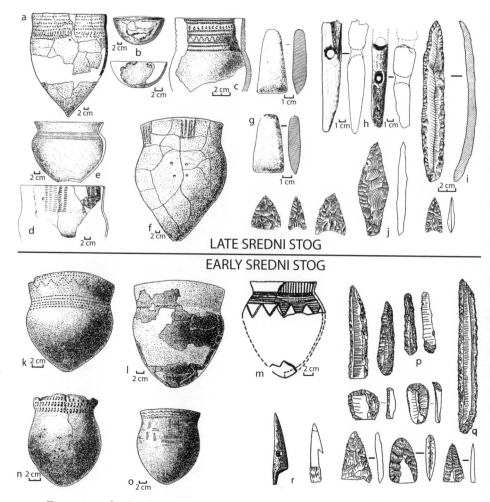

Figure 11.7 Sredni Stog pottery and tools, early and late. Perforated bone or antler artifacts like (h) were identified as cheekpieces for horse bits, but this identification is speculative. After Telegin 2002, figure 3.1.

Sredni Stog and Strilcha Skelya on the Dnieper represented an early phase (Ib), which Rassamakin called the Skelya culture. The pottery of this phase lacked cord-impressed decoration. The settlements at Dereivka (IIa) and Moliukhor Bugor (IIb) on the Dnieper represented the late phases, with braided cord impressions on the pottery (figure 11.7). Early Sredni Stog (phase I) was contemporary with the violent era of Tripolye B1 and the crisis in the Danube valley. Tripolye B1 painted pottery was found at

TABLE 11.2

Radiocarbon Dates for Late Eneolithic Cultures from the Lower Danube to the North Caucasus

Lab Number	BP Date	Sample	Calibrated Date
1. Sredni Stog culture			
Dereivka, Dnieper Valley			
Ki 2195	6240±100	settlement, shell	5270–5058 BCE
UCLA 1466a	5515±90	settlement, bone	4470–4240 BCE
Ki 2193	5400±100	settlement, shell	4360–4040 BCE
OxA 5030	5380±90	cemetery, grave 2	4350–4040 BCE
KI 6966	5370±70	settlement, bone	4340–4040 BCE
Ki 6960	5330±60	settlement, bone	4250–4040 BCE
KI 6964	5260±75	settlement, bone	4230–3990 BCE
Ki 2197	5230±95	settlement, bone	4230–3970 BCE
Ki 6965	5210±70	settlement, bone	4230–3960 BCE
UCLA 1671a	4900±100	settlement, bone	3900–3530 BCE
Ki 5488	4330±120	cult horse skull??	3300–2700 BCE
Ki 6962	2490±95	cult horse skull	790–520 BCE
OxA 7185	2295±60	cult horse tooth with bit wear	410–200 BCE
OxA 6577	1995±60	bone near cult horse	90 BCE–70CE
Aleksandriya, Donets Valley			
Ki-104	5470±300	?	4750–3900 BCE
2. North Caucasian Eneolithic			
Svobodnoe settlement			
Le-4531	5400±250	?	4500–3950 BCE
Le-4532	5475±100	?	4460–4160 BCE
3. Varna Culture, Bulgaria, lower Danube			
Durankulak tell settlement			
Bln-2122	5700±50	settlement, level 5	4600–4450 BCE
Bln-2111	5495±60	settlement, house 7	4450–4250 BCE
Bln-2121	5475±50	settlement, level 4	4360–4240 BCE
Pavelyanovo 1 tell settlement			
Bln-1141	5591±100	settlement	4540–4330 BCE

Table 11.2 (*continued*)

Lab Number	BP Date	Sample	Calibrated Date
4. Gumelnitsa culture, Romania, lower Danube			
Vulcanesti II, Bolgrad group			
MO-417	5110±150	settlement	4050–3700 BCE
Le-640	5300±60	settlement	4230–4000 BCE
Gumelnitsa, tell settlement			
GrN-3025	5715±70	settlement, charcoal	4680–4450 BCE
Bln-605	5675±80	settlement, charcoal	4620–4360 BCE
Bln-604	5580±100	settlement, charcoal	4540–4330 BCE
Bln-343	5485±120	settlement, charcoal	4460–4110 BCE
GrN-3028	5400±90	settlement, charred grain	4340–4050 BCE
5. Suvorovo Group, lower Danube			
Giurgiuleşti, cemetery, lower Prut/Danube			
Ki-7037	5398±69*	?	4340–4050 BCE

*This date was printed in Telegin et al. 2001 as 4398±69 BP, but I was told that this was a misprint and that the actual reported date was 5398+69 BP.

Strilcha Skelya. The stylistic changes that identified late Sredni Stog (phase II) probably began while the crisis in the Danube valley was going on, but then most of the late Sredni Stog period occurred after the collapse of Old Europe. Imported Tripolye B2 bowls were found in graves in the phase IIa cemeteries at Dereivka and Igren, and a Tripolye C1 vessel was found at the phase IIb Moliukhor Bugor settlement. The Dereivka settlement (phase IIa) is dated between 4200 and 3700 BCE by ten radiocarbon dates (table 11.2). The latest Sredni Stog period (IIb) is dated as late as 3600–3300 BCE by four radiocarbon dates at Petrovskaya Balka on the Dnieper. Early Sredni Stog probably began around 4400 BCE; late Sredni Stog probably lasted until 3400 BCE in some places on the Dnieper.

The origin of the Sredni Stog culture is poorly understood, but people from the east, perhaps from the Volga steppes, apparently played a role. Round-bottomed Sredni Stog shell-tempered pots were quite different from DDII pots of the Early Eneolithic, which were sand-tempered and flat-based (see figure 9.5). Almost all early Sredni Stog vessels had round or pointed bases and flaring, everted rims. Flat-based pots appeared only in the late period. Simple open bowls, probably food bowls, were the other common shape, usually undecorated. Sredni Stog pots were decorated just on the upper third of the vessel with rows of comb-stamped impressions, incised triangles, and cord impressions. Rows of U-shaped "caterpillar" impressions made with a U-shaped, cord-wrapped tool were typical (figure 11.7d). One pot shape, with a rounded body and a short vertical neck decorated with vertically combed lines (figure 11.7m) was copied directly from a common Tripolye B1 type. The round-based pots and shell temper seem to reflect influence from the east, from the Azov-Caspian or Volga regions, where there was a long tradition of shell-tempered, round-bottomed, everted-rim, impressed pottery beginning in the Neolithic and continuing through Eneolithic Khvalynsk.

Sredni Stog funeral rituals also were new. The new Sredni Stog burial posture (on the back with the knees raised) and standard orientation (head to the east-northeast) copied that of the Khvalynsk culture on the Volga (figure 11.8). The communal collective grave pits of DDII were abandoned. Individual single graves took their place. Cemeteries also became much smaller. The DDII cemetery near Dereivka had contained 173 individuals, most of them in large communal grave pits. The Sredni Stog cemetery near Dereivka contained only 12 graves, all single burials. Sredni Stog communities probably were smaller and more mobile. Graves had no surface marker, as at Dereivka, or exhibited a new surface treatment: some were surrounded by a small circle of stones and covered by a low stone or earth mound—a very modest kurgan—as at Kvityana or Maiorka. These probably were the earliest kurgans in the steppes. Stone circles and mounds were features that isolated and emphasized individuals. The shift from a communal funeral ritual to an individual ritual probably was a symptom of broader changes toward more openly self-aggrandizing social values, which were also reflected in a series of rich graves of the Suvorovo-Novodanilovka type discussed separately below.

Sredni Stog skull types also exhibited new traits. The DDII population had been a single homogeneous type, with a very broad, thick-boned face of the Proto-Europoid configuration. Sredni Stog populations included people with a more gracile bone structure and medium-width faces that showed the strongest statistical similarity to the Khvalynsk population. Immigrants from the Volga seem to have arrived in the Dnieper-Azov steppes at the beginning of the shift from DDII to Sredni Stog, instigating

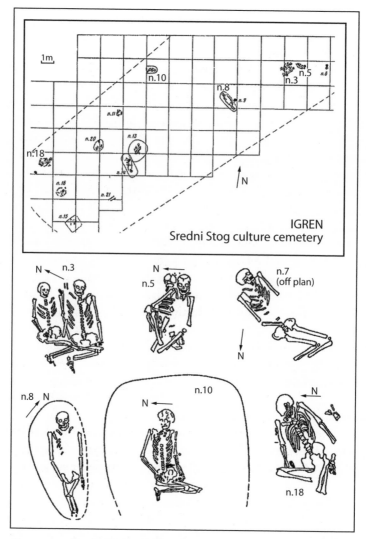

Figure 11.8 Sredni Stog graves, Igren cemetery, Dnieper Rapids. Graves were quite scattered. After Telegin et al. 2001.

changes in both funeral customs and pottery making. Perhaps they arrived on horseback.[35]

The places where people lived and put their cemeteries did not change markedly when Sredni Stog began. Sredni Stog settlements were stratified above DDII settlements at several sites near the Dnieper Rapids and on the Donets. Sredni Stog graves were located in or near DDII cemeteries at Mariupol, Igren, and Dereivka. Stone tools also showed continuity; lamellar flint blades, triangular flint points, and large almond-shaped flint points were made in both periods. Long unifacial flint blades were occasionally found in hoards in DDII sites but were found in much larger hoards in Sredni Stog sites, where some single hoards (Goncharovka) contained more than a hundred flint blades up to 20 cm long. These blades were typical grave gifts in Sredni Stog graves. Similar long flint blades became popular trade items across eastern Europe, appearing also in Funnel Beaker (TRB) sites in Poland and in Bodrogkeresztur sites in Hungary.

The Sredni Stog Economy: Horses and Agro-Pastoralism

Sredni Stog settlements had, on average, more than twice as many horse bones as DDII settlements in the Dnieper valley, where most of the studied sites are located. This increase in the use of horses for food could have been connected with the colder climate of the period 4200–3800 BCE, since domesticated horses are easier to maintain than cattle and sheep in snowy conditions (chapter 10). The maintenance advantage would, of course, have been gained only with domesticated horses. Horses were by far the most important source of meat at the Sredni Stog settlement of Dereivka. The 2,408 horse bones counted by Bibikova represented at least fifty-one animals (MNI)—more than half the mammals butchered at the site—and 9,000 kg of meat.[36]

Domesticated cattle, sheep, and pigs accounted for between 12% and 84% of the bones (NISP) from the settlements of Sredni Stog II, Dereivka, Aleksandriya, and Moliukhor Bugor (table 11.1). If horses are counted as domesticated animals, the percentage of domesticated animals at these settlements rises to 30–93%. The percentage of horse bones ranged from 7–63% of all bones found (average 54% NISP but with much variation). The highest percentage (63% of the mammal bones NISP, 52% of the individual mammals MNI) was at Dereivka, which was also the site with the largest sample of animal bones.[37] Sheep or goats were by far the most common animals (61% of mammals) in the southernmost site, Sredni Stog, in the driest steppe environment; and hunted game was most important (70% of mammals) at Moliukhor Bugor, the northenmost site, in the most forested environment. In the north, where forest resources were richer, deer hunting remained important, and in the steppe river valleys, where gallery forests were confined to the valley bottoms, sheep herding necessarily supplied a larger proportion of the diet.

TABLE 11.1
Mammal Bones from Sredni Stog Culture

	% horse	% cattle	% caprine	% pig	% dog	% wild
	(% of all bones, NISP/ % of individuals, MNI)*					
Sredni Stog II	7/12	21/12	61/47	2/6	3/11	7/22
Dereivka	63/52	16/8	2/7	3/4	1/2	17/45
Aleksandriya	29/24	37/20	7/12	—	—	27/44
Moliukhor BugorII	18/9	10/9	—	2/6	—	70/76

*NISP=number of identified species; MNI=minimum number of individuals.

Dereivka is the Sredni Stog settlement with the largest archaeological exposure, about 2000 m². It was located west of the Dnieper in the northern steppes. A scattered cemetery of twelve Sredni Stog graves was found half a kilometer upstream from the settlement.[38] Three shallow ovoid house pits, measuring about 12 m by 5 m, surrounded an open area used for ceramic manufacture, flint working, and other tasks (figure 11.9). A thick midden of river shellfish shells (*Unio* and *Paludinae*) enclosed one side. Only a part of the settlement was excavated, so we do not know how large it was. The mammal bones would have provided 1 kilo of meat per house, for the three houses, every day for more than eight years, indicating that Dereivka was occupied many times or for many years. On the other hand, the ephemeral nature of the Dereivka architectural remains and the small size of the nearby cemetery suggest that it was not a permanent settlement. Probably it was a favored living site that was revisited over many years by people who had large herds of horses (62% NISP) and cattle (16% NISP), hunted red deer (10% NISP), trapped or shot ducks (mallard and pintail), fished for wels catfish (*Silurus glanis*) and perch (*Lucioperca lucioperca*), and cultivated a little grain.

The ceramics from the Dereivka settlement have not been examined systematically for seed imprints, but Dereivka had flint blades with sickle gloss; three flat, ovoid grinding stones; and six polished schist mortars. Cultivated wheat, barley, and millet (*T. dicoccum, T. monococcum, H. vulgare, P. miliaceum*) have been identified in ceramic imprints at the phase IIb settlement of Moliukhor Bugor. Probably some grain cultivation occurred at Dereivka also, perhaps the first grain cultivation practiced east of the Dnieper.

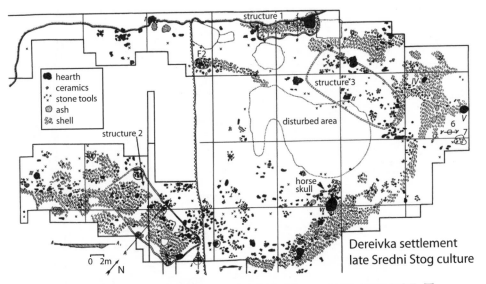

Figure 11.9 Dereivka settlement, Sredni Stog culture, 4200–3700 BCE. The location of the intrusive horse skull with bit wear is noted. The top edge is an eroded riverbank. After Telegin 1986.

Were the people of the Sredni Stog culture horse riders? Without bit wear or some other pathology associated with riding we cannot be certain. Objects from Dereivka tentatively identified as antler cheekpieces for bits (figure 11.7h) could have had other functions.[39] One way to approach this question is to ask if the steppe societies of the Late Eneolithic *behaved* like horseback riders. It looks to me like they did. Increased mobility (implied by smaller cemeteries), more long-distance trade, increased prestige and power for prominent individuals, status weapons appearing in graves, and heightened warfare against settled agricultural communities are all things we would expect to occur after horseback riding started, and we see them most clearly in cemeteries of the Suvorovo-Novodanilovka type.

MIGRATIONS INTO THE DANUBE VALLEY: THE SUVOROVO-NOVODANILOVKA COMPLEX

About 4200 BCE herders who probably came from the Dnieper valley appeared on the northern edge of the Danube delta. The lake country north of the delta was then occupied by Old European farmers of the Bolgrad culture. They left quickly after the steppe people showed up. The immigrants

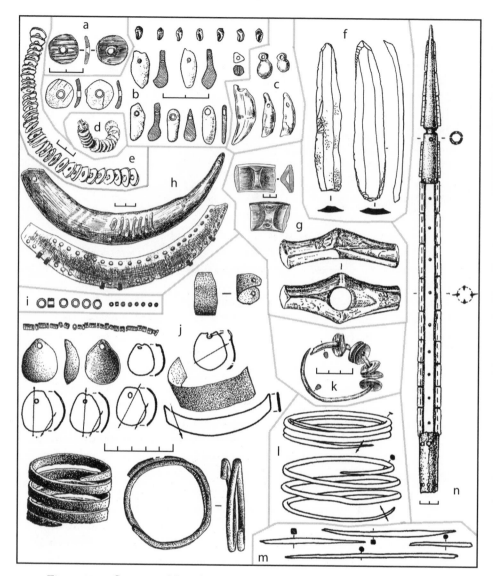

Figure 11.10 Suvorovo-Novodanilovka ornaments and weapons, about 4200–3900 BCE. (a, c) Vinogradni shell and canine tooth beads; (b) Suvorovo shell and deer tooth beads; (d) Decea Muresului shell beads; (e) Krivoy Rog shell beads; (f) Chapli lamellar flint blades; (g) Petro-Svistunovo, bone button and cast copper axe; (h) Petro-Svistunovo boar's tusk (*top*), Giurgiulesti copper-sheathed boar's tusk (*bottom*); (j) Chapli copper ornaments, including copper imitations of *Cardium* shells; (i) Utkonosovka bone beads; (k) Kainari copper "torque" with shell beads; (l) Petro-Svistunovo copper bracelet; (m) Suvorovo

built kurgan graves and carried maces with stone heads shaped like horse heads, objects that quickly appeared in a number of Old European towns. They acquired, either by trade or as loot, copper from the tell towns of the lower Danube valley, much of which they directed back into the steppes around the lower Dnieper. Their move into the lower Danube valley probably was the historical event that separated the Pre-Anatolian dialects, spoken by the migrants, from the archaic Proto-Indo-European language community back in the steppes.

The archaeology that documents this event emerged into the literature in small bits and pieces over the last fifty years, and it is still is not widely known. The steppe culture involved in the migration has been called variously the Skelya culture, the Suvorovo culture, the Utkonsonovka group, and the Novodanilovka culture. I will call it the Suvorovo-Novodanilovka complex (see figure 11.6). One cluster of graves, created by the migrants, is concentrated near the Danube delta. This was the Suvorovo group. Their relatives back home in the North Pontic steppes were the Novodanilovka group. Only graves are known for either group. About thirty-five to forty cemeteries are assigned to the complex, most containing fewer than ten graves and many, like Novodanilovka itself, represented by just a single rich burial. They first appeared during early Sredni Stog, around 4300–4200 BCE, and probably ceased before 3900 BCE.

In his earliest discussions Telegin interpreted the Novodanilovka graves (his term) as a wealthy elite element within the Sredni Stog culture. Later he changed his mind and made them a separate culture. I agree with his original position: the Suvorovo-Novodanilovka complex represents the chiefly elite within the Sredni Stog culture. Novodanilovka graves are distributed across the same territory as graves and settlements designated Sredni Stog, and many aspects of grave ritual and lithics are identical. The Suvorovo-Novodanilovka elite was involved in raiding and trading with the lower Danube valley during the Tripolye B1 period, just before the collapse of Old Europe.[40]

The people buried in these graves wore long belts and necklaces of shell disc beads, copper beads, and horse or deer tooth beads; copper rings; copper shell-shaped pendants; and copper spiral bracelets (figure 11.10). They bent thick pieces of copper wire into neckrings ("torques") decorated with shell beads, used copper awls, occasionally carried solid cast copper shaft-hole axes

---

Figure 11.10 (continued) and Aleksandriya copper awls; (n) Giurgiuleşti composite spear-head, bone with flint microblade edges and tubular copper fittings. After Ryndina 1998, figure 76; and Telegin et al. 2001.

(cast in a two-part mold), and put copper and gold fittings around the dark wood of their spears and javelins. In 1998 N. Ryndina counted 362 objects of copper and 1 of gold from thirty Suvorovo-Novodanilovka graves. They also carried polished stone mace heads made in several shapes, including horse heads (see figure 11.5). They used large triangular flint points, probably for spears/javelins; small round-butted flint axes with the cutting edge ground sharp; and long lamellar flint blades, often made of gray flint quarried from outcrops on the Donets River.

Most Suvorovo-Novodanilovka graves contained no pottery, and so they are difficult to link to a ceramic type. Imported ceramics were found in several graves: a Tripolye B1 pot in the Kainari kurgan, between the Prut and Dniester; a late Gumelniţa vessel in the Kopchak kurgan, not far from Kainari; another late Gumelniţa vessel in grave 2 at Giurgiuleşti, on the lower Prut; and a long-traveled pot of North Caucasian Svobodnoe type in the Novodanilovka grave in the Dnieper-Azov steppes. These imported pots were all the same age, dated roughly 4400–4000 BCE, and so are useful chronologically, but they throw no light on the cultural affiliation of the individuals in the graves. Only a few potsherds actually seem to have been made by the people who built the graves. One of the principal graves (gr. 1) at Suvorovo had two small sherds of a pot made of gray, shell-tempered clay, decorated with a small-toothed stamp and incised diagonal lines (figure 11.11). An analogous pot was found in Utkonosovka, kurgan 3, grave 2, near Suvorovo. These sherds resembled Cucuteni C ceramics: round body, round base, everted rim, shell-tempered, with diagonal incised and comb-stamped surface decoration.[41]

The Suvorovo graves around the Danube delta always were marked by the erection of a mound or kurgan, probably to increase their visibility on a disputed frontier, but possibly also as a visual response to the tells of the lower Danube valley (figure 11.11). Suvorovo kurgans were among the first erected in the steppes. Back in the Dnieper-Azov steppes, most Novodanilovka graves also had a surface marker of some kind, but earthen kurgans were less common than small stone cairns piled above the grave (Chapli, Yama). Kurgans in the Danube steppes rarely were more than 10 m in diameter, and often were surrounded by a ring of small stones or a cromlech (retaining wall) of large stones. The grave pit was usually rectangular but sometimes oval. The Sredni Stog burial posture (on the back with knees raised) appeared in most (Csongrad, Chapli, Novodanilovka, Giurgiuleşti, Suvorovo grave 7) but not all graves. In some the body was laid out extended (Suvorovo grave 1) or contracted on the side (Utkonosovka). Animal sacrifices occurred in some graves (cattle at Giurgiuleşti,

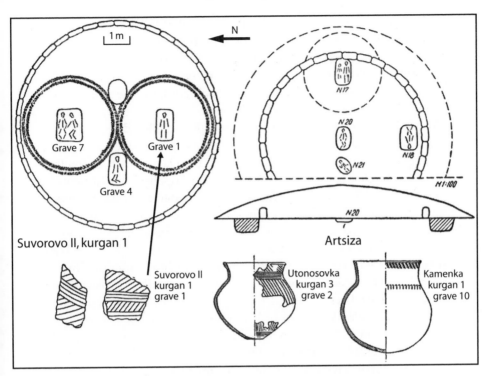

Figure 11.11 Suvorovo-type kurgan graves and pots. Most Suvorovo graves contained no pottery or contained pots made by other cultures, and so these few apparently self-made pots are important: *left*, Suvorovo cemetery II kurgan 1; *right*, Artsiza kurgan; *bottom*, sherds and pots from graves. After Alekseeva 1976, figure 1.

cattle and sheep at Chapli, and cattle at Krivoy Rog). The people buried in Novodanilovka graves in the Pontic steppes were wide-faced Proto-Europoid types, like the dominant element in Sredni Stog graves, whereas at least some of those buried in Suvorovo graves such as Giurgiuleşti had narrow faces and gracile skulls, suggesting intermarriage with local Old European people.[42]

The copper from Suvorovo-Novodanilovka graves helps to date them. Trace elements in the copper from Giurgiuleşti and Suvorovo in the lower Danube, and from Chapli and Novodanilovka in the Pontic steppes, are typical of the mines in the Bulgarian Balkans (Ai Bunar and/or Medni Rud) that abruptly ceased production when Old Europe collapsed. The eastern European copper trade shifted to chemically distinctive Hungarian and

Transylvanian ores during Tripolye B2, after 4000 BCE.[43] So Suvorovo-Novodanilovka is dated before 4000 BCE by its copper. On the other hand, Suvorovo kurgans replaced the settlements of the Bolgrad group north of the Danube delta, which were still occupied during early Tripolye B1, or after about 4400–4300 BCE. These two bookends (after the abandonment of Bolgrad, before the wider Old European collapse) restrict Suvorovo-Novodanilovka to a period between about 4300 and 4000 BCE.

Polished stone mace-heads shaped like horse heads were found in the main grave at Suvorovo and at Casimcea in the Danube delta region (figure 11.5). Similar mace-heads occurred at two Tripolye B1 settlements, at two late Karanovo VI settlements, and up the Danube valley at the settlement of Sălcuţa IV—all of them in Old European towns contemporary with the Suvorovo intrusion. Similar horse-head mace-heads were found in the Volga-Ural steppes and in the Kalmyk steppes north of the Terek River at Terekli-Mekteb.[44] "Eared" stone mace heads appeared first in several cemeteries of the Khvalynsk culture (Khvalynsk, Krivoluchie) and then somewhat later at several eastern steppe sites contemporary with Suvorovo-Novodanilovka (Novorsk, Arkhara, and Sliachovsko) and in two Tripolye B1 towns. Cruciform mace heads appeared first in the grave of a DDII chief at Nikol'skoe on the Dnieper (see figure 9.6), and then reappeared centuries later with the Suvorovo migration into Transylvania at Decea Mureşului and Ocna Sibiului; one example also appeared at a Tripolye settlement on the Prut (Bârlăleşti).

Polished stone maces were typical steppe prestige objects going back to Khvalynsk, Varfolomievka, and DDII, beginning ca. 5000–4800 BCE. They were not typical prestige objects for earlier Tripolye or Gumelniţa societies.[45] Maces shaped into horse-heads probably were made by people for whom the horse was a powerful symbol. Horse bones averaged only 3–6% of mammal bones in Tripolye B1 settlements and even less in Gumelniţa, and so horses were not important in Old European diets. The horse-head maces signaled a new iconic status for the horse just when the Suvorovo people appeared. If horses were *not* being ridden into the Danube valley, it is difficult to explain their sudden symbolic importance in Old European settlements.[46]

The Causes and Targets of the Migrations

Winters began to get colder in the interior steppes after about 4200 BCE. The marshlands of the Danube delta are the largest in Europe west of the Volga. Marshes were the preferred winter refuge for nomadic pastoralists in the Black Sea steppes during recorded history, because they offered good winter forage and cover for cattle. The Danube delta was richer in this resource than any other place on the Black Sea. The first Suvorovo herders who appeared on the northern edge of the Danube delta about 4200–4100 BCE might have brought some of their cattle south from the Dnieper steppes during a period of particularly cold winters.

Another attraction was the abundant copper that came from Old European towns. The archaeologist Susan Vehik argued that increased levels of conflict associated with climatic deterioration in the southwestern U.S. Plains around 1250 CE created an increased demand for gift-wealth (to attract and retain allies in tribal warfare) and therefore stimulated long-distance trade for prestige goods.[47] But the Suvorovo immigrants did not establish gift exchanges like those I have hypothesized for their relations with Cucuteni-Tripolye people. Instead, they seem to have chased the locals away.

The thirty settlements of the Bolgrad culture north of the Danube delta were abandoned and burned soon after the Suvorovo immigrants arrived. These small agricultural villages were composed of eight to ten semi-subterranean houses with fired clay hearths, benches, and large storage pots set in pits in the floor. Graphite-painted fine pottery and numerous female figurines show a mixture of Gumelniţa (Aldeni II type) and Tripolye A traits.[48] They were occupied mainly during Tripolye A, then were abandoned and burned during early Tripolye B1, probably around 4200–4100 BCE. Most of the abandonments apparently were planned, since almost everything was picked up. But at Vulcaneşti II, radiocarbon dated 4200–4100 BCE (5300 ± 60 BP), abandonment was quick, with many whole pots left to burn. This might date the arrival of the Suvorovo migrants.[49]

A second and seemingly smaller migration stream branched off from the first and ran westward to the Transylvanian plateau and then down the copper-rich Mureş River valley into eastern Hungary. These migrants left cemeteries at Decea Mureşului in the Mureş valley and at Csongrad in the plains of eastern Hungary. At Decea Mureşului, near important copper deposits, there were fifteen to twenty graves, posed on the back with the knees probably originally raised but fallen to the left or right, colored with red ochre, with Unio shell beads, long flint blades (up to 22 cm long), copper awls, a copper rod "torque," and two four-knobbed mace heads made of black polished stone (see figure 11.10). The migrants arrived at the end of the Tiszapolgar and the beginning of the Bodrogkeresztur periods, about 4000–3900 BCE, but seemed not to disrupt the local cultural traditions. Hoards of large golden and copper ornaments of

Old European types were hidden at Hencida and Mojgrad in eastern Hungary, probably indicating unsettled conditions, but otherwise there was a lot of cultural continuity between Tiszapolgar and Bodrogkeresztur.[50] This was no massive folk migration but a series of long-distance movements by small groups, exactly the kind of movement expected among horseback riders.

The Suvorovo Graves

The Suvorovo kurgan (Suvorovo II k.1) was 13 m in diameter and covered four Eneolithic graves (see figure 11.11).[51] Stones a meter tall formed a cromlech around the base of the mound. Within the cromlech two smaller stone circles were built on a north-south axis, each surrounding a central grave (gr. 7 and 1). Grave 7 was the double grave of an adult male and female buried supine with raised legs, heads to the east. The floor of the grave was covered with red ochre, white chalk, and black fragments of charcoal. A magnificent polished stone mace shaped like the head of a horse lay on the pelvis of the male (see figure 11.5). Belts of shell disk beads draped the female's hips. The grave also contained two copper awls made of Balkan copper, three lamellar flint blades, and a flint end scraper. Grave 1, in the other stone circle, contained an adult male in an extended position and two sherds of a shell-tempered pot.

The Suvorovo cemetery at Giurgiuleşti, near the mouth of the Prut, contained five graves grouped around a hearth full of burned animal bones.[52] Above grave 4, that of the adult male, was another deposit of cattle skulls and bones. Graves 4 and 5 were those of an adult male and female; graves 1, 2, and 3, contained three children, apparently a family group. The graves were covered by a mound, but the excavators were uncertain if the mound was built for these graves or was made later. The pose in four of the five graves was on the back with raised knees (grave 2 contained disarticulated bones), and the grave floors were painted with red ochre. Two children (gr. 1 and 3) and the adult woman (gr. 5) together wore nineteen copper spiral bracelets and five boars-tusk pendants, one of which was covered in sheet copper (see figure 11.10:h). Grave 2 contained a late Gumelniţa pot. The children and adult female also had great numbers (exact count not published) of copper beads, shell disc beads, beads of red deer teeth, two beads made of Aegean coral, flint blades, and a flint core. Six of eight metal objects analyzed by N. Ryndina were made from typical Varna-Gumelniţa Balkan ores. One bracelet and one ring were made of an intentional arsenic-copper alloy (respectively, 1.9% and 1.2% arsenic) that had never occurred in Varna or Gumelniṭa metals. The adult male buried in grave 4 had two gold rings and two composite projectile points, each more than 40 cm long, made with microlithic flint blades slotted along the edges of a bone point decorated with copper and gold tubular fittings (see figure 11.10:n). They probably were for two javelins, perhaps the preferred weapons of Suvorovo riders.

Kurgans also appeared south of the Danube River in the Dobruja at Casimcea, where an adult male was buried in an ochre-stained grave on his back with raised knees, accompanied by a polished stone horse-head mace (see figure 11.5), five triangular flint axes, fifteen triangular flint points, and three lamellar flint blades. Another Suvorovo grave was placed in an older Varna-culture cemetery at Devnya, near Varna. This single grave contained an adult male in an ochre-stained grave on his back with raised knees, accompanied by thirty-two golden rings, a copper axe, a copper decorative pin, a copper square-sectioned chisel 27 cm long, a bent copper wire 1.64 m long, thirty-six flint lamellar blades, and five triangular flint points.

A separate (about 80–90 km distant) but contemporary cluster of kurgans was located between the Prut and Dniester valleys near the Tripolye frontier (Kainari, Artsiza, and Kopchak). At Kainari, only a dozen kilometers from the Tripolye B1 settlement of Novi Ruşeşti, a kurgan was erected over a grave with a copper "torque" strung with *Unio* shell disc beads (see figure 11.10:k); long lamellar flint blades, red ochre, and a Tripolye B1 pot.

The Novodanilovka Group

Back in the steppes north of the Black Sea the elite were buried with copper spiral bracelets, rings, and bangles; copper beads of several types; copper shell-shaped pendants; and copper awls, all containing Balkan trace elements and made technologically just like the objects at Giurgiuleşti and Suvorovo.[53] Copper shell-shaped pendants, a very distinctive steppe ornament type, occurred in both Novodanilovka (Chapli) and Suvorovo (Giurgiuleşti) graves (see figure 11.10:j): The grave floors were strewn with red ochre or with a chunk of red ochre. The body was positioned on the back with raised knees and the head oriented toward the east or northeast. Surface markers were a small kurgan or stone cairn, often surrounded by a stone circle or cromlech. The following were among the richest:

Novodanilovka, a single stone-lined cist grave containing two adults at Novodanilovka in the dry hills between the Dnieper and the Sea of Azov with two copper spiral bracelets, more than a hundred

Unio shell beads, fifteen lamellar flint blades, and a pot imported from the North Caucasian Svobodnoe culture;

Krivoy Rog, in the Ingulets valley, west of the Dnieper, a kurgan covering two graves (1 and 2) with flint axes, flint lamellar blades, a copper spiral bracelet, two copper spiral rings, hundreds of copper beads, a gold tubular shaft fitting, *Unio* disc beads, and other objects;

Chapli (see figure 11.10) at the north end of the Dnieper Rapids, with five rich graves. The richest of these (1a and 3a) were children's graves with two copper spiral bracelets, thirteen shell-shaped copper pendants, more than three hundred copper beads, a copper foil headband, more than two hundred *Unio* shell beads, one lamellar flint blade, and one boars-tusk pendant like those at Giurgiuleşti; and

Petro-Svistunovo (see figure 11.10), a cemetery of twelve cromlechs at the south end of the Dnieper Rapids largely destroyed by erosion, with Grave 1 alone yielding two copper spiral bracelets, more than a hundred copper beads, three flint axes, and a flint lamellar blade, and the other graves yielding three more spiral bracelets, a massive cast copper axe comparable to some from Varna, and boars-tusk pendants like those at Chapli and Giurgiuleşti.

About eighty Sredni Stog cemeteries looked very similar in ritual and occurred in the same region but did not contain the prestige goods that appeared in the Novodanilovka graves, which probably were the graves of clan chiefs. The chiefs redistributed some of their imported Balkan wealth. For example, in the small Sredni Stog cemetery at Dereivka, grave 1 contained three small copper beads and grave 4 contained an imported Tripolye B1 bowl. The other graves contained no grave gifts at all.

WARFARE, CLIMATE CHANGE, AND LANGUAGE SHIFT IN THE LOWER DANUBE VALLEY

The colder climate of 4200–3800 BCE probably weakened the agricultural economies of Old Europe at the same time that steppe herders pushed into the marshes and plains around the mouth of the Danube. Climate change probably played a significant role in the ensuing crisis, because virtually all the cultures that occupied tell settlements in southeastern Europe abandoned them about 4000 BCE—in the lower Danube valley, the Balkans, on the Aegean coast (the end of Sitagroi III), and even in Greece (the end of Late Neolithic II in Thessaly).[54]

But even if climatic cooling and crop failures must have been significant causes of these widespread tell abandonments, they were not the only cause. The massacres at Yunatsite and Hotnitsa testify to conflict. Polished stone mace heads were status weapons that glorified the cracking of heads. Many Suvorovo-Novodanilovka graves contained sets of lanceolate flint projectile points, flint axes, and, in the Giurgiuleşti chief's grave, two fearsome 40-cm javelin heads decorated with copper and gold. Persistent raiding and warfare would have made fixed settlements a strategic liability. Raids by Slavic tribes caused the abandonment of all the Greek-Byzantine cities in this same region over the course of less than a hundred years in the sixth century CE. Crop failures exacerbated by warfare would have encouraged a shift to a more mobile economy.[55] As that shift happened, the pastoral tribes of the steppes were transformed from scruffy immigrants or despised raiders to chiefs and patrons who were rich in the animal resources that the new economy required, and who knew how to manage larger herds in new ways, most important among these that herders were mounted on horseback.

The Suvorovo chiefs displayed many of the behaviors that fostered language shift among the Acholi in East Africa: they imported a new funeral cult with an associated new mortuary ideology; they sponsored funeral feasts, always events to build alliances and recruit allies; they displayed icons of power (stone maces); they seem to have glorified war (they were buried with status weapons); and it was probably their economic example that prompted the shift to pastoral economies in the Danube valley. Proto-Indo-European religion and social structure were both based on oath-bound promises that obligated patrons (or the gods) to provide protection and gifts of cattle and horses to their clients (or humans). The oath (*$h_1óitos$*) that secured these obligations could, in principle, be extended to clients from the Old European tells.

An archaic Proto-Indo-European language, probably ancestral to Anatolian, spread into southeastern Europe during this era of warfare, dislocation, migration, and economic change, around 4200–3900 BCE. In a similar situation, in a context of chronic warfare on the Pathan/Baluch border in western Pakistan, Frederik Barth described a steady stream of agricultural Pathans who had lost their land and then crossed over and joined the pastoral Baluch. Landless Pathan could not regain their status in other Pathan villages, where land was necessary for respectable status. Tells and their fixed field systems might have played a similar limiting role in Old European status hierarchies. Becoming the client of a pastoral patron who offered protection and rewards in exchange for service was an alternative that held the promise of vertical social mobility for the children. The speakers of Proto-Indo-European talked about gifts and honors awarded for great deeds and loot/booty acquired unexpectedly, suggesting that achievement-based honor and wealth could be acquired.[56] Under conditions of chronic warfare, displaced tell dwellers may well have adopted an Indo-European patron and language as they adopted a pastoral economy.

AFTER THE COLLAPSE

In the centuries after 4000 BCE, sites of the Cernavoda I type spread through the lower Danube valley (figure 11.12). Cernavoda I was a settlement on a promontory overlooking the lower Danube. Cernavoda I material culture probably represented the assimilation of migrants from the steppes with local people who had abandoned their tells. Cernavoda I ceramics appeared at Pevec and Hotnitsa-Vodopada in north-central Bulgaria, and at Renie II in the lower Prut region. These settlements were small, with five to ten pit-houses, and were fortified. Cernavoda I pottery also occurred in settlements of other cultural types, as at Telish IV in northwestern Bulgaria. Cernavoda I pottery included simplified versions of late Gumelniţa shapes, usually dark-surfaced and undecorated but made in shell-tempered fabrics. The U-shaped "caterpillar" cord impressions (figure 11.12i), dark surfaces, and shell tempering were typical of Sredni Stog or Cucuteni C.[57]

Prominent among these new dark-surfaced, shell-tempered pottery assemblages were loop-handled drinking cups and tankards called "Scheibenhenkel," a new style of liquid containers and servers that appeared throughout the middle and lower Danube valley. Andrew Sherratt interpreted the Scheibenhenkel horizon as the first clear indicator of a new custom of drinking intoxicating beverages.[58] The replacement of highly decorated storage and serving vessels by plain drinking cups could indicate that new elite drinking rituals had replaced or nudged aside older household feasts.

The Cernavoda I economy was based primarily on the herding of sheep and goats. Many horse bones were found at Cernavoda I, and, for the first time, domesticated horses became a regular element in the animal herds of the middle and lower Danube valley.[59] Greenfield's zoological studies in the middle Danube showed that, also for the first time, animals were

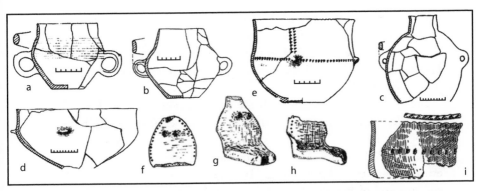

Figure 11.12 Black- or grey-surfaced ceramics from the Cernavoda I settlement, lower Danube valley, about 3900–3600 BCE, including two-handled tankards. After Morintz and Roman 1968.

butchered at different ages in upland and lowland sites. This suggested that herders moved animals seasonally between upland and lowland pastures, a form of herding called "transhumant pastoralism." The new pastoral economy might have been practiced in a new, more mobile way, perhaps aided by horseback riding.[60]

Kurgan graves were created only during the initial Suvorovo penetration. Afterward the immigrants' descendants stopped making kurgans. The flat-grave cemetery of Ostrovul Corbului probably dates to this settling-in period, with sixty-three graves, some displaying a posture on the back with raised knees, others contracted on the side, on the ruins of an abandoned tell. Cernavoda I flat graves also appeared at the Brailiţa cemetery, where the males had wide Proto-Europoid skulls and faces like the steppe Novo-danilovka population, and the females had gracile Mediterranean faces, like the Old European Gumelnitsa population.

By about 3600 BCE the Cernavoda I culture developed into Cernavoda III. Cernavoda III was, in turn, connected with one of the largest and most influential cultural horizons of eastern Europe, the Baden-Boleraz horizon, centered in the middle Danube (Hungary) and dated about 3600–3200 BCE. Drinking cups of this culture featured very high strap handles and were made in burnished grey-black fabrics with channeled flutes decorating their shoulders. Somewhat similar drinking sets were made from eastern Austria and Moravia to the mouth of the Danube and south to the Aegean coast (Dikili Tash IIIA–Sitagroi IV). Horse bones appeared almost everywhere, with larger sheep interpreted as wool sheep. At lowland sites in the middle Danube region, 60–91% of the sheep-goat lived to adult ages, suggesting management for secondary products, probably wool. Similarly 40–50% of the caprids were adults in two late TRB sites of this same era (Schalkenburg and Bronocice) in upland southern Poland. After 3600 BCE horses and wool sheep were increasingly common in eastern Europe.

Pre-Anatolian languages probably were introduced to the lower Danube valley and perhaps to the Balkans about 4200–4000 BCE by the Suvorovo migrants. We do not know when their descendants moved into Anatolia. Perhaps pre-Anatolian speakers founded Troy I in northwestern Anatolia around 3000 BCE. In prayers recited by the later Hittites, the sun god of heaven, *Sius* (cognate with Greek *Zeus*), was described as rising from the sea. This has always been taken as a fossilized ritual phrase retained from some earlier pre-Hittite homeland located west of a large sea.[61] The graves of Suvorovo were located west of the Black Sea. Did the Suvorovo people ride their horses down to the shore and pray to the rising sun?

CHAPTER TWELVE

Seeds of Change on the Steppe Borders
Maikop Chiefs and Tripolye Towns

After Old Europe collapsed, the dedication of copper objects in North Pontic graves declined by almost 80%.[1] Beginning in about 3800 BCE and until about 3300 BCE the varied tribes and regional cultures of the Pontic-Caspian steppes seem to have turned their attention away from the Danube valley and toward their other borders, where significant social and economic changes were now occurring.

On the southeast, in the North Caucasus Mountains, spectacularly ostentatious chiefs suddenly appeared among what had been very ordinary small-scale farmers. They displayed gold-covered clothing, gold and silver staffs, and great quantities of bronze weapons obtained from what must have seemed beyond the rim of the earth—in fact, from the newly formed cities of Middle Uruk Mesopotamia, through Anatolian middlemen. The first contact between southern urban civilizations and the people of the steppe margins occurred in about 3700–3500 BCE. It caused a social and political transformation that was expressed archaeologically as the Maikop culture of the North Caucasus piedmont. Maikop was the filter through which southern innovations—including possibly wagons—first entered the steppes. Sheep bred to grow long wool might have passed from north to south in return, a little considered possibility. The Maikop chiefs used a tomb type that looked like an elaborated copy of the Suvorovo-Novodanilovka kurgan graves of the steppes, and some of them seem to have moved north into the steppes. A few Maikop traders might have lived inside steppe settlements on the lower Don River. But, oddly, very little southern wealth was shared with the steppe clans. The gold, turquoise, and carnelian stayed in the North Caucasus. Maikop people might have driven the first wagons into the Eurasian steppes, and they certainly introduced new metal alloys that made a more sophisticated metallurgy possible. We do not know what they took in return—possibly wool, possibly horses, possibly even *Cannabis* or saiga antelope hides, though there is only circumstantial evidence for any of these. But in most parts of the Pontic-Caspian steppes the evidence for contact with Maikop is slight—a pot here, an arsenical bronze axe-head there.

On the west, Tripolye (C1) agricultural towns on the middle Dnieper began to bury their dead in cemeteries—the first Tripolye communities to accept the ritual of cemetery burial—and their coarse pottery began to look more and more like late Sredni Stog pottery. This was the first stage in the breakdown of the Dnieper frontier, a cultural border that had existed for two thousand years, and it seems to have signaled a gradual process of cross-border assimilation in the middle Dnieper forest-steppe zone. But while assimilation and incremental change characterized Tripolye towns on the middle Dnieper frontier, Tripolye towns closer to the steppe border on the South Bug River ballooned to enormous sizes, more than 350 ha, and, between about 3600 and 3400 BCE, briefly became the largest human settlements in the world. The super towns of Tripolye C1 were more than 1 km across but had no palaces, temples, town walls, cemeteries, or irrigation systems. They were not cities, as they lacked the centralized political authority and specialized economy associated with cities, but they were actually bigger than the earliest cities in Uruk Mesopotamia. Most Ukrainian archaeologists agree that warfare and defense probably were the underlying reasons why the Tripolye population aggregated in this way, and so the super towns are seen as a defensive strategy in a situation of confrontation and conflict, either between the Tripolye towns or between those towns and the people of the steppes, or both. But the strategy failed. By 3300 BCE all the big towns were gone, and the entire South Bug valley was abandoned by Tripolye farmers.

Finally, on the east, on the Ural River, a section of the Volga-Ural steppe population decided, about 3500 BCE, to migrate eastward across Kazakhstan more than 2000 km to the Altai Mountains. We do not know why they did this, but their incredible trek across the Kazakh steppes led to the appearance of the Afanasievo culture in the western Gorny Altai. The Afanasievo culture was intrusive in the Altai, and it introduced a suite of domesticated animals, metal types, pottery types, and funeral customs that were derived from the Volga-Ural steppes. This long-distance migration almost certainly separated the dialect group that later developed into the Indo-European languages of the Tocharian branch, spoken in Xinjiang in the caravan cities of the Silk Road around 500 CE but divided at that time into two or three quite different languages, all exhibiting archaic Indo-European traits. Most studies of Indo-European sequencing put the separation of Tocharian after that of Anatolian and before any other branch. The Afanasievo migration meets that expectation. The migrants might also have been responsible for introducing horseback riding to the pedestrian foragers of the northern Kazakh steppes, who were quickly transformed into the horse-riding, wild-horse–hunting Botai culture just when the Afanasievo migration began.

By this time, early Proto-Indo-European dialects must have been spoken in the Pontic-Caspian steppes, tongues revealing the innovations that separated all later Indo-European languages from the archaic Proto-Indo-European of the Anatolian type. The archaeological evidence indicates that a variety of different regional cultures still existed in the steppes, as they had throughout the Eneolithic. This regional variability in material culture, though not very robust, suggests that early Proto-Indo-European probably still was a regional language spoken in one part of the Pontic-Caspian steppes—possibly in the eastern part, since this was where the migration that led to the Tocharian branch began. Groups that distinguished themselves by using eastern innovations in their speech probably were engaging in a political act—allying themselves with specific clans, their political institutions, and their prestige—and in a religious act—accepting rituals, songs, and prayers uttered in that eastern dialect. Songs, prayers, and poetry were central aspects of life in all early Indo-European societies; they were the vehicle through which the right way of speaking reproduced itself publicly.

THE FIVE CULTURES OF THE FINAL ENEOLITHIC IN THE STEPPES

Much regional diversity and relatively little wealth existed in the Pontic-Caspian steppes between about 3800 and 3300 BCE (table 12.1). Regional variants as defined by grave and pot types, which is how archaeologists define them, had no clearly defined borders; on the contrary, there was a lot of border shifting and inter-penetration. At least five Final Eneolithic archaeological cultures have been identified in the Pontic-Caspian steppes (figure 12.1). Sites of these five groups are sometimes found in the same regions, occasionally in the same cemeteries; overlapped in time; shared a number of similarities; and were, in any case, fairly variable. In these circumstances, we cannot be sure that they all deserve recognition as

TABLE 12.1
Selected Radiocarbon Dates for Final Eneolithic Sites in the Steppes and Early Bronze Age Sites in the North Caucasus Piedmont

Lab Number	BP Date	Sample		Calibrated Date

1. Maikop culture

Klady kurgan cemetery, Farsa River valley near Maikop

Lab Number	BP Date	Sample		Calibrated Date
Le 4529	4960±120	Klady k29/1 late	bone	3940–3640 BCE
OxA 5059	4835±60	Klady k11/50 early	bone	3700–3520 BCE
OxA 5061	4765±65	Klady k11/55 early	bone	3640–3380 BCE
OxA 5058	4675±70	Klady k11/43 early	bone	3620–3360 BCE
OxA 5060	4665±60	Klady k11/48 early		3520–3360 BCE
Le 4528	4620±40	Klady k30/1 late	bone	3500–3350 BCE

Galugai settlement, upper Tersek River

OxA 3779	4930±120	Galugai I		3940–3540 BCE
OxA 3778	4650±80	Galugai I	bone	3630–3340 BCE
OxA 3777	4480±70	Galugai I		3340–3030 BCE

2. Tripolye C1 settlements

BM-495	4940±105	Soroki-Ozero		3940–3630 BCE
UCLA-1642F	4904±300	Novorozanovka 2		4100–3300 BCE
Bln-2087	4890±50	Maidanets'ke	charcoal	3710–3635 BCE
UCLA-1671B	4890±60	Evminka		3760–3630 BCE
BM-494	4792±105	Soroki-Ozero		3690–3370 BCE
UCLA-1466B	4790±100	Evminka		3670–3370 BCE
Bln-631	4870±100	Chapaevka		3780–3520 BCE
Ki-880	4810±140	Chapaevka	charcoal	3760–3370 BCE
Ki-1212	4600±80	Maidanets'ke		3520–3100 BCE

3. Repin culture

Kyzyl-Khak II settlement, North Caspian desert, lower Volga

?	4900±40	house 2	charcoal	3705–3645 BCE

Mikhailovka II settlement, lower part of level II

Ki-8010	4710±80	square 14, 2.06m depth	bone	3630–3370 BCE

Podgorovka settlement, Aidar River, Donets River tributary

Ki-7843	4560±50	?		3490–3100 BCE
Ki-7841	4370±55	?		3090–2900 BCE
Ki-7842	4330±50	?		3020–2880 BCE

4. Late Khvalynsk culture

Kara-Khuduk settlement, North Caspian desert, lower Volga

UPI-431	5100±45	pit-house	charcoal	3970–3800 BCE

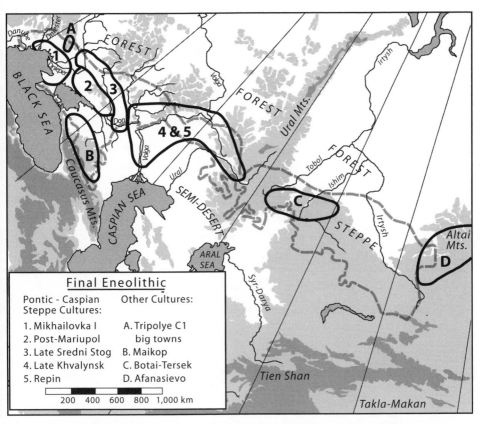

Figure 12.1 Final Eneolithic culture areas from the Carpathians to the Altai, 3800–3300 BCE.

different archaeological cultures. But we cannot understand the archaeological descriptions of this period without them, and together they provide a good picture of what was happening in the Pontic-Caspian steppes between 3800 and 3300 BCE. The western groups were engaged in a sort of two-pronged death dance, as it turned out, with the Cucuteni-Tripolye culture. The southern groups interacted with Maikop traders. And the eastern groups cast off a set of migrants who rode across Kazakhstan to a new home in the Altai, a subject reserved for the next chapter. Horseback riding is documented archaeologically in Botai-Tersek sites in Kazakhstan during this period (chapter 10) and probably appeared earlier, and so we proceed on the assumption that most steppe tribes were now equestrian.

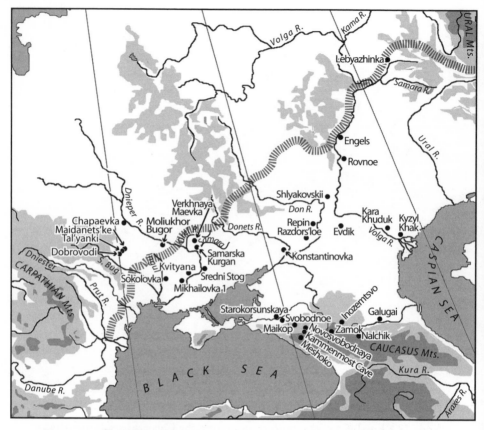

Figure 12.2 Final Eneolithic sites in the steppes and Early Bronze Age sites in the North Caucasus piedmont.

The Mikhailovka I Culture

The westernmost of the five Final Eneolithic cultures of the Pontic-Caspian steppes was the Mikhailovka I culture, also called the Lower Mikhailovka or Nizhnimikhailovkskii culture, named after a stratified settlement on the Dnieper located below the Dnieper Rapids (figure 12.2).[2] Below the last cascade, the river spread out over a broad basin in the steppes. Braided channels crisscrossed a sandy, marshy, forested low-land 10–20 km wide and 100 km long, a rich place for hunting and fishing and a good winter refuge for cattle, now inundated by hydroelectric dams. Mikhailovka overlooked this protected depression at a strategic river crossing. Its initial establishment probably was an outgrowth of increased east-west traffic across the river. It was the most important settlement on the lower Dnieper from the Late Eneolithic through the Early Bronze Age, about 3700–2500 BCE. Mikhailovka I, the original settlement, was occupied about 3700–3400 BCE, contemporary with late Tripolye B2 and early C1, late Sredni Stog, and early Maikop. A few late Sredni Stog and Maikop pottery sherds occurred in the occupation layer at Mikhailovka I. A whole Maikop pot was found in a grave with Mikhailovka I sherds at Sokolovka on the Ingul River, in kurgan 1, grave 6a. Tripolye B2 and C1 pots also are found in Mikhailovka I graves. These exchanges of pottery show that the Mikhailovka I culture had at least sporadic contacts with Tripolye B2/C1 towns, the Maikop culture, and late Sredni Stog communities.[3]

The people of Mikhailovka I cultivated cereal crops. At Mikhailovka I, imprints of cultivated seeds were found on 9 pottery sherds of 2,461 examined, or 1 imprint in 273 sherds.[4] The grain included emmer wheat, barley, millet, and 1 imprint of a bitter vetch seed (*Vicia ervilia*), a crop grown today for animal fodder. Zoologists identified 1,166 animal bones (NISP) from Mikhailovka I, of which 65% were sheep-goat, 19% cattle, 9% horse, and less than 2% pig. Wild boar, aurochs, and saiga antelope were hunted occasionally, accounting for less than 5 percent of the animal bones.

The high number of sheep-goat at Mikhailovka I might suggest that long-wool sheep were present. Wool sheep probably were present in the North Caucasus at Svobodnoe (see below) by 4000 BCE, and almost certainly were in the Danube valley during the Cernavoda III–Boleraz period around 3600–3200 BCE, so wool sheep could have been kept at Mikhailovka I. But even if long-wool sheep *were* bred in the steppes during this period, they clearly were not yet the basis for a widespread new wool economy, because cattle or even deer bones still outnumbered sheep in other steppe settlements.[5]

Mikhailovka I pottery was shell-tempered and had dark burnished surfaces, usually unornamented (figure 12.3). Common shapes were egg-shaped pots or flat-based, wide-shouldered tankards with everted rims. A few silver ornaments and one gold ring, quite rare in the Pontic steppes of this era, were found in Mikhailovka I graves.

Mikhailovka I kurgans were distributed from the lower Dnieper westward to the Danube delta and south to the Crimean peninsula, north and northwest of the Black Sea. Near the Danube they were interspersed with cemeteries that contained Danubian Cernavoda I–III ceramics.[6] Most

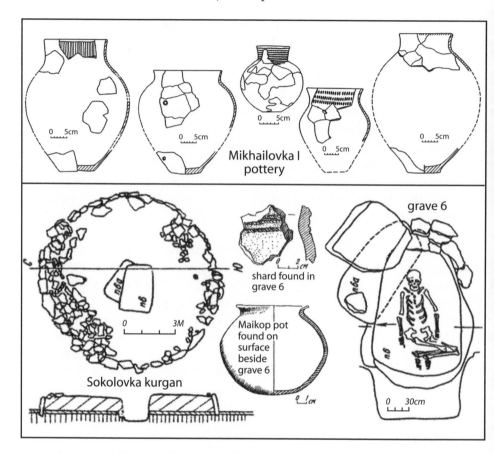

Figure 12.3 Ceramics from the Mikhailovka I settlement, after Lagodovskaya, Shaposhnikova, and Makarevich 1959; and a Mikhailovka I grave (gr. 6) stratified above an older Eneolithic grave (gr. 6a) at Sokolovka kurgan on the Ingul River west of the Dnieper, after Sharafutdinova 1980.

Mikhailovka I kurgans were low mounds of black earth covered by a layer of clay, surrounded by a ditch and a stone cromlech, often with an opening on the southwest side. The graves frequently were in cists lined with stone slabs. The body could be in an extended supine position or contracted on the side or supine with raised knees, although the most common pose was contracted on the side. Occasionally (e.g., Olaneshti, k. 2, gr. 1, on the lower Dniester) the grave was covered by a stone anthropomorphic stela—a large stone slab carved at the top into the shape of a head projecting above rounded shoulders (see figure 13.11). This was the beginning of a long and important North Pontic tradition of decorating some graves with carved stone stelae.[7]

The skulls and faces of some Mikhailovka I people were delicate and narrow. The skeletal anthropologist Ina Potekhina established that another North Pontic culture, the Post-Mariupol culture, looked most like the old wide-faced Suvorovo-Novodanilovka population. The Mikhailovka I people, who lived in the westernmost steppes closest to the Tripolye culture and to the lower Danube valley, seem to have intermarried more with people from Tripolye towns or people whose ancestors had lived in Danubian tells.[8]

The Mikhailovka I culture was replaced by the Usatovo culture in the steppes northwest of the Black Sea after about 3300 BCE. Usatovo retained some Mikhailovka I customs, such as making a kurgan with a surrounding stone cromlech that was open to the southwest. The Usatovo culture was led by a warrior aristocracy centered on the lower Dniester estuary that probably regarded Tripolye agricultural townspeople as tribute-paying clients, and that might have begun to engage in sea trade along the coast. People in the Crimean peninsula retained many Mikhailovka I customs and developed into the Kemi-Oba culture of the Early Bronze Age after about 3300 BCE. These EBA cultures will be described in a later chapter.

The Post-Mariupol Culture

The clumsiest culture name of the Final Eneolithic is the "Post-Mariupol" or "Extended-Position-Grave" culture, both names conveying a hint of definitional uncertainty. Rassamakin called it the "Kvityana" culture. I will use the name "Post-Mariupol." All these names refer to a grave type recognized in the steppes just above the Dnieper Rapids in the 1970s but defined in various ways since then. N. Ryndina counted about three hundred graves of the Post-Mariupol type in the steppes from the Dnieper valley eastward to the Donets. They were covered by low kurgans, occasionally surrounded by a stone cromlech. Burial was in an extended supine position in a narrow oblong or rectangular pit, often lined with stone and covered with wooden beams or stone slabs. Usually there were no ceramics in the grave (although this rule was fortunately broken in a few graves), but a fire was built above the grave; red ochre was strewn heavily on the grave floor; and lamellar flint blades, bone beads, or a few small copper

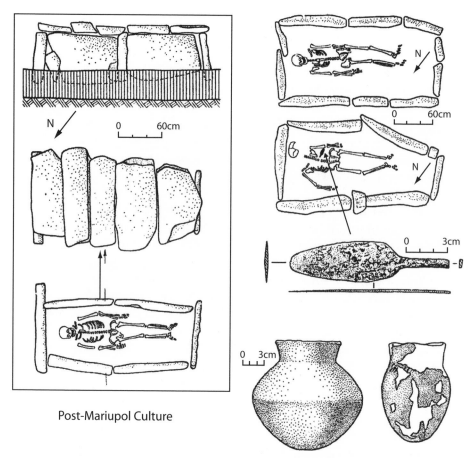

Post-Mariupol Culture

Figure 12.4 Post-Mariupol ceramics and graves: *left*, Marievka kurgan 14, grave 7; *upper right*, Bogdanovskogo Karera Kurgan 2, graves 2 and 17; *lower right*, pots from Chkalovskaya kurgan 3. After Nikolova and Rassamakin 1985, figure 7.

beads or twists were included (figure 12.4). Three cattle skulls, presumably sacrificed at the funeral, were placed at the edge of one grave at Chkalovska kurgan 3. The largest cluster is just north of the Dnieper Rapids on the east side of the Dnieper, between two tributary rivers, the Samara (smaller than the Volga-region Samara River) and the Orel. Two chronological phases are identified: an early (Final Eneolithic) phase contemporary with Tripolye B2/C1, about 3800–3300 BCE; and a later (Early Bronze Age) phase contemporary with Tripolye C2 and the Early Yamnaya horizon, about 3300–2800 BCE.[9]

About 40 percent of the Post-Mariupol graves in the core Orel-Samara region contained copper ornaments, usually just one or two. All forty-six of the copper objects examined by Ryndina from early-phase graves were made from "clean" Transylvanian ores, the same ores used in Tripolye B2 and C1 sites. The copper in the second phase, however, was from two sources: ten objects still were made of "clean" Transylvanian copper but twenty-three were made of arsenical bronze. They were most similar to the arsenical bronzes of the Ustatovo settlement or the late Maikop culture. Only one Post-Mariupol object (a small willow-leaf pendant from Bulakhovka kurgan cemetery I, k. 3, gr. 9) looked metallurgically like a direct import from late Maikop.[10]

Two Post-Mariupol graves were metalsmiths' graves. They contained three bivalve molds for making sleeved axes. (A sleeved axe had a single blade with a cast sleeve hole for the handle on one side.) The molds copied a late Maikop axe type but were locally made.[11] They probably were late Post-Mariupol, after 3300 BCE. They are the oldest known two-sided ceramic molds in the steppes, and they were buried with stone hammers, clay tubes or *tulieres* for bellows attachments, and abrading stones. These kits suggest a new level of technological skill among steppe metalsmiths and the graves began a long tradition of the smith being buried with his tools.

The Late Sredni Stog Culture

The third and final culture group in the *western* part of the Pontic-Caspian steppes was the late Sredni Stog culture. Late Sredni Stog pottery was shell-tempered and often decorated with cord-impressed geometric designs (see figure 11.7), quite unlike the plain, dark-surfaced pots of Mikhailovka I and the Post-Mariupol culture. The late Sredni Stog settlement of Moliukhor Bugor was located on the Dnieper in the forest-steppe zone. A Tripolye C1 vessel was found there. The people of Moliukhor Bugor lived in a house 15 m by 12 m with three internal hearths, hunted red deer and wild boar, fished, kept a lot of horses and a few domesticated cattle and sheep, and grew grain. Eight grain impressions were found among 372 sherds (one imprint in 47 sherds), a higher frequency than at Mikhailovka I. They included emmer wheat, einkorn wheat, millet, and barley. The well-known Sredni Stog settlement at Dereivka was occupied somewhat earlier, about 4000 BCE, but also produced many flint blades with sickle gloss and six stone querns for grinding grain, and so also probably included some grain cultivation.

Horses represented 63% of the animal bones at Dereivka (see chapter 10). The Sredni Stog societies on the Dnieper, like the other western steppe groups, had a mixed economy that combined grain cultivation, stock-breeding, horseback riding, and hunting and fishing.

Late Sredni Stog sites were located in the northern steppe and southern forest-steppe zones on the middle Dnieper, north of the Post-Mariupol and Mikhailovka I groups. Sredni Stog sites also extended from the Dnieper eastward across the middle Donets to the lower Don. The most important stratified settlement on the lower Don was Razdorskoe [raz-DOR-sko-ye]. Level 4 at Razdorskoe contained an early Khvalynsk component, level 5 above it had an early Sredni Stog (Novodanilovka period) occupation, and, after that, levels 6 and 7 had pottery that resembled late Sredni Stog mixed with imported Maikop pottery. A radiocarbon date said to be associated with level 6, on organic material in a core removed for pollen studies, produced a date of 3500–2900 BCE (4490±180 BP). Near Razdorskoe was the fortified settlement at Konstantinovka. Here, in a place occupied by people who made similar lower-Don varieties of late Sredni Stog pottery, there might actually have been a small Maikop colony.[12]

Bodies buried in Sredni Stog graves usually were in the supine-with-raised knees position that was such a distinctive aspect of steppe burials beginning with Khvalynsk. The grave floor was strewn with red ochre, and the body often was accompanied by a unifacial flint blade or a broken pot. Small mounds sometimes were raised over late Sredni Stog graves, but in many cases they were flat.

Repin and Late Khvalynsk in the Lower Don-Volga Steppes

The two eastern groups can be discussed together. They are identified with two quite different kinds of pottery. One type clearly resembled a late variety of Khvalynsk pottery. The other type, called Repin, probably began on the middle Don, and is identified by round-based pots with cord-impressed decoration and decorated rims.

Repin, excavated in the 1950s, was located 250 km upstream from Razdorskoe, on the middle Don at the edge of the feather-grass steppe. At Repin 55% of the animal bones were horse bones. Horse meat was much more important in the diet than the meat of cattle (18%), sheep-goat (9%), pigs (9%), or red deer (9%).[13] Perhaps Repin specialized in raising horses for export to North Caucasian traders (?).The pottery from

Repin defined a type that has been found at many sites in the Don-Volga region. Repin pottery sometimes is found stratified beneath Yamnaya pottery, as at the Cherkasskaya settlement on the middle Don in the Voronezh oblast.[14] Repin components occur as far north as the Samara oblast in the middle Volga region, at sites such as Lebyazhinka I on the Sok River, in contexts also thought to predate early Yamnaya. The Afanasievo migration to the Altai was carried out by people with a Repin-type material culture, probably from the middle Volga-Ural region. On the lower Volga, a Repin antelope hunters' camp was excavated at Kyzyl Khak, where 62% of the bones were saiga antelope (figure 12.5). Cattle were 13%, sheep 9%, and horses and onagers each about 7%. A radiocarbon date (4900±40 BP) put the Repin occupation at Kyzyl-Khak at about 3700–3600 BCE.

Kara Khuduk was another antelope hunters' camp on the lower Volga but was occupied by people who made late Khvalynsk-type pottery (figure 12.5). A radiocarbon date (5100±45 BP, UPI 430) indicated that it was occupied in about 3950–3800 BCE, earlier than the Repin occupation at Kyzyl-Khak nearby. Many large scrapers, possibly for hide processing, were found among the flint tools. Saiga antelope hides seem to have been highly desired, perhaps for trade. The animal bones were 70% saiga antelope, 13% cattle, and 6% sheep. The ceramics (670 sherds from 30–35 vessels) were typical Khvalynsk ceramics: shell-tempered, round-bottomed vessels with thick, everted lips, covered with comb stamps and corded-impressed U-shaped "caterpillar" impressions.

Late Khvalynsk graves without kurgans were found in the 1990s at three sites on the lower Volga: Shlyakovskii, Engels, and Rovnoe. The bodies were positioned on the back with knees raised, strewn with red ochre, and accompanied by lamellar flint blades, flint axes with polished edges, polished stone mace heads of Khvalynsk type, and bone beads. Late Khvalynsk populations lived in scattered enclaves on the lower Volga. Some of them crossed the northern Caspian, perhaps by boat, and established a group of camps on its eastern side, in the Mangyshlak peninsula.

The Volga-Don late Khvalynsk and Repin societies played a central role in the evolution of the Early Bronze Age Yamnaya horizon beginning around 3300 BCE (discussed in the next chapter). One kind of early Yamnaya pottery was really a Repin type, and the other kind was actually a late Khvalynsk type; so, if no other clues are present, it can be difficult to separate Repin or late Khvalynsk pottery from early Yamnaya pottery. The Yamnaya horizon probably was the medium through which late

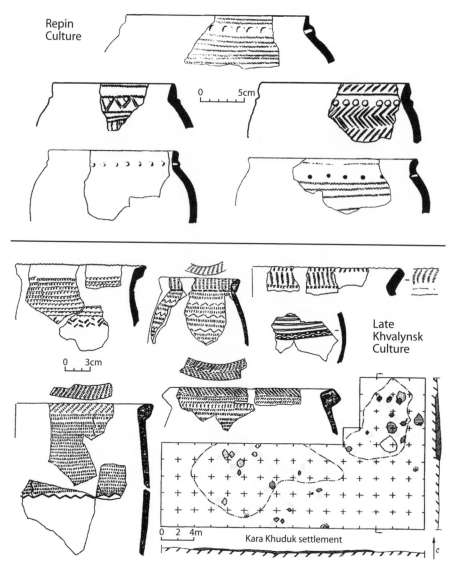

Figure 12.5 Repin pottery from Kyzl-Khak (*top*) and late Khvalynsk pottery and settlement plan from Kara-Khuduk (*bottom*) on the lower Volga. After Barynkin, Vasiliev, and Vybornov 1998, figures 5 and 6.

Proto-Indo-European languages spread across the steppes. This implies that classic Proto-Indo-European dialects were spoken among the Repin and late Khvalynsk groups.[15]

CRISIS AND CHANGE ON THE TRIPOLYE FRONTIER: TOWNS BIGGER THAN CITIES

Two notable and quite different kinds of changes affected the Tripolye culture between about 3700 and 3400 BCE. First, the Tripolye settlements in the forest-steppe zone on the middle Dnieper began to make pottery that looked like Pontic-Caspian ceramics (dark, occasionally shell-tempered wares) and adopted Pontic-Caspian–style inhumation funerals. The Dnieper frontier became more porous, probably through gradual assimilation. But Tripolye settlements on the South Bug River, near the steppe border, changed in very different ways. They mushroomed to enormous sizes, more than 400 ha, twice the size of the biggest cities in Mesopotamia. Simply put, they were the biggest human settlements in the world. And yet, instead of evolving into cities, they were abruptly abandoned.

Contact with Sredni Stog on the Dnieper Frontier

Chapaevka was a Tripolye B2/C1 settlement of eleven dwellings located on a promontory west of the Dnieper valley in the northern forest-steppe zone. It was occupied about 3700–3400 BCE.[16] Chapaevka is the earliest known Tripolye community to adopt cemetery burial (figure 12.6). A cemetery of thirty-two graves appeared on the edge of settlement. The form of burial, in an extended supine position, usually with a pot, sometimes with a piece of red ochre under the head or chest, was not exactly like any of the steppe grave types, but just the acceptance of the burial of the body was a notable change from the Old European funeral customs of the Tripolye culture. Chapaevka also had lightly built houses with dug-out floors rather than houses with plastered log floors (*ploshchadka*). Tripolye C1 pottery was found at Moliukhor Bugor, about 150 km to the south, perhaps the source of some of these new customs.

Most of the ceramics in the Chapaevka houses were well-fired fine wares with fine sand temper or very fine clay fabrics (50–70%), of which a small percentage (1–10%) were painted with standard Tripolye designs; but generally they were black to grey in color, with burnished surfaces, and

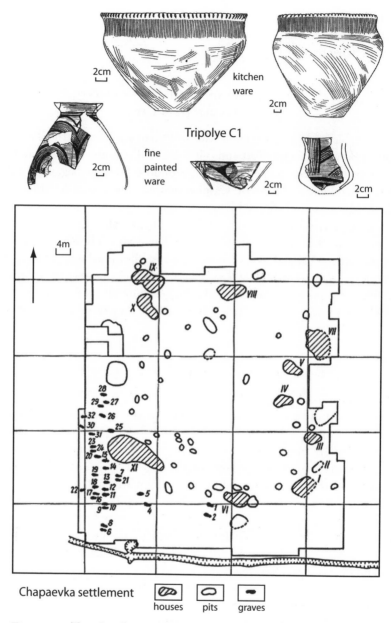

kitchen
ware

Tripolye C1

fine
painted
ware

Chapaevka settlement — houses — pits — graves

Figure 12.6 Tripolye C1 settlement at Chapaevka on the Dnieper with eleven houses (features I–XI) and cemetery (gr. 1–32) and ceramics. After Kruts 1977, figures 5 and 16.

were often undecorated. They were quite different from the orange wares that had typified earlier Tripolye ceramics. Undecorated grey-to-black ware also was typical of the Mikhailovka I and Post-Mariupol cultures, although their shapes and clay fabrics differed from most of those of the Tripolye C1 culture. One class of Chapaevka kitchen-ware pots with vertical combed decoration on the collars looked so much like late Sredni Stog pots that it is unclear whether this kind of ware was borrowed from Tripolye by late Sredni Stog potters or by Tripolye C1 potters from late Sredni Stog.[17] Around 3700–3500 BCE the Dnieper frontier was becoming a zone of gradual, probably peaceful assimilation between Tripolye villagers and indigenous Sredni Stog societies east of the Dnieper.

Towns Bigger Than Cities: The Tripolye C1 Super Towns

Closer to the steppe border things were quite different. All the Tripolye settlements located between the Dnieper and South Bug rivers, including Chapaevka, were oval, with houses arranged around an open central plaza. Some villages occupied less than 1 ha, many were towns of 8–15 ha, some were more than 100 ha, and a group of three Tripolye C1 sites located within 20 km of one another reached sizes of 250–450 ha between about 3700 and 3400 BCE. These super sites were located in the hills east of the South Bug River, near the edge of the steppe in the southern forest-steppe zone. They were the largest communities not just in Europe but in the world.[18]

The three known super-sites—Dobrovodi (250 ha), Maidanets'ke (250 ha), and Tal'yanki (450 ha)—perhaps were occupied sequentially in that order. None of these sites contained an obvious administrative center, palace, storehouse, or temple. They had no surrounding fortification wall or moat, although the excavators Videiko and Shmagli described the houses in the outer ring as joined in a way that presented an unbroken two-story-high wall pierced only by easily defended radial streets. The most thoroughly investigated of the three, Maidanets'ke, covered 250 ha. Magnetometer testing revealed 1,575 structures (figure 12.7). Most were inhabited simultaneously (there was almost no overbuilding of newer houses over older ones) by a population estimated at fifty-five hundred to seventy-seven hundred people. Using Bibikov's estimate of 0.6 ha of cultivated wheat per person per year, a population of that magnitude would have required 3,300–4,620 ha of cultivated fields each year, which would have necessitated cultivating fields more than 3 km from the town.[18] The houses were built close to one another in concentric oval rings, on a common plan, oriented toward a cen-

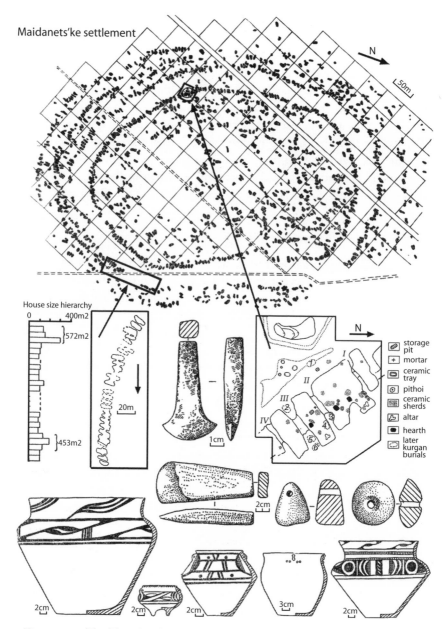

Figure 12.7 The Tripolye C1 Maidanets'ke settlement, with 1,575 structures mapped by magnetometers: *left*: smaller houses cluster around larger houses, thought to be clan or sub-clan centers; *right*: a house group very well preserved by the Yamnaya kurgan built on top of it, showing six inserted late

tral plaza. The excavated houses were large, 5–8 m wide and 20–30 m long, and many were two-storied. Videiko and Shmagli suggested a political organization based on clan segments. They documented the presence of one larger house for each five to ten smaller houses. The larger houses usually contained more female figurines (rare in most houses), more fine painted pots, and sometimes facilities such as warp-weighted looms. Each large house could have been a community center for a segment of five to ten houses, perhaps an extended family (or a "super-family collective," in Videiko's words). If the super towns were organized in this way, a council of 150–300 segment leaders would have made decisions for the entire town. Such an unwieldy system of political management could have contributed to its own collapse. After Maidanests'ke and Tal'yanki were abandoned, the largest town in the South Bug hills was Kasenovka (120 ha, with seven to nine concentric rings of houses), dated to the Tripolye C1/C2 transition, perhaps 3400–3300 BCE. When Kasenovka was abandoned, Tripolye people evacuated most of the South Bug valley.

Specialized craft centers appeared in Tripolye C1 communities for making flint tools, weaving, and manufacturing ceramics. These crafts became spatially segregated both within and between towns.[20] A hierarchy appeared in settlement sizes, comprised of two and perhaps three tiers. These kinds of changes usually are interpreted as signs of an emerging political hierarchy and increasing centralization of political power. But, as noted, instead of developing into cities, the towns were abandoned.

Population concentration is a standard response to increased warfare among tribal agriculturalists, and the subsequent abandonment of these places suggests that warfare and raiding was at the root of the crisis. The aggressors could have been steppe people of Mikhailovka I or late Sredni Stog type. A settlement at Novorozanovka on the Ingul, west of the Dnieper, produced a lot of late Sredni Stog cord-impressed pottery, some Mikhailovka I pottery, and a few imported Tripolye C1 painted fine pots. Mounted raiding might have made it impossible to cultivate fields more than 3 km from the town. Raiding for cattle or captives could have caused the fragmentation and dispersal of the Tripolye population and the abandonment of town-based craft traditions just as it had in the Danube valley

---

Figure 12.7(continued) Yamnaya graves. Artifacts from the settlement: *top center*, a cast copper axe; *central row*, a polished stone axe and two clay loom weights; *bottom row*, selected painted ceramics. After Shmagli and Videiko 1987; and Videiko 1990.

some five hundred years earlier. Farther north, in the forest-steppe zone on the middle Dnieper, assimilation and exchange led ultimately in the same direction but more gradually.

THE FIRST CITIES AND THEIR CONNECTION TO THE STEPPES

Steppe contact with the civilizations of Mesopotamia was, of course, much less direct than contact with Tripolye societies, but the southern door might have been the avenue through which wheeled vehicles first appeared in the steppes, so it was important. Our understanding of these contacts with the south has been completely rewritten in recent years.

Between 3700 and 3500 BCE the first cities in the world appeared among the irrigated lowlands of Mesopotamia. Old temple centers like Uruk and Ur had always been able to attract thousands of laborers from the farms of southern Iraq for building projects, but we are not certain why they began to live around the temples permanently (figure 12.8). This shift in population from the rural villages to the major temples created the first cities. During the Middle and Late Uruk periods (3700–3100 BCE) trade into and out of the new cities increased tremendously in the form of tribute, gift exchange, treaty making, and the glorification of the city temple and its earthly authorities. Precious stones, metals, timber, and raw wool (see chapter 4) were among the imports. Woven textiles and manufactured metal objects probably were among the exports. During the Late Uruk period, wheeled vehicles pulled by oxen appeared as a new technology for land transport. New accounting methods were developed to keep track of imports, exports, and tax payments—cylinder seals for marking sealed packages and the sealed doors of storerooms, clay tokens indicating package contents, and, ultimately, writing.

The new cities had enormous appetites for copper, gold, and silver. Their agents began an extraordinary campaign, or perhaps competing campaigns by different cities, to obtain metals and semiprecious stones. The native chiefdoms of Eastern Anatolia already had access to rich deposits of copper ore, and had long been producing metal tools and weapons. Emissaries from Uruk and other Sumerian cities began to appear in northern cities like Tell Brak and Tepe Gawra. South Mesopotamian garrisons built and occupied caravan forts on the Euphrates in Syria at Habubu Kabira. The "Uruk expansion" began during the Middle Uruk period about 3700 BCE and greatly intensified during Late Uruk, about 3350–3100 BCE. The city of Susa in southwestern Iran might have become an Uruk colony. East of Susa on the Iranian plateau a series of large mudbrick edifices

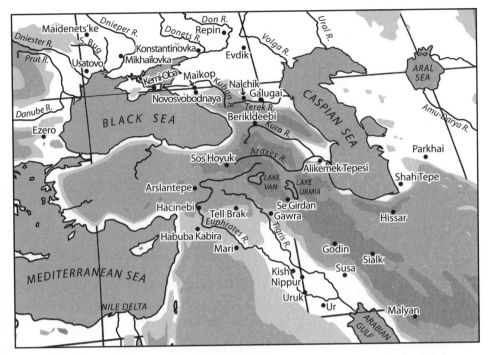

Figure 12.8 Maikop culture and selected sites associated with the Uruk expansion.

rose above the plains, protecting specialized copper production facilities that operated partly for the Uruk trade, regulated by local chiefs who used the urban tools of trade management: seals, sealed packages, sealed storerooms, and, finally, writing. Copper, lapis lazuli, turquoise, chlorite, and carnelian moved under their seals to Mesopotamia. Uruk-related trade centers on the Iranian plateau included Sialk IV_1, Tal-i-Iblis V–VI, and Hissar II in central Iran. The tentacles of trade reached as far northeast as the settlement of Sarazm in the Zerafshan Valley of modern Tajikistan, probably established to control turquoise deposits in the deserts nearby.

The Uruk expansion to the northwest, toward the gold, silver, and copper sources in the Caucasus Mountains, is documented at two important local strongholds on the upper Euphrates. Hacinebi was a fortified center with a large-scale copper production industry. Its chiefs began to deal with Middle Uruk traders during its phase B2, dated about 3700–3300 BCE. More than 250 km farther up the Euphrates, high in the mountains of Eastern Anatolia, the stronghold at Arslantepe expanded in wealth and size at about the same time (Phase VII), although it retained its own native system of seals, architecture, and administration. It also had its own large-scale copper production facilities based on local ores. Phase VIA, beginning about 3350 BCE, was dominated by two new pillared buildings similar to Late Uruk temples. In them officials regulated trade using some Uruk-style seals (among many local-style seals) and gave out stored food in Uruk-type, mass-produced ration bowls. The herds of Arslantepe VII had been dominated by cattle and goats, but in phase VIA sheep rose suddenly to become the most numerous and important animal, probably for the new industry of wool production. Horses also appeared, in very small numbers, at Arslantepe VII and VIA and Hacinebi phase B, but they seem not to have been traded southward into Mesopotamia. The Uruk expansion ended abruptly about 3100 BCE for reasons that remain obscure. Arslantepe and Hacinebi were burned and destroyed, and in the mountains of eastern Anatolia local Early Trans-Caucasian (ETC) cultures built their humble homes over the ruins of the grand temple buildings.[21]

Societies in the mountains to the north of Arslantepe responded in various ways to the general increase in regional trade that began about 3700–3500 BCE. Novel kinds of public architecture appeared. At Berikldeebi, northwest of modern Tbilisi in Georgia, a settlement that had earlier consisted of a few flimsy dwellings and pits was transformed about 3700–3500 BCE by the construction of a massive mudbrick wall that enclosed a public building, perhaps a temple, measuring 14.5×7.5 m (50×25 ft). At Sos level Va near Erzerum in northeastern Turkey there were similar architectural hints of increasing scale and power.[22] But neither prepares us for the funerary splendor of the Maikop culture.

The Maikop culture appeared about 3700–3500 BCE in the piedmont north of the North Caucasus Mountains, overlooking the Pontic-Caspian steppes. The semi-royal figure buried under the giant Maikop chieftan's kurgan acquired and wore Mesopotamian ornaments in an ostentatious funeral display that had no parallel that has been preserved even in Mesopotamia. Into the grave went a tunic covered with golden lions and bulls, silver-sheathed staffs mounted with solid gold and silver bulls, and silver sheet-metal cups. Wheel-made pottery was imported from the south, and the new technique was used to make Maikop ceramics similar to some of the vessels found at Berikldeebi and at Arslantepe VII/VIA.[23] New high-nickel arsenical bronzes and new kinds of bronze weapons (sleeved axes, tanged daggers) also spread into the North Caucasus from the south, and a cylinder seal from the south was worn as a bead in another Maikop grave. What kinds of societies lived in the North Caucasus when this contact began?

The North Caucasus Piedmont: Eneolithic Farmers before Maikop

The North Caucasian piedmont separates naturally into three geographic parts. The western part is drained by the Kuban River, which flows into the Sea of Azov. The central part is a plateau famous for its bubbling hot springs, with resort towns like Mineralnyi Vody (Mineral Water) and Kislovodsk (Sweet Water). The eastern part is drained by the Terek River, which flows into the Caspian Sea. The southern skyline is dominated by the permanently glaciated North Caucasus Mountains, which rise to icy peaks more than 5,600 m (18,000 ft) high; and off to the north are the rolling brown plains of the steppes.

Herding, copper-using cultures lived here by 5000 BCE. The Early Eneolithic cemetery at Nalchik and the cave occupation at Kammenomost Cave (chapter 9) date to this period. Beginning about 4400–4300 BCE the people of the North Caucasus began to settle in fortified agricultural villages such as Svobodnoe and Meshoko (level 1) in the west, Zamok on the central plateau, and Ginchi in Dagestan in the east, near the Caspian. About ten settlements of the Svobodnoe type, of thirty to forty houses each, are known in the Kuban River drainage, apparently the most densely settled region. Their earthen or stone walls enclosed central plazas surrounded by solid wattle-and-daub houses. Svobodnoe, excavated by A. Nekhaev, is the best-reported site (figure 12.9). Half the animal bones from Svobodnoe were from wild red deer and boar, so hunting was important. Sheep were the most important domesticated animal, and the proportion of sheep to goats was 5:1, which suggests that sheep were kept for wool. But pig keeping also was important, and pigs were the most important meat animals at the settlement of Meshoko.

Svobodnoe pots were brown to orange in color and globular with everted rims, but decorative styles varied greatly between sites (e.g., Zamok, Svobodnoe, and Meshoko are said to have had quite different domestic pottery types). Female ceramic figurines suggest female-centered domestic rituals. Bracelets carved and polished of local serpentine were manufactured in the hundreds at some sites. Cemeteries are almost unknown, but a few individual graves found among later graves under kurgans in the Kuban region have been ascribed to the Late Eneolithic. The Svobodnoe culture differed from Repin or late Khvalynsk steppe cultures in its house

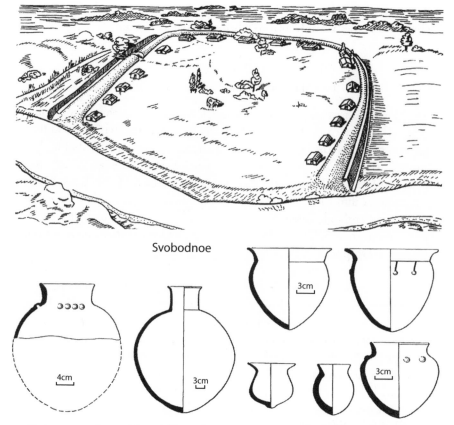

Svobodnoe

Figure 12.9 Svobodnoe settlement and ceramics, North Caucasus. After Nekhaev 1992.

forms, settlement types, pottery, stone tools, and ceramic female figurines. Probably it was distinct ethnically and linguistically.[24]

Nevertheless, the Svobodnoe culture was in contact with the steppes. A Svobodnoe pot was deposited in the rich grave at Novodanilovka in the Azov steppes, and a copper ring made of Balkan copper, traded through the Novodanilovka network, was found at Svobodnoe. Potsherds that look like early Sredni Stog types were noted at Svobodnoe and Meshoko 1. Green serpentine axes from the Caucasus appeared in several steppe graves and in settlements of the early Sredni Stog culture (Strilcha Skelya, Aleksandriya, Yama). The Svobodnoe-era settlements in the Kuban River valley participated in the eastern fringe of the steppe Suvorovo-Novodanilovka activities around 4000 BCE.

THE MAIKOP CULTURE

The shift from Svobodnoe to Maikop was accompanied by a sudden change in funeral customs—the clear and widespread adoption of kurgan graves—but there was continuity in settlement locations and settlement types, lithics, and some aspects of ceramics. Early Maikop ceramics showed some similarities with Svobodnoe pot shapes and clay fabrics, and some similarities with the ceramics of the Early Trans-Caucasian (ETC) culture south of the North Caucasus Mountains. These analogies indicate that Maikop developed from local Caucasian origins. But some Maikop pots were wheel-made, a new technology introduced from the south, and this new method of manufacture probably encouraged new vessel shapes.

The Maikop chieftain's grave, discovered on the Belaya River, a tributary of the Kuban River, was the first Maikop-culture tomb to be excavated, and it remains the most important early Maikop site. When excavated in 1897 by N. I. Veselovskii, the kurgan was almost 11 m high and more than 100 m in diameter. The earthen center was surrounded by a cromlech of large undressed stones. Externally it looked like the smaller Mikhailovka I and Post-Mariupol kurgans (and, before them, the Suvorovo kurgans), which also had earthen mounds surrounded by stone cromlechs. Internally, however, the Maikop chieftan's grave was quite different. The grave chamber was more than 5 m long and 4 m wide, 1.5 m deep, and was lined with large timbers. It was divided by timber partitions into two northern chambers and one southern chamber. The two northern chambers each held an adult female, presumably sacrificed, each lying in a contracted position on her right side, oriented southwest, stained with red ochre, with one to four pottery vessels and wearing twisted silver foil ornaments.[25]

The southern chamber contained an adult male. He also probably was positioned on his right side, contracted, with his head oriented southwest, the pose of most Maikop burials. He also lay on ground deeply stained with red ochre. With him were eight red-burnished, globular pottery vessels, the type collection for Early Maikop; a polished stone cup with a sheet-gold cover; two arsenical bronze, sheet-metal cauldrons; two small cups of sheet gold; and fourteen sheet-silver cups, two of which were decorated with impressed scenes of animal processions including a Caucasian spotted panther, a southern lion, bulls, a horse, birds, and a shaggy animal (bear? goat?) mounting a tree (figure 12.10). The engraved horse is the oldest clear image of a post-glacial horse, and it looked like a modern Przewalski: thick neck, big head, erect mane, and thick, strong legs. The chieftan also had arsenical

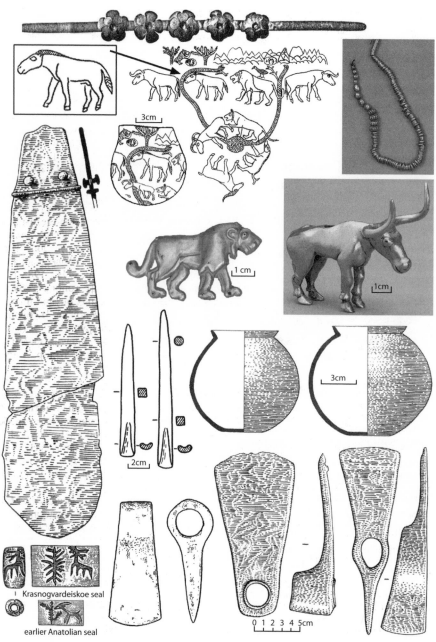

Krasnogvardeiskoe seal

earlier Anatolian seal

Figure 12.10 Early Maikop objects from the chieftain's grave at Maikop, the State Hermitage Museum, St. Petersburg; and a seal at lower left from the early Maikop Krasnogvardeiskoe kurgan, with a comparative seal from Chalcolithic Degirmentepe in eastern Anatolia. The lion, bull, necklace, and diadem are gold; the cup with engraved design is silver; the two pots are ceramic; and

bronze tools and weapons. They included a sleeved axe, a hoe-like adze, an axe-adze, a broad spatula-shaped metal blade 47 cm long with rivets for the attachment of a handle, and two square-sectioned bronze chisels with round-sectioned butts. Beside him was a bundle of six (or possibly eight) hollow silver tubes about 1 m long. They might have been silver casings for a set of six (or eight) wooden staffs, perhaps for holding up a tent that shaded the chief. Long-horned bulls, two of solid silver and two of solid gold, were slipped over four of the silver casings through holes in the middle of the bulls, so that when the staffs were erect the bulls looked out at the visitor. Each bull figure was sculpted first in wax; very fine clay was then pressed around the wax figure; this clay was next wrapped in a heavier clay envelope; and, finally, the clay was fired and the wax burned off—the lost wax method for making a complicated metal-casting mold. The Maikop chieftain's grave contained the first objects made this way in the North Caucasus. Like the potters wheel, the arsenical bronze, and the animal procession motifs engraved on two silver cups, these innovations came from the south.[26]

The Maikop chieftan was buried wearing Mesopotamian symbols of power—the lion paired with the bull—although he probably never saw a lion. Lion bones are not found in the North Caucasus. His tunic had sixty-eight golden lions and nineteen golden bulls applied to its surface. Lion and bull figures were prominent in the iconography of Uruk Mesopotamia, Hacinebi, and Arslantepe. Around his neck and shoulders were 60 beads of turquoise, 1,272 beads of carnelian, and 122 golden beads. Under his skull was a diadem with five golden rosettes of five petals each on a band of gold pierced at the ends. The rosettes on the Maikop diadem had no local prototypes or parallels but closely resemble the eight-petaled rosette seen in Uruk art. The turquoise almost certainly came from northeastern Iran near Nishapur or from the Amu Darya near the trade settlement of Sarazm in modern Tajikistan, two regions famous in antiquity for their turquoise. The red carnelian came from western Pakistan and the lapis lazuli from eastern Afghanistan. Because of the absence of cemeteries in Uruk Mesopotamia, we do not know much about the decorations worn there. The abundant personal ornaments at Maikop, many of them traded up the Euphrates through eastern Anatolia, probably were not made just for the barbarians. They provide an eye-opening glimpse of the kinds of styles that must have been seen in the streets and temples of Uruk.

Figure 12.10 (*continued*) the other objects are arsenical bronze. The bronze blade with silver rivets is 47 cm long and had sharp edges. After Munchaev 1994 and the Metropolitan Museum of Art, New York.

The Age and Development of the Maikop Culture

The relationship between Maikop and Mesopotamia was misunderstood until just recently. The extraordinary wealth of the Maikop culture seemed to fit comfortably in an age of ostentation that peaked around 2500 BCE, typified by the gold treasures of Troy II and the royal "death-pits" of Ur in Mesopotamia. But since the 1980s it has slowly become clear that the Maikop chieftain's grave probably was constructed about 3700–3400 BCE, during the Middle Uruk period in Mesopotamia—a thousand years *before* Troy II. The archaic style of the Maikop artifacts was recognized in the 1920s by Rostovtseff, but it took radiocarbon dates to prove him right. Rezepkin's excavations at Klady in 1979–80 yielded six radiocarbon dates averaging between 3700 and 3200 BCE (on human bone, so possibly a couple of centuries too old because of old carbon contamination from fish in the diet). These dates were confirmed by three radiocarbon dates also averaging between 3700 and 3200 BCE at the early Maikop-culture settlement of Galugai, excavated by S. Korenevskii between 1985 and 1991 (on animal bone and charcoal, so probably accurate). Galugai's pot types and metal types were exactly like those in the Maikop chieftain's grave, the type site for early Maikop. Graves in kurgan 32 at Ust-Dzhegutinskaya that were stylistically post-Maikop were radiocarbon dated about 3000–2800 BCE. These dates showed that Maikop was contemporary with the first cities of Middle and Late Uruk-period Mesopotamia, 3700–3100 BCE, an extremely surprising discovery.[27]

The radiocarbon dates were confirmed by an archaic cylinder seal found in an early Maikop grave excavated in 1984 at Krasnogvardeiskoe, about 60 km north of the Maikop chieftain's grave. This grave contained an east Anatolian agate cylinder seal engraved with a deer and a tree of life. Similar images appeared on stamp seals at Degirmentepe in eastern Anatolia before 4000 BCE, but cylinder seals were a later invention, appearing first in Middle Uruk Mesopotamia. The one from the kurgan at Kransogvardeiskoe (Red Guards), perhaps worn as a bead, is among the oldest of the type (see figure 12:10).[28]

The Maikop chieftain's grave is the type site for the early Maikop period, dated between 3700 and 3400 BCE. All the richest graves and hoards of the early period were in the Kuban River region, but the innovations in funeral ceremonies, arsenical bronze metallurgy, and ceramics that defined the Maikop culture were shared across the North Caucasus piedmont to the central plateau and as far as the middle Terek River valley. Galugai on the middle Terek River was an early Maikop settlement, with round houses

6–8 m in diameter scattered 10–20 m apart along the top of a linear ridge. The estimated population was less than 100 people. Clay, bell-shaped loom weights indicated vertical looms; four were found in House 2. The ceramic inventory consisted largely of open bowls (probably food bowls) and globular or elongated, round-bodied pots with everted rims, fired to a reddish color; some of these were made on a slow wheel. Cattle were 49% of the animal bones, sheep-goats were 44%, pigs were 3%, and horses (presumably horses that looked like the one engraved on the Maikop silver cup) were 3%. Wild boar and onagers were hunted only occasionally. Horse bones appeared in other Maikop settlements, in Maikop graves (Inozemstvo kurgan contained a horse jaw), and in Maikop art, including a frieze of nineteen horses painted in black and red colors on a stone wall slab inside a late Maikop grave at Klady kurgan 28 (figure 12.11). The widespread appearance of horse bones and images in Maikop sites suggested to Chernykh that horseback riding began in the Maikop period.[29]

The late phase of the Maikop culture probably should be dated about 3400–3000 BCE, and the radiocarbon dates from Klady might support this if they were corrected for reservoir effects. Having no ^{15}N measurements from Klady, I don't know if this correction is justified. The type sites for the late Maikop phase are Novosvobodnaya kurgan 2, located southeast of Maikop in the Farsa River valley, excavated by N. I. Veselovskii in 1898; and Klady (figure 12.11), another kurgan cemetery near Novosvobodnaya, excavated by A. D. Rezepkin in 1979–80. Rich graves containing metals, pottery, and beads like Novosvobodnaya and Klady occurred across the North Caucasus piedmont, including the central plateau (Inozemtsvo kurgan, near Mineralnyi Vody) and in the Terek drainage (Nalchik kurgan). Unlike the sunken grave chamber at Maikop, most of these graves were built on the ground surface (although Nalchik had a sunken grave chamber); and, unlike the timber-roofed Maikop grave, their chambers were constructed entirely of huge stones. In Novosvobodnaya-type graves the central and attendant/gift grave compartments were divided, as at Maikop, but the stone dividing wall was pierced by a round hole. The stone walls of the Nalchik grave chamber incorporated carved stone stelae like those of the Mikhailovka I and Kemi-Oba cultures (see figure 13.11).

Arsenical bronze tools and weapons were much more abundant in the richest late Maikop graves of the Klady-Novosvobodnaya type than they were in the Maikop chieftain's grave. Grave 5 in Klady kurgan 31 alone contained fifteen heavy bronze daggers, a sword 61 cm long (the oldest sword in the world), three sleeved axes and two cast bronze hammer-axes, among many other objects, for one adult male and a seven-year-old child

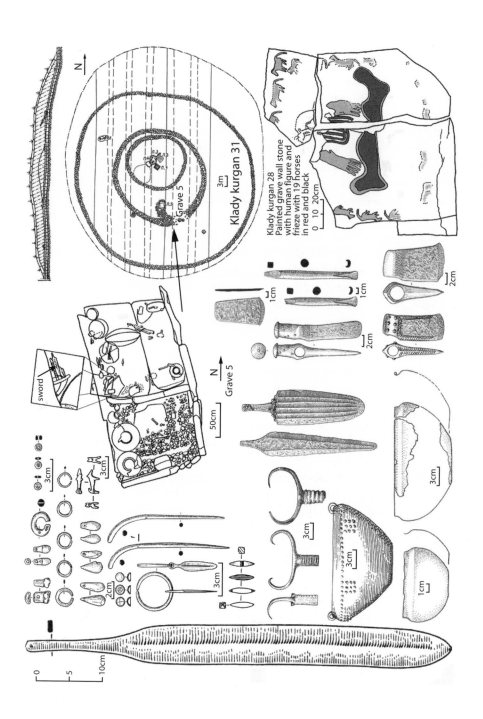

Klady kurgan 31

Grave 5

3m

N

Klady kurgan 28
Painted grave wall stone
with human figure and
frieze with 19 horses
in red and black

0 10 20cm

sword

Grave 5

N

50cm

3cm

3cm

2cm

3cm

3cm

2cm

2cm

1cm

1cm

1cm

1cm

3cm

3cm

3cm

3cm

2cm

0

5

10cm

(see figure 12.11). The bronze tools and weapons in other Novosvobodnaya-phase graves included cast flat axes, sleeved axes, hammer-axes, heavy tanged daggers with multiple midribs, chisels, and spearheads. The chisels and spearheads were mounted to their handles the same way, with round shafts hammered into four-sided contracting bases that fit into a V-shaped rectangular hole on the handle or spear. Ceremonial objects included bronze cauldrons, long-handled bronze dippers, and two-pronged bidents (perhaps forks for retrieving cooked meats from the cauldrons). Ornaments included beads of carnelian from western Pakistan, lapis lazuli from Afghanistan, gold, rock crystal, and even a bead from Klady made of a human molar sheathed in gold (the first gold cap!). Late Maikop graves contained several late metal types—bidents, tanged daggers, metal hammer-axes, and a spearhead with a tetrahedral tang—that did not appear at Maikop or in other early sites. Flint arrowheads with deep concave bases also were a late type, and black burnished pots had not been in earlier Maikop graves.[30]

Textile fragments preserved in Novosvobodnaya-type graves included linen with dyed brown and red stripes (at Klady), a cotton-like textile, and a wool textile (both at Novosvobodnaya kurgan 2). Cotton cloth was invented in the Indian subcontinent by 5000 BCE; the piece tentatively identified in the Novosvobodnaya royal grave might have been imported from the south.[31]

The Road to the Southern Civilizations

The southern wealth that defined the Maikop culture appeared suddenly in the North Caucasus, and in large amounts. How did this happen, and why?

Figure 12.11 Late Maikop-Novosvobodnaya objects and graves at Klady, Kuban River drainage, North Caucasus: (*Right*) plan and section of Klady kurgan 31 and painted grave wall from Klady kurgan 28 with frieze of red-and-black horses surrounding a red-and-black humanlike figure; (*left and bottom*): objects from grave 5, kurgan 31. These included (*left*) arsenical bronze sword; (*top row, center*) two beads of human teeth sheathed in gold, a gold ring, and three carnelian beads; (*second row*) five gold rings; (*third row*) three rock crystal beads and a cast silver dog; (*fourth row*) three gold button caps on wooden cores; (*fifth row*) gold ring-pendant and two bent silver pins; (*sixth row*) carved bone dice; (*seventh row*) two bronze bidents, two bronze daggers, a bronze hammer-axe, a flat bronze axe, and two bronze chisels; (*eighth row*) a bronze cauldron with repoussé decoration; (*ninth row*) two bronze cauldrons and two sleeved axes. After Rezepkin 1991, figures 1, 2, 4, 5, 6.

The valuables that seemed the most interesting to Mesopotamian urban traders were metals and precious stones. The upper Kuban River is a metal-rich zone. The Elbrusskyi mine on the headwaters of the Kuban, 35 km northwest of Elbruz Mountain (the highest peak in the North Caucasus) produces copper, silver, and lead. The Urup copper mine, on the upper Urup River, a Kuban tributary, had ancient workings that were visible in the early twentieth century. Granitic gold ores came from the upper Chegem River near Nalchik. As the metal prospectors who profited from the Uruk metal trade explored northward, they somehow learned of the copper, silver, and gold ores on the other side of the North Caucasus Mountains. Possibly they also pursued the source of textiles made of long-woolen thread.

It is possible that the initial contacts were made on the Black Sea coast, since the mountains are easy to cross between Maikop and Sochi on the coast, but much higher and more difficult in the central part of the North Caucasus farther east. Maikop ceramics have been found north of Sochi in the Vorontsovskaya and Akhshtyrskaya caves, just where the trail over the mountains meets the coast. This would also explain why the region around Maikop initially had the richest graves—if it was the terminal point for a trade route that passed through eastern Anatolia to western Georgia, up the coast to Sochi, and then to Maikop. The metal ores came from deposits located east of Maikop, so if the main trade route passed through the high passes in the center of the Caucasus ridge we would expect to see more southern wealth near the mines, not off to the west.

By the late Maikop (Novosvobodnaya) period, contemporary with Late Uruk, an eastern route was operating as well. Turquoise and carnelian beads were found at the walled town of Alikemek Tepesi in the Mil'sk steppe in Azerbaijan, near the mouth of the Kura River on the Caspian shore.[32] Alikemek Tepesi possibly was a transit station on a trade route that passed around the eastern end of the North Caucasian ridge. An eastern route through the Lake Urmia basin would explain the discovery in Iran, southwest of Lake Urmia, of a curious group of eleven conical, gravel-covered kurgans known collectively as Sé Girdan. Six of them, up to 8.2 m high and 60 m in diameter, were excavated by Oscar Muscarella in 1968 and 1970. Then thought to date to the Iron Age, they recently have been redated on the basis of their strong similarities to Novosvobodnaya-Klady graves in the North Caucasus.[33] The kurgans and grave chambers were made the same way as those of the Novosvobodnaya-Klady culture; the burial pose was the same; the arsenical bronze flat axes and short-nosed shaft-hole axes were similar in shape and manufacture to Novosvobodnaya-Klady types; and carnelian and gold beads were the same shapes, both containing silver vessels and fragments of silver tubes. The Sé Girdan kurgans could represent the migration southward of a Klady-type chief, perhaps to eliminate troublesome local middlemen. But the Lake Urmia chiefdom did not last. Moscarella counted almost ninety sites of the succeeding Early Trans-Caucasian Culture (ETC) around the southern Urmia Basin, but none of them had even small kurgans.

The power of the Maikop chiefs probably grew partly from the aura of the extraordinary that clung to the exotic objects they accumulated, which were palpable symbols of their personal connection with powers previously unknown.[34] Perhaps the extraordinary nature of these objects was one of the reasons why they were buried with their owners rather than inherited. Limited use and circulation were common characteristics of objects regarded as "primitive valuables." But the supply of new valuables dried up when the Late Uruk long-distance exchange system collapsed about 3100 BCE. Mesopotamian cities began to struggle with internal problems that we can perceive only dimly, their foreign agents retreated, and in the mountains the people of the ETC attacked and burned Arslantepe and Hacinebi on the upper Euphrates. Sé Girdan stood abandoned. This was also the end of the Maikop culture.

MAIKOP-NOVOSVOBODNAYA IN THE STEPPES: CONTACTS WITH THE NORTH

Valuables of gold, silver, lapis, turquoise, and carnelian were retained exclusively by the North Caucasian individuals in direct contact with the south and perhaps by those who lived near the silver and copper mines that fed the southern trade. But a revolutionary new technology for land transport—wagons—might have been given to the steppes by the Maikop culture. Traces of at least two solid wooden disc wheels were found in a late Maikop kurgan on the Kuban River at Starokorsunskaya kurgan 2, with Novosvobodnaya black-burnished pots. Although not dated directly, the wooden wheels in this kurgan might be among the oldest in Europe.[35] Another Novosvobodnaya grave contained a bronze cauldron with a schematic image that seems to portray a cart. It was found at Evdik.

Evdik kurgan 4 was raised by the shore of the Tsagan-Nur lake in the North Caspian Depression, 350 km north of the North Caucasus piedmont, in modern Kalmykia.[36] Many shallow lakes dotted the Sarpa Depression, an ancient channel of the Volga. At Evdik, grave 20 contained an adult male in a contracted position oriented southwest, the standard Maikop pose, stained with red ochre, with an early Maikop pot by his feet. This was

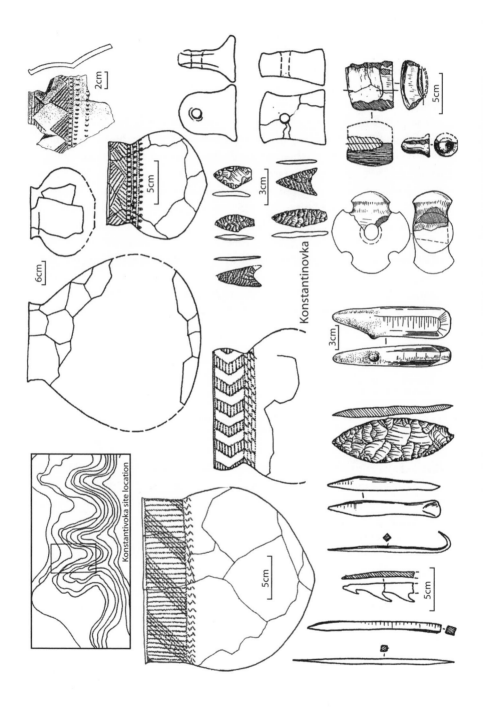

Konstantinovka

Konstantivoka site location

the original grave over which the kurgan was raised. Two other graves followed it, without diagnostic grave goods, after which grave 23 was dug into the kurgan. This was a late Maikop grave. It contained an adult male and a child buried together in sitting positions, an unusual pose, on a layer of white chalk and red ochre. In the grave was a bronze cauldron decorated with an image made in repoussé dots. The image seems to portray a yoke, a wheel, a vehicle body, and the head of an animal (see figure 4.3a). Grave 23 also contained a typical Novosvobodnaya bronze socketed bident, probably used with the cauldron. And it also had a bronze tanged dagger, a flat axe, a gold ring with 2.5 twists, a polished black stone pestle, a whetstone, and several flint tools, all typical Novosvobodnaya artifacts. Evdik kurgan 4 shows a deep penetration of the Novosvobodnaya culture into the lower Volga steppes. The image on the cauldron suggests that the people who raised the kurgan at Evdik also drove carts.

Evdik was the richest of the Maikop-Novosvobodnaya kurgans that appeared in the steppes north of the North Caucasus between 3700 and 3100 BCE. In such places, late Novosvobodnaya people whose speech would probably be assigned to a Caucasian language family met and spoke with individuals of the Repin and Late Khvalynsk cultures who probably spoke Proto-Indo-European dialects. The loans discussed in chapter 5 between archaic Caucasian and Proto-Indo-European languages probably were words spoken during these exchanges. The contact was most obvious, and therefore perhaps most direct, on the lower Don.

Trade across a Persistent Cultural Frontier

Konstantinovka, a settlement on the lower Don River, might have contained a resident group of Maikop people, and there were kurgan graves with Maikop artifacts around the settlement (figure 12.12). About 90% of the settlement ceramics were a local Don-steppe shell-tempered, cord-impressed type connected with the cultures of the Dnieper-Donets steppes to the west (late Sredni Stog, according to Telegin). The other 10% were red-burnished early Maikop wares. Konstantinovka was located on a steep-sided promontory overlooking the strategic lower Don valley, and was protected by a ditch and bank. The gallery forests below it were full of deer (31% of the bones) and

Figure 12.12 Konstantinovka settlement on the lower Don, with topographic location and artifacts. Plain pots are Maikop-like; cord-impressed pots are local. Loom-weights and asymmetrical flint points also are Maikop-like. *Lower right*: crucible and bellows fragments. After Kiashko 1994.

the plateau behind it was the edge of a vast grassland rich in horses (10%), onagers (2%), and herds of sheep/goats (25%). Maikop vistors probably imported the perforated clay loom weights similar to those at Galugai (unique in the steppes), copper chisels like those at Novosvobodnaya (again, unique except for two at Usatovo; see chapter 14), and asymmetrical shouldered flint projectile points very much like those of the Maikop-Novosvobodnaya graves. But polished stone axes and gouges, a drilled cruciform polished stone mace head, and boars-tusk pendants were steppe artifact types. Crucibles and slag show that copper working occurred at the site.

A. P. Nechitailo identified dozens of kurgans in the North Pontic steppes that contained single pots or tools or both that look like imports from Maikop-Novosvobodnaya, distributed from the Dniester River valley on the west to the lower Volga on the east. These widespread northern contacts seem to have been most numerous during the Novosvobodnaya/Late Uruk phase, 3350–3100 BCE. But most of the Caucasian imports appeared singly in local graves and settlements. The region that imported the largest number of Caucasian arsenical bronze tools and weapons was the Crimean Peninsula (the Kemi-Oba culture). The steppe cultures of the Volga-Ural region imported little or no Caucasian arsenical bronze; their metal tools and weapons were made from local "clean" copper. Sleeved, one-bladed metal axes and tanged daggers were made across the Pontic-Caspian steppes in emulation of Maikop-Novosvobodnaya types, but most were made locally by steppe metalsmiths.[37]

What did the Maikop chiefs want from the steppes? One possibility is drugs. Sherratt has suggested that narcotics in the form of *Cannabis* were one of the important exports of the steppes.[38] Another more conventional trade item could have been wool. We still do not know where wool sheep were first bred, although it makes sense that northern sheep from the coldest places would initially have had the thickest wool. Perhaps the Maikop-trained weavers at Konstantinovka were there with their looms to make some of the raw wool into large textiles for payment to the herders. Steppe people had felts or textiles made from narrow strips of cloth, produced on small, horizontal looms, then stitched together. Large textiles made in one piece on vertical looms were novelties.

Another possibility is horses. In most Neolithic and earlier Eneolithic sites across Transcaucasia there were no horse bones. After the evolution of the ETC culture beginning about 3300 BCE horses became widespread, appearing in many sites across Transcaucasia. S. Mezhlumian reported horse bones at ten of twelve examined sites in Armenia dated to the later fourth millennium BCE. At Mokhrablur one horse had severe wear on a P_2 consistent with bit wear. Horses were bitted at Botai and Kozhai 1 in

Kazakhstan during the same period, so bit wear at Mokhrablur would not be unique. At Alikemek Tepesi the horses of the ETC period were thought by Russian zoologists to be domesticated. Horses the same size as those of Dereivka appeared as far south as the Malatya-Elazig region in southeastern Turkey, as at Norşuntepe; and in northwestern Turkey at Demirci Höyük. Although horses were not traded into the lowlands of Mesopotamia this early, they might have been valuable in the steppe-Caucasian trade.[39]

PROTO-INDO-EUROPEAN AS A REGIONAL
LANGUAGE IN A CHANGING WORLD

During the middle centuries of the fourth millennium BCE the equestrian tribes of the Pontic-Caspian steppes exhibited a lot of material and probably linguistic variability. They absorbed into their conversations two quite different but equally surprising developments among their neighbors to the south, in the North Caucasus piedmont, and to their west, in the Cucuteni-Tripolye region. From the North Caucasus probably came wagons, and with them ostentatious displays of incredible wealth. In the west, some Tripolye populations retreated into huge planned towns larger than any settlements in the world, probably in response to raiding from the steppes. Other Tripolye towns farther north on the Dnieper began to change their customs in ceramics, funerals, and domestic architecture toward steppe styles in a slow process of assimilation.

Although regionally varied, steppe cultural habits and customs remained distinct from those of the Maikop culture. An imported Maikop or Novosvobodnaya potsherd is immediately obvious in a steppe grave. Lithics and weaving methods were different (no loom weights in the steppes), as were bead and other ornament types, economies and settlement forms, and metal types and sources. These distinctions persisted in spite of significant cross-frontier interaction. When Maikop traders came to Konstantinovka, they probably needed a translator.

The Yamnaya horizon, the material expression of the late Proto-Indo-European community, grew from an eastern origin in the Don-Volga steppes and spread across the Pontic-Caspian steppes after about 3300 BCE. Archaeology shows that this was a period of profound and rapid change along all the old ethnolinguistic frontiers surrounding the Pontic-Caspian steppes. Linguistically based reconstructions of Proto-Indo-European society often suggest a static, homogeneous ideal, but archaeology shows that Proto-Indo-European dialects and institutions spread through steppe societies that exhibited significant regional diversity, during a period of far-reaching social and economic change.

Chapter Thirteen

Wagon Dwellers of the Steppe
The Speakers of Proto-Indo-European

The sight of wagons creaking and swaying across the grasslands amid herds of wooly sheep changed from a weirdly fascinating vision to a normal part of steppe life between about 3300 and 3100 BCE. At about the same time the climate in the steppes became significantly drier and generally cooler than it had been during the Eneolithic. The shift to drier conditions is dated between 3500 and 3000 BCE in pollen cores in the lower Don, the middle Volga, and across the northern Kazakh steppes (table 13.1). As the steppes dried and expanded, people tried to keep their animal herds fed by moving them more frequently. They discovered that with a wagon you could keep moving indefinitely. Wagons and horseback riding made possible a new, more mobile form of pastoralism. With a wagon full of tents and supplies, herders could take their herds out of the river valleys and live for weeks or months out in the open steppes between the major rivers—the great majority of the Eurasian steppes. Land that had been open and wild became pasture that belonged to someone. Soon these more mobile herding clans realized that bigger pastures and a mobile home base permitted them to keep bigger herds. Amid the ensuing disputes over borders, pastures, and seasonal movements, new rules were needed to define what counted as an acceptable move—people began to manage local migratory behavior. Those who did not participate in these agreements or recognize the new rules became cultural Others, stimulating an awareness of a distinctive Yamnaya identity. That awareness probably elevated a few key behaviors into social signals. Those behaviors crystallized into a fairly stable set of variants in the steppes around the lower Don and Volga rivers. A set of dialects went with them, the speech patterns of late Proto-Indo-European. This is the sequence of changes that I believe created the new way of life expressed archaeologically in the Yamnaya horizon, dated about 3300–2500 BCE (figure 13.1). The spread

Table 13.1

Vegetation shifts in steppe pollen cores from the Don to the Irtysh

Site	Razdorskoe, *Lower Don* (Kremenetski 1997)	Buzuluk Forest Pobochnoye peat bog *Middle Volga* (Kremenetski et al. 1999)	Northern Kazakhstan Upper Tobol to *Upper Irtysh* (Kremenetski et al. 1997)
Type	Stratified settlement Pollen core	forest peat bog core	two lake cores and two peat bog cores
Dates Flora	*6500–3800 BCE* Birch-pine forest on sandy river terraces. On floodplain, elm and linden forest with hazelnut & black alder. Oak and hornbeam present after 4300 BCE.	*6000–3800 BCE* Oak trees appear, join elm, hazel, black alder forests around Pobochnoye lake. 4800–3800 BCE lake gets shallower, Typha reeds increase, forest expands.	*6500–3800 BCE* Birch-pine forest evolving to open pine forest in forest-steppe, with willow near waterways. In steppe, Artemesia and Chenopodia.
	3800–3300 BCE Slight reduction in deciduous trees, increase in Ephedra, hazel, lime, and pine on floodplain.	*3800–3300 BCE* Lake slowly converts to sedge-moss swamp. Typha reeds peak. Pine and lime trees peak. Probably warmer.	*3800–3300 BCE* Moist period, forests expand. Lime trees with oak, elm, and black alder also expand. Soils show increased moisture.
	Sub-Boreal 3300–2000 BCE Very dry. Sharp forest decline. Ceralia appears. Chenopodia sharp rise. Maximum aridity 2800–2000 BCE.	*3300–2000 BCE* Reduction in overall forest. In forest, pine down, birch up. Artemesia, an arid herb indicator, increases sharply. Lake is covered by alder shrubs by 2000 BCE.	*3300–2000 BCE* Forest retreats, broadleaf declines. Mokhove bog on the Tobol dries up about 2800 BCE. Steppe grows.

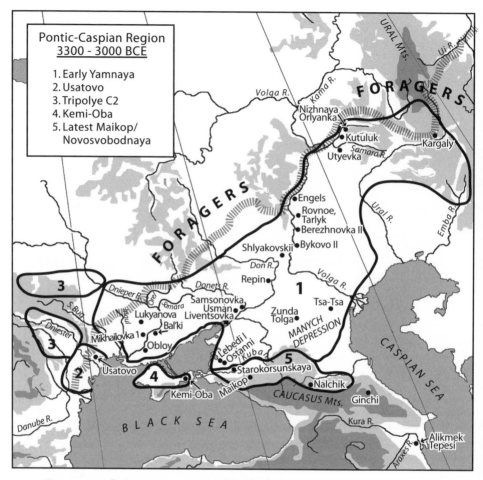

Figure 13.1 Culture areas in the Pontic-Caspian region about 3300–3000 BCE.

of the Yamnaya horizon was the material expression of the spread of late Proto-Indo-European across the Pontic-Caspian steppes.[1]

The behavior that really set the Yamnaya people apart was living on wheels. Their new economy took advantage of two kinds of mobility: wagons for slow bulk transport (water, shelter, and food) and horseback riding for rapid light transport (scouting for pastures, herding, trading and raiding expeditions). Together they greatly increased the potential scale of herding economies. Herders operating out of a wagon could stay with their herds out in the deep steppes, protected by mobile homes that carried tents, water, and food. A diet of meat, milk, yogurt, cheese, and soups made of wild *Chenopodium* seeds and wild greens can be deduced, with a little imagination, from the archaeological evidence. The reconstructed Proto-Indo-European vocabulary tells us that honey and honey-based mead also were consumed, probably on special occasions. Larger herds meant greater disparities in herd wealth, which is reflected in disparities in the wealth of Yamanaya graves. Mobile wagon camps are almost impossible to find archaeologically, so settlements became archaeologically invisible where the new economy took hold.

The Yamnaya horizon is the visible archaeological expression of a social adjustment to high mobility—the invention of the political infrastructure to manage larger herds from mobile homes based in the steppes. A linguistic echo of the same event might be preserved in the similarity between English *guest* and *host*. They are cognates, derived from one Proto-Indo-European root (*ghos-ti-*). (A "ghost" in English was originally a visitor or guest.) The two social roles opposed in English *guest* and *host* were originally two reciprocal aspects of the same relationship. The late Proto-Indo-European guest-host relationship required that "hospitality" (from the same root through Latin *hospes* 'foreigner, guest') and "friendship" (*keiwos-*) should be extended by hosts to guests (both *ghos-ti-*), in the knowledge that the receiver and giver of "hospitality" could later reverse roles. The social meaning of these words was then more demanding than modern customs would suggest. The guest-host relationship was bound by oaths and sacrifices so serious that Homer's warriors, Glaukos and Diomedes, stopped fighting and presented gifts to each other when they learned that their *grandfathers* had shared a guest-host relationship. This mutual obligation to provide "hospitality" functioned as a bridge between social units (tribes, clans) that had ordinarily restricted these obligations to their kin or co-residents (*h₄erós-*). Guest-host relationships would have been very useful in a mobile herding economy, as a way of separating people who were moving through your territory with your assent from those who were unwelcome, unregulated, and therefore unprotected. The guest-host institution might have been among the critical identity-defining innovations that spread with the Yamnaya horizon.[2]

It is difficult to document a shift to a more mobile residence pattern five thousand years after the fact, but a few clues survive. Increased mobility can be detected in a pattern of brief, episodic use, abandonment, and, much later, re-use at many Yamnaya kurgan cemeteries; the absence of degraded or overgrazed soils under early Yamnaya kurgans; and the first appearance of kurgan cemeteries in the deep steppe, on the dry plateaus between major river valleys. The principal indicator of increased mobility is a negative piece of evidence: the archaeological disappearance of long-term settlements east of the Don River. Yamnaya settlements are known west of the Don in Ukraine, but east of the Don in Russia there are no significant Yamnaya settlements in a huge territory extending to the Ural River containing many hundreds of excavated Yamnaya kurgan cemeteries and probably thousands of excavated Yamnaya graves (I have never seen a full count). The best explanation for the complete absence of settlements is that the eastern Yamnaya people spent much of their lives in wagons.

The Yamnaya horizon was the first more or less unified ritual, economic, and material culture to spread across the entire Pontic-Caspian steppe region, but it was never completely homogeneous even materially. At the beginning it already contained two major variants, on the lower Don and lower Volga, and, as it expanded, it developed other regional variants, which is why most archaeologists are reluctant to call it the Yamnaya "culture." But many broadly similar customs were shared. In addition to kurgan graves, wagons, and an increased emphasis on pastoralism, archaeological traits that defined the early Yamnaya horizon included shell-tempered, egg-shaped pots with everted rims, decorated with comb stamps and cord impressions; tanged bronze daggers; cast flat axes; bone pins of various types; the supine-with-raised-knees burial posture; ochre staining on grave floors near the feet, hips, and head; northeastern to eastern body orientation (usually); and the sacrifice at funerals of wagons, carts, sheep, cattle, and horses. The funeral ritual probably was connected with a cult of ancestors requiring specific rituals and prayers, a connection between language and cult that introduced late Proto-Indo-European to new speakers.

The most obvious material division within the early Yamnaya horizon was between east and west. The eastern (Volga–Ural–North Caucasian steppe) Yamnaya pastoral economy was more mobile than the western one (South Bug–lower Don). This contrast corresponds in an intriguing way to economic and cultural differences between eastern and western Indo-European language branches. For example, impressions of cultivated grain have been found in western Yamnaya pottery, in both settlements and graves, and Proto-Indo-European cognates related to cereal agriculture were well preserved in western Indo-European vocabularies. But grain imprints are absent in eastern Yamnaya pots, just as many of the cognates related to agriculture are missing from the eastern Indo-European languages.[3] Western Indo-European vocabularies contained a few roots that were borrowed from Afro-Asiatic languages, such as the word for the domesticated bull, **tawr-*, and the western Yamnaya groups lived next to the Tripolye culture, which might have spoken a language distantly derived from an Afro-Asiatic language of Anatolia. Eastern Indo-European generally lacked these borrowed Afro-Asiatic roots. Western Indo-European religious and ritual practices were female-inclusive, and western Yamnaya people shared a border with the female-figurine–making Tripolye culture: eastern Indo-European rituals and gods, however, were more male-centered, and eastern Yamnaya people shared borders with northern and eastern foragers who did not make female figurines. In western Indo-European branches the spirit of the domestic hearth was female (Hestia, the Vestal Virgins), and in Indo-Iranian it was male (Agni). Western Indo-European mythologies included strong female deities such as Queen Magb and the Valkyries, whereas in Indo-Iranian the furies of war were male Maruts. Eastern Yamnaya graves on the Volga contained a higher percentage (80%) of males than any other Yamnaya region. Perhaps this east-west tension in attitudes toward gender contributed to the separation of the feminine gender as a newly marked grammatical category in the dialects of the Volga-Ural region, one of the innovations that defined Proto-Indo-European grammar.[4]

Did the Yamnaya horizon spread into neighboring regions in a way that matches the known relationships and sequencing between the Indo-European branches? This also is a difficult subject to follow archaeologically, but the movements of the Yamnaya people match what we would expect surprisingly well. First, just before the Yamnaya horizon appeared, the Repin culture of the Volga-Ural region threw off a subgroup that migrated across the Kazakh steppes about 3700–3500 BCE and established itself in the western Altai, where it became the Afanasievo culture. The separation of the Afanasievo culture from Repin probably represented the separation of Pre-Tocharian from classic Proto-Indo-European. Second, some three to five centuries later, about 3300 BCE, the rapid diffusion of the early Yamnaya horizon across the Pontic-Caspian steppes scattered the speakers of late Proto-Indo-European dialects and sowed the seeds of regional differentiation. After a pause of only a century or two, about 3100–3000 BCE, a large migration stream erupted from within the western Yamnaya region and flowed up the Danube valley and into the Carpathian Basin during the Early Bronze Age. Literally thousands of kurgans can be assigned to this event, which could reasonably have incubated the ancestral dialects for several western Indo-European language branches, including Pre-Italic and Pre-Celtic. After this movement slowed or stopped, about 2800–2600 BCE, late Yamnaya people came face to face with people who made Corded Ware tumulus cemeteries in the east Carpathian foothills, a historic meeting through which dialects ancestral to the northern Indo-European languages (Germanic, Slavic, Baltic) began to spread among eastern Corded Ware groups. Finally, at the end of the Middle Bronze Age, about 2200–2000 BCE, a migration stream flowed from the late Yamnaya/Poltavka cultures of the Middle Volga–Ural region eastward around the southern Urals, creating the Sintashta culture, which almost certainly represented the ancestral Indo-Iranian–speaking community. These migrations are described in chapters 14 and 15.

The Yamnaya horizon meets the expectations for late Proto-Indo-European in many ways: chronologically (the right time), geographically (the right place), materially (wagons, horses, animal sacrifices, tribal pastoralism), and linguistically (bounded by persistent frontiers); and it generated migrations in the expected directions and in the expected sequence. Early Proto-Indo-European probably developed between 4000 and 3500 BCE in the Don–Volga–Ural region. Late Proto-Indo-European, with o-stems and the full wagon vocabulary, expanded rapidly across the Pontic-Caspian steppes with the appearance of the Yamnaya horizon beginning about 3300 BCE. By 2500 BCE the Yamnaya horizon had fragmented into daughter groups, beginning with the appearance of the Catacomb culture in the Don-Kuban region and the Poltavka culture in the Volga-Ural region about 2800 BCE. Late Proto-Indo-European also was so diversified by 2500 BCE that it probably no longer existed (chapter 3). Again, the linkage with the steppe archaeological evidence is compelling.

Why Not a Kurgan Culture?

Marija Gimbutas first articulated her concept of a "Kurgan culture" as the archaeological expression of the Proto-Indo-European language community in 1956.[5] The Kurgan culture combined two cultures first defined by V. A. Gorodtsov, who, in 1901, excavated 107 kurgans in the Don River valley. He divided his discoveries into three chronological groups. The oldest graves, stratified deepest in the oldest kurgans, were the Pit-graves (Yamnaya). They were followed by the Catacomb-graves (Katakombnaya), and above them were the timber-graves (Srubnaya). Gorodtsov's sequence still defines the Early (EBA), Middle (MBA), and Late Bronze Age (LBA) grave types of the western steppes.[6] Gimbutas combined the first two (EBA Pit-graves and MBA Catacomb-graves) into the Kurgan culture. But later she also began to include many other Late Neolithic and

Bronze Age cultures of Europe, including the Maikop culture and many of the Late Neolithic cultures of eastern Europe, as outgrowths or creations of Kurgan culture migrations. The Kurgan culture was so broadly defined that almost any culture with burial mounds, or even (like the Baden culture) without them could be included. Here we are discussing the steppe cultures of the Russian and Ukrainian EBA, just one part of the original core of Gimbutas's Kurgan culture concept. Russian and Ukrainian archaeologists do not generally use the term "Kurgan culture"; rather than lumping EBA Yamnaya and MBA Catacomb-graves together they tend to divide both groups and their associated time periods into ever finer slices. I will seek a middle ground.

The Yamnaya horizon is usually described by Slavic archaeologists not as a "culture" but as a "cultural-historical community." This phrase carries the implication that there was a thread of cultural identity or shared ethnic origin running through the Yamnaya social world, although one that diversified and evolved with the passage of time.[7] Although I agree that this probably was true in this case, I will use the Western term "horizon," which is neutral about cultural identity, in order to avoid using a term loaded toward that interpretation. As I explained in chapter 7, a horizon in archaeology is a style or fashion in material culture that is rapidly accepted by and superimposed on local cultures across a wide area. In this case, the five Pontic-Caspian cultures of the Final Eneolithic (chapter 12) were the local cultures that rapidly accepted, in varying degrees, the Yamnaya lifestyle.

Beyond the Eastern Frontier: The Afanasievo Migration to the Altai

In the last chapter I introduced the subject of the trans-continental, Repin-culture migration that created the Afanasievo culture in the western Altai Mountains and probably detached the Tocharian branch from common Proto-Indo-European. I describe it here because the process of migration and return migration that installed the early Afanasievo culture continued across the north Kazakh steppes during the Yamnaya period. In fact, it is usually discussed as an event connected with the Yamnaya horizon; it is only recently that early Afanasievo radiocarbon dates, and the broadening understanding of the age and geographic extent of the Repin culture, have pushed the beginning of the movement back into the pre-Yamnaya Repin period.

Two or three centuries before the Yamnaya horizon first appeared, the Repin-type communities of the middle Volga-Ural steppes experienced a

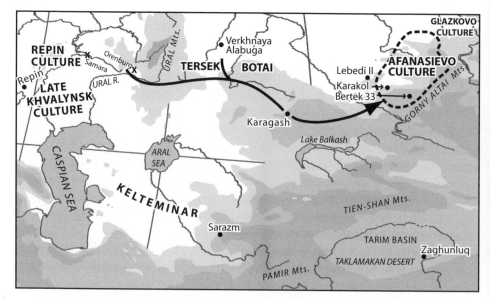

Figure 13.2 Culture areas in the steppes between the Volga and the Altai at the time of the Afanasievo migration, 3700–3300 BCE.

conflict that prompted some groups to move across the Ural River eastward into the Kazakh steppes (figure 13.2). I say a conflict because of the extraordinary distance the migrants eventually put between themselves and their relatives at home, implying a strongly negative push. On the other hand, connections with the Volga-Ural Repin-Yamnaya world were maintained by a continuing round of migrations moving in both directions, so some aspect of the destination must also have exerted a positive pull. It is remarkable that the intervening north Kazakh steppe was not settled, or at least that almost no kurgan cemeteries were constructed there. Instead, the indigenous horse-riding Botai-Tersek culture emerged in the north Kazakh steppe at just the time when the Repin-Afanasievo migration began.

The specific ecological target in this series of movements might have been the islands of pine forest that occur sporadically in the northern Kazakh steppes from the Tobol River in the west to the Altai Mountains in the east. I am not sure why these pine islands would have been targeted other than for the fuel and shelter they offered, but they do seem to correspond with the few site locations linked to Afanasievo in the steppes, and the same peculiar steppe-pine-forest islands occur also in the high mountain valleys of the western Altai where early Afanasievo sites appeared.[8] In the western Altai Mountains broad meadows and mountain steppes dip both westward toward the Irtysh River of western Siberia (probably the route of the first approach) and northward toward the Ob and Yenisei rivers (the later spread). The Afanasievo culture appeared in this beautiful setting, ideal for upland pastoralism, probably around 3700–3400 BCE, during the Repin–late Khvalynsk period.[9] It flourished there until about 2400 BCE, through the Yamnaya period in the Pontic-Caspian steppes.

The Altai Mountains were about 2000 km east of the Ural River frontier that defined the eastern edge of the early Proto-Indo-European world. Only three kurgan cemeteries old enough to be connected with the Afanasievo migrations have been found in the intervening 2000 km of steppes. All three are classified as Yamnaya kurgan cemeteries, although the pottery in some of the graves has Repin traits. Two were on the Tobol, not far east of the Ural River, at Ubagan I and Verkhnaya Alabuga, possibly an initial stopping place. The other, the Karagash kurgan cemetery, was found 1000 km east of the Tobol, southeast of Karaganda in central Kazakhstan. Karagash was on the elevated green slopes of an isolated mountain spur that rose prominently above the horizon, a very visible landmark near Karkaralinsk. The earthen mound of kurgan 2 at Karagash was 27 m in diameter. It covered a stone cromlech circle 23 m in diameter, made of oblong stones 1 m in length, projecting about 60–70 cm above the ground. Some stones had traces of paint on them. A pot was broken inside the southwestern edge of the cromlech on the original ground surface, before the mound was built. The kurgan contained three graves in stone-lined cists; the central grave and another under the southeastern part of the kurgan were later robbed. The lone intact grave was found under the northeastern part of the kurgan. In it were sherds from a shell-tempered pot, a fragment of a wooden bowl with a copper-covered lip, a tanged copper dagger, a copper four-sided awl, and a stone pestle. The skeleton was of a male forty to fifty years old laid on his back with his knees raised, oriented southwest, with pieces of black charcoal and red ochre on the grave floor. The metal artifacts were typical for the Yamnaya horizon; the stone cromlech, stone-lined cist, and pot were similar to Afansievo types. Directly east of Karagash and 900 km away, up the Bukhtarta River valley east of the Irtysh, were the peaks of the western Altai and the Ukok plateau, where the first Afanasievo graves appeared. The Karagash kurgan is unlikely to be a grave of the first migrants—it looks like a Yamnaya-Afanasievo kurgan built by later people still participating in a cross-Kazakhstan circulation of movements—but it probably does mark the initial route, since routes in long-distance migrations tend to be targeted and re-used.[10]

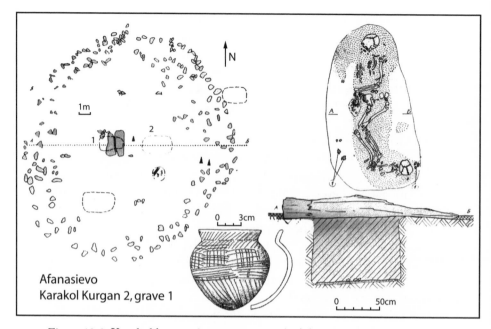

Figure 13.3 Karakol kurgan 2, grave 1, an early Afanasievo grave in the western Gorny Altai. After Kubarev 1988.

The early Afanasievo culture in the Altai introduced fully developed kurgan funeral rituals and Repin-Yamnaya material culture. At Karakol, kurgan 2 in the Gorny Altai, an early Afanasievo grave (gr. 1) contained a small pot similar to pots from the Ural River that are assigned to the Repin variant of early Yamnaya (figure 13.3).[11] Grave 1 was placed under a low kurgan in the center of a stone cromlech 20 m in diameter. Afanasievo kurgans always were marked by a ring of stones, and large stone slabs were used to cover grave pits (early) or to make stone-lined grave cists (late). Early Afanasievo skull types resembled those of Yamnaya and western populations. On the Ukok plateau, where the early Afanasievo cemetery at Bertek 33 was found, the Afanasievo immigrants occupied a virgin landscape—there were no earlier Mesolithic or Neolithic sites. Afanasievo sites also contained the earliest bones of domesticated cattle, sheep, and horses in the Altai. At the Afanasievo settlement of Balyktyul, domesticated sheep-goat were 61% of the bones, cattle were 12%, and horses 8%.[12]

Cemeteries of the local Kuznetsk-Altai foragers like Lebedi II were located in the forest and forest-meadow zone higher up on the slopes of the Altai, and contained a distinct set of ornaments (bear-teeth necklaces and bone carvings of elk and bear), lithics (asymmetrical curved flint knives), antler tools (harpoons), pottery (related to the Serovo-Glazkovo pottery tradition of the Baikal forager tradition), and funeral rituals (no kurgans, no stone slab over the grave). As time passed, Glazkovo forager sites located to the northeast began to show the influence of Afanasievo motifs on their ceramics, and metal objects began to appear in Glazkovo sites.[13]

It is clear that populations continued to circulate between the Ural frontier and the Altai well into the Yamnaya period in the Ural steppes, or after 3300 BCE, bringing many Yamnaya traits and practices to the Altai. About a hundred metal objects have been found in Afanasievo cemeteries in the Altai and Western Sayan Mountains, including three sleeved copper axes of a classic Volga-Ural Yamnaya type, a cast shaft-hole copper hammer-axe, and two tanged copper daggers of typical Yamnaya type. These artifacts are recognized by Chernykh as western types typical of Volga-Ural Yamnaya, with no native local precedents in the Altai region.[14]

Mallory and Mair have argued at book length that the Afanasievo migration detached the Tocharian branch from Proto-Indo-European. A material bridge between the Afanasievo culture and the Tarim Basin Tocharians could be represented by the long-known but recently famous Late Bronze Age Europoid "mummies" (not intentionally mummified but naturally freeze-dried) found in the northern Taklamakan Desert, the oldest of which are dated 1800–1200 BCE. In addition to the funeral ritual (on the back with raised knees, in ledged and roofed grave pits), there was a symbolic connection. On the stone walls of Late Afanasievo graves in the Altai (perhaps dated about 2500 BC) archaeologist V. D. Kubarev found paintings with "solar signs" and headdresses like the one painted on the cheek of one of the Tarim "mummies" found at Zaghunluq, dated about 1200 BCE. If Mallory and Mair were right, as seems likely, late Afanasievo pastoralists were among the first to take their herds from the Altai southward into the Tien Shan; and after 2000 BCE their descendants crossed the Tien Shan into the northern oases of the Tarim Basin.[15]

WAGON GRAVES IN THE STEPPES

We cannot say exactly when wagons first rolled into the Eurasian steppes. But an image of a wagon on a clay cup is securely dated to 3500–3300 BCE at Bronocice in southern Poland (chapter 4). The ceramic wagon models of the Baden culture in Hungary and the Novosvobodnaya wagon grave at Starokorsunskaya kurgan 2 on the Kuban River in the North Caucasus probably are about the same age. The oldest excavated wagon graves in the steppes are radiocarbon dated about 3100–3000 BCE, but it is unlikely that they actually were the first. Wagons probably appeared in the Pontic-Caspian steppes a couple of centuries before the Yamnaya horizon began. It would have taken some time for a new, wagon-dependent herding system to get organized and begin to succeed. The spread of the Yamnaya horizon was the signature of that success.

In a book published in 2000 Aleksandr Gei counted 257 Yamnaya and Catacomb-culture wagon and cart burials in the Pontic-Caspian steppes, dated by radiocarbon between about 3100 and 2200 BCE (see figures 4.4, 4.5, 4.6). Parts of wagons and carts were deposited in less than 5% of excavated Yamnaya-Catacomb graves, and the few graves that had them were concentrated in particular regions. The largest cluster of wagon-graves (120) was in the Kuban steppes north of the North Caucasus, not far from Maikop. Most of the Kuban wagons (115) were in graves of the Novotitorovskaya type, a local Kuban-region EBA culture that developed from early Yamnaya.[16]

Usually the vehicles used in funeral rituals were disassembled and the wheels were placed near the corners of the grave pit, as if the grave itself represented the wagon. But a whole wagon was buried west of the Dnieper in the Yamnaya grave at Lukyanova kurgan, grave 1; and whole wagons were found under nine Novotitorovskaya kurgans in the Kuban steppes. Many construction details can be reconstructed from these ten cases. All ten wagons had a fixed axle and revolving wheels. The wheels were made of two or three planks doweled together and cut in a circular shape about 50–80 cm in diameter. The wagon bed was about 1 m wide and 2–2.5 m long, and the gauge or track width between the wheels was 1.5–1.65 m. The Novotitorovskaya wagon at Lebedi kurgan 2, grave 116, is reconstructed by Gei with a box seat for the driver, supported on a cage of vertical struts doweled into a rectangular frame. Behind the driver was the interior of the wagon, the floor of which was braced with X-crossed planks (like the repoussé image on the Novosvobodnaya bronze cauldron from the Evdik kurgan) (see figure 4.3a). The Lukyanovka wagon frame also was braced with X-crossed planks. The passengers and cargo were protected under a "tilt," a wagon cover made of reed mats painted with red, white, and black stripes and curved designs, possibly sewn to a backing of felt. Similar painted reed mats with some kind of organic backing were placed on the floors of Yamnaya graves (figure 13.4).[17]

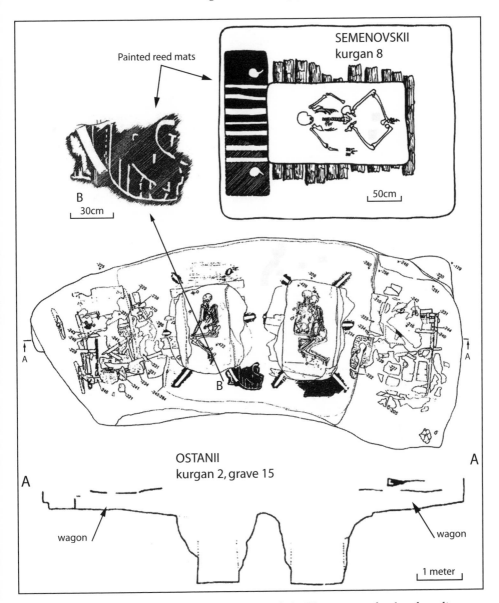

Figure 13.4 Painted reed mats in graves of the Yamnaya and related traditions. Top: Semenovskii kurgan 8, grave 9, late Yamnaya, lower Dniester steppes; bottom, Ostanni kurgan 2, double grave 15 with two wagons, Novotitorovskaya culture, Kuban River steppes. After Subbotin 1985, figure 7.7; and Gei 2000.

TABLE 13.3

Selected Radiocarbon Dates associated with the Afanasievo Migration and the Yamnaya Horizon

Lab nnumber	BP date	Sample	Calibrated date
1. Afanasievo culture, Altai Mountains (from Parzinger 2002, Figure 10)			
Unidentified sites			
Bln4764	4409±70	?	3310–2910 BCE
Bln4765	4259±36	?	2920–2780 BCE
Bln4767	4253±36	?	2920–3780 BCE
Bln4766	4205±44	?	2890–2690 BCE
Bln4769	4022+40	?	2580–2470 BCE
Bln4919	3936±35	?	2490–2340 BCE
Kara-Koba I enclosure 3			
?	5100±50	?	3970–3800 BCE
Elo-bashi enclosure 5			
?	4920±50	?	3760–3640 BCE

2. Yamnaya horizon kurgan cemeteries with multiple kurgans built together and long gaps between construction phases

A. Yamnaya horizon cemeteries in Ukraine (from Telegin et al. 2003)

Avgustnivka cemetery				
Phase 1	Ki2118	4800±55	k 1/gr2	3650–3520 BCE
Phase 2	Ki7110	4130±55	k 5/gr2	2870–2590 BCE
	Ki7111	4190±60	k 4/gr2	2890–2670 BCE
	Ki7116	4120±60	k 4/gr1	2870–2570 BCE
Verkhnetarasovka cemetery				
Phase 1	Ki602	4070±120	k 9/18	2870–2460 BCE
	Ki957	4090±95	k 70/13	2870–2490 BCE
Phase 2	Ki581	3820±190	k 17/3	2600–1950 BCE
	Ki582	3740±150	k 21/11	2400–1940 BCE
Vinogradnoe cemetery				
Phase 1	Ki9414	4340±70	k 3/10	3090–2880 BCE
Phase 2	Ki9402	3970±70	k 3/25	2580–2340 BCE
	Ki987	3950±80	k 2/11	2580–2300 BCE
	Ki9413	3930±70	k 24/37	2560–2300 BCE

TABLE 13.3 (*continued*)

Lab number		BP date	Sample	Calibrated date
Golovkovka cemetery				
Phase 1	Ki6722	3980±60	k 7/4	2580–2350 BCE
	Ki6719	3970±55	k 6/8	2580–2350 BCE
	Ki6730	3960±60	k 5/3	2570–2350 BCE
	Ki6724	3950±50	k 12/3	2560–2340 BCE
	Ki6729	3920±50	k 14/9	2560–2340 BCE
	Ki6727	3910±15	k 14/2	2460–2350 BCE
	Ki6728	3905±55	k 14/7	2470–2300 BCE
	Ki6721	3850±55	k 6/11	2460–2200 BCE
	Ki2726	3840±50	k 4/4	2400–2200 BCE
Dobrovody cemetery				
Phase 1	Ki2129	4160±55	k 2/4	2880–2630 BCE
Phase 2	Ki2107	3980±45	k 2/6	2580–2450 BCE
	Ki7090	3960±60	k 1/6	2570–2350 BCE
Minovka cemetery				
Phase 1	Ki8296	4030±70	k 2/5	2840–2460 BCE
	Ki 421	3970±80	k 1/3	2620–2340 BCE
Novoseltsy cemetery				
Phase 1	Ki1219	4520±70	k 19/7	3360–3100 BCE
Phase 2	Ki1712	4350±70	k 19/15	3090–2880 BCE
Phase 3	Ki7127	4055±65	k 19/19	2840–2470 BCE
	Ki7128	4005±50	k 20/8	2580–2460 BCE
Otradnoe cemetery				
Phase 1	Ki478	3990±100	k 26/9	2850–2300 BCE
Phase 2	Ki 431	3890±105	k 1/17	2550–2200 BCE
	Ki 470	3860±105	k 24/1	2470–2140 BCE
	Ki452	3830±120	k 1/21	2470–2070 BCE
Pereshchepyno cemetery				
Phase 1	Ki9980	4150±70	k 4/13	2880–2620 BCE
	Ki9982	4105±70	k 1/7	2870–2500 BCE
	Ki9981	4080±70	k 1/6	2860–2490 BCE
Svatove cemetery				
Phase 1	Ki585	4000±190	k 1/1	2900–2200 BCE
	Ki586	4010±180	k 2/1	2900–2250 BCE

TABLE 13.3 *(continued)*

Lab number		BP date	Sample	Calibrated date
Talyanki cemetery				
Phase 1	Ki6714	3990±50	k 1/1	2580–2460 BCE
	Ki6716	3950±50	k 1/3	2560–2340 BCE
Phase 2	Ki2612	3760±70	k 2/3	2290–2030 BCE

B. Yamnaya horizon cemeteries in the middle Volga region (Samara Valley Project)

Nizhnaya Orlyanka 1				
Phase 1	AA1257	4520±75	k 4/2	3360–3090 BCE
	OxA**	4510±75	k 1/15	3360–3090 BCE
Grachevka II				
Phase 1	AA53805	4342±56	k 5/2	3020–2890 BCE
	AA53807	4361±65	k 7/1	3090–2890 BCE

C. Poltavka cemetery in the middle Volga region, three kurgans built in a single phase.

Krasnosamarskoe IV cemetery			
AA37034	4306±53	kurgan 1, grave 4	2929–2877 BCE
AA37031	4284±79	kurgan 1, grave 1	3027–2700 BCE
AA37033	4241±70	kurgan 1, grave 3 central	2913–2697 BCE
AA37036	4327±59	kurgan 2, grave 2 central	3031–2883 BCE
AA37041	4236±47	kurgan 3, grave 9 central	2906–2700 BCE
AA37040	4239±49	kurgan 3, grave 8	2910–2701 BCE

The Yamnaya-Poltavka dates show that multiple kurgans were constructed almost simultaneously with long gaps of time between episodes, perhaps indicating episodic use of the associated pastures.

The oldest radiocarbon dates from steppe vehicle graves bracket a century or two around 3000 BCE (table 13.3). One came from Ostannii kurgan 1, grave 160 in the Kuban, a grave of the third phase of the Novotitorovskaya culture dated 4440±40 BP, or 3320–2930 BCE. The other is from Bal'ki kurgan, grave 57, on the lower Dnieper, an early Yamnaya grave dated 4370±120 BP, or 3330–2880 BCE (see figures 4.4, 4.5). The probability distributions for both dates lie predominantly before 3000 BCE, which is why I use the figure 3100 BCE. But almost certainly these were not the first wagons in the steppes.[18]

Wagons probably appeared in the steppes between about 3500 and 3300 BCE, possibly from the west through Europe, or possibly through the late Maikop-Novosvobodnaya culture, from Mesopotamia. Since we cannot really say where the wheel-and-axle principle was invented, we do not know from which direction it first entered the steppes. But it had the greatest effect in the Don-Volga-Ural steppes, the eastern part of the early Proto-Indo-European world, and the Yamnaya horizon had its oldest roots there.

The subsequent spread of the Yamnaya horizon across the Pontic-Caspian steppes probably did not happen primarily through warfare, for which there is only minimal evidence. Rather, it spread because those who shared the agreements and institutions that made high mobility possible became potential allies, and those who did not share these institutions were separated as Others. Larger herds also probably brought increased prestige and economic power, because large herd-owners had more animals to loan or offer as sacrifices at public feasts. Larger herds translated into richer bride-prices for the daughters of big herd owners, which would have intensified social competition between them. A similar competitive dynamic was partly responsible for the Nuer expansion in east Africa (chapter 6). The Don-Volga dialect associated with the biggest and therefore most mobile herd owners probably was late Proto-Indo-European.

Where Did the Yamnaya Horizon Begin?

Why, as I just stated, did the Yamnaya horizon have its oldest roots in the eastern part of the Proto-Indo-European world? The artifact styles and funeral rituals that defined the early Yamnaya horizon appeared earliest in the east. Most archaeologists accept Nikolai Merpert's judgment that the oldest Yamnaya variants appeared in the Volga-Don steppes, the driest and easternmost part of the Pontic-Caspian steppe zone.

The Yamnaya horizon was divided into nine regional groups in Merpert's classic 1974 study. His regions have been chopped into finer and finer pieces by younger scholars.[19] These regional groups, however defined, did not pass through the same chronological stages at the same time. The pottery of the earliest Yamnaya phase (A) is divided by Telegin into two variants, A1 and A2 (figure 13.5).[20] Type A1 pots had a longer collar, decoration was mainly in horizontal panels on the upper third of the vessel, and "pearl" protrusions often appeared on and beneath the collar. Type A1 was like Repin pottery from the Don. Type A2 pots had decorations all over the vessel body, often in vertical panels, and had shorter, thicker, more everted

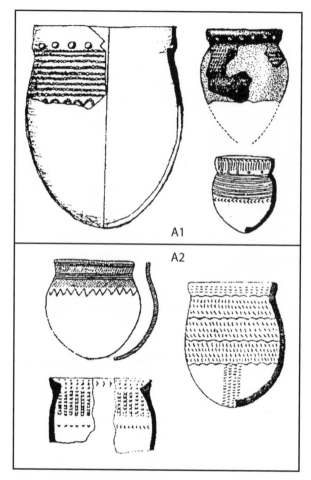

Figure 13.5 Early Yamnaya ceramic types A1 (Repin-related) and A2 (Khvalynsk-related). After Telegin et al. 2003.

rims. Type A2 was like late Khvalynsk pottery from the lower Volga. Repin vessels were made by coiling strips of clay; Type A2 Yamnaya vessels were usually made by pounding strips of clay into bag-shaped depressions or moulds to build up the walls, a very specific technological style. Pots of both subtypes were made of clays mixed with shell. Some of the shell temper seems to have been intentionally added, and some, particularly in Type A2 vessels, came from lake-bottom clays that naturally contained bits of shell and lake snails. Both the A1 and A2 types appeared across the Pontic-Caspian steppes in the earliest Yamnaya graves.

Early Yamnaya on the Lower Volga and Lower Don

Archaeological surveys led by I. V. Sinitsyn on the lower Volga between 1951 and 1953 revealed a regular series of Bronze Age kurgan cemeteries spaced 15–20 km apart along the level plains on the eastern bank between Saratov and Volgograd (then Stalingrad). Some of these kurgans contained stratified sequences of graves, and this stratigraphic evidence was employed to identify the earliest Yamnaya monuments. Important stratified kurgans included Bykovo cemetery II, kurgan 2, grave 1 (with a pot of Telegin's Type A1 stratified beneath later Yamnaya graves) and Berezhnovka cemetery I, kurgans 5 and 32, graves 22 and 2, respectively (with pots of Telegin's Type A2 stratified beneath later graves). In 1956 Gimbutas suggested that the "Kurgan Culture" began on the lower Volga. Merpert's synthesis of the Yamnaya horizon in 1974 supported Gimbutas. Recent excavations have reconfirmed the antiquity of Yamnaya traditions on the lower Volga. Archaic antecedents of both the A1 and A2 types of early Yamnaya pottery have been found in settlements on the lower Volga at Kyzyl Khak and Kara Khuduk (see figure 12.5), dated by radiocarbon between 4000 and 3500 BCE. Graves that seem intermediate between late Khvalynsk and Yamnaya in style and ritual have also been found at Shlyakovskii kurgan, Engels and Tarlyk between Saratov and Volgograd on the lower Volga.

The A1 or Repin style was made earliest in the middle Don–middle Volga region. Repin pottery is stratified beneath Yamnaya pottery at Cherkassky on the middle Don and is dated between 3950 and 3600 BCE at an antelope hunters' camp on the lower Volga at Kyzyl-Khak. The earliest Repin pottery was somewhat similar in form and decoration to the late Sredni Stog–Konstantinovka types on the lower Don, and it is now thought that contact with the late Maikop-Novosvobodnaya culture on the lower Don at places like Konstantinovka stimulated the emergence and spread of the early Repin culture and, through Repin, early Yamnaya. The metal-tanged daggers and sleeved axes of the early Yamnaya horizon certainly were copied after Maikop-Novosvobodnaya types.

The A2 or Khvalynsk style began on the lower Volga among late Khvalynsk populations. This bag-shaped kind of pottery remained the most common type in lower Volga Yamnaya graves, and later spread up the Volga into the middle Volga-Ural steppes, where the A2 style gradually replaced Repin-style Yamnaya pottery. Again, contact with people from the late Maikop-Novosvobodnaya culture, such as the makers of the kurgan at Evdik on the lower Volga, might have stimulated the change from late Khvalynsk to early Yamnaya. One of the stimuli introduced from the North Caucasus might have been wagons and wagon-making skills.[21]

Early Yamnaya on the Dnieper

The type site for early Yamnaya in Ukraine is a settlement, Mikhailovka. That Mikhailovka is a settlement, not a kurgan cemetery, immediately identifies the western Yamnaya way of life as more residentially stable than that of eastern Yamnaya. The strategic hill fort at Mikhailovka (level I) on the lower Dnieper was occupied before 3400 BCE by people who had connections in the coastal steppes to the west (the Mikhailovka I culture). After 3400–3300 BCE Mikhailovka (level II) was occupied by people who made pottery of the Repin-A1 type, and therefore had connections to the east. While Repin-style pottery had deep roots on the middle Don, it was intrusive on the Dnieper, and quite different from the pottery of Mikhailovka I. Mikhailovka II is itself divided into a lower level and an upper level. Lower II was contemporary with late Tripolye C1 and probably should be dated 3400–3300 BCE, whereas upper II was contemporary with early Tripolye C2 and should be dated 3300–3000 BCE. Repin-style pottery was found in both levels. The Mikhailovka II archaeological layer was about 60–70 cm thick. Houses included both dug-outs and surface houses with one or two hearths, tamped clay floors, partial stone wall foundations, and roofs of reed thatch, judging by thick deposits of reed ashes on the floors. This settlement was occupied by people who were newly allied to or intermarried with the Repin-style early Yamnaya communities of the Volga-Don region.

The people of Mikhailovka II farmed much less than those of Mikhailovka I. The frequency of cultivated grain imprints was 1 imprint per 273 sherds at Mikhailovka I but declined to 1 in 604 sherds for early Yamnaya Mikhailovka II, and 1 in 4,065 sherds for late Yamnaya Mikhailovka III, fifteen times fewer than in Mikhailovka I. At the same time food remains in the form of animal bones were forty-five times greater in the Yamnaya levels than in Mikhailovka I.[22] So although the total amount of food debris increased greatly during the Yamnaya period, the contribution of grain to the diet decreased. Grain imprints did occur in late Yamnaya funeral pottery from western Ukraine, as at Belyaevka kurgan 1, grave 20 and Glubokoe kurgan 2, grave 8, kurgans on the lower Dniester. These imprints included einkorn wheat, bread wheat (*Triticum aestivum*), millet (*Panicum miliaceum*), and barley (*Hordeum vulgare*). Some

Yamnaya groups in the Dnieper-Dniester steppes occasionally cultivated small plots of grain, as pastoralists have always done in the steppes. But cultivation declined in importance at Mikhailovka even as the Yamnaya settlement grew larger.[23]

When Did the Yamnaya Horizon Begin?

Dimitri Telegin and his colleagues used 210 radiocarbon dates from Yamnaya graves to establish the outlines of a general Yamnaya chronology. The earliest time interval with a substantial number of Yamnaya graves is about 3400–3200 BCE. Almost all the early dates are on wood taken from graves, so they do not need to be corrected for old carbon reservoir effects that can affect human bone. Graves dated in this interval can be found across the Pontic-Caspian steppes: in the northwestern Pontic steppes (Novoseltsy k. 19 gr. 7, Odessa region), the lower Dnieper steppes (Obloy k. 1, gr. 7, Kherson region), the Donets steppes (Volonterivka k. 1, gr. 4, Donetsk region), the lower Don steppes (Usman k. 1, gr. 13, Rostov region), the middle Volga steppes (Nizhnaya Orlyanka I, k. 1, gr. 5 and k. 4, gr. 1), and the Kalmyk steppes south of the lower Volga (Zunda Tolga, k. 1, gr. 15). Early Yamnaya must have spread rapidly across all the Pontic-Caspian steppes between about 3400 and 3200 BCE. The rapidity of the spread is interesting, suggesting both a competitive advantage and an aggressive exploitation of it. Other local cultures survived in pockets for centuries, since radiocarbon dates from Usatovo sites on the Dniester, late Post-Mariupol sites on the Dnieper and Kemi-Oba on the Crimean peninsula overlap with early Yamnaya radiocarbon dates between about 3300 and 2800 BCE. All three groups were replaced by late Yamnaya variants after 2800 BCE.[24]

Were the Yamnaya People Nomads?

Steppe nomads have fascinated and horrified agricultural civilizations since the Scythians looted their way through Assyria in 627 BCE. We still tend to stereotype all steppe nomads as people without towns, living in tents or wagons hung with brilliant carpets, riding shaggy horses among their cattle and sheep, and able to combine their fractious clans into vast pitiless armies that poured out of the steppes at unpredictable intervals for no apparent reason other than pillage. Their peculiar kind of mobile pastoral economy, nomadic pastoralism, is often interpreted by historians as a parasitic adaptation that depended on agriculturally based states. Nomads needed states, according to this *dependency hypothesis*, for grain, metals, and loot. They needed enormous amounts of food and weapons to feed and arm their armies, and huge quantities of loot to maintain their loyalty, and that volume of food and wealth could only be acquired from agricultural states. Eurasian nomadic pastoralism has been interpreted as an opportunistic response to the evolution of centralized states like China and Persia on the borders of the steppe zone. Yamnaya pastoralism, whatever it was, could not have been nomadic pastoralism, because it appeared before there were any states for the Yamnaya people to depend on.[25]

But the dependency model of Eurasian nomadic pastoralism really explains only the *political* and *military* organization of Iron Age and Medieval nomads. The historian Nicola DiCosmo has shown that political and military organizations among nomads were transformed by the evolution of large standing armies that protected the leader—essentially a permanent royal bodyguard that ballooned into an army, with all the costs that implied. As for the *economic* basis of nomadic pastoralism, Sergei Vainshtein, the Soviet ethnographer, and DiCosmo both recognized that many nomads raised a little barley or millet, leaving a few people to tend small valley-bottom fields during the summer migrations. Nomads also mined their own metal ores, abundant in the Eurasian steppes, and made their own metal tools and weapons in their own styles. The metal crafts and subsistence economy that made Eurasian nomadic pastoralism possible did not depend on imported metal or agricultural subsidies from neighboring farmers. Centralized agricultural states like those of Uruk-period Mesopotamia were very good at concentrating wealth, and if steppe pastoralists could siphon off part of that wealth it could radically transform tribal steppe military and political structures, but the everyday subsistence economics of nomadic pastoralism did not require outside support from states.[26]

If nomadic pastoralism is an economic term, referring not to political organization and military confederacies but simply to a form of pastoral economy dependent on high residential mobility, it appeared during the Yamnaya horizon. After the EBA Yamnaya period an increasingly bifurcated economy appeared, with both mobile and settled elements, in the MBA Catacomb culture. This sedentarizing trend then intensified with the appearance of permanent, year-round settlements across the northern Eurasian steppes during the Late Bronze Age (LBA) with the Srubnaya culture. Finally mobile pastoral nomadism of a new militaristic type appeared in the Iron Age with the Scythians. But the Scythians did not invent the first pastoral economy based on mobility. That seems to have been the great innovation of the Yamnaya horizon.

Yamnaya Herding Patterns

An important clue to how the Yamnaya herding system worked is the location of Yamnaya kurgan cemeteries. Most Yamnaya kurgan cemeteries across the Pontic-Caspian region were located in the major river valleys, often on the lowest river terrace overlooking riverine forests and marshes. But at the beginning of the Yamnaya period kurgan cemeteries also began to appear for the first time in the deep steppes, on the plateaus between the major river valleys. If a cemetery can be interpreted as an ancestral claim to property ("here are the graves of my ancestors"), then the appearance of kurgan cemeteries in the deep steppes signaled that deep-steppe pastures had shifted from wild and free to cultured and owned resources. In 1985 V. Shilov made a count of the excavated kurgans located in the deep steppes, on inter-valley plateaus, in the steppe region between the lower Don, the lower Volga, and the North Caucasus. He counted 799 excavated graves in 316 kurgans located in the deep steppes, outside major river valleys. The earliest graves, the first ones to appear in these locations, were Yamnaya graves. Yamnaya accounted for 10% (78) of the graves, and 45% (359) were from MBA cultures related to the Catacomb culture, 7% (58) were from the LBA Srubnaya culture, 29% (230) were of Scytho-Sarmatian origin, and 9% (71) were historical-Medieval. The exploitation of pastures on the plateaus between the river valleys began during the EBA and rapidly reached its all-time peak during the MBA.[27]

N. Shishlina collected seasonal botanical data from kurgan graves in the Kalmyk steppes, north of the North Caucasus, part of the same region that Shilov had studied. Shishlina found that Yamnaya people moved seasonally between valley-bottom pastures (occupied during all seasons) and deep-steppe plateau pastures (probably in the spring and summer) located within 15–50 km of the river valleys. Shishlina emphasized the localized nature of these migratory cycles. Repetitive movements between the valleys and plateau steppes created overgrazed areas with degraded soils (preserved today under MBA kurgan mounds) by the end of the Yamnaya period.

What was the composition of Bronze Age herds in the Don-Volga steppes? Because there are no Yamnaya settlements east of the Don, faunal information has to be extracted from human graves. Of 2,096 kurgan graves reviewed by Shilov in both the river valleys and the inter-valley plateaus—a much bigger sample than just the graves on the plateaus—just 15.2% of Yamnaya graves contained sacrifices of domesticated

TABLE 13.2

Domesticated Animals in Early Bronze Age Graves and Settlements in the Pontic–Caspian Steppes

Culture	Cattle	Sheep/gt	Horse	Pig	Dog
Don–Volga steppe, Yamnaya graves	15%	65%	8%	—	5%
Mikhailovka II/III, Yamnaya settlement	59%	29%	11%	9%	0.7%
Repin (lower Don), settlement	18%	9%	55%	9%	—

Note: Missing % were unidentifiable as to species.

animals. Most of these contained the bones of sheep or goats (65%), with cattle a distant second (15%), horses third (8%) and dogs fourth (5%) (table 13.2).[28]

Yamnaya herding patterns were different in the west, between the Dnieper and Don valleys. One difference was the presence of Yamnaya settlements, implying a less mobile, more settled herding pattern. At Mikhailovka levels II and III, which define early and late Yamnaya in the Dnieper valley, cattle (60%) were more numerous than sheep (29%), unlike the sheep-dominant herds of the east. Kurgan cemeteries penetrated only a few kilometers into the plateaus; most cemeteries were located in the Dnieper valley or its larger tributaries. This riverine cattle-herding economy was tethered to fortified strongholds like Mikhailovka, supported by occasional small grain fields. About a dozen small Yamnaya settlements have been excavated in the Dnieper-Don steppes at places such as Liventsovka and Samsonovka on the lower Don. Most occupy less than 1 ha and were relatively low-intensity occupations, although fortification ditches protected Samsonovka and Mikhailovka, and a stone fortification wall was excavated at Skelya-Kamenolomnya. Cattle are said to predominate in the animal bones from all these places.[29]

East of Repin no Yamnaya settlements have been found. Occasional wind-eroded scatters of microliths and Yamnaya pottery sherds have been observed in valley bottoms and near lakes in the Manych and North Caspian desert-steppes and deserts, but without intact cultural layers. In the lusher grasslands where it is more difficult to see small surface sites, even Yamnaya surface scatters are almost unknown. For example, the Samara oblast on the middle Volga was dotted with known settlements of the Mesolithic, Neolithic, Eneolithic, and Late Bronze Ages, but it had no EBA Yamnaya settlements. In 1996, during the Samara Valley Project, we attempted to find ephemeral Bronze Age camps by digging test pits at twelve favorable-looking places along the bottom of a stream valley, Peschanyi Dol, that had four Yamnaya kurgan cemeteries clustered near its mouth around the village of Utyevka (see figure 16.11 for a map). The Peschanyi Dol valley is today used as a summer pasturing place for cattle herds from three nearby Russian rural villages. We discovered seven ephemeral LBA Srubnaya ceramic scatters in this pleasant valley and a larger Srubnaya settlement, Barinovka, at its mouth. The LBA settlement and one camp also had been occupied during the MBA; each yielded a small handful of MBA ceramic sherds. But we found no EBA sherds—no Yamnaya settlements.

If we cannot find the camps that Yamnaya herders occupied through the winter, when they had to retreat with their herds to the protection of riverine forests and marshes (where most Yamnaya cemeteries were located), then their herds were so large that they had to keep moving even in winter. In a similar northern grassland environment with very cold winters, the fifty bands of the Blackfoot Indians of Canada and Montana had to move a few miles several times each winter just to provide fresh forage for their horses. And the Blackfeet did not have to worry about feeding cattle or sheep. Mongolian herders move their tents and animal herds about once a month throughout the winter. The Yamnaya herding system probably was equally mobile.[30]

Yamnaya herders watched over their herds on horseback. At Repin on the Don, 55% of the animal bones were horse bones. A horse skull was placed in a Yamnaya grave in a kurgan cemetery overlooking the Caspian Depression near Tsa-Tsa, south of the Volga, in kurgan 7, grave 12. Forty horses were sacrificed in a Catacomb-period grave in the same cemetery in kurgan 1, grave 5.[31] The grave probably was dug around 2500 BCE. An adult male was buried in a contracted position on his left side, oriented northeast. Fragments of red ochre and white chalk were placed by his hip. A bronze dagger blade was found under his skull. Above his grave were forty horse skulls arranged in two neat rows. Three ram skulls lay on the floor of the grave. The amount of meat forty horses would have yielded—assuming they were slightly bigger than Przewalskis, or about 400 kg live weight—would be roughly 8,000 k, enough for four thousand portions of 2 k each. This suggests a funeral feast of amazing size. Horses were suitable animals for extraordinary ritual sacrifices.

Wild Seeds and Dairy Foods in the Don-Volga Steppes

A ceramics lab in Samara has microscopically examined many Yamnaya pot-sherds from graves, but no cultivated grain imprints appeared on Yamnaya pottery here or anywhere else east of the Don. Yamnaya people from the middle Volga region had teeth that were entirely free of caries (no caries in 428 adult Yamnaya-Poltavka teeth from Samara oblast [see figure 16.12]), which indicates a diet very low in starchy carbohydrates, like the teeth of foragers.[32] Eastern Yamnaya people might have eaten wild *Chenopodium* and *Amaranthus* seeds and even *Phragmites* reed tubers and rhizomes. Analysis of pollen grains and phytoliths (silica bodies that form inside plant cells) by N. Shishlina from Yamnaya grave floors in the eastern Manych depression, in the steppes north of the North Caucasus, found pollen and phytoliths of *Chenopodium* (goosefoot) and amaranths, which can produce seed yields greater in weight per hectare than einkorn wheat, and without cultivation.[33] Cultivated grain played a small role, if any, in the eastern Yamnaya diet.

Although they were very tall and robust and showed few signs of systemic infections, the Yamnaya people of the middle Volga region exhibited significantly more childhood iron-deficiency anemia (bone lesions called *cribra orbitalia*) than did the skeletons from any earlier or later period (figure 13.6). A childhood diet *too* rich in dairy foods can lead to anemia, since the high phosphorus content of milk can block the absorption of iron.[34] Health often declines in the early phases of a significant dietary change, before the optimal mix of new foods has been established. The anomalous Yamnaya peak in *cribra orbitalia* could also have resulted from an increased parasite load among children, which again would be consistent with a living pattern involving closer contact between animals and people. Recent genetic research on the worldwide distribution of the mutation that created lactose tolerance, which made a dairy-based diet possible, indicates that it probably emerged first in the steppes west of the Ural Mountains between about 4600 and 2800 BCE—the Late Eneolithic (Mikhailovka I) and the EBA Yamnaya periods.[35] Selection for this mutation, now carried by all adults who can tolerate dairy foods, would have been strong in a population that had recently shifted to a mobile herding economy.

The importance of dairy foods might explain the importance of the cow in Proto-Indo-European myth and ritual, even among people who depended largely on sheep. Cattle were sacred because cows gave more milk

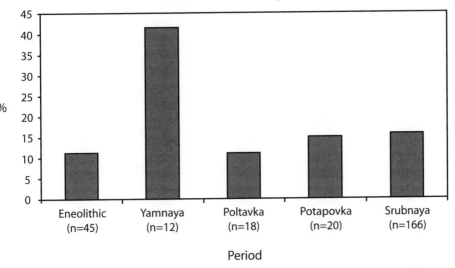

Figure 13.6 Frequencies of cribra orbitalia, associated with anemia, in cultures of the Samara oblast, middle Volga region. After Murphy and Khokhlov 2004.

than any other herd animal in the Eurasian steppe—twice as much as mares and five times more than goats, according to the Soviet ethnographer Vainshtein. He noted that, even among the sheep herders of Tuva in Siberia, an impoverished family of nomads that had lost all its sheep would try to keep at least one cow because that meant they could eat. The cow was the ultimate milk producer, even where herders counted their wealth in sheep.[36]

The Yamnaya wagon-based herding economy seems to have evolved in the steppes east of the Don, like the earliest Yamnaya pottery styles. Unlike the pottery and grave styles, the high-mobility, sheep-herding strategy of eastern Yamnaya pastoralism did not spread westward into the Dnieper steppes or northward into the middle Volga-Ural steppes, where cattle breeding remained the dominant aspect of the herding economies. Instead, it seems that social, religious, and political institutions (guest-host agreements, patron-client contracts, and ancestor cults) spread with the Yamnaya horizon. Some new chiefs from the east probably migrated into the Dnieper steppes, but in the west they added cattle to their herds and lived in fortified home bases.

Yamnaya Social Organization

The speakers of late Proto-Indo-European expressed thanks for sons, fat cattle, and swift horses to Sky Father, *dyew pater*, a male god whose prominence probably reflected the importance of fathers and brothers in the herding units that composed the core of earthly social organization. The vocabulary for kin relations in Proto-Indo-European was that of a people who lived in a patrilineal, patrilocal social world, meaning that rights, possessions, and responsibilities were inherited only from the father (not the mother), and residence after marriage was with or near the husband's family. Kinship terms referring to grandfather, father, brother, and husband's brother survive in clearly corresponding roots in nearly all Indo-European languages, whereas those relating to wife and wife's family are few, uncertain, and variable. Kinship structure is only one aspect of social organization, but in tribal societies it was the glue that held social units together. We will see, however, that where the linguistic evidence suggests a homogeneous patri-centered Proto-Indo-European kinship system, the archaeological evidence of actual behavior is more variable.

As Jim Mallory admitted years ago, we know very little about the social meanings of kurgan cemeteries, and kurgan cemeteries are all the archaeological evidence left to us over much of the Yamnaya world.[37] We can presume that they were visible claims to territory, but we do not know the rules by which they were first established or who had the right to be buried there or how long they were used before they were abandoned. Archaeologists tend to write about them as static finished objects, but when they were first made they were dynamic, evolving monuments to specific people, clans, and events.

Gender and the Meaning of Kurgan Burial

We can be confident that kurgans were not used as family cemeteries. Mallory's review of 2,216 Yamnaya graves showed that the median Yamnaya kurgan contained fewer than 3 Yamnaya graves. About 25% contained just 1 grave. Children never were buried alone in the central or principal grave—that status was limited to adults. A count of kurgans per century in the well-studied and well-dated Samara River valley, in the middle Volga region, indicated that Yamnaya kurgans were built rarely, only one every five years or so even in regions with many Yamnaya cemeteries. So kurgans commemorated the deaths of special adults, not of everyone in the social group or even of everyone in the distinguished person's family. In the lower Volga, 80% of the Yamnaya graves contained males. E. Murphy and A. Khokhlov have confirmed that 80% of the sexable Yamnaya-Poltavka graves in the middle Volga region also contained males. In Ukraine, males predominated but not as strongly. In the steppes north of the North Caucasus, both in the eastern Manych steppes and in the western Kuban-Azov steppes, females and males appeared about equally in central graves and in kurgan graves generally. Mallory described the near-equal gender distribution in 165 Yamnaya graves in the eastern Manych region, and Gei gave similar gender statistics for 400 Novotitorovskaya graves in the Kuban-Azov steppes. Even in the middle Volga region some kurgans have central graves containing adult females, as at Krasnosamarskoe IV. Males were not always given the central place under kurgans even in regions where they strongly tended to occupy the central grave, and in the steppes north of the North Caucasus (where Maikop influence was strongest before the Yamnaya period) males and females were buried equally.[38]

The male-centered funerals of the Volga-Ural region suggest a more male-centered eastern social variant within the Yamnaya horizon, an archaeological parallel to the male-centered deities reconstructed for eastern Indo-European mythological traditions. But even on the Volga the people buried in central graves were not *exclusively* males. In the patrilocal, patrilineal society reconstructed by linguists for Proto-Indo-European speakers, *all* lineage heads would have been males. The appearance of adult females in one out of five kurgan graves, including central graves, suggests that gender was not the only factor that determined who was buried under a kurgan. Why were adult females buried in central graves under kurgans even on the Volga? Among later steppe societies women could occupy social positions normally assigned to men. About 20% of Scythian-Sarmatian "warrior graves" on the lower Don and lower Volga contained females dressed for battle as if they were men, a phenomenon that probably inspired the Greek tales about the Amazons. It is at least interesting that the frequency of adult females in central graves under Yamnaya kurgans in the same region, but two thousand years earlier, was about the same. Perhaps the people of this region customarily assigned some women leadership roles that were traditionally male.[39]

Kurgan Cemeteries and Mobility

Were the kurgans in a cemetery built together in a rapid sequence and then abandoned, or did people stay around them and use them regularly for longer periods of time? For interval dating *between* kurgans it would be ideal to obtain radiocarbon dates from all the kurgans in a cemetery. In a Yamnaya cemetery, that would usually be from three to as many as forty or fifty kurgans. Very few kurgan cemeteries have been subjected to this intensity of radiocarbon dating.

We can try to approximate the time interval between kurgans from the 210 radiocarbon dates on Yamnaya graves published in 2003 by Telegin and his colleagues. In his list we find nineteen Yamnaya kurgan cemeteries for which there are radiocarbon dates from at least two kurgans in the same cemetery. In eleven of these nineteen, more than half, at least two kurgans yielded radiocarbon dates that are statistically indistinguishable (see table 13.3 for radiocarbon dates). This suggests that kurgans were built rapidly in clusters. In many cases, the cemetery was then abandoned for a period of centuries before it was reused. For example, at the Poltavka cemetery of Krasnosamarskoe IV in the middle Volga region we can show this pattern, because we excavated all three kurgans in a small kurgan group and obtained multiple radiocarbon dates from each (figure 13.7). Like many kurgan groups in Ukraine, all three kurgans here were built within an indistinguishably brief time. The central graves all dated about 2700–2600 BCE (dates reduced by 200 radiocarbon years to account for the measured ^{15}N in the human bone used for the date), and then the cemetery was abandoned. Cemeteries like Krasnosamarskoe IV were used intensively for very short periods.

If pastures were like the cemeteries that marked them, then they were used briefly and abandoned. This episodic pasturing pattern, similar to swidden horticulture, possibly was encouraged by similar conditions—a low-productivity environment demanding frequent relocation. But herding, unlike swidden horticulture, required large pastures for each animal, and it could produce trade commodities (wool, felt, leather) if the herds were sufficiently large. To "rest" pastures under these circumstances would have been attractive only at low population densities.[40] It could have happened when the new Yamnaya economy was expanding into the previously unexploited pastures between the river valleys. But as the population of wagon-driving herders grew during the Early Bronze Age, some pastures began to show signs of overuse. A. A. Golyeva established that EBA Yamnaya kurgans in the Manych steppes were built on pristine soils and grasses, but many MBA Catacomb-culture kurgans were built on soils that had already been overgrazed.[41] Yamnaya kurgan cemeteries were dynamic aspects of a new herding system during its initial expansionary phase.

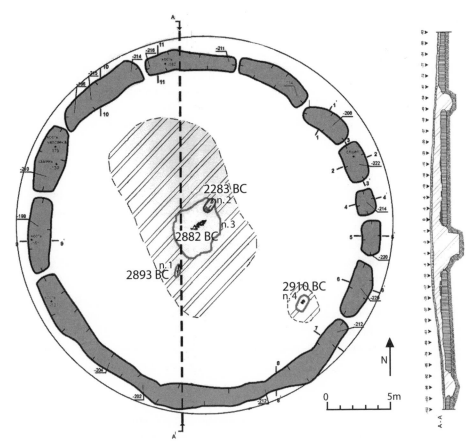

Figure 13.7 Krasnosamarskoe cemetery IV, kurgan 1, early Poltavka culture on the middle Volga. Three graves were created simultaneously when the kurgan was raised, about 2800 BCE: the central grave, covered by a layer of clay, a peripheral grave to its southeast, and an overlying grave in the kurgan. Author's excavation.

Proto-Indo-European Chiefs

The speakers of Proto-Indo-European followed chiefs (**weik-potis*) who sponsored feasts and ceremonies and were immortalized in praise poetry. The richer Yamnaya graves probably commemorated such individuals. The dim outlines of a social hierarchy can be extracted from the amount of labor required to build kurgans. A larger kurgan probably meant that a larger number of people felt obligated to respond to the death of the person buried in the central grave. Most graves contained nothing but the body, or in some cases just the head, with clothing, perhaps a bead or two, reed mats, and wooden beams. The skin of a domestic animal with a few leg or head bones attached was an unusual gift, appearing in about 15% of graves, and a copper dagger or axe was very rare, appearing in less than 5%. Sometimes a few sherds of pottery were thrown into the grave. It is difficult to define social roles on the basis of such slight evidence.

Do big kurgans contain the richest graves? Kurgan size and grave wealth have been compared in at least two regions, in the Ingul River valley west of the Dnieper in Ukraine (a sample of 37 excavated Yamnaya kurgans), and in the Volga-Ural region (a sample of more than 90 kurgans).[42] In both regions kurgans were easily divided into widely disparate size classes—three classes in Ukraine and four on the Volga. In both regions the class 1 kurgans were 50 m or more in diameter, about the width of a standard American football field (or two-thirds the width of a European soccer field), and their construction required more than five hundred man-days, meaning that five hundred people might have worked for one day to build them, or one hundred people for five days, or some other combination totaling five hundred.

The biggest kurgans were not built over the richest central graves in either region. Although the largest class 1 kurgans did contain rich graves, so did smaller kurgans. In both regions wealthy graves occurred both in the central position under a kurgan and in peripheral graves. In the Ingul valley, where there were no metal-rich graves in the study sample, more objects were found in peripheral graves than in central graves. In some cases, where we have radiocarbon dates for many graves under a single kurgan, we can establish through overlapping radiocarbon dates that the central grave and a *richer* peripheral grave were dug simultaneously in a single funeral ceremony, as at Krasnosamarskoe IV. The richest graves in some Novosvobodnaya kurgans, including the Klady cemetery, were peripheral graves, located off-center under the mound. It could be misleading to count the objects in peripheral graves, including some wheeled vehicle sacrifices, as separate from the central grave. In at least some cases, a richer peripheral grave accompanied the central grave in the same funeral ceremony.

Elite status was marked by artifacts as well as architecture, and the most widespread indication of status was the presence of metal grave goods. The largest metal artifact found in any Yamnaya grave was laid on the left arm of a male buried in Kutuluk cemetery I, kurgan 4, overlooking the Kinel River, a tributary of the Samara River in the Samara oblast east of the

Figure 13.8 Kutuluk cemetery I, kurgan 4, grave 1, middle Volga region. An Early Yamnaya male with a large copper mace or club, the heaviest metal object of the Yamnaya horizon. Photograph and excavation by P. Kuznetsov; see Kuznetsov 2005.

Volga (figure 13.8). A solid copper club or mace weighing 750 gm, it was 48.7 cm long and more than 1 cm thick, with a diamond cross-section. The kurgan was medium-sized, 21 m in diameter and less than 1 m high, but the central grave pit (gr. 1) was large. The male was oriented east, positioned supine with raised knees, with ochre at his head, hips, and feet—a classic early Yamnaya grave type. Two samples of bone taken from his skeleton were dated about 3100–2900 BCE (4370±75 AA12570 and 4400±70 BP OxA 4262), but ^{15}N levels suggest that the date probably was too old and should be revised to about 2900–2700 BCE.

In the Samara River valley, near the village of Utyevka on the flood-plain of the Samara River, was the richest steppe grave of the Yamnaya-Poltavka period. Utyevka cemetery I, kurgan 1 was 110 m in diameter. Central grave 1 was a Yamnaya-Poltavka grave containing an adult male, positioned supine with legs in an uncertain position. He was buried with two golden rings with granulated decoration, unique objects with analogies in the North Caucasus or Anatolia; also a copper tanged dagger, a copper pin with a forged iron head, a flat copper axe, a copper awl, a copper sleeved axe of the classic Volga-Ural type IIa with a slightly rising blade, and a polished stone pestle[43] (figure 13.9). In the Volga-Ural region numerous Yamnaya graves contained metal daggers, chisels, and cast shaft-hole axes.

Overall, the wide disparities in labor invested in kurgans of different sizes, from 10 m to more than 110 m in diameter, indicate a broad sociopolitical hierarchy, though one not always correlated with grave wealth. The class 1 kurgans tended to contain rich graves but they were not always the central grave, and rich graves frequently occurred in smaller kurgans. Chernykh observed that kurgans seem to have been bigger, as a rule, in the North Pontic steppes, where many also had additional stone elements including cromlechs or curbs, carved stone stelae, and even coverings of stone or gravel, whereas the graves of the Volga-Ural region were richer in metal but had simpler earthen monuments.[44]

The Identity of the Metalworker

The craft of the steppe metalsmith improved and became more sophisticated under Yamnaya chiefs. Metalworkers in the Pontic-Caspian steppes made cast-copper objects regularly for the first time, and in late Yamnaya they even experimented with forged iron. Thin seams of copper ore (azurite, malachite) are interbedded with iron-bearing sandstones between the central North Caucasus region (Krasnodar) and the Ural Mountains (Kargaly), including the entire Volga-Ural region. These ores are exposed by erosion on the sides of many stream valleys, and were mined by Yamnaya metalworkers. A Yamnaya grave at Pershin in Orenburg oblast, near the enormous copper deposits and mines at Kargaly on the middle Ural River, contained a male buried with a two-piece mold for a sleeved, one-bladed axe of Chernykh's type 1. The grave is dated about 2900–2700 BCE

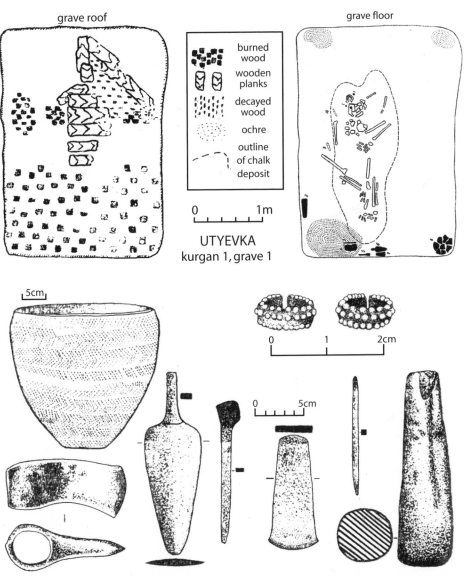

Figure 13.9 Utyevka cemetery I, kurgan 1, grave 1, between 2800 and 2500 BCE, middle Volga region. The richest grave and among the largest kurgans (more than 100 m in diameter) of the Yamnaya-Poltavka horizon. Gold rings with granulated decoration, ceramic vessel, copper shaft-hole axe, copper dagger, copper pin with iron head, copper flat axe, copper awl, and stone pestle. After Vasiliev 1980.

(4200±60, BM-3157). A Yamnaya mining pit has been found at Kargaly with radiocarbon dates of the same era. Almost all the copper objects from the Volga-Ural region were made of "clean" copper from these local sources. Although the cast sleeved single-bladed axes and tanged daggers of the early Yamnaya period imitated Novosvobodnaya originals, they were made locally from local copper ores. North Caucasian arsenical bronze was imported by people buried in graves in the Kalmyk steppe south of the lower Volga and in Kemi-Oba sites on the Crimean peninsula, but not in the Volga-Ural steppes.[45]

The grave at Pershin was not the only smith's grave of the period. Metalworkers were clearly identified in several Yamnaya-period graves, perhaps because metalworking was still a form of shamanic magic, and the tools remained dangerously polluted by the spirit of the dead smith. Two Post-Mariupol smith's graves on the Dnieper (chapter 12) probably were contemporary with early Yamnaya, as was a smith's grave with axe molds, crucibles, and *tulieres* in a Novotitorovskaya-culture grave in the Kuban steppes at Lebedi I (figure 13.10). Copper slag, the residue of metalworking, was included in other graves, as at Utyevka I kurgan 2.[46]

One unappreciated aspect of EBA and MBA steppe metallurgy was its experimentation with iron. The copper pin in Utyevka kurgan 1 with a forged iron head was not unique. A Catacomb-period grave at Gerasimovka on the Donets, probably dated around 2500 BCE, contained a knife with a handle made of arsenical bronze and a blade made of iron. The iron did not contain magnetite or nickel, as would be expected in meteoric iron, so it is thought to have been forged. Iron objects were rare, but they were part of the experiments conducted by steppe metalsmiths during the Early and Middle Bronze Ages, long before iron began to be used in Hittite Anatolia or the Near East.[47]

THE STONE STELAE OF THE NORTH PONTIC STEPPES

The Yamnaya horizon developed in the Pontic-Caspian steppes largely because an innovation in land transport, wagons, was added to horseback riding to make a new kind of herding economy possible. At the same time an innovation in sea transport, the introduction of the multi-oared longboat, probably was responsible for the permanent occupation of the Cycladic Islands by Grotta-Pelos mariners about 3300–3200 BCE, and for the initial development of the northwest Anatolian trading communities such as Kum Tepe that preceded the founding of Troy.[48] These two horizons, one on the sea and the other on a sea of grass, came into contact around the shores of the Black Sea.

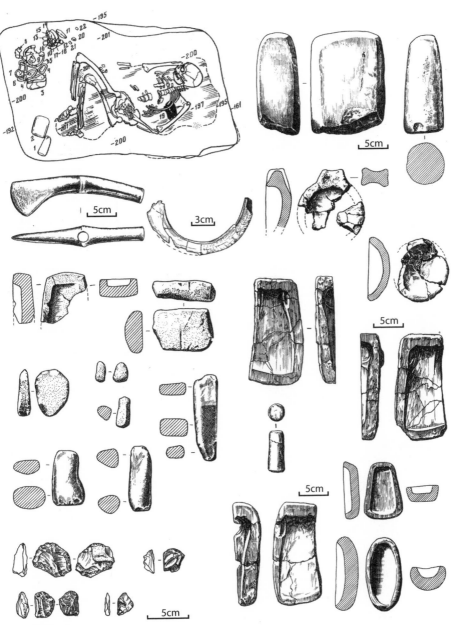

Figure 13.10 Lebedi cemetery I, kurgan 3, grave 10, a metal worker's gave of the late Novotitorovskaya culture, perhaps 2800–2500 BCE, Kuban River steppes. He wore a boars-tusk pendant. Under his arm was a serpentine hammer-axe (*upper left*). By his feet was a complete smithing kit: heavy stone hammers and abraders, sharp-edged flint tools, a round clay crucible (*upper right*), and axe molds for both flat and sleeved axes. After Gei 1986, figures 1, 4, 6, 7, and 9.

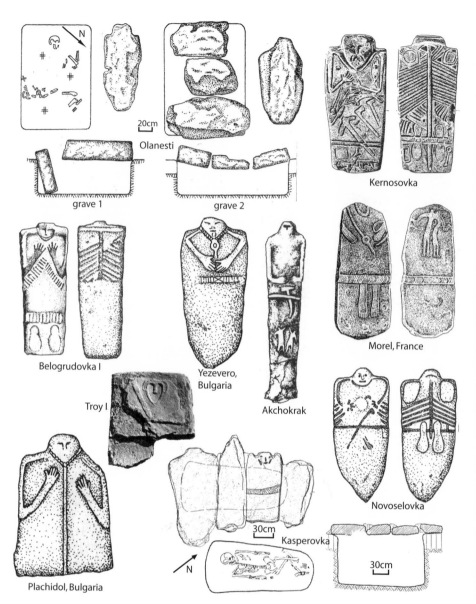

Figure 13.11 Carved stone anthropomorphic stelae of the Pontic steppes, Bulgaria, Troy I, and southeastern France. Graves 1 and 2 of Olaneşti kurgan 2 (*upper left*), located in the lower Dniester steppes, are pre-Usatovo, so before 3300 BCE. The Yamnaya stelae of Ukraine and Crimea (Kernosovka, Belogrudovka, Akchorak, Novoselovka, and Kasperovka) and Bulgaria (Plachidol, Yezerovo) probably date 3300–2500 BCE. Parallels at Troy I and in the mountains of southeastern France (Morel) are striking. After Telegin and Mallory 1994; and Yarovoy 1985.

The Kemi-Oba culture was a kurgan-building culture dated 3200–2600 BCE centered in the Crimean peninsula. Its dark-surfaced pottery was a continuation of Mikhailovka I ceramic traditions. Kemi-Oba grave cists were lined with flat-shaped stones, some painted in geometric designs, a custom shared with Novosvobodnaya royal graves (e.g., the Tsar kurgan at Nalchik). Kemi-Oba graves also contained large, stone funeral stelae, many with human heads carved at the top and arms, hands, belts, tunics, weapons, crooks, sandals, and even animal scenes sometimes carved on one or both faces (figure 13.11) This custom spread from the Crimean peninsula into both the Caucasus (where only a few stelae appeared) and the western Pontic steppes. At least three hundred stelae have been found in Yamnaya and Catacomb graves in the North Pontic steppes, usually re-used as grave-pit covers, with more than half concentrated between the South Bug and Ingul rivers.[49] The carving of funeral stelae seems to have expanded in frequency and elaboration in the Crimean and Pontic steppes after about 3300 BCE. Their original purpose is unknown. Perhaps they marked the future site of a kurgan cemetery before the first kurgan was built, or maybe they marked the first kurgan until the second one was built. In any case, they are usually found re-used as stone covers over grave pits, sealed beneath kurgans.

Eerily similar stelae, with carved heads, bent arms, hands, weapons, and even specific objects such as crooks, were carved in northern Tuscany and the Italian piedmont at about the same time, and a fragment of a similar-looking stela was built into a stone building in Troy I. It is difficult to imagine that these widely separated but strikingly similar and contemporaneous funeral stelae were unconnected. A newly invigorated maritime trade probably was responsible for carrying ideas and technologies across the sea. The Yamnaya horizon spread across the Pontic-Caspian steppes while an invigorated sea trade spread across the eastern Mediterranean. A full understanding of the significance of the Yamnaya horizon requires an understanding of its external relations—the subject of the next chapter.

CHAPTER FOURTEEN

The Western Indo-European Languages

"A wild river full of possibilities flowed from my new tongue."
—Andrew Lam, *Learning a Language, Inventing a Future* 2006

We will not understand the early expansion of the Proto-Indo-European dialects by trying to equate language simply with artifact types. Material culture often has little relationship to language. I have proposed an exception to that rule in the case of robust and persistent frontiers, but that does seem to be an exception. The essence of language expansion is psychological. The initial expansion of the Indo-European languages was the result of widespread cultural shifts in group self-perception. Language replacement always is accompanied by revised self-perceptions, a restructuring of the cultural classifications within which the self is defined and reproduced. Negative evaluations associated with the dying language lead to a descending series of reclassifications by succeeding generations, until no one wants to speak like Grandpa any more. Language shift and the stigmatization of old identities go hand in hand.

The pre–Indo-European languages of Europe were abandoned because they were linked to membership in social groups that became stigmatized. How that process of stigmatization happened is a fascinating question, and the possibilities are much more varied than just invasion and conquest. Increased out-marriage, for example, can lead to language shift. The Gaelic spoken by Scottish "fisher" folk was abandoned after World War II, when increased mobility and new economic opportunities led to out-marriage between Gaelic "fishers" and the surrounding English-speaking population, and the formerly tightly closed and egalitarian "fisher" community became intensely aware both of its low ranking in a larger world and of alternative economic opportunities. Gaelic rapidly disappeared, although only a few people—soldiers, professionals, teachers—moved very far. Similarly, the general situation in Europe after 3300 BCE was one of increased mobility, new pastoral economies, explicitly status-ranked political systems, and inter-regional connectivity—exactly the kind of context that might have led to the stigmatization of the tightly closed identities associated with languages spoken by localized groups of village farmers.[1]

The other side of understanding language shift is to ask why the identities associated with Indo-European languages were emulated and admired. It cannot have been because of some essential quality or inner potential in Indo-European languages or people. Usually language shift flows in the direction of paramount prestige and power. Paramount status can attach to one ethnic group (Celt, Roman, Scythian, Turk, American) for centuries, but eventually it flows away. So we want to know what in this particular era attached prestige and power to the identities associated with Proto-Indo-European speech—Yamnaya identities, principally. At the beginning of this period, Indo-European languages still were spoken principally by pastoral societies from the Pontic-Caspian steppes. Five factors probably were important in enhancing their status:

1. Pontic-Caspian steppe societies were more familiar with horse breeding and riding than anyone outside the steppes. They had many more horses than anywhere else, and measurements show that their steppe horses were larger than the native marsh and mountain ponies of central and western Europe. Larger horses appeared in Baden, Cernavoda III, and Cham sites in central Europe and the Danube valley about 3300–3000 BCE, probably imported from the steppes.[2] Horses began to appear commonly in most sites of the ETC culture in Transcaucasia at the same time, and larger horses appeared among them, as in southeastern Anatolia at Norşuntepe. Steppe horse-breeders might also have had the most manageable male bloodline—the genetic lineage of the original domesticated male founder was preserved even in places with native wild populations (see chapter 10). If they had the largest, strongest, *and* most manageable horses, and they had more than anyone else, steppe societies could have grown rich by trading horses. In the sixteenth century the Bukhara khanate in Central Asia, drawing on horse-breeding grounds in the Ferghana valley, exported one hundred thousand horses *annually* just to one group of customers: the Mughal rulers of India and Pakistan. Although I am not suggesting anything near that scale, the annual demand for steppe horses in Late Eneolithic/Early Bronze Age Europe could easily have totaled thousands of animals during the initial expansion of horseback riding beyond the steppes. That would have made some steppe horse dealers wealthy.[3]

2. Horseback riding shortened distances, so riders traveled farther than walkers. In addition to the conceptual changes in human geography this caused, riders gained two functional advantages. First, they could manage herds larger than those tended by pedestrian herders, and could move those larger herds more easily from one pasture to another. Any single herder became more productive on horseback. Second, they could advance to and retreat from raids faster than pedestrian warriors. Riders could show up unexpectedly, dismount and attack people in their fields, run back to their horses and get away quickly. The decline in the economic importance of cultivation across Europe after 3300 BCE occurred in a social setting of increased levels of warfare almost everywhere. Riding probably added to the general increase in insecurity, making riding more necessary, and expanding the market for horses (see paragraph above).

3. Proto-Indo-European institutions included a belief in the sanctity of verbal contracts bound by oaths (*$h_1óitos$*), and in the obligation of patrons (or gods) to protect clients (or humans) in return for loyalty and service. "Let this racehorse bring us good cattle and good horses, male children and all-nourishing wealth," said a prayer accompanying the sacrifice of a horse in the *Rig Veda* (I.162), a clear statement of the contract that bound humans to the gods. In Proto-Indo-European religion generally the chasm between gods and humans was bridged by the sanctity of oath-bound contracts and reciprocal obligations, so these were undoubtedly important tools regulating the daily behavior of the powerful toward the weak, at least for people who belonged under the social umbrella. Patron-client systems like this could incorporate outsiders as clients who enjoyed rights and protection. This way of legitimizing inequality probably was an old part of steppe social institutions, going back to the initial appearance of differences in wealth when domesticated animals were accepted.[4]

4. With the evolution of the Yamnaya horizon, steppe societies must have developed a political infrastructure to manage migratory behavior. The change in living patterns and mobility described in the previous chapter cannot have happened without social effects. One of those might have been the creation of mutual obligations of "hospitality" between guest-hosts (*$ghos-ti-$*). This institution, discussed in the last chapter, redefined who belonged under the social umbrella, and extended protection to new groups. It would have been very useful as a new way to incorporate outsiders as people with clearly defined rights and protections, as it was used from *The Odyssey* to medieval Europe.[5] The apparent absence of this root in Anatolian and Tocharian suggests that this might have been a new development connected with the migratory behavior of the early Yamnaya horizon.

5. Finally, steppe societies had created an elaborate political theater around their funerals, and perhaps on more cheerful public occasions as well. Proto-Indo-European contained a vocabulary related to gift giving and gift taking that is interpreted as referring to potlatch-like feasts meant to build prestige and display wealth. The public performance of praise poetry, animal sacrifices, and the distribution of meat and mead were central elements of the show. Calvert Watkins found a special kind of song he called the "praise of the gift" in Vedic, Greek, Celtic, and Germanic, and therefore almost certainly in late Proto-Indo-European. Praise poems proclaimed the generosity of a patron and enumerated his gifts. These performances were both acclamations of identity and recruiting events.[6]

Wealth, military power, and a more productive herding system probably brought prestige and power to the identities associated with Proto-Indo-European dialects after 3300 BCE. The guest-host institution extended the protections of oath-bound obligations to new social groups. An Indo-European–speaking patron could accept and integrate outsiders as clients without shaming them or assigning them permanently to submissive roles, as long as they conducted the sacrifices properly. Praise poetry at public feasts encouraged patrons to be generous, and validated the language of the songs as a vehicle for communicating with the gods who regulated everything. All these factors taken together suggest that the spread of Proto-Indo-European probably was more like a franchising operation than an invasion. Although the initial penetration of a new region (or "market" in the franchising metaphor) often involved an actual migration from the steppes and military confrontations, once it began to reproduce new patron-client agreements (franchises) its connection to the original steppe immigrants became genetically remote, whereas the myths, rituals, and institutions that maintained the system were reproduced down the generations.[7]

The End of the Cucuteni-Tripolye Culture and the Roots of the Western Branches

In this chapter we examine the archaeological evidence associated with the initial expansion of the western Indo-European languages, including the separation of Pre-Germanic, the ultimate ancestor of English. It is possible to connect prehistoric languages with archaeological cultures in this particular time and place *only* because the possibilities are already constrained by three critical parameters. These are (1) that the late Proto-Indo-European dialects did expand; (2) that they expanded into eastern and central Europe from a homeland in the Pontic-Caspian steppes; and (3) that the separations of Pre-Italic, Pre-Celtic, and Pre-Germanic, at least, from late Proto-Indo-European probably happened at about this time, between 3300 and 2500 BCE (see the conclusions of chapters 3 and 4).

The Roots of the Oldest Western Indo-European Branches

These constraints oblige us to turn our attention to the region just to the west of the early Yamnaya territory, or west of the South Bug River valley, beginning about 3300 BCE. On this frontier we can identify three archaeological cases of cross-cultural contact in which people from the western Pontic steppes established long-term relationships with people outside the steppe zone to their west during the steppe Early Bronze Age, 3300–2800 BCE. Each of these new intercultural meetings provided a context in which language expansion might have occurred, and, given the constraints just described, probably did. But each case happened differently.

The first occurrence involved close integration, noted particularly in pottery but evident in other customs as well, between the steppe Usatovo culture and the late Tripolye villages of the upper Dniester and Prut valleys (figure 14.1). It is fairly clear from the archaeological evidence that the steppe aspect of the integrated culture had separate origins and stood in a position of military dominance over the upland farmers, a situation that would have encouraged the spread of the steppe language into the uplands. In the second case, people of the Yamnaya horizon moved in significant numbers into the lower Danube valley and the Carpathian Basin. This was a true "folk migration," a massive and sustained flow of outsiders into a previously settled landscape. Again there are archaeological signs, in pottery particularly, of integration with the local Cotsofeni culture. Integration with the locals would have provided a medium for language shift. In the third case, the Yamnaya horizon expanded toward the border with the Corded Ware horizon on the headwaters of the Dniester in far northwestern Ukraine. In some places it appears there was no integration at all, but on the east flank of this contact zone, near the middle Dnieper, a hybrid border culture emerged. It is probably safe to assume that the separations of several western Indo-European branches were associated somehow with these events. The linguistic evidence suggests that Italic, Celtic, and Germanic, at least, separated next after Tocharian (discussed in the previous chapter). The probable timing of separations suggests that they happened around this time, and these are the visible events that seem like good candidates.

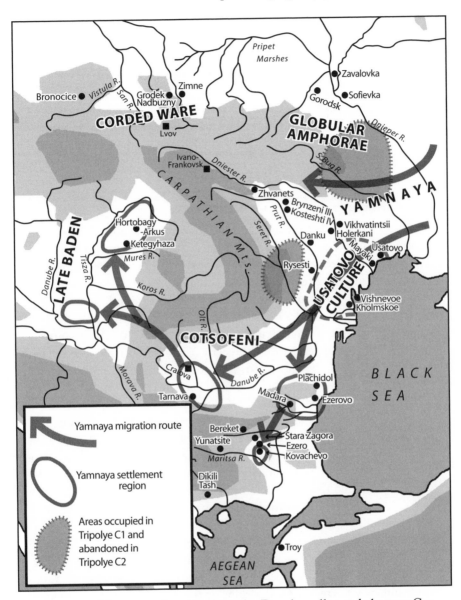

Figure 14.1 Yamnaya migrations into the Danube valley and the east Carpathian piedmont, 3100–2600 BCE. The older western IE branches probably evolved from dialects scattered by these migrations.

The End of the Cucuteni-Tripolye Culture

The people whose dialects would separate to become the root speech communities for the northwestern Indo-European language branches (Pre-Germanic, Pre-Baltic, and Pre-Slavic) probably moved initially toward the northwest. That would mean moving through or into Late Tripolye territory if it happened between 3300 and 2600 BCE, the time span of the final, staggering C2 phase of the Tripolye culture, after which all Tripolye traditions disappeared entirely. The period began with the sudden abandonment of large regions near the steppe border, including almost the entire South Bug valley. In the regions where the Tripolye culture survived, no Tripolye C2 towns had more than thirty to forty houses. The houses themselves were smaller and less substantial. Painted fine ceramics declined in frequency, while clinging to old motifs and styles. Domestic rituals utilizing clay female figurines became less frequent, the female traits became stylized and abstract, and then the rituals disappeared entirely. Two major episodes of change can be seen. The first major shock came at the transition from Tripolye C1 to C2 about 3300 BCE, simultaneously with the appearance of the early Yamnaya horizon. The second and final sweep of change erased the last remnants of Tripolye customs around 2800–2600 BCE, when the early Yamnaya period ended.

The first crisis, at the Tripolye C1/C2 transition about 3300 BCE (table 14.1), is evident in the abandonment of large regions that had contained hundreds of Tripolye C1 towns and villages. The vacated regions included the Ros' River valley, a western tributary of the Dnieper south of Kiev, near the steppe border; all of the middle and lower South Bug valley, near the steppe border; and the southern Siret and Prut valleys in southeastern Romania (between Iasi and Bîrlad), also near the steppe border. After this event almost no Cucuteni-Tripolye sites survived in what is now Romania, so after two thousand years the Cucuteni sequence came to an end. All these regions had been densely occupied during Cucuteni B2/Tripolye C1. We do not know what happened to the evacuated populations. A Yamnaya kurgan was erected on the ruins of the Tripolye C1 super town at Maidanetsk'e (see figure 12.7) in the South Bug valley, but this seems to have happened centuries after its abandonment. Other kurgans in the South Bug valley (Serezlievka) contained Tripolye C2 figurines and pots, so it is clear that kurgan-building people occupied the South Bug valley, but their population seems to have been sparse, and their use of Tripolye pottery has led to arguments over their origins.[8] With the disappearance of agricultural towns from most of the South Bug valley, surviving Tripolye populations

TABLE 14.1

Selected Radiocarbon Dates for the Usatovo Culture, other Tripolye C2 groups, and Yamnaya graves in the Danube valley.

Lab Number	BP Date	Sample	Calibrated Date
1. Usatovo culture			
Mayaki settlement, lower Dniester			
Ki–282	4580±120	charcoal from fortification ditch	3520–3090 BCE
Ki–281	4475±130	same	3360–2930 BCE
Bln–629	4400±100	same	3320–2900 BCE
UCLA 1642B	4375±60	same	3090–2900 BCE
Le–645	4340±65	same	3080–2880 BCE
Usatovo, flat cemetery II, unrecorded grave number			
UCLA–1642A	4330±60	?bone	3020–2880 BCE
2. Tripolye C2 sites on the middle Dnieper			
Gorodsk settlement, fortified promontory, Teterev River			
GrN–5090	4551±35	?bone	3370–3110 BCE
Ki–6752	4495±45	shell	3340–3090 BCE
Sofievka cemetery, Borispol district, Kiev region			
Ki–5012	4320±70	grave 1, cremated bone	3080–2870 BCE
Ki–5029	4300±45	charcoal	3020–2870 BCE
Ki–5013	4270±90	square M11, cremated bone	3020–2690 BCE
3. Tripolye C2 sites on the upper Dniester			
Zhvanets settlement, early C2, upper Dniester, Kamianets–Podolsky region			
Ki–6745	4530±50	animal bone, pit–house 1	3360–3100 BCE
Ki–6743	4480±40	animal bone, surface house 2	3340–3090 BCE
Ki–6754	4380±60	charcoal	3100–2910 BCE
Ki–6744	4355±60	animal bone, pit–house 6	3080–2890 BCE
4. Yamnaya graves in the Danube valley			
Poruchik–Geshanovo kurgan cemetery, northeast Bulgaria			
Bln–3302	4360±50	charcoal from unpublished grave	3080–2900 BCE
Bln–3303	4110±50	same	2860–2550 BCE
Bln–3301	4080±50	same	2860–2490 BCE

TABLE 14.1 (*continued*)

Lab Number	BP Date	Sample	Calibrated Date
Plachidol kurgan cemetery 1, northeast Bulgaria			
Bln-2504	4269±60	charcoal, grave 2 with stela	3010–2700 BCE
Bln-2501	4170±50	charcoal, grave 1 with wagon	2880–2670 BCE
Baia Hamangia, Danube delta, Romania			
GrN-1995	4280±65	charcoal from grave	3020–2700 BCE
Bln-29	4090±160	charcoal from grave	2880–2460 BCE
Ketegyhaza kurgan 3, grave 4 (latest grave in kurgan 3), eastern Hungary			
Bln-609	4265±80	charcoal from grave	3020–2690 BCE

resolved into two geographic groups north and south of the South Bug (see figure 13.1).

The northern Tripolye C2 group was located on the middle Dnieper and its tributaries around Kiev, where the forest-steppe graded into the closed northern forest. Cross-border assimilation with steppe cultures had begun on the middle Dnieper during Tripolye C1, as at Chapaevka (see figures 12.2, 12.6), and this process continued during Tripolye C2. At towns like Gorodsk, west of the Dnieper, and cemeteries like Sofievka, east of the Dnieper, the mix of cultural elements included late Sredni Stog, early Yamnaya, late Tripolye, and various influences from southern Poland (late Baden, late TRB). The hybrid that emerged from all these intercultural meetings slowly became its own distinct culture.

The southern Tripolye C2 group, centered in the Dniester valley, was closely integrated with a steppe culture, the Usatovo culture, described in detail below. The two surviving late Tripolye settlement centers on the Dnieper and Dniester continued to interact—Dniester flint continued to appear in Dnieper sites—but they also slowly grew apart. For reasons that will be clear in the next chapter, I believe that the emerging hybrid culture on the middle Dnieper played an important role in the evolution of both the Pre-Baltic and Pre-Slavic language communities after 2800–2600 BCE. Pre-Germanic is usually assigned an earlier position in branching diagrams. If early Pre-Germanic speakers moved away from the Proto-Indo-European homeland toward the northwest, as seems likely, they moved through one of these Tripolye settlement centers before 2800 BCE. Perhaps it was the other one in the Dniester valley. Its steppe partner was the Usatovo culture.

STEPPE OVERLORDS AND TRIPOLYE CLIENTS: THE USATOVO CULTURE

The Usatovo culture appeared about 3300–3200 BCE in the steppes around the mouth of the Dniester River, a strategic corridor that reached northwest into southern Poland. The rainfall-farming zone in the Dniester valley had been densely occupied by Cucuteni-Tripolye communities for millennia, but they never established settlements in the steppes. Kurgans had overlooked the Dniester estuary in the steppes since the Suvorovo migration about 4000 BCE; these are assigned to various groups including Mikhailovka I and the Cernavoda I–III cultures. Usatovo represented the rapid evolution of a new level of social and political integration between lowland steppe and upland farming communities. The steppe element used Tripolye material culture but clearly declared its greater prestige, wealth, and military power. The upland farmers who lived on the border itself adopted the steppe custom of inhumation burial in a cemetery, but they did not erect kurgans or take weapons to their graves. This integrated culture appeared in the Dniester valley just after the abandonment of all the Tripolye C1 towns in the South Bug valley on one side and the final Cucuteni B2 towns in southern Romania on the other. The chaos caused by the dissolution of hundreds of Cucuteni-Tripolye farming communities probably convinced the Tripolye townspeople of the middle Dniester valley to accept the status of clients. Explicit patronage defined the Usatovo culture.[9]

Cultural Integration between Usatovo and Upland Tripolye Towns

The stone-walled houses of the Usatovo settlement occupied the brow of a grassy ridge overlooking a bay near modern Odessa, the best seaport on the northwest coast of the Black Sea. Usatovo covered about 4–5 ha. A stone defensive wall probably defended the town on its seaward side. The settlement was largely destroyed by modern village construction and limestone quarrying prior to the first excavation by M. F. Boltenko in 1921, but parts of it survived (figure 14.2). Behind the ancient town four separate cemeteries crowned the hillcrest, all of them broadly contemporary. Two were kurgan cemeteries and two were flat-grave cemeteries. In one of the kurgan cemeteries, the one closest to the town, half the central graves contained men buried with bronze daggers and axes. These bronze weapons occurred in no other graves, not even in the second kurgan cemetery. Female figurines were limited to the flat-grave cemeteries and the settlement, never occurring in the kurgan

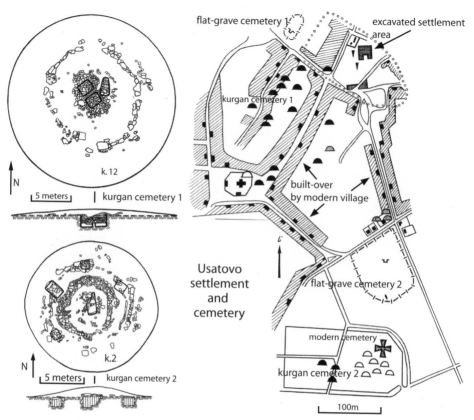

Figure 14.2 The Usatovo settlement (inside dotted line), kurgan cemeteries, and flat-grave cemeteries within the modern bay-side village of Usatovo, at the northeastern edge of the city of Odessa. After Patovka 1976 (village plan) and Zbenovich 1974 (kurgans).

graves. The flat-grave cemeteries were similar to flat-grave cemeteries that appeared outside Tripolye villages in the uplands, notably at Vikhvatinskii on the Dniester, where excavation of perhaps one-third of the cemetery yielded sixty-one graves of people with a gracile Mediterranean skull-and-face configuration. Upland cemeteries appeared at several other Tripolye sites (Holerkani, Ryşeşti, and Danku) located at the border between the steppes and the rainfall agriculture zone in the forest-steppe.

Clearly segregated funeral rituals (kurgan or flat grave) for different social groups appeared also at Mayaki, another Usatovo settlement on the

Dniester. The dagger chiefs of Usatovo probably dominated a hierarchy of steppe chiefs. Their relationship with the Tripolye villages in the Prut and Dniester forest-steppe seems unequal. Kurgan graves and graves containing weapons occurred only in the steppe. The upland Vikhvatinskii cemetery contained female figurines, but no metal weapons and only one copper object, a simple awl. Probably the Usatovo chiefs were patrons who received tribute, including fine painted pottery, from upland Tripolye clients. This relationship would have provided a prestige and status gradient that encouraged the adoption of the Usatovo language by late Tripolye villagers.

Usatovo is classified in all eastern European accounts as a Tripolye C2 culture. All eastern European archaeological cultures are defined first (sometimes only!) by ceramic types. Tripolye C2 pottery was a defining feature of Usatovo graves and settlements (figure 14.3). But the Usatovo culture was different from any Tripolye variant in that all the approximately fifty known Usatovo sites appeared exclusively in the steppe zone, at first around the mouth of the Dniester and later spreading to the Prut and Danube estuaries. Its funeral rituals were entirely derived from steppe traditions. Its coarse pottery, although made in standard Tripolye shapes, was shell-tempered and decorated with cord-impressed geometric designs like those of Yamnaya pottery. If the settlements were not so disturbed, we might be able to say whether they included compounds where Tripolye craftspeople worked as specialists. To explore how the Tripolye element was integrated in Usatovo society we have to look at other kinds of evidence.

The Usatovo economy was based primarily on sheep and goats (58–76% of bones at the Usatovo and Mayaki settlements, respectively). Sheep clearly predominated over goats, suggesting a wool butchering pattern.[10] At the same time, during Tripolye C2, clay loom weights and conical spindle whorls increased in frequency in upland towns in both the middle Dnieper and the Dniester regions, as if the Tripolye textile industry had accelerated. Usatovo settlements contained comparatively few spindle-whorls.[11] Perhaps upland Tripolye weavers made the wool from steppe sheep into finished textiles in a reciprocal exchange arrangement. Usatovo herders also kept cattle (28–13%) and horses (14–11%). Horse images were incised on two stone kurgan stelae at Usatovo (kurgan cemetery I, k. 11 and 3) and on a pot from an Usatovo grave at Tudorovo (figure 14.3n). Horses were important symbolically probably because riding was important in herding and raiding, and possibly because horses were important trade commodities.

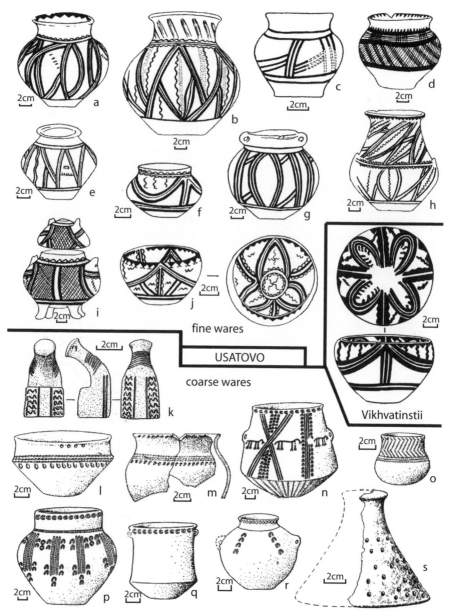

fine wares

USATOVO

coarse wares

Vikhvatinstii

Figure 14.3 Usatovo-culture ceramics (a, e, h, p, q, r) Usatovo kurgan cemetery I; (b) Tudorovo flat grave; (c) Sarata kurgan; (d) Shabablat kurgan; (f)
Parkany kurgan 182; (g, j, l) Usatovo kurgan cemetery II; (i) Parkany kurgan
91; (k) abstract figurine from Usatovo flat grave cemetery II; (m) Mayaki
settlement; (n) Tudorovo kurgan; (o) Usatovo flat grave cemetery II; (s) Mayaki settlement, probably a cheese strainer. Also shown: a painted fine bowl
from the Tripolye C2 cemetery at Vikhvatintsii. After Zbenovich 1968.

Impressions in pottery at the Usatovo settlement showed cultivated wheat (mostly emmer and bread wheats), barley, millet (frequent), oats (frequent), and peas.[12] The settlement also contained grinding stones and flint sickle teeth with characteristic edge gloss from cereal harvesting. This was the first evidence for cereal cultivation in the Dniester steppes, and, in fact, it is surprising, since rainfall agriculture is risky where precipitation is less than 350 mm per year. The grain would have been grown more easily in the upland settlements, perhaps cultivated by Tripolye people who resided part-time at Usatovo.

Tripolye C2 fine pots were particularly valued as grave gifts for the chiefs who died at Usatovo. Tripolye pots with an orange clay fabric, fired at almost 900°C, constituted 18% of the ceramics at the Usatovo settlement but 30% in the kurgan graves (figure 14.3, top). About 80% of the pottery at Usatovo and at other Usatovo-culture settlements was shell-tempered gray or brown ware, undecorated or decorated with cord impressions, and fired at only 700°C. This ware was made like steppe pottery. Though the shapes were like those made in the uplands by late Tripolye potters, some decorative motifs resembled those seen on Yamnaya Mikhailovka II–style pottery. A few of these shell-tempered gray pots at Usatovo were coated with a thick orange slip to make them *look* like fine Tripolye pots, indicating that the two kinds of pottery really were regarded as different.[13]

The painted Tripolye pots in Usatovo kurgan graves were most similar to those of the Tripolye C2 settlements at Brynzeny III on the Prut and Vikhvatintsii on the Dniester. Vikhvatinskii was 175 km up the Dniester from Usatovo near the steppe border, and Brynzeny III was about 350 km distant, hidden in the steep forested valleys of the East Carpathian piedmont. A fine painted pot of Brynzeny type was buried in the central grave of kurgan cemetery I, kurgan 12, at Usatovo, with an imported Maikop pot and a riveted bronze dagger. At this time Brynzeny III still had thirty-seven two-story *ploshchadka* houses, clay ovens, loom weights for large vertical looms, and female figurines. These traditional Tripolye customs survived in towns that showed ceramic connections with Usatovo, perhaps because patron-client agreements protected them. As the identities associated with the dying Tripolye culture were stigmatized and those associated with the Usatovo chiefs were emulated, people who lived at places like Brynzeny III and Vikhvatintsii might well have become bilingual. Their children then shifted to the Usatovo language.

Although fine Tripolye pots were preferred grave gifts for the Usatovo elite, the Tripolye culture itself occupied a secondary position of power and prestige. This is clearest in funeral customs. At Usatovo the chiefs buried under the kurgan graves were richer and more important than the people buried in the flat graves, and the flat graves were exactly reproduced in the upland Tripolye cemeteries at Vikhvatinskii and Holerkani.

The Usatovo Chiefs and Long-distance Trade

Another aspect of the Usatovo economy was long-distance trade, probably conducted by sea. All six known Usatovo settlements overlooked shallow coastal river mouths that would have made good harbors. These river mouths are today closed off from the sea by siltation, creating brackish lakes called *limans*, but they would have been more open to the sea in 3000 BCE. The sherds of small ceramic jugs and bowls of the Cernavoda III and Cernavoda II types from the lower Danube valley made up 1–2% of the broken crockery in the settlement at Usatovo, perhaps carried in by longboat rowers engaged in coastal trade down to Bulgaria. But these Cernavoda vessels never were offered as gifts in Usatovo graves. Whole imported late Maikop-Novosvobodnaya pots were included as grave gifts in the two central graves in kurgans 12 and 13 in kurgan cemetery I at Usatovo, two of the largest kurgans; but Maikop pottery never occurred in the settlement. Imported Maikop pots had a very different social meaning from Cernavoda pots.

Trade might have linked Usatovo to the emerging Aegean maritime chiefdoms of the EBI period, including Troy I. A white glass bead recovered from Usatovo kurgan cemetery II, kurgan 2, grave, 1 is the oldest known glass in the Black Sea region and perhaps in the ancient world. Glaze, the simplest form of glass, was applied to ceramics by about 4500–4000 BCE in northern Mesopotamia and Egypt. Glazes were made by mixing powdered quartz sand, lime, and either soda or ash and then heating the mixture to about 900°C, when it fused into a viscous state and could be dipped or poured. Faience beads were made of the same materials, molded into bead shapes, and glazed, beginning about the same time. But translucent glass, which required a higher temperature, has not been securely dated before the fifth dynasty of Egypt, or before 2450 BCE. The Usatovo bead and two others from Tripolye C2 Sofievka on the middle Dnieper are probably four hundred to seven hundred years older than that, equivalent to the first dynasty or the late Pre-Dynastic period. The Tripolye culture had no glazed ceramics or faience, so this vitreous technology was exotic. Almost certainly the Usatovo and Sofievka glass beads were made somewhere in the Eastern Mediterranean and imported. Another Tripolye C2 cemetery near Sofievka at Zavalovka, radiocarbon dated 2900–2800 BCE and similar to Sofievka in grave types and pottery, contained beads made of amber from the Baltic, perhaps the earliest expression of the exchange of northern amber for Mediterranean luxuries.[14]

In addition, two of the central dagger graves (k. 1 and 3) at Usatovo and an Usatovo grave at Sukleya on the lower Dniester contained daggers with rivet holes for the handle, cast in bivalve molds with a midrib on the blade. [see figure 14.4, top]. This kind of blade appeared also in Anatolia at Troy II and contemporary sites in Greece and Crete (David Stronach's Type 4 daggers). Like the glass, the Usatovo examples seem older than the Aegean ones—they should date to the equivalent of Troy I. But, in this case, the type might well have been locally invented in southeastern Europe and spread to the Aegean. Daggers with rivet holes but with a simpler lenticular-sectioned blade (without a midrib) certainly were made locally across southeastern Europe. They appeared in at least seven other Usatovo-culture graves, in graves at Sofievka on the middle Dnieper, and in Cotsofeni sites in the lower Danube valley, radiocarbon dated just before and after 3000 BCE [see figure 14.4, middle]. Regardless of the direction of borrowing, the shared riveted dagger types of Usatovo and the Aegean point to long-distance contacts between the two regions, perhaps in oared longboats.[15]

Patrons and Clients: Graves of the Warrior Chiefs at Usatovo

Usatovo kurgan cemetery I was quite near the Usatovo settlement (see figure 14.2). It originally contained about twenty kurgans. Fifteen were excavated between 1921 and 1973. They were complex constructions. Each kurgan had an earth core built up inside a stone cromlech made of large rectangular stones laid horizontally. All the cromlechs were covered by earth when the kurgans were enlarged; whether this was part of the original funeral or an entirely unconnected later event is unknown. The central grave was a deep shaft (up to 2 m deep) dug in the center of the cromlech circle, and in most kurgans it was accompanied by several (1–3) other graves also located inside the cromlech circle, in shallow pits covered by stone lids. At least five kurgans in cemetery I (3, 9, 11, 13, 14) were guarded by standing stone stelae on the southwestern sector of the mound. One stela (k. 13) was shaped at its top into a head, making an anthropomorphic shape, like many contemporary Yamnaya stelae in the South Bug–Dnieper steppes (see figure 13.11). Kurgan 3 (31 m in diameter) had two stelae standing side by side. The larger one (1.1 m tall) was inscribed with the images of a man, a deer, and three horses; the smaller one had just one horse. Kurgan 11 (40 m in diameter, the largest at Usatovo) covered a cromlech circle and inner mound 26 m in diameter surfaced with eighty-five hundred stones. On its southwest border were three stelae, one 2.7 m tall (!) with inscribed images of either dogs or horses. The central grave was robbed.

Only adult men were buried in the central graves of kurgan cemetery I, in a contracted position on the left side oriented east-northeast. Only the central graves and the peripheral graves on the southwestern sector contained red ochre. Seven of the fifteen central graves (k. 1, 3, 4, 6, 9, 12, and 14) had arsenical bronze dagger blades with two to four rivet holes for the handle. No other graves at Usatovo contained daggers (figure 14.4). Bronze daggers emerged as new symbols of status here and in the graves of the Yamnaya horizon at this time, but Yamnaya daggers had long tangs for the handle, like Novosvobodnaya daggers and unlike the Usatovo and Sofievka daggers with rivet holes for the handle. The central graves at Usatovo also contained fine Tripolye pots, arsenical bronze awls, flat axes, two Novosvobodnaya-style chisels, adzes, silver rings and spiral twists, flint microlithic blades, and flint hollow-based arrowheads. Bronze weapons and tools appeared only in the central graves.

Kurgan cemetery II was about 400 m away from kurgan cemetery I. It originally contained probably ten kurgans, most of them smaller than those in kurgan cemetery I; three were excavated. They yielded no daggers, no weapons, only small metal objects (awls, rings), and only a few fine painted Tripolye ceramic vessels. Six individuals had designs painted on their skulls with red ochre (figure 14.5). Three of these were men who had been killed by hammer blows to the head. Hammer wounds did not appear in kurgan cemetery I. Kurgan cemetery II was used for a distinct social group or status, perhaps warriors. But similar red designs were painted on the head of one male in kurgan cemetery I, in a peripheral grave under kurgan 12, grave 2, in the southwestern sector; similar designs were painted on the skulls of some Yamnaya graves at the Popilnaya kurgan cemetery on the South Bug.[16]

The flat graves at Usatovo were shallow pits covered by large flat stones, usually containing a body in a contracted position on the left side, oriented east or northeast. The peripheral graves under the kurgans had the same form as flat graves, and two cemeteries contained just flat graves, without

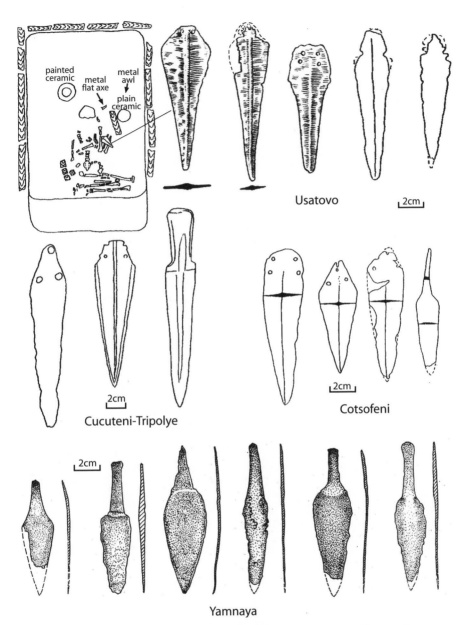

Figure 14.4 Daggers of the EBA, 3300–2800 BCE. *Top row*: Usatovo kurgan cemetery I, kurgan 3, central grave, with midrib dagger; kurgan 1, midrib dagger; Sukleya kurgan, midrib dagger; kurgan 9, lenticular-sectioned dagger; kurgan 6, lenticular-sectioned dagger. *Middle row left*: Werteba Cave, upper Dniester, riveted dagger; Cucuteni B, Moldova, midrib dagger; Werteba Cave, bone dagger carved in the shape of a metal dagger. *Middle row right*, Cotsofeni daggers from the lower Danube valley. *Bottom row*, Yamnaya tanged daggers from the North Pontic steppes. After Anthony 1996; and Nechitailo 1991.

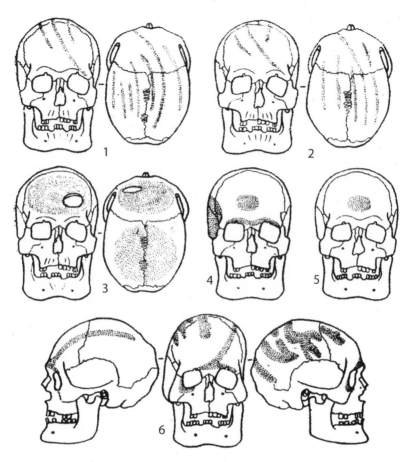

Usatovo (1-5) and Mayaki (6) painted skulls

Figure 14.5 Skulls painted with red ochre designs from the Usatovo and Mayaki cemeteries. Number 3 was killed by the hammer wound in the forehead. After Zin'kovskii and Petrenko 1987.

kurgans (thirty-six graves in flat cemetery I; thirty graves in flat cemetery II). Whereas just seven of the fifty-one graves (14%) in the kurgan cemeteries contained children, and two of these were buried with adults, twelve of the thirty-six graves (33%) in flat cemetery I contained children. Most of the adults in the flat graves were males, with a few old females. Each grave had from one to five pottery vessels but no metal, and only 4% of the pottery was fine painted ware. They did have ceramic female figurines (principally in children's graves), flint tools, and projectile points, and fifteen skulls were painted in the same red ochre designs as those in the kurgan graves, but none had hammer wounds.

Kurgan cemetery I was reserved for leaders who displayed arsenical bronze riveted daggers and axes and wore silver rings but suffered no hammer wounds, perhaps patrons. Kurgan cemetery II honored old men, old women, young men, and children who did not have bronze daggers or metal weapons of any kind but sometimes died of hammer wounds to the head, perhaps those who died in battle and their close kin. The flat cemeteries contained many children, a few women, and old men who had plain pots and no daggers. All were connected to one another, and to external Yamnaya groups, by linear red designs painted on some skulls. The social organization of Usatovo has been interpreted as a male-centered military aristocracy, but it could also be read as remarkably like the tripartite social system suggested by Dumezil for the speakers of Proto-Indo-European, with priest-patrons (kurgan cemetery I), warriors (kurgan cemetery II), and ordinary producers (flat graves).

The Ancestor of English: The Origin and Spread of the Usatovo Dialect

The Usatovo culture was exclusively a steppe culture, and it appeared simultaneously with the rapid expansion of the Yamnaya horizon across the steppes, after the permanent dissolution of many Tripolye towns near the steppe border. Usatovo is often interpreted as a Tripolye population that migrated into the steppes, but Tripolye farmers had never done this during the previous two thousand years, and in neighboring valleys (the lower Siret, lower Prut, the entire South Bug valley, the Ros') they were retreating from the steppe border, not advancing across it. The funeral customs of Usatovo were starkly hierarchical, with a typical steppe kurgan ritual reserved for the elite. Although Usatovo ceramics were almost entirely borrowed from and made by Tripolye potters, even here there were similarities with Yamnaya ceramics in some cord-impressed ornament on the coarse wares. Usatovo is not counted as a part of the Yamnaya horizon because of its close integration with the Tripolye culture, but it appeared at the same time as the Yamnaya horizon, in the steppes, with kurgan funeral rituals that repeated many old steppe customs; sacrifices and broken pottery also were placed on the southwestern side of the kurgan in Yamnaya and even Afanasievo graves. The painted skulls were also repeated in Yamnaya graves. Usatovo probably began with steppe clans connected with the early

Yamnaya horizon who were able to impose a patron-client relationship on Tripolye farming villages because of the protection that client status offered in a time of great insecurity. The pastoral patrons quickly became closely integrated with the farmers.

Tripolye clients of the Usatovo chiefs could have been the agents through which the Usatovo language spread northward into central Europe. After a few generations of clientage, the people of the upper Dniester might have wanted to acquire their own clients. Nested hierarchies in which clients are themselves patrons of other clients are characteristic of the growth of patron-client systems. The archaeological evidence for some kind of northward spread of people or political relationships consists of pottery exchanges between Tripolye sites on the upper Dniester and late TRB (Trichterbecker or Funnel-Beaker culture) sites in southeastern Poland. Substantial quantities of fine painted Tripolye C2 pottery of the Brynzeny III type occurred in southern Polish settlements of the late TRB culture dated 3000–2800 BCE, importantly at Gródek Nadbużny and Zimne, and late TRB pots were imported into the Tripolye C2 sites of Zhvanets and Brynzeny III.[17] Zhvanets was a production center for fine Tripolye pottery, with seven large two-chambered kilns, a possible source of local economic and political prestige. Conflict accompanied or alternated with exchange, since both the Polish sites and the Tripolye C2 sites closest to southeastern Poland were heavily fortified. The Tripolye C2 settlement of Kosteshti IV had a stone wall 6 m wide and a fortification ditch 5 m wide, and Zhvanets had three lines of fortification walls faced with stone, and both were located on high promontories.[18] Tripolye C2 community leaders whose parents had already adopted the Usatovo language could have attempted to extend to the late TRB communities of southern Poland the same kind of patron-client relationships that the Usatovo chiefs had offered them, an extension that might well have been encouraged or even backed up by paramount Usatovo chiefs.

If I had to hazard a guess I would say that this was how the Proto-Indo-European dialects that would ultimately form the root of Pre-Germanic first became established in central Europe: they spread up the Dniester from the Usatovo culture through a nested series of patrons and clients, and eventually were spoken in some of the late TRB communities between the Dniester and the Vistula. These late TRB communities later evolved into early Corded Ware communities, and it was the Corded Ware horizon (see below) that provided the medium through which the Pre-Germanic dialects spread over a wider area.

The Yamnaya Migration up the Danube Valley

About 3100 BCE, during the initial rapid spread of the Yamnaya horizon across the Pontic-Caspian steppes, and while the Usatovo culture was still in its early phase, Yamnaya herders began to move through the steppes past Usatovo and into the lower Danube valley. The initial groups were followed by a regular stream of people that continued for perhaps three hundred years, between 3100 and 2800 BCE.[19] The passage through the Usatovo chiefdoms probably was managed through guest-host relationships. The migrants did not claim any Usatovo territory—at least they did not create their own cemeteries there. Instead, they kept going into the Danube valley, a minimum distance of 600–800 km from where they began in the steppes east of Usatovo—in the South Bug valley and farther east. The largest number of Yamnaya migrants ended up in eastern Hungary, an amazing distance (800–1,300 km depending on the route taken). This was a major, sustained population movement, and, like all such movements, it must have been preceded by scouts who collected information while on some other kind of business, possibly horse trading. The scouts knew just a few areas, and these became the targets of the migrants.[20]

The Yamnaya migrations into the Danube valley were targeted toward at least five specific destinations (see figure 14.1). One cluster of Yamnaya kurgan cemeteries, probably the earliest, appeared on the elevated plain northwest of Varna bay in Bulgaria (kurgan cemeteries at Plachidol, Madara, and other nearby places). This cluster overlooked the fortified coastal settlement at Ezerovo, an important local Early Bronze Age center. The second cluster of kurgan cemeteries appeared in the Balkan uplands 200 km to the southwest (the Kovachevo and Troyanovo cemeteries). They overlooked a fertile plain between the Balkan peaks and the Maritsa River, where many old tells such as Ezero and Mihailich had just been reoccupied and fortified. The third target was 300 km farther up the Danube valley in northwestern Bulgaria (Tarnava), on low ridges overlooking the broad plain of the Danube. These three widely separated clusters in Bulgaria contained at least seventeen Yamnaya cemeteries, each with five to twenty kurgans. Across the Danube and just 100 km west of the northwestern Bulgarian cluster, a larger group of kurgan cemeteries appeared in southwestern Romania, where at least a hundred Yamnaya kurgans dotted the low plains overlooking the Danube around Rast in southern Oltenia, south of Craiova. The Tarnava and Rast kurgans were in the same terrain and can be counted as one group, separated by the Danube River (and a modern international border).

Pushing westward through Cotsofeni-culture territory, Yamnaya migrants found their way over the mountains around the Iron Gates, where the Danube sweeps through a long, steep set of gorges, and into the wide plains on the Serbian side. A few kurgan groups were erected in a fourth cluster west of the Iron Gates in the plains of northern Serbia (Jabuka). Finally, the fifth and largest group of kurgans appeared in the eastern Hungarian plains north of the Körös and east of the Tisza rivers.[21] The number of kurgans raised in the east Hungarian cluster is unknown, but Ecsedy estimated at least three thousand, spread over about 6000–8000 km^2. Archaeologists have mapped forty-five Yamnaya cemeteries, each of which contained five to thirty-five kurgans. One kurgan at Kétegyháza was built on top of the remains of a Cernavoda III settlement. The east Hungarian Yamnaya population seems to have been the largest that accumulated in any of the five target areas. Some of them wore leather caps, silver temple rings, and dog-canine-tooth necklaces in their graves.

The first three clusters near Varna, Ezero, and the Cotsofeni territory seem to have been chosen for their proximity to settled areas, perhaps by ambitious men seeking clients, whereas the last two clusters seem to have been chosen for their pastures, perhaps by others who wanted to increase their herds. In all places the Yamnaya funeral ritual was similar, and it was not native but intrusive. Kurgans were 15–60 m in diameter. The grave pit floors often had traces of organic mats, some painted with designs, as in the steppes (figure 14.6). The central graves contained an adult (80% are males in Bulgaria) buried supine with raised knees (some were contracted on the side), with the head oriented toward the west (or, in Bulgaria, sometimes to the south). Most had Proto-Europoid skull-face shapes, like the predominant element in the Pontic steppe Yamnaya population. Most graves contained no grave goods. A few contained a flint tool, beads of pierced dog teeth, or a temple ring with one and a half twists of copper, silver, or gold. In Hungary a lump of red ochre was placed near the head; in Romania and Bulgaria, in addition to a lump placed near the head, red ochre covered the floor or stained the skull, feet, legs, and hands. At Kétegyháza, where there was no local source of hematite from which to make red ochre, a lump of clay was painted red to imitate true ochre, a clear indication of a cult practice imported from a region with different minerals. One grave at Gurbaneşti in Romania contained a clay vessel with carbonized hemp seeds, the earliest evidence for the burning of *Cannabis*. Sherratt suggested that *Cannabis* smoking was introduced to the Danube valley by the Yamnaya immigrants. In northeast Bulgaria at Plachidol, one Yamnaya grave (k. 1, gr. 1) had four wooden wagon wheels placed at the corners

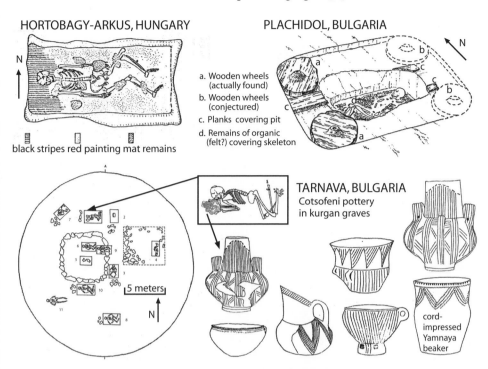

HORTOBAGY-ARKUS, HUNGARY

PLACHIDOL, BULGARIA

a. Wooden wheels
   (actually found)
b. Wooden wheels
   (conjectured)
c. Planks covering pit
d. Remains of organic
   (felt?) covering skeleton

black stripes   red painting   mat remains

TARNAVA, BULGARIA
Cotsofeni pottery
in kurgan graves

5 meters

cord-
impressed
Yamnaya
beaker

Figure 14.6 Kurgan graves and ceramics from Bulgaria and eastern Hungary associated with the Yamnaya migration about 3000 BCE. The graves under Tarnava kurgan 1 in northwestern Bulgaria contained principally Cotsofeni pottery, but one grave under kurgan 2 contained a typical Yamnaya beaker. After Ecsedy 1979; Panaiotov 1989; and Sherratt 1986.

just as in many wagon graves in the steppes (figure 14.6). Cemeteries in this cluster near Varna contained anthropomorphic stone stelae like the Yamnaya and Kemi-Oba stelae in the steppes.

The source of the Yamnaya migration is commonly said to have been in the lower Dniester steppes, where Yamnaya graves also were consistently oriented to the west. But the lower Dniester steppes were occupied by the Usatovo culture between 3100 and 2800 BCE. Yamnaya graves in the Dniester steppes are consistently stratified above Usatovo graves, and most of them are radiocarbon dated between 2800 and 2400 BCE, so most of them postdated the Danube valley migration. The Dniester variant of Yamnaya might instead represent a return migration *from* the Danube valley back into the steppes, since almost all significant migration streams produce a flowback of return migration. The Yamnaya wagon graves (Kholmskoe, Vishnevoe, and others) located in the steppes just north of the Danube delta are stratified above Usatovo graves, so probably were made later than the Yamnaya wagon grave in Bulgaria at Plachidol. The Danube valley migration probably originated east of the Usatovo area, in the steppes around the South Bug, Ingul, and Dnieper valleys. Western-oriented Yamnaya graves are found as a minor variant in Yamnaya cemeteries in the Dnieper-South Bug region. The oldest dated Yamnaya wagon grave (ca. 3000 BCE) at Bal'ki (k. 1 gr. 57) on the lower Dnieper was oriented to the west.[22]

What started this movement? A popular candidate has been a shortage of pasture in the steppes, but I find it hard to believe that there was any absolute shortage of pasture during the initial expansion of a new wagon-based economy. If the migration into the Danube valley began with raiding that then developed into a migration, we have to ask what caused the raiding. In the discussion of the causes of steppe warfare, in chapter 11, I mentioned the Proto-Indo-European *Trito* myth, which legitimized the cattle raid; the likelihood that competition between high-status families would lead to escalating bride-prices calculated in livestock, which might *create* a consumer shortage of animals and pastures in places where no absolute shortage existed; and the Proto-Indo-European initiation ritual that sent all young men out raiding.

The institution of the *Männerbünde* or *korios*, the warrior brotherhood of young men bound by oath to one another and to their ancestors during a ritually mandated raid, has been reconstructed as a central part of Proto-Indo-European initiation rituals.[23] One material trait linked to these ceremonies was the dog or wolf; the young initiates were symbolized by the dog or wolf and in some Indo-European traditions wore dog or wolf skins during their initiation. The canine teeth of dogs were frequently worn as pendants in Yamnaya graves in the western Pontic steppes, particularly in the Ingul valley, one probable region of origin for the Yamnaya migration.[24] A second material trait linked to the *korios* was the belt. The *korios* raiders wore a belt and little else (like the warrior figures in some later Germanic and Celtic art, e.g., the Anglo-Saxon Finglesham belt buckle). The initiates on a raid wore two belts, their leader one, symbolizing that the leader was bound by a single oath to the god of war/ancestors, and the initiates were double-bound to the god/ancestors and to the leader. Stone anthropomorphic stelae were erected over hundreds of Yamnaya graves between the Ingul and the South Bug valleys, in the same region where dog-canine pendants were common. The most common clothing element carved or painted on the stelae was a belt, often with an axe or a pair of sandals attached to it. Usually it was a single belt, perhaps symbolizing the leader of a raid. That stone stelae with belts were erected also by the Yamnaya migrants in Bulgaria near Plachidol provides another link between the migrants and the symbolism of the *korios* raid.[25]

There must also have been other pulls, positive rumors about opportunities in the Danube valley, because the migrants did not just raid but decided to live in the target region. These attractions are difficult to identify now, although the opportunity to acquire clients might have been a powerful pull.

Language Shift and the Yamnaya Migration

The Yamnaya migration occurred at a time of great fluidity and change throughout southeastern Europe. In Bulgaria, the tells in the upland plains of the Balkans at Ezero, Yunatsite, and Dubene-Sarovka were reoccupied about 3300–3200 BCE at the beginning of the Early Bronze Age (EBI) after almost a millennium of abandonment. The reoccupied tell settlements were fortified with substantial stone walls or ditches and palisades. One target of the Yamnaya migration was precisely this region. Yamnaya kurgan cemeteries could be seen for many miles; visually, they dominated the landscapes around them. In contrast, local cemeteries in the lower Danube valley and the Balkans, like the EBI cemetery at the Bereket tell settlement near Stara Zagora, usually had no visible surface monuments.[26]

A series of new artifact types diffused very widely across the lower and middle Danube valleys in connection with the Yamnaya migration. Concave-based arrowheads similar to steppe arrowheads appeared in the newly occupied tell sites in Bulgaria (Ezero) and in Aegean Macedonia (Dikili–Tash IIIB). These possibly were a sign of warfare with intrusive Yamnaya raiding groups. A new ceramic style spread across the entire middle and lower Danube, including the Morava and Struma valleys leading to Greece and the Aegean, and in Aegean Macedonia. The defining trait of this style was cord-impressed pottery encrusted with white paint.[27] White-encrusted, cord-impressed pottery appeared also in the Yamnaya graves. The Yamnaya immigrants could, perhaps, have played a role in joining one region to another and helping to spread this new style. But the pottery styles they spread were not their own. The Yamnaya immigrants usually deposited no pottery in their graves, and, when they did, they borrowed local ceramic styles, so their ceramic footprint is almost invisible.

Many Yamnaya kurgans in the lower Danube valley contained Cotsofeni ceramic vessels. The Cotsofeni culture evolved in mountain refuges in western Romania and Transylvania beginning about 3500 BCE, probably from Old European roots. Cotsofeni settlements were small agricultural hamlets of a few houses. Their owners cremated their dead and buried the ashes in flat graves, some of which contained riveted daggers like Usatovo daggers.[28] When Yamnaya herders reached the plains around Craiova, they probably realized that control over this region was the key to movement up and down the Danube valley through the mountain passes around the Iron Gates. They established alliances or patron-client contracts with the leaders of the Cotsofeni communities, through which they obtained Cotsofeni pottery (and probably other less visible Cotsofeni products), as Usatovo patrons obtained Tripolye pottery. Cotsofeni pottery then was carried into other regions by Yamnaya people. A Cotsofeni vessel was found in a Yamnaya kurgan as far afield as Tarakliya, Moldova, probably in the grave of a returned migrant. In northwestern Bulgaria, kurgan 1 at Tarnava (figure 14.6) contained an unusual concentration of six Cotsofeni pots in six Yamnaya graves.[29] Most of the Yamnaya kurgans in Bulgaria contained no ceramics, but, when they did, they were often Cotsofeni ceramics.

The situation of the Yamnaya chiefs might have been similar to that described by Barth in his account of the Yusufai Pathan invasion of the Swat valley in Pakistan in the sixteenth century. The invader, "faced with the sea of politically undifferentiated villagers proceeds to organize a central island of authority, and from this island he attempts to exercise authority over the surrounding sea. Other landowners establish similar islands, some with overlapping spheres of influence, others having unadministered gaps between them."[30] The mechanism through which the immigrant chief made himself indispensable to the villagers and tied them to him was the creation of a contract in which he guaranteed protection, hospitality, and the recognition of the villagers' rights to agricultural production in exchange for their loyalty, service, and best land. Yamnaya herding groups needed more land for pastures than did farming groups of equal population, and this could have provided a rationale for the Yamnaya people to claim use-rights over most of the available pasture lands and the migration routes that linked them, eventually creating a web of landownership that covered much of southeastern Europe. The reestablishment of tell settlements in the Balkans might have been part of a newly bifurcated economy in which farmers settled on fortified tells and increased grain production in response to reductions in their pastures, taken by their Yamnaya patrons.

The widely separated pockets of Yamnaya settlement in the lower Danube valley and the Balkans established speakers of late Proto-Indo-European dialects in scattered islands where, if they remained isolated from one another, they could have differentiated over centuries into various Indo-European languages. The many thousands of Yamnaya kurgans in eastern Hungary suggest a more continuous occupation of the landscape by a larger population of immigrants, one that could have acquired power and prestige partly just through its numerical weight. This regional group could have spawned both pre-Italic and pre-Celtic. Bell Beaker sites of the Csepel type around Budapest, west of the Yamnaya settlement region, are dated about 2800–2600 BCE. They could have been a bridge between Yamnaya on their east and Austria/Southern Germany to their west, through which Yamnaya dialects spread from Hungary into Austria and Bavaria, where they later developed into Proto-Celtic.[31] Pre-Italic could have developed among the dialects that remained in Hungary, ultimately spreading into Italy through the Urnfield and Villanovan cultures. Eric Hamp and others have revived the argument that Italic and Celtic shared a common parent, so a single migration stream could have contained dialects that later were ancestral to both.[32] Archaeologically, however, the Yamnaya immigrants here, as elsewhere, left no lasting material impression except their kurgans.

YAMNAYA CONTACTS WITH THE CORDED WARE HORIZON

The Corded Ware horizon is often invoked as the archaeological manifestation of the cultures that introduced the northern Indo-European languages to Europe: Germanic, Baltic, and Slavic. The Corded Ware horizon spread across most of northern Europe, from Ukraine to Belgium, after 3000 BCE, with the initial rapid spread happening mainly between 2900 and 2700 BCE. The defining traits of the Corded Ware horizon were a pastoral, mobile economy that resulted in the near disappearance of settlement sites (much like Yamnaya in the steppes), the almost universal adoption of funeral rituals involving single graves under mounds (like Yamnaya), the diffusion of stone hammer-axes probably derived from Polish TRB styles, and the spread of a drinking culture linked to particular kinds of cord-decorated cups and beakers, many of which had local stylistic prototypes in variants of TRB ceramics. The material culture of the Corded Ware horizon was mostly native to northern Europe, but the underlying behaviors were very similar to those of the Yamnaya horizon— the broad adoption of a herding economy based on mobility (using ox-drawn wagons and horses), and a corresponding rise in the ritual prestige and value of livestock.[33] The economy and political structure of the Corded Ware horizon certainly was influenced by what had emerged earlier in the steppes, and, as I just argued, some Corded Ware groups in south-eastern Poland might have evolved from Indo-European–speaking late TRB societies through connections with Usatovo and late Tripolye. The Corded Ware horizon established the material foundation for the evolution of most of the Bronze Age cultures of the northern European plain, so most discussions of Germanic, Baltic, or Slavic origins look back to the Corded Ware horizon.

The Yamnaya and Corded Ware horizons bordered each other in the hills between Lvov and Ivano-Frankovsk, Ukraine, in the upper Dniester piedmont around 2800–2600 BCE (see figure 14.1). At that time early Corded Ware cemeteries were confined to the uppermost headwaters of the Dniester west of Lvov, the same territory that had earlier been occupied by the late TRB communities infiltrated by late Tripolye groups. If Corded Ware societies in this region evolved from local late TRB origins, as many believe, they might already have spoken an Indo-European language. Between 2700 and 2600 BCE Corded Ware and late Yamnaya herders met each other on the upper Dniester over cups of mead or beer.[34] This meeting was another opportunity for language shift, and it is possible that Pre-Germanic dialects either originated here or were enriched by this additional contact.

The wide-ranging pattern of interaction that the Corded Ware horizon inaugurated across northern Europe provided an optimal medium for language spread. Late Proto-Indo-European languages penetrated the eastern end of this medium, either through the incorporation of Indo-European dialects in the TRB base population before the Corded Ware horizon evolved, or through Corded Ware–Yamnaya contacts later, or both. Indo-European speech probably was emulated because the chiefs who spoke it had larger herds of cattle and sheep and more horses than could be raised in northern Europe, and they had a politico-religious culture already adapted to territorial expansion. The dialects that were ancestral to Germanic probably were initially adopted in a small territory between the Dniester and the Vistula and then spread slowly. As we will see in the next chapter, Slavic and Baltic probably evolved from dialects spoken on the middle Dnieper.[35]

THE ORIGINS OF GREEK

The only major post-Anatolian branch that is difficult to derive from the steppes is Greek. One reason for this is chronological: Pre-Greek probably split away from a later set of developing Indo-European dialects and languages, not from Proto-Indo-European itself. Greek shared traits with Armenian and Phrygian, both of which probably descended from languages spoken in southeastern Europe before 1200 BCE, so Greek shared a common background with some southeastern European languages that might have evolved from the speech of the Yamnaya immigrants in Bulgaria. As noted in chapter 3, Pre-Greek also shared many traits with pre–Indo-Iranian. This linguistic evidence suggests that Pre-Greek should have been spoken on the eastern border of southeastern Europe, where it could have shared some traits with Pre-Armenian and Pre-Phrygian on the west and pre–Indo-Iranian on the east. The early western Catacomb culture would fit these requirements (see figure 15.5), as it was in touch with southeastern Europe on one side and with the developing Indo-Iranian world of the east on the other. But it is impossible, as far as I know, to identify a Catacomb-culture migration that moved directly from the western steppes into Greece.

A number of artifact types and customs connect the Mycenaean Shaft Grave princes, the first definite Greek speakers at about 1650 BCE, with steppe or southeastern European cultures. These parallels included specific types of cheekpieces for chariot horses, specific types of socketed spearheads, and even the custom of making masks for the dead, which was common on the Ingul River during the late Catacomb culture, between about 2500 and 2000 BCE. It is very difficult, however, to define the specific source of the migration stream that brought the Shaft Grave princes into Greece. The people who imported Greek or Proto-Greek to Greece might have moved several times, perhaps by sea, from the western Pontic steppes to southeastern Europe to western Anatolia to Greece, making their trail hard to find. The EHII/III transition about 2400–2200 BCE has long been seen as a time of radical change in Greece when new people might have arrived, but the resolution of this problem is outside the scope of this book.[36]

Conclusion: The Early Western Indo-European Languages Disperse

There was no Indo-European invasion of Europe. The spread of the Usatovo dialect up the Dniester valley, if it happened as I have suggested, was quite different from the Yamnaya migration into the Danube valley. But even that migration was not a coordinated military invasion. Instead, a succession of Pontic steppe tribal segments fissioned from their home clans and moved toward what they perceived as places with good pastures and opportunities for acquiring clients. The migrating Yamnaya chiefs then organized islands of authority and used their ritual and political institutions to establish control over the lands they appropriated for their herds, which required granting legal status to the local populations nearby, under patron-client contracts. Western Indo-European languages might well have remained confined to scattered islands across eastern and central Europe until after 2000 BCE, as Mallory has suggested.[37] Nevertheless, the movements into the East Carpathians and up the Danube valley occurred in the right sequence, at the right time, and in the right directions to be connected with the detachment of Pre-Italic, Pre-Celtic, and Pre-Germanic—the branch that ultimately gave birth to English.

CHAPTER FIFTEEN

Chariot Warriors of the Northern Steppes

The publication of the book *Sintashta* in 1992 (in Russian) opened a new era in steppe archaeology.[1] Sintashta was a settlement east of the Ural Mountains in the northern steppes. The settlement and the cemeteries around it had been excavated by various archaeologists between 1972 and 1987. But only after 1992 did the significance of the site begin to become clear. Sintashta was a fortified circular town 140 m in diameter, surrounded by a timber-reinforced earthen wall with timber gate towers (figure 15.1). Outside the wall was a V-shaped ditch as deep as a man's shoulders. The Sintashta River, a western tributary of the upper Tobol, had washed away half of it, but the ruins of thirty-one houses remained. The original town probably contained fifty or sixty. Fortified strongholds like this were unprecedented in the steppes. A few smaller fortified settlements had appeared west of the Don (Mikhailovka, for example) during the Yamnaya period. But the walls, gates, and houses of Sintashta were much more substantial than at any earlier fortified site in the steppes. And inside each and every house were the remains of metallurgical activity: slag, ovens, hearths, and copper. Sintashta was a fortified metallurgical industrial center.

Outside the settlement were five funerary complexes that produced spectacular finds (figure 15.2). The most surprising discoveries were the remains of chariots, which radiocarbon dates showed were the oldest chariots known anywhere. They came from a cemetery of forty rectangular grave pits without an obvious kurgan labeled SM for *Sintashta mogila*, or *Sintashta cemetery*. The other four mortuary complexes were a mid-size kurgan (SI, for *Sintashta I*), 32 m in diameter and only 1 m high, that covered sixteen graves; a second flat or non-kurgan cemetery (SII) with ten graves; a second small kurgan (SIII), 16 m in diameter, that covered a single grave containing the partial remains of five individuals; and finally a huge kurgan, 85 m in diameter and 4.5 m high (SB, for *Sintashta bolshoi kurgan*), built over a central grave (robbed in antiquity) constructed of logs and sod on the

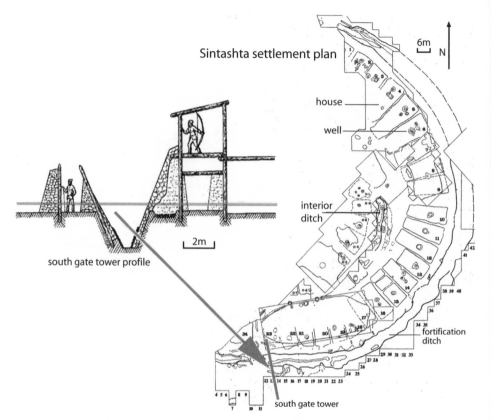

Figure 15.1 The Sintashta settlement: rectangular houses arranged in a circle within a timber-reinforced earthen wall, with excavators' reconstruction of south gate tower and outer defense wall. After Gening, Zdanovich, and Gening 1992, figures 7 and 12.

original ground surface. The southern skirt of the SB kurgan covered, and so was later than, the northern edge of the SM cemetery, although the radiocarbon dates suggest that SM was only slightly older than SB. The forty SM graves contained astounding sacrifices that included whole horses, up to eight in and on a single grave (gr. 5), with bone disc-shaped cheekpieces, chariots with spoked wheels, copper and arsenical bronze axes and daggers, flint and bone projectile points, arsenical bronze socketed spearheads, polished stone mace heads, many ceramic pots, and a few small silver and gold ornaments (figure 15.3). What was impressive in these graves was weaponry, vehicles, and animal sacrifices, not crowns or jewelry.

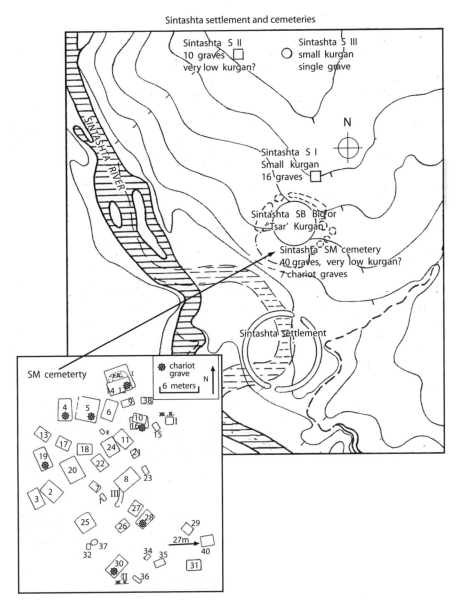

Figure 15.2 The Sintashta settlement landscape, with associated cemeteries, and detail of the SM cemetery. After Gening, Zdanovich, and Gening 1992, figures 2 and 42.

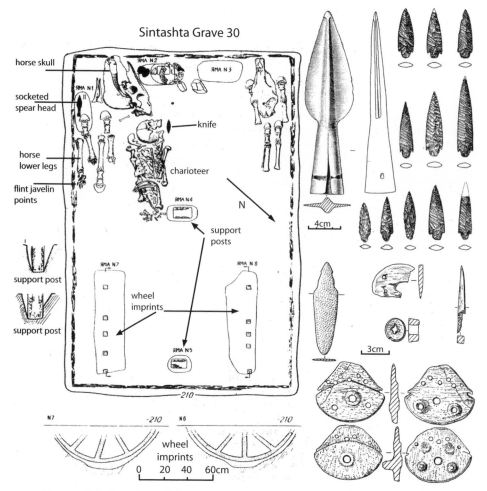

Figure 15.3 Sintashta SM cemetery, grave 30, with chariot wheel impressions, skulls and lower leg bones of horse team, cheekpieces for bits, and weapons. After Gening, Zdanovich, and Gening, figures 111, 113, and 114.

The radiocarbon dates for both the cemeteries and the settlement at Sintashta were worryingly diverse, from about 2800–2700 BCE (4200+100 BP), for wood from grave 11 in the SM cemetery, to about 1800–1600 BCE (3340+60BP), for wood from grave 5 in the SII cemetery. Probably there was an older Poltavka component at Sintashta, as later was found at many other sites of the Sintashta type, accounting for the older dates. Wood from the central grave of the large kurgan (SB) yielded consistent dates (3520+65, 3570+60, and 3720+120), or about 2100–1800 BCE. The same age range was produced by radiocarbon dates from the similar settlement at Arkaim, from several Sintashta cemeteries (Krivoe Ozero, Kammeny Ambar), and from the closely related graves of the Potapovka type in the middle Volga region (table 15.1).

The details of the funeral sacrifices at Sintashta showed startling parallels with the sacrificial funeral rituals of the *Rig Veda*. The industrial scale of metallurgical production suggested a new organization of steppe mining and metallurgy and a greatly heightened demand for copper and bronze. The substantial fortifications implied surprisingly large and determined attacking forces. And the appearance of Pontic-Caspian kurgan rituals, vehicle burials, and weapon types in the steppes east of the Ural River indicated that the Ural frontier had finally been erased.

After 1992 the flow of information about the Sintashta culture grew to a torrent, almost all of it in Russian and much of it still undigested or actively debated as I write.[2] Sintashta was just one of more than twenty related fortified settlements located in a compact region of rolling steppes between the upper Ural River on the west and the upper Tobol River on the east, southeast of the Ural Mountains. The settlement at Arkaim, excavated by G. B. Zdanovich, was not damaged by erosion, and twenty-seven of its fifty to sixty structures were exposed (figure 15.4). All the houses at Arkaim contained metallurgical production facilities. It has become a conference center and national historic monument. Sintashta and Arkaim raised many intriguing questions. Why did these fortified metal-producing towns appear in that place at that time? Why the heavy fortifications—who were they afraid of? Was there an increased demand for copper or just a new organization of copper working and mining or both? Did the people who built these strongholds invent chariots? And were they the original Aryans, the ancestors of the people who later composed the *Rig Veda* and the *Avesta*?[3]

The End of the Forest Frontier: Corded Ware Herders in the Forest

To understand the origins of the Sintashta culture we have to begin far to the west. In what had been the Tripolye region between the Dniester and Dnieper rivers, the interaction between Corded Ware, Globular Amphorae, and Yamnaya populations between 2800 and 2600 BCE produced a complicated checkerboard of regional cultures covering the rolling hills and valleys of the forest-steppe zone (figure 15.5). To the south, in the

TABLE 15.1
Selected radiocarbon dates for the Sintashta–Arkaim (S) and Potapovka (P) cultures in the
south Ural steppes and middle Volga steppes.

Lab Number	BP Date	Sample Source	C, K	Calibrated Date
Sintashta SB Big Kurgan (S)				
GIN–6186	3670±40	birch log		2140–1970 BCE
GIN–6187	3510±40	"		1890–1740 BCE
GIN–6188	3510±40	"		1890–1740 BCE
GIN–6189	3260±40	"		1610–1450 BCE
Sintashta SM cemetery (S)				
Ki–653	4200±100	grave 11, wood	K	2900–2620 BC
Ki–658	4100±170	grave 39, wood	K	2900–2450 BC
Ki–657	3760±120	grave 28, wood	C	2400–1970 BC
Ki–864	3560±180	grave 19, wood	C	2200–1650 BCE
Ki–862	3360±70	grave 5, wood	C, K	1740–1520 BC
Krivoe Ozero cemetery, kurgan 9, grave 1 (S)				
AA–9874b	3740±50	horse 1 bone	C, K	2270–2030 BC
AA–9875a	3700±60	horse 2 bone		2200–1970 BC
AA–9874a	3580±50	horse 1 bone		2030–1780 BC
AA–9875b	3525±50	horse 2 bone		1920–1750 BC
Kammeny Ambar 5 (S)				
OxA–12532	3604±31	k2: grave 12, human bone		2020–1890 BCE
OxA–12530	3572±29	k2: grave 6, "	K	1950–1830 BCE
OxA–12533	3555±31	k2: grave 15, "		1950–1780 BCE
OxA–12531	3549±49	k2: grave 8, "	C, K	1950–1770 BCE
OxA–12534	3529±31	k4: grave 3, "		1920–1770 BCE
OxA–12560	3521±28	k4: grave 1, "		1890–1770 BCE
OxA–12535	3498±35	k4: grave 15, "		1880–1740 BCE
Utyevka cemetery VI (P)				
AA–12568	3760±100	k6: grave 4, human bone	K	2340–1980 BC
OxA–4264	3585±80	k6: grave 6, human bone		2110–1770 BC
OxA–4306	3510±80	k6: grave 4, human bone	K	1940–1690 BC
OxA–4263	3470±80	k6: grave 6, human bone	K	1890–1680 BC
Potapovka cemetery I (P)				
AA–12569	4180±85	k5: grave 6, dog bone*		2890–2620 BC

TABLE 15.1 (*continued*)

Lab Number	BP Date	Sample Source	C, K	Calibrated Date
AA–47803	4153±59	k.3: grave 1, human bone*		2880–2620 BC
OxA–4265	3710±80	k5: grave 13, human bone		2270–1960 BC
OxA–4266	3510±80	k5: grave 3, human bone		1940–1690 BC
AA–47802	3536±57	k.3: grave 1, horse skull*		1950–1770 BC
Other Potapovka cemeteries (P)				
AA–53803	4081±54	Kutuluk I, k1:1, human bone		2860–2490 BC
AA–53806	3752±52	Grachevka II k5:3, human bone		2280–2030 BC

*See note 17

Graves that contained chariots are marked C; graves that contained studded disc cheekpieces are marked K.

steppes, late Yamnaya and a few late Usatovo groups continued to erect kurgan cemeteries. Some late Yamnaya groups penetrated northward into the forest-steppe, up the Dniester, South Bug, and Dnieper valleys. Eastern Carpathian groups making Globular Amphorae pottery moved from the upper Dniester region around Lvov eastward into the forest-steppe around Kiev, and then retreated back to the Dniester. Corded Ware groups from southern Poland replaced them around Kiev. Under the influence of this combined Globular Amphorae and Corded Ware expansion to the east, the already complex mixture of Yamnaya-influenced Late Tripolye people in the Middle Dnieper valley created the Middle Dnieper culture in the forest-steppe region around Kiev. This was the first food-producing, herding culture to push into the Russian forests north of Kiev.[4]

The Middle Dnieper and Fatyanovo Cultures

The people of the Middle Dnieper culture carried stockbreeding economies (cattle, sheep, and pigs, depending on the region) north into the forest zone, up the Dnieper and Desna into what is now Belarus (figure 15.5). They followed marshes, open lakes, and riverine floodplains where there were natural openings in the forest. These open places had grass and reeds for the animals, and the rivers supplied plentiful fish. The earliest Middle Dnieper sites are dated about 2800–2600 BCE; the latest ones continued to about 1900–1800 BCE.[5] Early Middle Dnieper pottery showed clear similarities with Carpathian and eastern Polish Corded

Arkaim settlement and finds

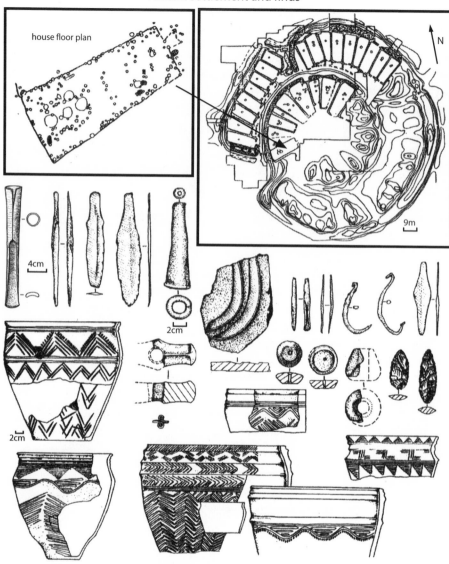

Figure 15.4 Arkaim settlement, house plan, and artifacts, including a mold for casting curved sickle or knife blades. After Zdanovich 1995, figure 6.

Figure 15.5 Culture groups of the Middle Bronze Age, 2800–2200 BCE.

Ware pottery, and Middle Dnieper pots have been found in Corded Ware graves near Grzeda Sokalska between the upper Dniester and the upper Vistula.[6] Some late Sredni Stog or Yamnaya elements also appeared in Middle Dnieper ceramics (figure 15.6). Middle Dnieper cemeteries contained both kurgans and flat-graves, both inhumation burials and cremations, with hollow-based flint arrowheads like those of the Yamnaya and Catacomb cultures, large trapezoidal flint axes like Globular Amphorae, and drilled stone "battle-axes" like those of the Corded Ware cultures. The Middle Dnieper culture clearly emerged from a series of encounters and exchanges between steppe and forest-steppe groups around Kiev, near the strategic fords over the Dnieper.[7]

A second culture, Fatyanovo, emerged at the northeastern edge of the Middle Dnieper culture. After the cattle herders moved out of the south-flowing Dnieper drainage and into the north-flowing rivers such as the Oka that coursed through the pine-oak-birch forests to the Upper Volga, they began to make pottery in distinctive Fatyanovo forms. But Fatyanovo pottery still showed mixed Corded Ware/Globular Amphorae traits, and the Fatyanovo culture probably was derived from an early variant of the Middle Dnieper culture. Ultimately Fatyanovo-type pottery, graves, and the cattle-raising economy spread over almost the entire Upper Volga basin. In the enormous western part of the Fatyanovo territory, from the Dvina to the Oka, very few Fatyanovo settlements are known, but more than three hundred large Fatyanovo flat-grave cemeteries, without kurgans, have been found on hills overlooking rivers or marshes. The Late Eneolithic Volosovo culture of the indigenous forest foragers was quite different in its pottery, economy, and mortuary customs. It disappeared when the Fatyanovo pioneers pushed into the Upper and Middle Volga basin.

The Middle Dnieper and Fatyanovo migrations overlapped the region where river and lake names in Baltic dialects, related to Latvian and Lithuanian, have been mapped by linguists: through the upper and middle Dnieper basin and the upper Volga as far east as the Oka. These names indicate the former extent of Baltic-speaking populations, which once occupied an area much larger than the area they occupy today. The Middle Dnieper and Fatyanovo migrations probably established the populations that spoke pre-Baltic dialects in the Upper Volga basin. Pre-Slavic probably developed between the middle Dnieper and upper Dniester among the populations that stayed behind.[8]

As Fatyanovo groups spread eastward down the Volga they discovered the copper ores of the western Ural foothills, and in this region, around the lower Kama River, they created long-term settlements. The Volga-Kama region,

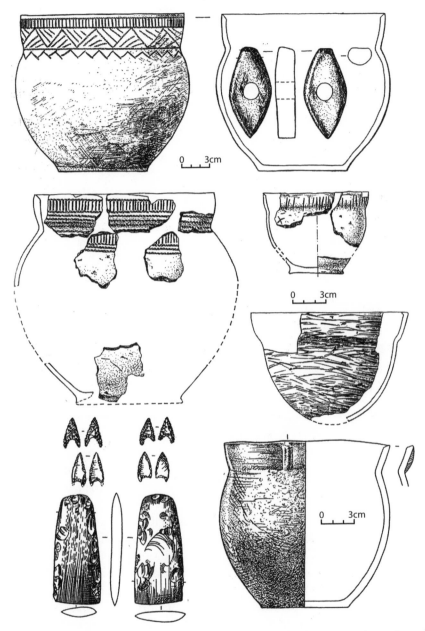

Figure 15.6 Ceramics and stone tools of the Middle Dnieper culture from sites in Belarus. After Kryvaltsevich and Kovalyukh 1999, figures 2 and 3.

which became the metallurgical heartland for almost all Fatyanovo metallurgy, has been separated from the rest of Fatyanovo and designated the Balanovo culture. Balanovo seems to be the settled, metal-working aspect of eastern Fatyanovo. At the southern fringe of Balanovo territory, in the forest-steppe zone of the middle Volga and upper Don where the rivers again flowed south, a fourth group emerged (after Middle Dnieper, Fatyanovo, and Balanovo). This was Abashevo, the easternmost of the Russian forest-zone cultures that were descended from Corded Ware ceramic traditions. The Abashevo culture played an important role in the origin of Sintashta.

The Abashevo Culture

Abashevo probably began about 2500 BCE or a little later. A late Abashevo kurgan at Pepkino on the middle Volga is dated 2400–2200 BCE (3850±95, Ki-7665); I would guess that the grave actually was created closer to 2200 BCE. Late Abashevo traditions persisted west of the Urals probably as late as 1900 BCE, definitely into the Sintashta period, since late Abashevo vessels are found in Sintashta and Potapovka graves. Early Abashevo ceramic styles strongly influenced Sintashta ceramics.

Abashevo sites are found predominantly in the forest-steppe zone, although a few extended into the northern steppes of the middle Volga. Within the forest-steppe, they are distributed between the upper Don on the west, a region with many Abashevo settlements (e.g., Kondrashovka); the middle Volga region in the center, represented largely by kurgan cemeteries (including the type-site, the Abashevo kurgan cemetery); and up the Belaya River into the copper-rich southwestern foothills of the Urals on the east, again with many settlements (like Balanbash, with plentiful evidence of copper smelting). More than two hundred Abashevo settlements are recorded; only two were clearly fortified, and many seem to have been occupied briefly. The easternmost Abashevo sites wrapped around the southern slopes of the Urals and extended into the Upper Ural basin, and it is these sites in particular that played a role in the origins of Sintashta.[9]

Some of the Volosovo foragers who had occupied these regions before 2500 BCE were absorbed into the Abashevo population, and others moved north. At the northern border of Abashevo territory, cord-impressed Abashevo and comb-stamped Volosovo ceramics are occasionally found inside the same structures at sites such as Bolshaya Gora.[10] Contact between late Volosovo and Abashevo populations west of the Urals probably helped to spread cattle-breeding economies and metallurgy into transitional northern forest cultures such as Chirkovska.

Whereas early Abashevo pottery looked somewhat like Fatyanovo/Balanovo Corded Ware, early Abashevo graves were covered by kurgans, unlike Fatyanovo flat cemeteries. Abashevo kurgans were surrounded by a circular ditch, the grave pit had ledges at the edges, and the body position was either contracted on the side or supine with raised knees—funeral customs derived from the Poltavka culture on the Volga. Abashevo ceramics also showed increasing decorative influences from steppe Catacomb-culture ceramic traditions, in both motifs (horizontal line-and-dot, horizontal fluting) and technology (shell tempering). Some Abashevo metal types such as waisted knives copied Catacomb and Poltavka types. A. D. Pryakhin, the preeminent expert on the Abashevo culture, concluded that it originated from contacts between Fatyanovo/Balanovo and Catacomb/Poltavka populations in the southern forest-steppe. In many ways, the Abashevo culture was a conduit through which steppe customs spread northward into the forest-steppe. Most Russian archaeologists interpret the Abashevo culture as a border culture associated with Indo-Iranian speakers, unlike Fatyanovo.[11]

Abashevo settlements in the Belaya River valley such as Balanbash contained crucibles, slag, and casting waste. Cast shaft-hole axes, knives, socketed spears, and socketed chisels were made by Abashevo metalsmiths. About half of all analyzed Abashevo metal objects were made of pure copper from southwestern Ural sandstone ores (particularly ornaments), and about half were arsenical bronze thought to have been made from southeastern Ural quartzitic ores (particularly tools and weapons), the same ores later exploited by Sintashta miners. High-status Abashevo graves contained copper and silver ornaments, semicircular solid copper and silver bracelets, cast shaft-hole axes, and waisted knives (figure 15.7). High-status Abashevo women wore distinctive headbands decorated with rows of flat and tubular beads interspersed with suspended double-spiral and cast rosette pendants, made of copper and silver. These headbands were unique to the Abashevo culture and probably were signals of ethnic as well as political status.[12]

The clear signaling of identity seen in Abashevo womens' headbands occurred in a context of intense warfare—not just raiding but actual warfare. At the cemetery of Pepkino, near the northern limit of Abashevo territory on the lower Sura River, a single grave pit 11 m long contained the bodies of twenty-eight young men, eighteen of them decapitated, others with axe wounds to the head, axe wounds on the arms, and dismembered extremities. This mass grave, probably dated about 2200 BCE, also contained Abashevo pottery, a two-part mold for making a shaft-hole axe

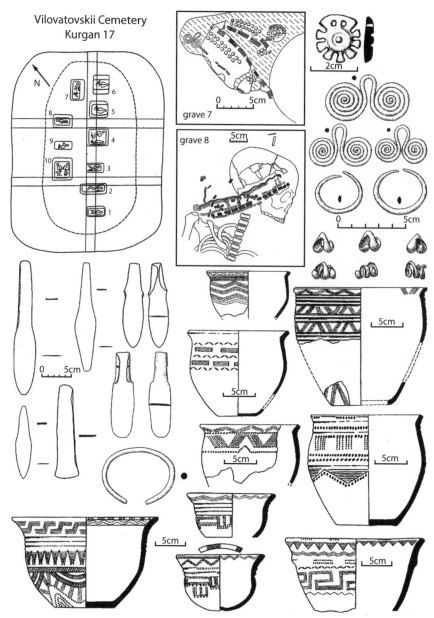

Figure 15.7 Abashevo culture graves and metal objects from the middle Volga forest-steppe (*upper left*), including distinctive cast copper rosettes; and ceramics from the south Ural region (*lower right*). After O. V. Kuzmina 1999, figures 23 and 24 (ceramics); and Bol'shov 1995, figure 13 (grave goods).

of Chernykh's Type V, and a crucible. It was covered by a single kurgan and so probably reflected a single event, clearly a serious battle or massacre. The absence of women or children in the grave indicates that it was not a settlement massacre. If it was the result of a battle, it implies a force of 280 to 560 on the Abashevo side alone, because deaths in tribal battles rarely reached 10% of the fighting force and usually were more like 5%.[13] Forces this size would require a considerable degree of inter-regional political integration. Intense warfare, perhaps on a surprising scale, was part of the political landscape during the late Abashevo era. In this context, the fortifications around Sintashta settlements and the invention of new fighting technologies—including the chariot—begin to make sense.

Linguists have identified loans that were adopted into the early Finno-Ugric (F-U) languages from Pre-Indo-Iranian and Proto-Indo-Iranian (Proto-I-I). Archaeological evidence for Volosovo-Abashevo contacts around the southern Urals probably were the medium through which these loans occurred. Early Proto-Indo-Iranian words that were borrowed into common Finno-Ugric included Proto-I-I *asura-* 'lord, god' > F-U *asera*; Proto-I-I *med^hu-* 'honey' > F-U *mete*; Proto-I-I *čekro-* 'wheel' > F-U *kekrä*; and Proto-I-I *arya-* 'Aryan' > F-U *orya*. Proto-Indo-Iranian *arya-*, the self designation "Aryan," was borrowed into Pre-Saami as *orja-*, the root of *oarji*, meaning "southwest," and of *ārjel*, meaning "southerner," confirming that the Proto-Aryan world lay south of the early Uralic region. The same borrowed *arya-* root developed into words with the meaning "slave" in the Finnish and Permic branches (Finnish, Komi, and Udmurt), a hint of ancient hostility between the speakers of Proto-Indo-Iranian and Finno-Ugric.[14]

PRE-SINTASHTA CULTURES OF THE EASTERN STEPPES

Who lived in the Ural-Tobol steppes during the late Abashevo era, before the Sintashta strongholds appeared there? There are two local antecedents and several unrelated neighbors.

Sintashta Antecedents

Just to the north of the steppe zone later occupied by Sintashta settlements, the southern forest-steppe zone contained scattered settlements of the late Abashevo culture. Abashevo miners regularly worked the quartzitic arsenic-rich copper ores of the Ural-Tobol region. Small settlements of the Ural variant of late Abashevo appeared in the upper Ural River valley and perhaps as far east as the upper Tobol. Geometric meanders first became a significant new decorative motif on Abashevo pottery made in the Ural region [see figure 15.7], and the geometric meander remained popular in Sintashta motifs. Some early Sintashta graves contained late Abashevo pots, and some late Abashevo sites west of the Urals contained Sintashta-type metal weapons and chariot gear such as disc-shaped cheekpieces that might have originated in the Sintashta culture. But Ural Abashevo people did not conduct mortuary animal sacrifices on a large scale, many of their metal types and ornaments were different, and, even though a few of their settlements were surrounded by small ditches, this was unusual. They were not fortified like the Sintashta settlements in the steppes.

Poltavka-culture herders had earlier occupied the northern steppe zone just where Sintashta appeared. The Poltavka culture was essentially a Volga-Ural continuation of the early Yamnaya horizon. Poltavka herding groups moved east into the Ural-Tobol steppes probably between 2800 and 2600 BCE. Poltavka decorative motifs on ceramics (vertical columns of chevrons) were very common on Sintashta pottery. A Poltavka kurgan cemetery (undated) stood on a low ridge 400 m south of the future site of Arkaim before that fortified settlement was built near the marshy bottom of the valley.[15] The cemetery, Aleksandrovska IV, contained twenty-one small (10–20 m in diameter) kurgans, a relatively large Poltavka cemetery (figure 15.8). Six were excavated. All conformed to the typical Poltavka rite: a kurgan surrounded by a circular ditch, with a single grave with ledges, the body tightly contracted on the left or right side, lying on an organic mat, red ochre or white chalk by the head and occasionally around the whole body, with a pot or a flint tool or nothing. A few animal bones occasionally were dropped in the perimeter ditch. A Poltavka settlement was stratified beneath the Sintashta settlement of Kuisak, which is intriguing because Poltavka settlements, like Yamnaya settlements, are generally unknown. Unfortunately this one was badly disturbed by the Sintashta settlement that was built on top of it.[16]

In the middle Volga region, the Potapovka culture was a contemporary sister of Sintashta, with similar graves, metal types, weapons, horse sacrifices, and chariot-driving gear (bone cheekpieces and whip handles), dated by radiocarbon to the same period, 2100–1800 BCE. Potapovka pottery, like Sintashta, retained many Poltavka decorative traits, and Potapovka graves were occasionally situated directly on top of older Poltavka monuments. Some Potapovka graves were dug right through preexisting Poltavka graves, destroying them, as some Sintashta strongholds were built on top of and incorporated older Poltavka settlements.[17] It is difficult to

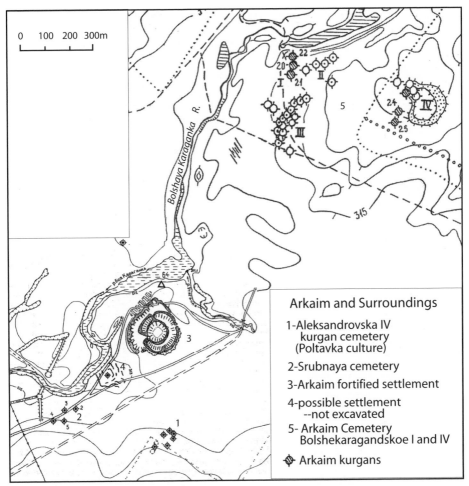

Figure 15.8 Arkaim settlement landscape with the kurgan cemeteries of Aleksandrovka IV (1), an older Poltavka cemetery of six kurgans; and Bolshekaragandskoe I and IV (5), with two excavated Sintashta-culture kurgans (24 and 25). Composite of Zdanovich 2002, Figure 3; and Batanina and Ivanova 1995, figure 2.

imagine that this was accidental. A symbolic connection with old Poltavka clans must have guided these choices.

Poltavka herders might have begun to explore across the vast Kazakh plains toward Sarazm, an outpost of Central Asian urban civilization established before 3000 BCE near modern Samarkand in the Zeravshan valley (see figure 16.1). Its northern location placed it just beyond the range of steppe herders who pushed east of the Urals around 2500 BCE.[18]

Hunters and Traders in Central Asia and the Forest Zone

Between the Poltavka territory in the upper Tobol steppes and Sarazm in the Zeravshan Valley lived at least two distinct groups of foragers. In the south, around the southern, western, and eastern margins of the Aral Sea, was the Kelteminar culture, a culture of relatively sedentary hunters and gatherers who built large reed-covered houses near the marshes and lakes in the steppes and in the riverbank thickets (called *tugai* forest) of the Amu Darya (Oxus) and lower Zeravshan rivers, where huge Siberian tigers still prowled. Kelteminar hunters pursued bison and wild pigs in the *tugai*, and gazelle, onagers, and Bactrian camels in the steppes and deserts. No wild horses ranged south of the Kyzl Kum desert, so Kelteminar hunters never saw horses, but they caught lots of fish, and collected wild pomegranates and apricots. They made a distinctive incised and stamped pottery. Early Kelteminar sites such as Dingil'dzhe 6 had microlithic flint industries much like those of Dzhebel Cave layer IV, dated about 5000 BCE. Kelteminar foragers probably began making pottery about this time, toward the end of the sixth millennium BCE. Late Kelteminar lasted until around 2000 BCE. Kelteminar pottery was found at Sarazm (level II), but the Kyzl Kum desert, north of the Amu Darya River, seems to have been an effective barrier to north-south communication with the northern steppes. Turquoise, which outcropped on the lower Zeravshan and in the desert southeast of the Aral Sea, was traded southward across Iran but not into the northern steppes. Turquoise ornaments appeared at Sarazm, at many early cities on the Iranian plateau, and even in the Maikop chieftain's grave (chapter 12), but not among the residents of the northern steppes.[19]

A second and quite different network of foragers lived in the northern steppes, north of the Aral Sea and the Syr Darya river (the ancient Jaxartes). Here the desert faded into the steppes of central and northern Kazakhstan, where the biggest predators were wolves and the largest grazing mammals were wild horses and saiga antelope (both absent in the Kelteminar region). In the lusher northern steppes, the descendants of the late Botai-Tersek culture still rode horses, hunted, and fished, but some of them now kept a few domesticated cattle and sheep and also worked metal. The post-Botai settlement of Sergeivka on the middle Ishim River is dated by radiocarbon about 2800–2600 BCE (4160±80

BP, OxA-4439). It contained pottery similar to late Botai-Tersek pottery, stone tools typical for late Botai-Tersek, and about 390 bones of horses (87%) but also 60 bones of cattle and sheep (13%), a new element in the economy of this region. Fireplaces, slag, and copper ore also were found. Very few sites like Sergeivka have been recognized in northern Kazakhstan. But Sergeivka shows that by 2800–2600 BCE an indigenous metallurgy and a little herding had begun in northern Kazakhstan. The impetus for these innovations probably was the arrival of Poltavka herders in the Tobol steppes. Pottery similar to that at Sergeivka was found in the Poltavka graves at Aleksandrovska IV, confirming contact between the two.[20]

North of the Ural-Tobol steppes, the foragers who occupied the forested eastern slopes of the Ural Mountains had little effect on the early Sintashta culture. Their natural environment was rich enough to permit them to live in relatively long-term settlements on river banks while still depending just on hunting and fishing. They had no formal cemeteries. Their pottery had complex comb-stamped geometric motifs all over the exterior surface. Ceramic decorations and shapes were somewhat similar between the forest-zone Ayatskii and Lipchinskii cultures on one side and the steppe zone Botai-Tersek cultures on the other. But in most material ways the forest-zone cultures remained distinct from Poltavka and Abashevo, until the appearance of the Sintashta culture, when this relationship changed. Forest-zone cultures adopted many Sintashta customs after about 2200–2100 BCE. Crucibles, slag, and copper rods interpreted as ingots appeared at Tashkovo II and Iska III, forager settlements located on the Tobol River north of Sintashta. The animal bones from these settlements were still from wild game—elk, bear, and fish. Some Tashkovo II ceramics displayed geometric meander designs borrowed from late Abashevo or Sintashta. And the houses at Tashkovo II and Andreevskoe Ozero XIII were built in a circle around an open central plaza, as at Sintashta or Arkaim, a settlement plan atypical of the forest zone.

THE ORIGIN OF THE SINTASHTA CULTURE

A cooler, more arid climate affected the Eurasian steppes after about 2500 BCE, reaching a peak of aridity around 2000 BCE. Ancient pollen grains cored from bogs and lake floors across the Eurasian continent show the effects this event had on wetland plant communities.[21] Forests retreated, open grassland expanded, and marshes dwindled. The steppes southeast of the Ural Mountains, already drier and colder than the Middle Volga grasslands southwest of the Urals, became drier still. Around 2100 BCE a

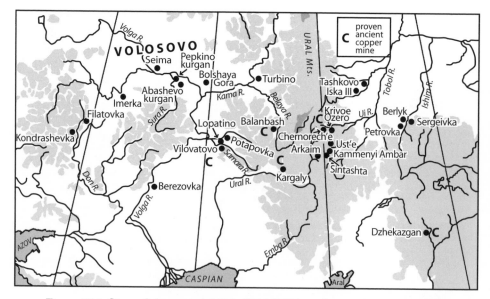

Figure 15.9 Sites of the period 2100–1800 BCE in the northern steppe and southern forest-steppe between the Don and the Ishim, with the locations of proven Bronze Age copper mines. The Sintashta-Potapovka-Filatovka complex probably is the archaeological manifestation of the Indo-Iranian language group.

mixed population of Poltavka and Abashevo herders began to settle in fortified strongholds between the upper Tobol and Ural River valleys, near the shrinking marshes that were vital for wintering their herds (see figure 15.9). Eurasian steppe pastoralists have generally favored marshy regions as winter refuges because of the winter forage and protection offered by stands of *Phragmites* reeds up to three meters tall. In a study of mobility among Late Mesolithic foragers in the Near East, Michael Rosenberg found that mobile populations tended to settle near critical resources when threatened with increased competition and declining productivity. He compared the process to a game of musical chairs,[22] in which the risk of losing a critical resource, in this case, winter marshlands for the cattle, was the impetus for settling down. Most Sintashta settlements were built on the first terrace overlooking the floodplain of a marshy, meandering stream. Although heavily fortified, these settlements were put in marshy, low places rather than on more easily defended hills nearby (see figures 15.2 and 15.8).

More than twenty Sintashta-type walled settlements were erected in the Ural-Tobol steppes between about 2100 and 1800 BCE. Their impressive fortifications indicate that concentrating people and herds near a critical wintering place was not sufficient in itself to protect it. Walls and towers also were required. Raiding must have been endemic. Intensified fighting encouraged tactical innovations, most important the invention of the light war chariot. This escalation of conflict and competition between rival tribal groups in the northern steppes was accompanied by elaborate ceremonies and feasts at funerals conducted within sight of the walls. Competition between rival hosts led to potlatch-type excesses such as the sacrifice of chariots and whole horses.

The geographic position of Sintashta societies at the eastern border of the Pontic-Caspian steppe world exposed them to many new cultures, from foragers to urban civilizations. Contact with the latter probably was most responsible for the escalation in metal production, funeral sacrifices, and warfare that characterized the Sintashta culture. The brick-walled towns of the Bactria-Margiana Archaeological Complex (BMAC) in Central Asia connected the metal miners of the northern steppes with an almost bottomless market for copper. One text from the city of Ur in present-day Iraq, dated to the reign of Rim-Sin of Larsa (1822–1763 BCE), recorded the receipt of 18,333 kg (40,417 lb, or 20 tons) of copper in a single shipment, most of it earmarked for only one merchant.[23] This old and well-oiled Asian trade network was connected to the northern Eurasian steppes for the first time around 2100–2000 BCE (see chapter 16 for the contact between Sintashta and BMAC sites).

The unprecedented increase in demand for metal is documented most clearly on the floors of Sintashta houses. Sintashta settlements were industrial centers that specialized in metal production. Every excavated structure at Sintashta, Arkaim, and Ust'e contained the remains of smelting ovens and slag from processing copper ore. The metal in the majority of finished objects was arsenical bronze, usually in alloys of 1–2.5% arsenic; tin-bronzes comprised only 2% or less of metal objects. At Sintashta, 36% of tested objects were made of copper with elevated arsenic (from 0.1–1% arsenic), and 48% were classified as arsenical bronze (over 1% arsenic). Unalloyed copper objects were more frequent at Arkaim, where they constituted almost half the tested objects, than at Sintashta, where they made up only 10% of tested objects. Clay tubular pipes probably for the mouths of the bellows, or *tulieres*, occurred in graves and settlements (see figure 15.4). Pieces of crucibles were found in graves at Krivoe Ozero. Closed two-piece molds were required to cast bronze shaft-hole axes and spear blades (see figure 15.10). Open single-piece molds for casting curved sickles and rod-like copper ingots were found in the Arkaim settlement.

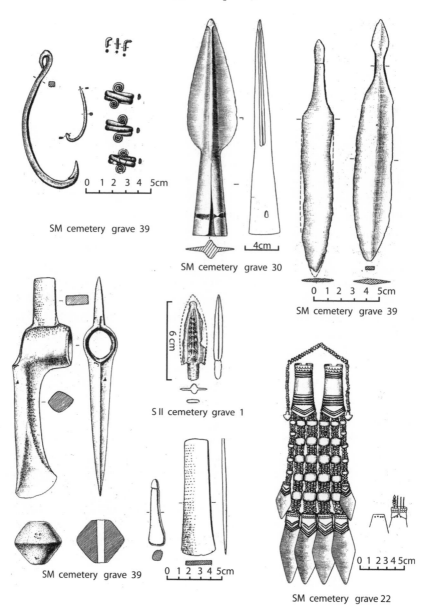

Figure 15.10 Weapons, tools, and ornaments from graves at Sintashta. After Gening, Zdanovich, and Gening 1992, figures 99, 113, 126, and 127.

Ingots or rods of metal weighing 50–130 g might have been produced for export. An estimated six thousand tons of quartzitic rock bearing 2–3% copper was mined from the single excavated mining site of Vorovskaya Yama east of the upper Ural River.[24]

Warfare, a powerful stimulus to social and political change, also shaped the Sintashta culture, for a heightened threat of conflict dissolves the old social order and creates new opportunities for the acquisition of power. Nicola DiCosmo has recently argued that complex political structures arose among steppe nomads in the Iron Age largely because intensified warfare led to the establishment of permanent bodyguards around rival chiefs, and these grew in size until they became armies, which engendered state-like institutions designed to organize, feed, reward, and control them. Susan Vehik studied political change in the deserts and grasslands of the North American Southwest after 1200 CE, during a period of increased aridity and climatic volatility comparable to the early Sintashta era in the steppes. Warfare increased sharply during this climatic downturn in the Southwest. Vehik found that long-distance trade increased greatly at the same time; trade after 1350 CE was more than forty times greater than it had been before then. To succeed in war, chiefs needed wealth to fund alliance-building ceremonies before the conflict and to reward allies afterward. Similarly, during the climatic crisis of the late MBA in the steppes, competing steppe chiefs searching for new sources of prestige valuables probably discovered the merchants of Sarazm in the Zeravshan valley, the northernmost outpost of Central Asian civilization. Although the connection with Central Asia began as an extension of old competitions between tribal chiefs, it created a relationship that fundamentally altered warfare, metal production, and ritual competition among the steppe cultures.[25]

Warfare in the Sintashta Culture: Fortifications and Weapons

A significant increase in the intensity of warfare in the southern Ural steppes is apparent from three factors: the regular appearance of large fortified towns; increased deposits of weapons in graves; and the development of new weapons and tactics. All the Sintashta settlements excavated to date, even relatively small ones like Chernorech'ye III, with perhaps six structures (see figure 15.11), and Ust'e, with fourteen to eighteen structures, were fortified with V-shaped ditches and timber-reinforced earthen walls.[26] Wooden palisade posts were preserved inside the earthen walls at

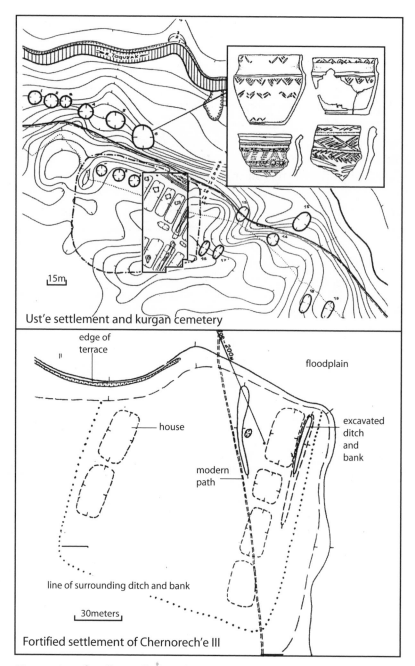

Ust'e settlement and kurgan cemetery

15m

edge of
terrace

floodplain

house

excavated
ditch
and
bank

modern
path

line of surrounding ditch and bank

30meters

Fortified settlement of Chernorech'e III

Figure 15.11 Smaller walled settlements of the Sintashta type at Ust'e and Chernorech'e III. After Vinogradov 2003, figure 3.

Ust'ye, Arkaim, and Sintashta. Communities build high walls and gates when they have reason to fear that their homes will come under attack.

The graves outside the walls now also contained many more weapons than in earlier times. The Russian archaeologist A. Epimakhov published a catalogue of excavated graves from five cemeteries of the Sintashta culture: Bol'shekaragandskoe (the cemetery for the Arkaim citadel), Kammeny Ambar 5, Krivoe Ozero, Sintashta, and Solntse II.[27] The catalogue listed 242 individuals in 181 graves. Of these, 65 graves contained weapons. Only 79 of the 242 individuals were adults, but 43 of these, or 54% of all adults, were buried with weapons. Most of the adults in the weapon graves were not assigned a gender, but of the 13 that were, 11 were males. Most adult males of the Sintashta culture probably were buried with weapons. In graves of the Poltavka, Catacomb, or Abashevo cultures, weapons had been unusual. They were more frequent in Abashevo than in the steppe graves, but the great majority of Abashevo graves did not contain weapons of any kind, and, when they did, usually it was a single axe or a projectile point. My reading of reports on kurgan graves of the earlier EBA and MBA suggests to me that less than 10% contained weapons. The frequency of weapons in adult graves of the Sintashta culture (54%) was much higher.

New types of weapons also appeared. Most of the weapon types in Sintashta graves had appeared earlier—bronze or copper daggers, flat axes, shaft-hole axes, socketed spears, polished stone mace heads, and flint or bone projectile points. In Sintashta-culture graves, however, longer, heavier projectile point types appeared, and they were deposited in greater numbers. One new projectile was a spearhead made of heavy bronze or copper with a socketed base for a thick wooden spear handle. Smaller, lighter-socketed spearheads had been used occasionally in the Fatyanovo culture, but the Sintashta spear was larger (see figure 15.3). Sintashta graves also contained two varieties of chipped flint projectile points: lanceolate and stemmed (see figure 15.12). Short lanceolate points with flat or slightly hollow bases became longer in the Sintashta period, and these were deposited in groups for the first time. They might have been for arrows, since prehistoric arrow points were light in weight and usually had flat or hollow bases. Lanceolate flint points with a hollow or flat base occurred in seven graves at Sintashta, with up to ten points in one grave (SM gr. 39). A set of five lanceolate points was deposited in the chariot grave of Berlyk II, kurgan 10.

More interesting were flint points of an entirely new type, with a contracting stem, defined shoulders, and a long, narrow blade with a thick medial ridge, 4–10 cm long. These new stemmed points might have been for javelins. Their narrow, thick blades were ideal for javelin points because the

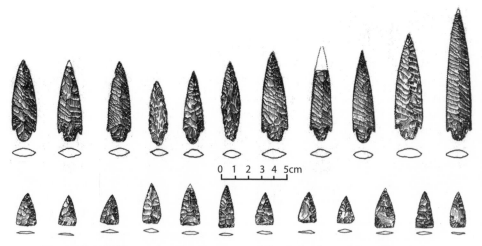

Figure 15.12 Flint projectile point types of the Sintashta culture. The top row was a new type for steppe cultures, possibly related to the introduction of the javelin. The bottom row was an old type in the steppes, possibly used for arrows, although in older EBA and MBA graves it was more triangular. After Gening, Zdanovich, and Gening 1992.

heavier shaft of a javelin (compared to an arrow) causes greater torque stress on the embedded point at the moment of impact; moreover, a narrow, thick point could penetrate deeper before breaking than a thin point could.[28] A stemmed point, by definition, is mounted in a socketed foreshaft, a complex type of attachment usually found on spears or javelins rather than arrows. Smaller stemmed points had existed earlier in Fatyanovo and Balanovo tool kits and were included in occasional graves, as at the Fatyanovo cemetery of Volosovo-Danilovskii, where 1 grave out of 107 contained a stemmed point, but it was shorter than the Sintashta type (only 3–4 cm long). Sintashta stemmed points appeared in sets of up to twenty in a single grave (chariot gr. 20 at the Sintashta SM cemetery), as well as in a few Potapovka graves on the middle Volga. Stemmed points made of cast bronze, perhaps imitations of the flint stemmed ones, occurred in one chariot grave (SM gr. 16) and in two other graves at Sintashta (see figure 15.10).

Weapons were deposited more frequently in Sintashta graves. New kinds of weapons appeared, among them long points probably intended for javelins, and they were deposited in sets that appear to represent warriors' equipment for battle. Another signal of increased conflict is the most hotly debated artifact of this period in the steppes—the light, horse-drawn chariot.

Sintashta Chariots: Engines of War

A chariot is a two-wheeled vehicle with spoked wheels and a standing driver, pulled by bitted horses, and usually driven at a gallop. A two-wheeler with solid wheels or a seated driver is a cart, not a chariot. Carts, like wagons, were work vehicles. Chariots were the first wheeled vehicles designed for speed, an innovation that changed land transport forever. The spoked wheel was the central element that made speed possible. The earliest spoked wheels were wonders of bent-wood joinery and fine carpentry. The rim had to be a perfect circle of joined wood, firmly attached to individually carved spokes inserted into mortices in the outer wheel and a multi-socketed central nave, all carved and planed out of wood with hand tools. The cars also were stripped down to just a few wooden struts. Later Egyptian chariots had wicker walls and a floor of leather straps for shock absorption, with only the frame made of wood. Perhaps originally designed for racing at funerals, the chariot quickly became a weapon and, in that capacity, changed history.

Today most authorities credit the invention of the chariot to Near Eastern societies around 1900–1800 BCE. Until recently, scholars believed that the chariots of the steppes post-dated those of the Near East. Carvings or petroglyphs showing chariots on rock outcrops in the mountains of eastern Kazakhstan and the Russian Altai were ascribed to the Late Bronze Age Andronovo horizon, thought to date after 1650 BCE. Disk-shaped cheekpieces made of antler or bone found in steppe graves were considered copies of older Mycenaean Greek cheekpieces designed for the bridles of chariot teams. Because the Mycenaean civilization began about 1650 BCE, the steppe cheekpieces also were assumed to date after 1650 BCE.[29]

The increasing amount of information about chariot graves in the steppes since about 1992 has challenged this orthodox view. The archaeological evidence of steppe chariots survives only in graves where the wheels were placed in slots that had been dug into the grave floors. The lower parts of the wheels left stains in the earth as they rotted (see figure 15.13). These stains show an outer circle of bent wood 1–1.2 m in diameter with ten to twelve square-sectioned spokes. There is disagreement as to the number of clearly identified chariot graves because the spoke imprints are faint, but even the conservative estimate yields sixteen chariot graves in nine cemeteries. All belonged to either the Sintashta culture in the Ural-Tobol steppes or the Petrovka culture east of Sintashta in northern Kazakhstan. Petrovka was contemporary with late Sintashta, perhaps 1900–1750 BC, and developed directly from it.[30]

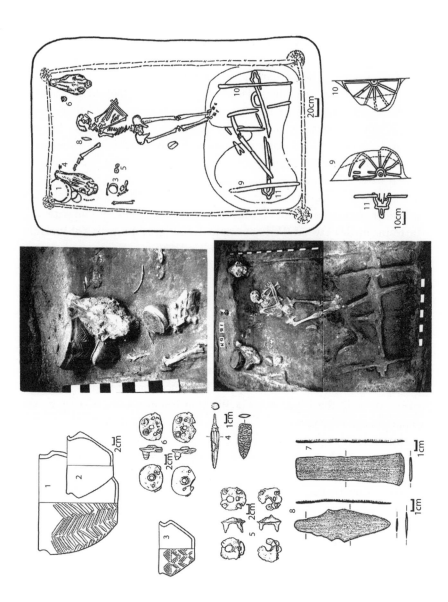

Scholars disagree as to whether steppe chariots were effective instruments of war or merely symbolic vehicles designed only for parade or ritual use, made in barbaric imitation of superior Near Eastern originals.[31] This debate has focused, surprisingly, on the distance between the chariots' wheels. Near Eastern war chariots had crews of two or even three—a driver and an archer, and occasionally a shield-bearer to protect the other two from incoming missiles. The gauge or track width of Egyptian chariots of ca. 1400–1300 BCE, the oldest Near Eastern chariots preserved well enough to measure, was 1.54–1.80 m. The hub or nave of the wheel, a necessary part that stabilized the chariot, projected at least 20 cm along the axle on each side. A gauge around 1.4–1.5 m would seem the minimum to provide enough room between the wheels for the two inner hubs or naves (20 + 20 cm) and a car at least 1 m wide to carry two men. Sintashta and Petrovka-culture chariots with less than 1.4–1.5 m between their wheels were interpreted as parade or ritual vehicles unfit for war.

This dismissal of the functional utility of steppe chariots is unconvincing for six reasons. First, steppe chariots were made in many sizes, including two at Kammeny Ambar 5, two at Sintashta (SM gr. 4, 28) and two at Berlyk (Petrovka culture) with a gauge between 1.4 and 1.6 m, big enough for a crew of two. The first examples published in English, which were from Sintashta (SM gr. 19) and Krivoe Ozero (k. 9, gr. 1), had gauges of only about 1.2–1.3 m, as did three other Sintashta chariots (SM gr. 5, 12, 30) and one other Krivoe Ozero chariot. The argument against the utility of steppe chariots focused on these six vehicles, most of which, in spite of their narrow gauges, were buried with weapons. However, six other steppe vehicles were as wide as some Egyptian war chariots. One (Sintashta SM gr. 28) with a gauge of about 1.5 m was placed in a grave that also contained the partial remains of two adults, possibly its crew. Even if we accept the doubtful assumption that war chariots needed a crew of two, many steppe chariots were big enough.[32]

Second, steppe chariots were not necessarily used as platforms for archers. The preferred weapon in the steppes might have been the javelin. A single

Figure 15.13 Chariot grave at Krivoe Ozero, kurgan 9, grave 1, dated about 2000 BCE: (1–3) three typical Sintashta pots; (5–6) two pairs of studded disk cheekpieces made of antler; (4) a bone and a flint projectile point; (7–8) a waisted bronze dagger and a flat bronze axe; (9–10) spoked wheel impressions from wheels set into slots in the floor of the grave; (11) detail of artist's reconstruction of the remains of the nave or hub on the left wheel. After Anthony and Vinogradov 1995, photos by Vinogradov.

warrior-driver could hold the reins in one hand and hurl a javelin with the other. From a standing position in a chariot, a driver-warrior could use his entire body to throw, whereas a man on horseback without stirrups (invented after 300 CE) could use only his arm and shoulder. A javelin-hurling charioteer could strike a man on horseback before the rider could strike him. Unlike a charioteer, a man on horseback could not carry a large sheath full of javelins and so would be at a double disadvantage if his first cast missed. A rider armed with a bow would fare only slightly better. Archers of the steppe Bronze Age seem to have used bows 1.2–1.5 m long, judging by bow remains found at Berezovka (k. 3, gr. 2) and Svatove (k. 12, gr. 12).[33] Bows this long could be fired from horseback only to the side (the left side, for a right-handed archer), which made riders with long bows vulnerable. A charioteer armed with javelins could therefore intimidate a Bronze Age rider on horseback. Many long-stemmed points, suitable for javelins, were found in some chariot graves (Sintashta SM gr. 4, 5, 30). If steppe charioteers used javelins, a single man could use narrower cars in warfare.

Third, if a single driver-warrior needed to switch to a bow in battle, he could fire arrows while guiding the horses with the reins around his hips. Tomb paintings depicted the Egyptian pharaoh driving and shooting a bow in this way. Although it may have been a convention to include only the pharaoh in these illustrations, Littauer noted that a royal Egyptian scribe was also shown driving and shooting in this way, and in paintings of Ramses III fighting the Libyans the archers in the Egyptian two-man chariots had the reins around their hips. Their car-mates helped to drive with one hand and used a shield with the other. Etruscan and Roman charioteers also frequently drove with the reins wrapped around their hips.[34] A single driver-warrior might have used a bow in this manner, although it would have been safer to shift the reins to one hand and cast a javelin.

The fourth reason not to dismiss the functionality of steppe chariots is that most of these chariots, including the narrow-gauge ones, were buried with weapons. I have seen complete inventories for twelve Sintashta and Petrovka chariot graves, and ten contained weapons. The most frequent weapons were projectile points, but chariot graves also contained metal-waisted daggers, flat metal axes, metal shaft-hole axes, polished stone mace heads, and one metal-socketed spearhead 20 cm long (from Sintashta SM gr. 30; see figure 15.3). According to Epimakhov's catalogue of Sintashta graves, cited earlier, all chariot graves where the skeleton could be assigned a gender contained an adult male. If steppe chariots were not designed for war, why were most of them buried with a male driver and weapons?

Fifth, a new kind of bridle cheekpiece appeared in the steppes at the very time that chariots did (see figure 15.14). It was made of antler or bone

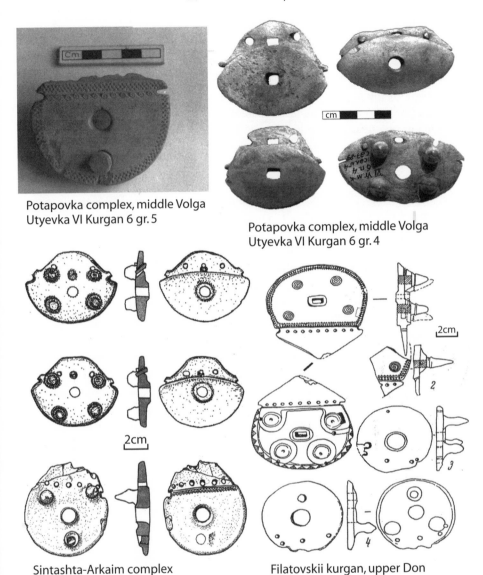

Potapovka complex, middle Volga
Utyevka VI Kurgan 6 gr. 5

Potapovka complex, middle Volga
Utyevka VI Kurgan 6 gr. 4

Sintashta-Arkaim complex
Kamennyi Ambar 5 Kurgan 2 Grave 8

Filatovskii kurgan, upper Don
Grave 1, 2 pairs of cheekpieces

Figure 15.14 Studded disk cheekpieces from graves of the Sintashta, Potapovka, and Filatovka types. The band of running spirals beneath the checkerboard panel on the upper left specimen from Utyevka VI was once thought to be derived from Mycenae. But the steppe examples like this one were older than Mycenae. Photos by the author; drawings after Epimakhov 2002; and Siniuk and Kosmirchuk 1995.

and shaped like an oblong disk or a shield, perforated in the center so that cords could pass through to connect the bit to the bridle and in various other places to allow for attachments to the noseband and cheek-strap. Pointed studs or prongs on its inner face pressed into the soft flesh at the corners of the horse's mouth when the driver pulled the reins on the opposite side, prompting an immediate response from the horse. The development of a new, more severe form of driving control suggests that rapid, precise maneuvers by the driving team were necessary. When disk cheekpieces are found in pairs, different shapes with different kinds of wear are often found together, as if the right and left sides of the horse, or the right and left horses, needed slightly different kinds of control. For example, at Krivoe Ozero (k. 9, gr. 1), the cheekpieces with the left horse had a slot located above the central hole, angled upward, toward the noseband (see figure 15.13). The cheekpieces with the right horse had no such upward-angled slot. A similar unmatched pair, with and without an upward-angled slot, were buried with a chariot team at Kamennyi Ambar 5 (see figure 15.14). The angled slot may have been for a noseband attached to the reins that would pull down on the inside (left) horse's nose, acting as a brake, when the reins were pulled, while the outside (right) horse was allowed to run free—just what a left-turning racing team would need. The chariot race, as described in the *Rig Veda*, was a frequent metaphor for life's challenges, and Vedic races turned to the left. Chariot cheekpieces of the same general design, a bone disk with sharp prongs on its inner face, appeared later in Shaft Grave IV at Mycenae and in the Levant at Tel Haror, made of metal. The oldest examples appeared in the steppes.[35]

Finally, the sixth flaw in the argument that steppe chariots were poorly designed imitations of superior Near Eastern originals is that the oldest examples of the former predate any of the dated chariot images in the Near East. Eight radiocarbon dates have been obtained from five Sintashta-culture graves containing the impressions of spoked wheels, including three at Sintashta (SM cemetery, gr. 5, 19, 28), one at Krivoe Ozero (k. 9, gr. 1), and one at Kammeny Ambar 5 (k. 2, gr. 8). Three of these (3760±120 BP, 3740±50 BP, and 3700±60 BP), with probability distributions that fall predominantly before 2000 BCE, suggest that the earliest chariots *probably* appeared in the steppes before 2000 BCE (table 15.1). Disk-shaped cheekpieces, usually interpreted as specialized chariot gear, also occur in steppe graves of the Sintashta and Potapovka types dated by radiocarbon before 2000 BCE. In contrast, in the Near East the oldest images of true chariots—vehicles with *two spoked* wheels, pulled by *horses* rather than asses or onagers, controlled with *bits* rather than lip- or nose-rings, and guided by a *standing warrior*, not a seated driver—first appeared about 1800 BCE, on Old Syrian seals. The oldest images in Near Eastern art of vehicles with two spoked wheels appeared on seals from Karum Kanesh II, dated about 1900 BCE, but the equids were of an uncertain type (possibly native asses or onagers) and they were controlled by nose-rings (see figure 15.15). Excavations at Tell Brak in northern Syria recovered 102 cart models and 191 equid figurines from the parts of this ancient walled caravan city dated to the late Akkadian and Ur III periods, 2350–2000 BCE by the standard or "middle" chronology. None of the equid figurines was clearly a horse. Two-wheeled carts were common among the vehicle models, but they had built-in seats and solid wheels. No chariot models were found. Chariots were unknown here as they were elsewhere in the Near East before about 1800 BCE.[36]

Chariots were invented earliest in the steppes, where they were used in warfare. They were introduced to the Near East through Central Asia, with steppe horses and studded disk cheekpieces (see chapter 16). The horse-drawn chariot was faster and more maneuverable than the old solid-wheeled battle-cart or battle-wagon that had been pulled into inter-urban battles by ass-onager hybrids in the armies of Early Dynastic, Akkadian, and Ur III kings between 2900 and 2000 BCE. These heavy, clumsy vehicles, mistakenly described as chariots in many books and catalogues, were similar to steppe chariots in one way: they were consistently depicted carrying javelin-hurling warriors, not archers. When horse-drawn chariots appeared in the Near East they quickly came to dominate inter-urban battles as swift platforms for archers, perhaps a Near Eastern innovation. Their wheels also were made differently, with just four or six spokes, apparently another improvement on the steppe design.

Among the Mitanni of northern Syria, in 1500–1350 BC, whose chariot tactics might have been imported with their Old Indic chariot terminology from a source somewhere in the steppes, chariots were organized into squadrons of five or six; six such units (thirty to thirty-six chariots) were combined with infantry under a brigade commander. A similar organization appeared in Chou China a millennium later: five chariots in a squadron, five squadrons in a brigade (twenty-five), with ten to twenty-five support infantry for each chariot.[37] Steppe chariots might also have operated in squadrons supported by individuals on foot or even on horseback, who could have run forward to pursue the enemy with hand weapons or to rescue the charioteer if he were thrown.

Chariots were effective in tribal wars in the steppes: they were noisy, fast, and intimidating, and provided an elevated platform from which a skilled

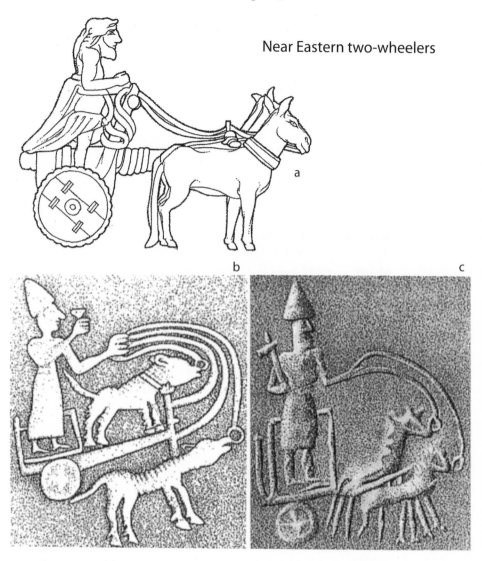

Near Eastern two-wheelers

Figure 15.15 Two-wheeled, high-speed vehicles of the ancient Near East prior to the appearance of the chariot: (a) cast copper model of a straddle-car with solid wheels pulled by a team of ass-onager–type equids from Tell Agrab, 2700–2500 BCE; (b and c) engraved seal images of vehicles with four-spoked wheels, pulled by equids (?) controlled with lip- or nose-rings from *karum* Kanesh II, 1900 BCE. After Raulwing 2000, figures 7.2 and 10.1.

driver could hurl a sheath full of javelins. As the car hit uneven ground at high speed, the driver's legs had to absorb each bounce, and the driver's weight had to shift to the bouncing side. To drive through a turn, the inside horse had to be pulled in while the outside horse was given rein. Doing this well and hurling a javelin at the same time required a lot of practice. Chariots were supreme advertisements of wealth; difficult to make and requiring great athletic skill *and* a team of specially trained horses to drive, they were available only to those who could delegate much of their daily labor to hired herders. A chariot was material proof that the driver was able to fund a substantial alliance or was supported by someone who had the means. Taken together, the evidence from fortifications, weapon types, and numbers, and the tactical innovation of chariot warfare, all indicate that conflict increased in both scale and intensity in the northern steppes during the early Sintashta period, after about 2100 BCE. It is also apparent that chariots played an important role in this new kind of conflict.

Tournaments of Value

Parallels between the funerals of the Sintashta chiefs and the funeral hymns of the *Rig Veda* (see below) suggest that poetry surrounded chariot burials. Archaeology reveals that feasts on a surprising scale also accompanied chiefly funerals. Poetry and feasting were central to a mortuary performance that emphasized exclusivity, hierarchy, and power—what the anthropologist A. Appadurai called "tournaments of value," ceremonies meant to define membership in the elite and to channel political competition within clear boundaries that excluded most people. In order to understand the nature of these sacrificial dramas, we first have to understand the everyday secular diet.[38]

Flotation of seeds and charcoal from the soils excavated at Arkaim recovered only a few charred grains of barley, too few, in fact, to be certain that they came from the Sintashta-culture site rather than a later occupation. The people buried at Arkaim had no dental caries, indicating that they ate a very low-starch diet, not starchy cereals.[39] Their teeth were like those of hunter-gatherers. Charred millet was found in test excavations at the walled Alands'koe stronghold, indicating that some millet cultivation probably occurred at some sites, and dental decay *was* found in the Krivoe Ozero cemetery population, so some communities might have consumed cultivated grain. Gathering wild seeds from *Chenopodium* and *Amaranthus*, plants that still played an important role in the LBA steppe diet centuries later (see chapter 16 for LBA wild plants), could have supplemented occasional cereal cultivation. Cultivated cereals seem to have played a minor role in the Sintashta diet.[40]

The scale of animal sacrifices in Sintashta cemeteries implies very large funerals. One example was Sacrificial Complex 1 at the northern edge of the Sintashta SM cemetery (see figure 15.16). In a pit 50 cm deep, the heads and hooves of six horses, four cattle, and two rams lay in two rows facing one another around an overturned pot. This single sacrifice provided about six thousand pounds (2,700 kg) of meat, enough to supply each of three thousand participants with two pounds (.9 kg). The Bolshoi Kurgan, built just a few meters to the north, required, by one estimate, three thousand man-days.[41] The workforce required to build the kurgan matched the amount of food provided by Sacrificial Complex 1. However, the Bolshoi Kurgan was unique; the other burial mounds at Sintashta were small and low. If the sacrifices that accompanied the other burials at Sintashta were meant to feed work parties, what they built is not obvious. It seems more likely that most sacrifices were intended to provide food for the funeral guests. With up to eight horses sacrificed for a single funeral, Sintashta feasts would have fed hundreds, even thousands of guests. Feast-hosting behavior is the most common and consistently used avenue to prestige and power in tribal societies.[42]

The central role of horses in Sintashta funeral sacrifices was unprecedented in the steppes. Horse bones had appeared in EBA and earlier MBA graves but not in great numbers, and not as frequently as those of sheep or cattle. The animal bones from the Sintashta and Arkaim settlement refuse middens were 60% cattle, 26% sheep-goat, and 13% horse. Although beef supplied the preponderance of the meat diet, the funeral sacrifices in the cemeteries contained just 23% cattle, 37% sheep-goat, and 39% horse. Horses were sacrificed more than any other animal, and horse bones were three times more frequent in funeral sacrifices than in settlement middens. The zoologist L. Gaiduchenko suggested that the Arkaim citadel specialized in horse breeding for export because the high level of ^{15}N isotopes in human bone suggested that horses, very low in ^{15}N, were not eaten frequently. Foods derived from cattle and sheep, significantly higher in ^{15}N than the horses from these sites, probably composed most of the diet.[43] According to Epimakhov's catalogue of five Sintashta cemeteries, the most frequent animal sacrifices were horses but they were sacrificed in no more than 48 of the 181 graves catalogued, or 27%; multiple horses were sacrificed in just 13% of graves. About one-third of the graves contained weapons, but, among these, two-thirds of graves with horse sacrifices contained weapons, and 83% of graves with multiple horse sacrifices contained weapons. Only a minority of Sintashta graves contained

Sintashta cemetery SM sacrificial complex 1

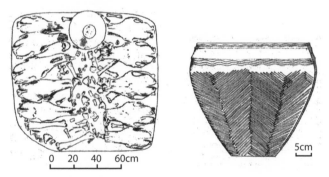

Figure 15.16 Sacrificial complex number 1 at the northern edge of the Sintashta SM cemetery. After Gening, Zdanovich, and Gening 1992, figure 130.

horse sacrifices, but those that did usually also contained weapons, a symbolic association between the ownership of large horse herds, the hosting of feasts, and the warrior's identity.

There is little jewelry or ornaments in Sintashta graves, and no large houses or storage facilities in the settlements. The signs of craft specialization, a signal of social hierarchy, are weak in all crafts except metallurgy, but even in that craft, every household in every settlement seems to have worked metal. The absence of large houses, storage facilities, or craft specialists has led some experts to doubt whether the Sintashta culture had a strong social hierarchy.[44] Sintashta cemeteries contained the graves of a cross-section of the entire age and sex spectrum, including many children, apparently a more inclusive funeral ritual than had been normal in EBA and earlier MBA mortuary ceremonies in the steppes. On the other hand, most Sintashta cemeteries did not contain enough graves to account for more than a small segment of the population of the associated walled settlements. The Sintashta citadel included about fifty to sixty structures, and its associated cemeteries had just sixty-six graves, most of them the graves of children. If the settlement contained 250 people for six generations (150 years), it should have generated more than fifteen hundred graves. Only a few exceptional families were given funerals in Sintashta cemeteries, but the entire family, including children, was honored in this way. This privilege, like the sacrifice of horses and chariots, was not one that everyone could claim. Horses, chariots, weapons, and multiple animal sacrifices identified the graves of the Sintashta chiefs.

The funeral sacrifices of the Simtashta culture are a critical link between archaeology and history. They closely resembled the rituals described in the *Rig Veda*, the oldest text preserved in an Indo-Iranian language.

Sintashta and the Origins of the Aryans

The oldest texts in Old Indic are the "family books," books 2 through 7, of the *Rig Veda* (RV). These hymns and prayers were compiled into "books" or mandalas about 1500–1300 BCE, but many had been composed earlier. The oldest parts of the *Avesta* (AV), the Gathas, the oldest texts in Iranian, were composed by Zarathustra probably about 1200–1000 BCE. The undocumented language that was the parent of both, common Indo-Iranian, must be dated well before 1500 BCE, because, by this date, Old Indic had already appeared in the documents of the Mitanni in North Syria (see chapter 3). Common Indo-Iranian probably was spoken during the Sintashta period, 2100–1800 BCE. Archaic Old Indic probably emerged as a separate tongue from archaic Iranian about 1800–1600 BCE (see chapter 16). The RV and AV agreed that the essence of their shared parental Indo-Iranian identity was linguistic and ritual, not racial. If a person sacrificed to the right gods in the right way using the correct forms of the traditional hymns and poems, that person was an Aryan.[45] Otherwise the individual was a *Dasyu*, again not a racial or ethnic label but a ritual and linguistic one—a person who interrupted the cycle of giving between gods and humans, and therefore a person who threatened cosmic order, *r'ta* (RV) or *aša* (AV). Rituals performed *in the right words* were the core of being an Aryan.

Similarities between the rituals excavated at Sintashta and Arkaim and those described later in the RV have solved, for many, the problem of Indo-Iranian origins.[46] The parallels include a reference in RV 10.18 to a kurgan ("let them . . . bury death in this hill"), a roofed burial chamber supported with posts ("let the fathers hold up this pillar for you"), and with shored walls ("I shore up the earth all around you; let me not injure you as I lay down this clod of earth"). This is a precise description of Sintashta and Potapovka-Filatovka grave pits, which had wooden plank roofs supported by timber posts and plank shoring walls. The horse sacrifice at a royal funeral is described in RV 1.162: "Keep the limbs undamaged and place them in the proper pattern. Cut them apart, calling out piece by piece." The horse sacrifices in Sintashta, Potapovka, and Filatovka graves match this description, with the lower legs of horses carefully cut apart at the joints and placed in and over the grave. The preference for horses as sacrificial animals in Sintashta funeral rituals, a species choice setting Sintashta apart from earlier steppe cultures, was again paralleled in the RV. Another verse in the same hymn read: "Those who see that the racehorse is cooked, who say, 'It smells good! Take it away!' and who wait for the doling out of the flesh of the charger—let their approval encourage us." These lines describe the public feasting that surrounded the funeral of an important person, exactly like the feasting implied by head-and-hoof deposits of horses, cattle, goats, and sheep in Sintashta graves that would have yielded hundreds or even thousands of kilos of meat. In RV 5.85, Varuna released the rain by overturning a pot: "Varuna has poured out the cask, turning its mouth downward. With it the king of the whole universe waters the soil." In Sacrificial Deposit 1 at Sintashta an overturned pot was placed between two rows of sacrificed animals—in a ritual possibly associated with the construction of the enormous Bolshoi Kurgan.[47] Finally, the RV eloquently documents the importance of the poetry and speech making that accompanied all these events. "Let us speak great words as men of power in the sacrificial gathering" was the standard closing attached repeatedly to several different hymns (RV 2.12, 2.23, 2.28) in one of the "family books." These public performances played an important role in attracting and converting celebrants to the Indo-Iranian ritual system and language.

The explosion of Sintashta innovations in rituals, politics, and warfare had a long-lasting impact on the later cultures of the Eurasian steppes. This is another reason why the Sintashta culture is the best and clearest candidate for the crucible of Indo-Iranian identity and language. Both the Srubnaya and the Andronovo horizons, the principal cultural groups of the Late Bronze Age in the Eurasian steppes (see chapter 16), grew from origins in the Potapovka-Sintashta complex.

A Srubnaya site excavated by this author contained surprising evidence for one more parallel between Indo-Iranian (and perhaps even Proto-Indo-European) ritual and archaeological evidence in the steppes: the midwinter New Year's sacrifice and initiation ceremony, held on the winter solstice. Many Indo-European myths and rituals contained references to this event. One of its functions was to initiate young men into the warrior category (*Männerbünde, korios*), and its principal symbol was the dog or wolf. Dogs represented death; multiple dogs or a multi-headed dog (*Cerberus, Saranyu*) guarded the entrance to the Afterworld. At initiation, death came to both the old year and boyhood identities, and as boys became warriors they would feed the dogs of death. In the RV the oath brotherhood of warriors that performed sacrifices at midwinter were called the Vrâtyas, who also were called dog-priests. The ceremonies associated with them featured many contests, including poetry recitation and chariot races.[48]

At the Srubnaya settlement of Krasnosamarskoe (Krasno-sa-MAR-sko-yeh) in the Samara River valley, we found the remains of an LBA midwinter dog sacrifice, a remarkable parallel to the reconstructed midwinter New Year ritual, dated about 1750 BCE. The dogs were butchered only at midwinter, many of them near the winter solstice, whereas the cattle and sheep at this site were butchered throughout the year. Dogs accounted for 40% of all the animal bones from the site. At least eighteen dogs were butchered, probably more. Nerissa Russell's studies showed that each dog head was burned and then carefully chopped into ten to twelve small, neat, almost identical segments with axe blows. The postcranial remains were not chopped into ritually standardized little pieces, and none of the cattle or sheep was butchered like this. The excavated structure at Krasnosamarskoe probably was the place where the dog remains from a midwinter sacrifice were discarded after the event. They were found in an archaeological context assigned to the early Srubnaya culture, but early Srubnaya was a direct outgrowth from Potapovka and Abashevo, the same circle as Sintashta, and nearly the same date. Krasnosamarskoe shows that midwinter dog sacrifices were practiced in the middle Volga steppes, as in the dog-priest initiation rituals described in the RV. Although such direct evidence for midwinter dog rituals has not yet been recognized in Sintashta settlements, many individuals buried in Sintashta graves wore necklaces of dog canine teeth. Nineteen dog canine pendants were found in a single collective grave with eight youths—probably of initiation age—under a Sintashta kurgan at Kammenyi Ambar 5, kurgan 4, grave 2.[49]

In many small ways the cultures between the upper Don and Tobol rivers in the northern steppes showed a common kinship with the Aryans of the *Rig Veda* and *Avesta*. Between 2100 and 1800 BCE they invented the chariot, organized themselves into stronghold-based chiefdoms, armed themselves with new kinds of weapons, created a new style of funeral rituals that involved spectacular public displays of wealth and generosity, and began to mine and produce metals on a scale previously unimagined in the steppes. Their actions reverberated across the Eurasian continent. The northern forest frontier began to dissolve east of the Urals as it had earlier west of the Urals; metallurgy and some aspects of Sintashta settlement designs spread north into the Siberian forests. Chariotry spread west through the Ukrainian steppe MVK culture into southeastern Europe's Monteoru (phase Ic1-Ib), Vatin, and Otomani cultures, perhaps with the *satəm* dialects that later popped up in Armenian, Albanian, and Phrygian, all of which are thought to have evolved in southeastern Europe. (Pre-Greek must have departed before this, as it did not share in the *satəm* innovations.) And the Ural frontier was finally broken—herding economies spread eastward across the steppes. With them went the eastern daughters of Sintashta, the offspring who would later emerge into history as the Iranian and Vedic Aryans. These eastern and southern connections finally brought northern steppe cultures into face-to-face contact with the old civilizations of Asia.

CHAPTER SIXTEEN

The Opening of the Eurasian Steppes

Between about 2300 and 2000 BCE the sinews of trade and conquest began to pull the far-flung pieces of the ancient world together into a single interacting system. The mainspring that drove inter-regional trade was the voracious demand of the Asiatic cities for metal, gems, ornamental stones, exotic woods, leather goods, animals, slaves, and power. Participants gained access to and control over knowledge of the urban centers and their power-attracting abilities—a source of social prestige in most societies.[1] Ultimately, whether through cultural means of emulation and resistance or political means of treaty and alliance, a variety of regional centers linked their fortunes to those of the paramount cities of the Near East, Iran, and South Asia. Regional centers in turn extended their influence outward, partly in a search for raw materials for trade, and partly to feed their own internal appetites for power. On the edges of this expanding, uncoordinated system of consumption and competition were tribal cultures that probably had little awareness of its urban core, at least initially (figures 16.1 and 16.2). But eventually they were drawn in. By 1500 BCE chariot-driving mercenaries not too far removed from the Eurasian steppes, speaking an Old Indic language, created the Mitanni dynasty in northern Syria in the heart of the urban Near East.[2]

How did tribal chiefs from the steppes intrude into the dynastic politics of the Near East? Where else did they go? To understand the crucial role that Eurasian steppe cultures played in the knitting together of the ancient world during the Bronze Age, we should begin in the heartland of cities, where the demand for raw materials was greatest.

BRONZE AGE EMPIRES AND THE HORSE TRADE

About 2350 BCE Sargon of Akkad conquered and united the feuding kingdoms of Mesopotamia and northern Syria into a single super-state—

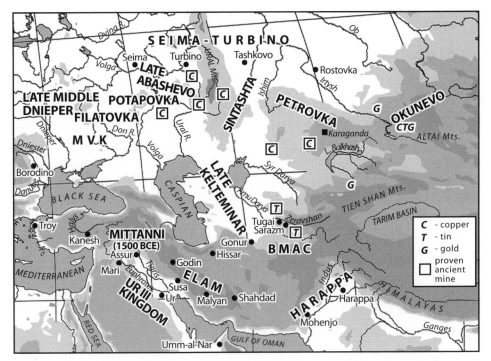

Figure 16.1 Cultures of the steppes and the Asian civilizations between about 2200 and 1800 BCE, with the locations of proven Bronze Age mines in the steppes and the Zeravshan valley.

the first time the world's oldest cities were ruled by one king. The Akkadian state lasted about 170 years. It had economic and political interests in western and central Iran, leading to increased trade, occasionally backed up by military expeditions. Images of horses, distinguished from asses and onagers by their hanging manes, short ears, and bushy tails, began to appear in Near Eastern art during the Akkadian period, although they still were rare and exotic animals. Some Akkadian seals had images of men riding equids in violent scenes of conflict (figure 16.3). Perhaps a few Akkadian horses were acquired from the chiefs and princes of western Iran known to the Akkadians as the Elamites.

Elamite was a non–Indo-European language, now extinct, then spoken across western Iran. A string of walled cities and trade centers stood on the Iranian plateau, revealed by excavations at Godin, Malyan, Konar Sandal, Hissar, Shar-i-Sokhta, Shahdad, and other places. Malyan, the ancient city of Anshan, the largest city on the plateau, certainly was an

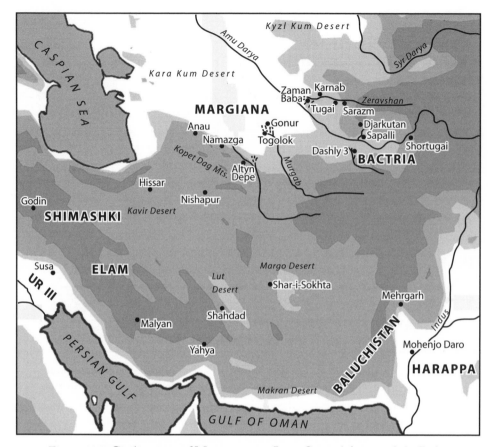

Figure 16.2 Civilizations of Mesopotamia, Iran, Central Asia, and the Indus valley about 2200–1800 BCE.

Elamite city allied to the Elamite king in Susa. Some of the other brick-built towns, almost all of them smaller than Malyan, were part of an alliance called Shimashki, located north of Malyan and south of the Caspian Sea. Among the fifty-nine personal names recorded in the Shimashki alliance, only twelve can be classified as Elamite; the others are from unknown non–Indo-European languages. East of the Iranian plateau, the Harappan civilization of Indo-Pakistan, centered in huge mudbrick cities on the Indus River, used its own script to record a language that has not been definitively deciphered but might have been related to modern Dravidian. The Harappan cities exported precious stones, tropical woods, and metals westward on ships that sailed up the Persian Gulf, through a chain

Figure 16.3 Early images of men riding equids in the Near East and Central Asia: (*top*) Akkadian seal impression from Kish, 2350–2200 BCE (after Buchanan 1966); (*middle*) seal impression of the BMAC from a looted grave in Afghanistan, 2100–1800 BCE (after Sarianidi 1986); (*bottom*) Ur III seal impression of Abbakalla, animal disburser for king Shu-Sin, 2050–2040 BCE (after Owen 1991).

of coastal kingdoms scattered from Oman to Kuwait. Harappa probably was the country referred to as "Melukkha" in the Mesopotamian cuneiform records.[3]

Akkadian armies and trade networks reached far and wide, but inside Akkad was an enemy it could not conquer with arms: crop failure. During the Akkadian era the climate became cooler and drier, and the agricultural economy of the empire suffered. Harvey Weiss of Yale has argued that some northern Akkadian cities were entirely abandoned, and their populations might have moved south into the irrigated floodplains of southern Mesopotamia.[4] The Gutians, a coalition of chiefs from the western Iranian uplands (perhaps Azerbaijan?) defeated the Akkadian army and overran the city of Akkad in 2170 BCE. Its ruins have never been found.

About 2100 BCE the first king of the Third Dynasty of Ur, even then an ancient Sumerian city in what is now southern Iraq, expelled the Gutians and reestablished the power of southern Mesopotamia. The brief Ur III period, 2100–2000 BCE, was the last time that Sumerian, the language of the first cities, was a language of royal administration. A century of bitter wars erupted between the Sumerian Ur III kings and the Elamite city-states of the Iranian plateau, occasionally interrupted by negotiations and marriage exchanges. King Shu-Sin of Ur bragged that he conquered a path across Elam and through Shimashki until his armies finally were stopped only by the Caspian Sea.

During this period of struggle and empire, 2100–2000 BCE, the bones of horses appeared for the first time at important sites on the Iranian plateau such as the large city of Malyan in Fars and the fortified administrative center at Godin Tepe in western Iran. Bit wear made with a hard bit, probably metal, appeared on the teeth of some of the equids (both mules and horses) from Malyan. Excavated by Bill Sumner and brought by Mindy Zeder to the collections of the Smithsonian Museum of Natural History in Washington, D.C. these teeth were the first archaeological specimens that we examined when we started our bit wear project in 1985. Now we know what then we only suspected: the horses and mules of the Kaftari phase at Malyan were bitted with hard bits. Bits were a new technology for controlling equids in Iran, different from the lip- and nose-rings that had appeared before this in Mesopotamian works of art. Of course bits and bit-wear were very old in the steppes by 2000 BCE.[5]

Horses also appeared in significant numbers in the cities of Mesopotamia for the first time during the Ur III period; this was when the word for *horse* first appeared in written records. It meant "ass of the mountains," showing that horses were flowing into Mesopotamia from western Iran and eastern Anatolia. The Ur III kings fed horses to lions for exotic entertainment. They did not use horse-drawn chariots, which had not yet appeared in Near Eastern warfare. But they did have solid-wheeled battle wagons and battle carts armed with javelins, pulled by teams of their smaller native equids—asses, which were manageable but small, and onagers or hemiones, which were almost untamable but larger. Ass-onager hybrids probably pulled Sumerian battle carts and battle wagons. Horses could have been used initially as breeding stock to make a larger, stronger ass-horse hybrid—a mule. Mules were bitted at Malyan.

The Sumerians recognized in horses an arched-neck pride that asses and onagers simply did not possess. King Shulgi compared himself in one inscription to "a horse of the highway that swishes his tail." We are not sure exactly what horses were doing on Ur III highways, but a seal impression of one Abbakalla, the royal animal disburser for king Shu-Sin, showed a man riding a galloping equid that looks like a horse (see figure 16.3).[6] Ceramic figurines of the same age showed humans astride schematic animals that have equine proportions; and ceramic plaques dated at the time of Ur III or just afterward showed men astride equids that probably were horses, some riding in awkward poses on the rump and others in more natural forward seats. No Ur III images showed a chariot, so the first clear images of horses in Mesopotamia show men riding them.[7]

About 2000 BCE an Elamite and Shimashki alliance defeated the last of the Ur III kings, Ibbi-Sin, and dragged him to Elam in chains. After this stunning event the kings of Elam and Shimashki played a controlling role in Mesopotamian politics for several centuries. Between 2000 and 1700 BCE the power, independence, and wealth of the Old Elamite (Malyan) and Shimashkian (Hissar? Godin?) overlords of the Iranian plateau was at its height. The treaties they negotiated for the Ur III wars were sealed by gifts and trade agreements that channeled lapis lazuli, carved steatite vessels, copper, tin, and horses from one prince to another. The Sintashta culture appeared at just the same time, but showed up 2000 km to the north in the remote grasslands of the Ural-Tobol steppes. The metal trade and the horse trade might have tied the two worlds together. Could the Elamite defeat of Ibbi-Sin have been aided by chariot-driving Sintashta mercenaries from the steppes? It is possible. Vehicles like chariots, with two spoked wheels and a standing driver, but guided by equids with lip- or nose-rings, began to appear on seal images in Anatolia just after the defeat of Ibbi-Sin. They were not yet common, but that was about to change.

The metal trade might have provided the initial incentive for prospectors to explore across the Central Asian deserts that had previously separated the northern Eurasian steppe cultures from those of Iran. Vast amounts of metal were demanded by Near Eastern merchants during the heyday of the Old Elamite kings. Zimri-Lim, king of the powerful city-state of Mari in northern Syria between 1776 and 1761 BCE, distributed gifts totaling more than 410 kg (905 lb) of tin—not bronze, but tin—to his allies during a single tour in his eighth year. Zimri-Lim also was chided by an adviser for riding a horse in public, an activity still considered insulting to the honor of an Assyrian king:[8]

> May my lord honor his kingship. You may be the king of the Haneans, but you are also the king of the Akkadians. May my Lord not ride horses; (instead) let him ride either a chariot or *kudanu*-mule so that he would honor his kingship.

Zimri-Lim's advisers accepted the fact that kings could ride in chariots—Near Eastern monarchs had by then ridden in wheeled vehicles of other kinds for more than a thousand years. But only rude barbarians actually rode on the backs of the large, sweaty, smelly animals that pulled them. Horses, in Zimri-Lim's day, were still exotic animals associated with crude foreigners. A steady supply of horses first began between 2100 and 2000 BCE. Chariots appeared across the Near East after 2000 BCE. How?

The Tin Trade and the Gateway to the North

Tin was the most important trade commodity in the Bronze Age Near East. In the palace records of Mari it was said to be worth ten times its weight in silver. A copper-tin alloy was easier for the metal smith to cast, and it made a harder, lighter-colored metal than either pure copper or arsenical bronze, the older alternatives. But the source of Near Eastern tin remains an enigma. Large tin deposits existed in England and Malaysia, but these places were far beyond the reach of Near Eastern traders in the Bronze Age. There were small tin deposits in western Serbia—and a scatter of Old European copper objects from the Danube valley contained elevated tin, perhaps derived from this source—but no ancient mines have been found there. Ancient mines in eastern Anatolia near Goltepe might have supplied a trickle of tin before 2000 BCE, but their proven tin content is very low, and tin was *imported* at great cost to Anatolia from northern Syria after 2000 BCE. It was imported into northern Syria from somewhere far to the east. The letters of king Zimri-Lim of Mari said flatly that he acquired his tin from Elam, through merchants at Malyan (Anshan) and Susa. An inscription on a statue of Gudea of Lagash, ca.

2100 BCE, was thought to refer to the "tin of Melukkha," implying that tin came up the Arabian Gulf in ships sent by Harappan merchants; but the passage might have been mistranslated. Intentional tin-bronze alloys occurred in about 30% of the objects tested from the Indus-valley cities of Mohenjo-Daro and Harappa, although most had such a low tin content (70% of them had only 1% tin, 99% copper) that it seems the best recipe for tin bronze (8–12% tin, 92–88% copper) was not yet known in Harappa. Still, "Melukkha" could have been one source of Mesopotamian tin. Tin-bronzes have been found in sites in Oman, at the entrance to the Arabian Gulf, in association with imported pottery and beads from Harappa and bone combs and seals made in Bactria. Oman had no tin of its own but could have been a coastal port and trans-shipment point for tin that came from the Indus valley.[9]

Where were the tin mines? Could the tin exported by the Elamite kings and by Harappan merchants have come from the same sources? Quite possibly. The most probable sources were in western and northern Afghanistan, where tin ore has been found by modern mineral surveyors, although no ancient mines have been found there, and also in the Zeravshan River valley, where the oldest tin mines in the ancient world have been found near the site of Sarazm. Sarazm also was the portal through which horses, chariots, and steppe cultures first arrived at the edges of Central Asia.

Sarazm was founded before 3500 BCE (4880±30 BP, 4940±30 BP for phase I) as a northern colony of the Namazga I–II culture. The Namazga home settlements (Namazga, Anau, Altyn-Depe, Geoksur) were farming towns situated on alluvial fans where the rivers that flowed off the Iranian plateau emerged into the Central Asian deserts. Perhaps the lure that enticed Namazga farmers to venture north across the Kara Kum desert to Sarazm was the turquoise that outcropped in the desert near the lower Zeravshan River, a source they could have learned about from Kelteminar foragers. Sarazm probably was founded as a collection point for turquoise. It was situated on the middle Zeravshan more than 100 km upstream from the turquoise deposits at an elevation where the valley was lush and green and crops could be grown. It grew to a large town, eventually covering more than 30 ha (74 acres). Its people were buried with ornaments of turquoise, carnelian, silver, copper, and lapis lazuli. Late Kelteminar pottery was found at Sarazm in its phase II, dated about 3000–2600 BCE (4230±40BP), and turquoise workshops have been found in the late Kelteminar camps of Kaptarnikum and Lyavlyakan in the desert near the lower Zeravshan. Turquoise from the Zeravshan and from a second source near Nishapur in northeastern Iran was traded into Mesopotamia, the Indus valley, and perhaps even to Maikop (the Maikop chieftain was buried with a necklace of turquoise beads). But the Zeravshan also contained polymetallic deposits of copper, lead, silver—and tin.

Oddly, no tin has been found at Sarazm itself. Crucibles, slag, and smelting furnaces appeared at Sarazm at least as early as the phase III settlement (radiocarbon dated 2400–2000 BCE), probably for processing the rich copper deposits in the Zeravshan valley. Sarazm III yielded a variety of copper knives, daggers, mirrors, fishhooks, awls, and broad-headed pins. Most were made of pure copper, but a few objects contained 1.8–2.7% arsenic, probably an intentional arsenical bronze. Tin-bronzes began to appear in small amounts in the Kopet Dag home region, in Altyn-Depe and Namazga, during the Namazga IV period, equivalent to late Sarazm II and III. A small amount of tin, perhaps just placer minerals retrieved from the river, probably came from the Zeravshan before 2000 BCE, even if we cannot see it at Sarazm.[10]

The tin mines of the Zeravshan River valley were found and investigated by N. Boroffka and H. Parzinger between 1997 and 1999.[11] Two tin mines with Bronze Age workings were excavated. The largest was in the desert on the lower Zeravshan at Karnab (Uzbekistan), about 170 km west of Sarazm, exploiting cassiterite ores with a moderate tin content—probably ordinarily about 3%, although some samples yielded as much as 22% tin. The pottery and radiocarbon dates show that the Karnab mine was worked by people from the northern steppes, connected with the Andronovo horizon (see below). Dates ranged from 1900 to 1300 BCE (the oldest was Bln 5127, 3476 ± 32 BP, or 1900–1750 BCE; see table 16.1). A few pieces of Namazga V/VI pottery were found in the Andronovo mining camp at Karnab. The other mining complex was at Mushiston in the upper Zeravshan (Tajikistan), just 40 km east of Sarazm, working stannite, cassiterite and copper ores with a very high tin content (maximum 34%). Andronovo miners also left their pottery at Mushiston, where wood beams produced radiocarbon dates as old as Karnab. Sarazm probably was abandoned when these Andronovo mining operations began. Whether the Zeravshan tin mines were worked before the steppe cultures arrived is unknown.

Sarazm probably was abandoned around 2000 BCE, just at the Namazga V/VI transition. On the lower Zeravshan, the smaller villages of the Zaman Baba culture probably were abandoned about the same time as Sarazm.[12] The Zaman Baba culture had established small villages of pithouses supported by irrigation agriculture in the large oasis in the lower

Zeravshan delta just a couple of centuries earlier. Zaman Baba and Sarazm were abandoned when people from the northern steppes arrived in the Zeravshan.[13]

Sarazm exported both copper and turquoise southward during the Akkadian and Ur III periods. Could it have pulled steppe copper miners and horse traders into the chain of supply for the urban trade? Could that explain the sudden intensification of copper production in Sintashta settlements and the simultaneous appearance of horses in Iran and Mesopotamia beginning about 2100 BCE? The answer lies among the ruins of walled cities in Central Asia south of Sarazm, cities that interacted with the cultures of the northern steppes before the Andronovo tin miners appeared on the Zeravshan frontier.

The Bactria-Margiana Archaeological Complex

Around 2100 BCE a substantial population colonized the Murgab River delta north of the Iranian plateau. The Murgab River flowed down from the mountains of western Afghanistan, snaked across 180 km of desert, then fanned out into the sands, dropping deep loads of silt and creating a fertile island of vegetation about 80 by 100 km in size. This was Margiana, a region that quickly became and remained one of the richest oases in Central Asia. The immigrants built new walled towns, temples, and palaces (Gonur, Togolok) on virgin soil during the late Namazga V period, at the end of the regional Middle Bronze Age (figure 16.4). They might have been escaping from the military conflicts that raged periodically across the Iranian plateau, or they might have relocated to a larger river system with more reliable flows in a period of intensifying drought. Anthropological studies of their skeletons show that they came from the Iranian plateau, and their pottery types seem to have been derived from the Namazga V-type towns of the Kopet Dag.[14]

The colonization phase in Margiana, 2100–2000 BCE, was followed by a much richer period, 2000–1800 BCE, during Namazga VI, the beginning of the regional Late Bronze Age. New walled towns now spread to the upper Amu Darya valley, ancient Bactria, where Sapalli-Tepe, Dashly-3, and Djarkutan were erected on virgin soil. The towns of Bactria and Margiana shared a distinctive set of seal types, architectural styles, brick-lined tomb types, and pottery. The LBA civilization of Bactria and Margiana is called the Bactria-Margiana Archaeological Complex (BMAC). The irrigated countryside was dominated by large towns surrounded by thick yellow-brick walls with narrow gates and high corner towers. At the

Table 16.1

Selected Radiocarbon Dates from Earlier Late Bronze Age Cultures in the Steppes

Lab Number	BP Date	Kurgan		Grave	Mean Intercept BCE	BCE

1. Krasnosamarskoe kurgan cemetery IV, Samara oblast, LBA Pokrovka and Srubnaya graves

Lab Number	BP Date	Kurgan		Grave	Mean Intercept BCE	BCE
AA37038	3490±57	kurgan 3		1	1859, 1847, 1772	1881–1740
AA37039	3411±46	kurgan 3		6	1731, 1727, 1686	1747–1631
AA37042	3594±45	kurgan 3		10	1931	1981–1880
AA37043	3416±57	kurgan 3		11	1733, 1724, 1688	1769–1623
AA37044	3407±46	kurgan 3		13	1670, 1668, 1632	1685–1529
AA37045	3407±46	kurgan 3		16	1730, 1685	1744–1631
AA37046	3545±65	kurgan 3		17	1883	1940–1766
AA37047	3425±52	kurgan 3		23	1735, 1718, 1693	1772–1671

2. Krasnosamarskoe settlement, Samara oblast

Structure floor and cultural level outside structure, Pokrovka and Srubnaya occupations

Lab Number	BP Date	Square/quad		level	Mean Intercept BCE	BCE
AA41022	3531±43	L5	2	3	1879, 1832, 1826, 1790	1899–1771
AA41023	3445±51	M5	1	7	1741	1871–1678
AA41024	3453±43	M6	3	7	1743	1867–1685
AA41025	3469±45	N3	3	7	1748	1874–1690
AA41026	3491±52	N4	2	6	1860, 1846, 1772	1879–1743
AA41027	3460±52	O4	1	7	1745	1873–1685
AA41028	3450±57	O4	2	5	1742	1874–1679
AA41029	3470±43	P1	4	6	1748	1783–1735
AA41030	3477±39	S2	3	4	1752	1785–1738
AA41031	3476±38	R1	2	5	1750	1875–1706
AA41032	3448±47	N2	2	4	1742	1858–1685
AA47790	3311±54	O5	3	3	1598, 1567, 1530	1636–1518
AA47796	3416±59	Y2	2	4	1736, 1713, 1692	1857–1637
AA47797	3450±50	Y1	3	5	1742	1779–1681

Waterlogged Pokrovka artifacts from deep pit interpreted as a well inside the structure

Lab Number	BP Date	Square/quad		level	Mean Intercept BCE	BCE
AA47793	3615±41	M2	4	−276	1948	1984–1899
AA47794	3492±55	M2	4	−280	1860, 1846, 1773	1829–1742
AA47795	3550±54	M2	4	−300	1884	1946–1776

TABLE 16.1 (*continued*)

Lab Number	BP Date	Kurgan	Grave	Mean Intercept BCE	BCE
Srubnaya and Pokrovka artifacts from eroded part of settlement on the lake bottom					
AA47791	3494±56	Lake find 1	0	1862, 1845, 1774	1881–1742
AA47792	3492±55	Lake find 2	0	1860, 1846, 1773	1829–1742
Srubnaya herding camp at PD1 in the Peschanyi Dol valley					
AA47798	3480±52	A 16 3	3	1758	1789–1737
AA47799	3565±55	I 18 2	2	1889	1964–1872
3. Karnab mining camp, Zeravshan valley, Uzbekistan, Andronovo–Alakul occupation					
Bln-5127	3476±32				1880–1740
Bln-141274	3280±40				1620–1510
Bln-141275	3170±50				1520–1400
Bln-5126	3130±44				1490–1310
4. Alakul–Andronovo settlements and kurgan graves					
Alakul kurgan 15, grave 1					
Le-924	3360±50	charcoal			1740–1530
Subbotino kurgan 17, grave 3					
Le-1126	3460±50	wood			1880–1690
Subbotino kurgan 18, central grave					
Le-1196	3000±50	wood			1680–1510
Tasty-Butak settlement					
Rul-614	3550±65	wood, pit 14			2010–1770
Le-213	3190±80	wood, pit 11			1600–1320

center of the larger towns were walled palaces or citadels that contained temples. The brick houses and streets of Djarkutan covered almost 100 ha, commanded by a high-walled citadel about 100 by 100 m. Local lords ruled from smaller strongholds such as Togolok 1, just .5 ha (1.2 acres) in size but heavily walled with large corner turrets. Trade and crafts flourished in the crowded houses and alleys of these Central Asian walled towns and fortresses. Their rulers had relations with the civilizations of Mesopotamia, Elam, Harappa, and the Arabian Gulf.

Between 2000 and 1800 BCE, BMAC styles and exported objects (notably small jars made of carved steatite) appeared in many sites and

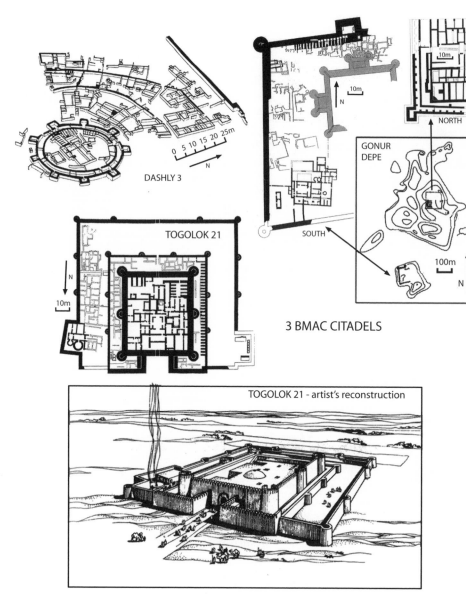

DASHLY 3

TOGOLOK 21

3 BMAC CITADELS

GONUR DEPE

NORTH

SOUTH

TOGOLOK 21 - artist's reconstruction

Figure 16.4 Three walled towns of the Bactria-Margiana Archaeological Complex (BMAC) in Central Asia, 2100–1800 BCE. Wall foundations of the central circular citadel/temple and town at Dashly 3, Bactria (after Sarianidi 1977, figure 13); wall foundations at Gonur Depe, Margiana (combined from Hiebert 1994; and Sarianidi 1995); wall foundations and artist's reconstruction of Togolok 21, Margiana (after Hiebert 1994; and Sarianidi 1987).

cemeteries across the Iranian plateau. Crested axes like those of the BMAC appeared at Shadad and other sites in eastern and central Iran. A cemetery at Mehrgarh VIII in Baluchistan, on the border between the Harappan and Elamite civilizations, contained so many BMAC artifacts that it suggests an actual movement of BMAC people into Baluchistan. BMAC-style sealings, ivory combs, steatite vessels, and pottery goblets appeared in the Arabian Gulf from Umm-al-Nar on the Oman peninsula up the Arabian coast to Falaika island in Kuwait. Beadmakers in BMAC towns used shells obtained from both the Indian Ocean (*Engina medicaria, Lambis truncate sebae*) and the Mediterranean Sea (*Nassarius gibbosulus*), as well as steatite, alabaster, lapis lazuli, turquoise, silver, and gold.[15]

The metalsmiths of the BMAC made beautiful objects of bronze, lead, silver, and gold. They cast delicate metal figures by the lost-wax process, which made it possible to cast very detailed metal objects. They made crested bronze shaft-hole axes with distinctive down-curved blades, tanged daggers, mirrors, pins decorated with cast animal and human figures, and a variety of distinctive metal compartmented seals (figure 16.5). The metals used in the first colonization period, late Namazga V, were unalloyed copper, arsenical bronze, and a copper-lead alloy with up to 8–10% lead.

About 2000 BCE, during the Namazga VI/BMAC period, tin-bronze suddenly appeared prominently in sites of the BMAC. Tin-bronzes were common at two BMAC sites, Sapalli and Djarkutan, reaching more than 50% of objects, although at neighboring Dashly-3, also in Bactria, tin-bronzes were just 9% of metal objects. Tin-bronzes were rare in Margiana (less than 10% of metal objects at Gonur, none at all at Togolok). Tin-bronze was abundant only in Bactria, closer to the Zeravshan. It looks like the tin mines of the Zeravshan were established or greatly expanded at the beginning of the mature BMAC period, about 2000 BCE.[16]

There were no wild horses in Central Asia. The native equids were onagers. Wild horses had not previously strayed south of what is today central Kazakhstan. Any horses found in BMAC sites must have been traded in from the steppes far off to the north. The animal bones discarded in and near BMAC settlements contained no horse bones. Hunters occasionally killed wild onagers but not horses. Most of the bones recovered from the settlement trash deposits were from sheep or goats. Asian zebu cattle and domesticated Bactrian camels also appeared. They were shown pulling wagons and carts in BMAC artwork. Small funeral wagons with solid wooden-plank wheels and bronze-studded tires were buried in royal graves associated with the first building phase, dated about 2100–2000 BCE, at

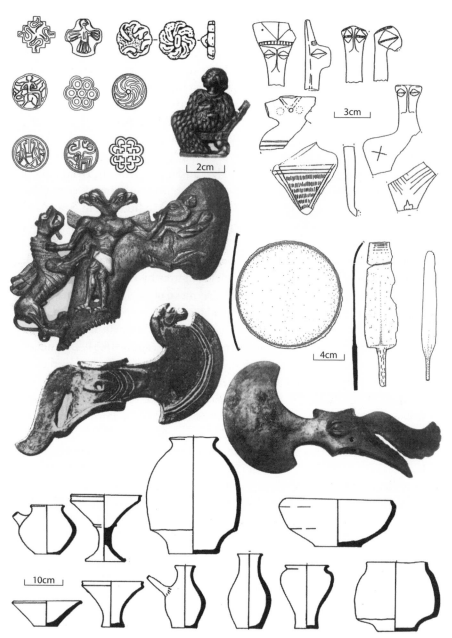

Figure 16.5 Artifacts of the Bactria-Margiana Archaeological Complex, 2100–1800 BCE: (*top left*) a sample of BMAC stamp seals, adapted after Salvatori 2000, and Hiebert 1994; (*top center*) cast silver pin head from Gonur North showing a goddess in a ritual dress, after Klochkov 1998, figure 3; (*top right*) ceramic female figurines from Gonur North, after Hiebert 1994; (*center left*)

Gonur in Margiana (called Gonur North, because the oldest phase was found at the northern end of the modern ruins).

In these graves at Gonur, associated with the early settlement of Gonur North, one horse was found. A brick-lined grave pit contained the contorted bodies of ten adult humans who were apparently killed in the grave itself, one of whom fell across a small funeral wagon with solid wooden wheels. The grave also contained a whole dog, a whole camel, and the decapitated body of a horse foal (the reverse of an Aryan horse sacrifice). This grave is thought to have been a sacrificial offering that accompanied a nearby "royal" tomb. The royal tomb contained funeral gifts that included a bronze image of a horse head, probably a pommel decoration on a wooden staff. Another horse head image appeared as a decoration on a crested copper axe of the BMAC type, unfortunately obtained on the art market and now housed in the Louvre. Finally, a BMAC-style seal probably looted from a BMAC cemetery in Bactria (Afghanistan) showed a man riding a galloping equid that looks very much like a horse (see figure 16.3). The design was similar to the contemporary galloping-horse-and-rider image on the Ur III seal of Abbakalla, dated 2040–2050 BCE. Both seals showed a galloping horse, a rider with a hair-knot on the back of his head, and a man walking.

These finds suggest that horses began to appear in Central Asia about 2100–2000 BCE but never were used for food. They appeared only as decorative symbols on high-status objects and, in one case, in a funeral sacrifice. Given their simultaneous appearance across Iran and Mesopotamia, and the position of BMAC between the steppes and the southern civilizations, horses were probably a trade commodity. After chariots were introduced to the princes of the BMAC, Iran, and the Near East around 2000–1900 BCE, the demand for horses could easily have been on the order of tens of thousands of animals annually.[17]

Steppe Immigrants in Central Asia

Fred Hiebert's excavations at the walled town of Gonur North in Margiana, dated 2100–2000 BCE, turned up a few sherds of strange pottery,

---

Figure 16.5 (continued) crested shaft-hole axes from the art market, probably from BMAC sites, with a possible horse-head on the lower one, after Aruz 1998, figure 24; and Amiet 1986, figure 167; (*center right*) a crested axe with eye amulet, and a copper mirror and dagger excavated from Gonur North, after Hiebert 1994; and Sarianidi 1995, figure 22; (*bottom*) ceramic vessel shapes from Gonur, after Hiebert 1994.

unlike any other pottery at Gonur. It was made with a paddle-and-anvil technique on a cloth-lined form—the clay was pounded over an upright cloth-covered pot to make the basic shape, and then was removed and finished. This is how Sintashta pottery was made. These strange sherds were imported from the steppe. At this stage (equivalent to early Sintashta) there was very little steppe pottery at Gonur, but it was there, at the same time a horse foal was thrown into a sacrificial pit in the Gonur North cemetery. Another possible trace of this early phase of contact were "Abashevo-like" pottery sherds decorated with horizontal channels, found at the tin miners' camp at Karnab on the lower Zeravshan. Late Abashevo was contemporary with Sintashta.

During the classic phase of the BMAC, 2000–1800 BCE, contact with steppe people became much more visible. Steppe pots were brought into the rural stronghold at Togolok 1 in Margiana, inside the larger palace/temple at Togolok 21, inside the central citadel at Gonur South, and inside the walled palace/temple at Djarkutan in Bactria (figure 16.6). These sherds were clearly from steppe cultures. Similar designs can be found on Sintashta pots at Krivoe Ozero (k. 9, gr. 3; k. 10, gr. 13) but were more common on pottery of early Andronovo (Alakul variant) type, dated after 1900–1800 BCE—pottery like that used by the Andronovo miners at Karnab. Although the amount of steppe pottery in classic BMAC sites is small, it is widespread, and there is no doubt that it derived from northern steppe cultures. In these contexts, dated 2000–1800 BCE, the most likely steppe sources were the Petrovka culture at Tugai or the first Alakul-Andronovo tin miners at Karnab, both located in the Zeravshan valley.[18]

The Petrovka settlement at Tugai appeared just 27 km downstream (west) of Sarazm, not far from the later site of Samarkand, the greatest caravan trading city of medieval Central Asia. Perhaps Tugai had a similar, if more modest, function in an early north-south trade network. The Petrovka culture (see below) was an eastern offshoot of Sintashta. The Petrovka people at Tugai constructed two copper-smelting ovens, crucibles with copper slag, and at least one dwelling. Their pottery included at least twenty-two pots made with the paddle-and-anvil technique on a cloth-lined form. Most of them were made of clay tempered with crushed shell, the standard mixture for Petrovka potters, but two were tempered with crushed talc/steatite minerals. Talc-tempered clays were typical of Sintashta, Abashevo, and even forest-zone pottery of Ural forager cultures, so these two pots probably were carried to the

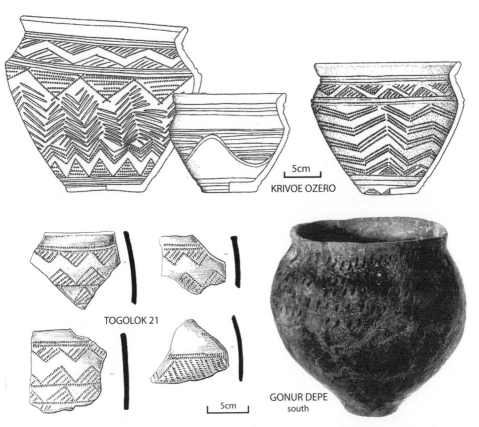

Figure 16.6 A whole steppe pot found inside the walls of the Gonur South town, after Hiebert 1994; steppe sherds with zig-zag decoration found inside the walls of Togolok 1, after Kuzmina 2003; and similar motifs on Sintashta sherds from graves at Krivoe Ozero, Ural steppes, after Vinogradov 2003, figures 39 and 74.

Zeravshan from the Ural steppes. The pottery shapes and impressed designs were classic early Petrovka (figure 16.7). A substantial group of Petrovka people apparently moved from the Ural-Ishim steppes to Tugai, probably in wagons loaded with pottery and other possessions. They left garbage middens with the bones of cattle, sheep, and goats, but they did not eat horses—although their Petrovka relatives in the northern steppes did. Tugai also contained sherds of wheel-made cups in red-polished and black-polished fabrics typical of the latest phase at Sarazm (IV). The

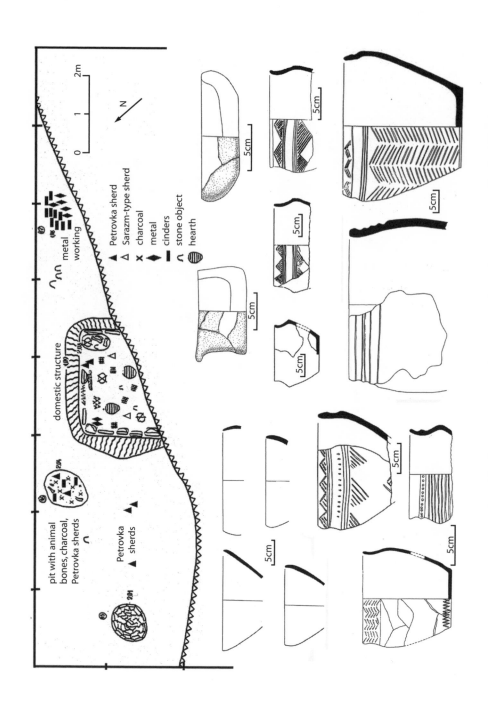

pit with animal
bones, charcoal,
Petrovka sherds

Petrovka
sherds

metal
working

domestic structure

N

0   1   2m

▲ Petrovka sherd
△ Sarazm-type sherd
✕ charcoal
◆ metal
▌ cinders
⌒ stone object
⬭ hearth

5cm

principal activity identified in the small excavated area was copper smelting.[19]

The steppe immigrants at Tugai brought chariots with them. A grave at Zardcha-Khalifa 1 km east of Sarazm contained a male buried in a contracted pose on his right side, head to the northwest, in a large oval pit, 3.2 m by 2.1 m, with the skeleton of a ram.[20] The grave gifts included three wheel-made Namazga VI ceramic pots, typical of the wares made in Bactrian sites of the BMAC such as Sappali and Dzharkutan; a trough-spouted bronze vessel (typical of BMAC) and fragments of two others; a pair of gold trumpet-shaped earrings; a gold button; a bronze straight-pin with a small cast horse on one end; a stone pestle; two bronze bar bits with looped ends; and two largely complete bone disc-shaped cheekpieces of the Sintashta type, with fragments of two others (figure 16.8). The two bronze bar bits are the oldest known metal bits anywhere. With the four cheekpieces they suggest equipment for a chariot team. The cheekpieces were a specific Sintashta type (the raised bump around the central hole is the key typological detail), though disc-shaped studded cheekpieces also appeared in many Petrovka graves. Stone pestles also frequently appeared in Sintashta and Petrovka graves. The Zardcha-Khalifa grave probably was that of an immigrant from the north who had acquired many BMAC luxury objects. He was buried with the only known BMAC-made pin with the figure of a horse—perhaps made just for him. The Zardcha-Khalifa chief may have been a horse dealer. The Zeravshan valley and the Ferghana valley just to the north might have become the breeding ground at this time for the fine horses for which they were known in later antiquity.

The fabric-impressed pottery and the sacrificed horse foal at Gonur North and perhaps the Abashevo (?) sherds at Karnab represent the exploratory phase of contact and trade between the northern steppes and the southern urban civilizations about 2100–2000 BCE, during the period when the kings of Ur III still dominated Elam. Information and perhaps even cult practices from the south flowed back to early Sintashta societies. On the eastern frontier in Kazakhstan, where Petrovka was budding off from Sintashta, the lure of the south prompted a migration across more

---

Figure 16.7 The Petrovka settlement at Tugai on the Zeravshan River: (*top*) plan of excavation; (*center left*) imported redware pottery like that of Sarazm IV; (*center right*) two coarse ceramic crucibles from the metal-working area; (*bottom*) Petrovka pottery. Adapted from Avanessova 1996.

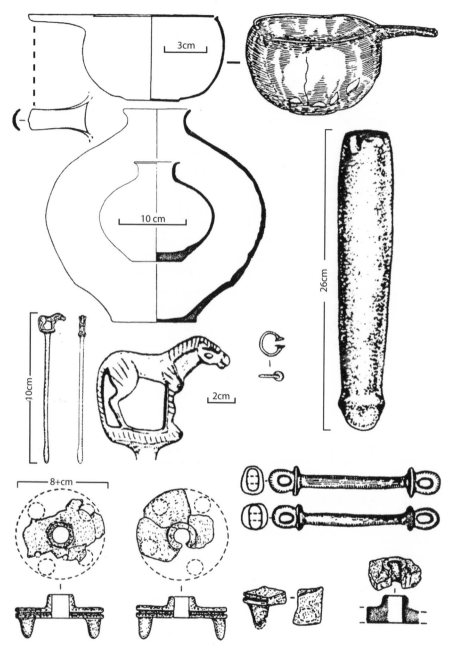

Figure 16.8 Objects from the grave at Zardcha-Khalifa on the Zeravshan River. The trough-spouted bronze vessel and ceramic pots are typical of the BMAC, 2000–1800 BCE; the cast copper horse pin shows BMAC casting methods; the bronze bar bits are the first ones dated this early; and the stone pestle, trumpet-shaped earring, and bone cheekpieces are steppe types. After Bobomulloev 1997, figures 2, 3, and 4.

than a thousand kilometers of hostile desert. The establishment of the Petrovka metal-working colony at Tugai, probably around 1900 BCE, was the beginning of the second phase, marked by the actual migration of chariot-driving tribes from the north into Central Asia. Sarazm and the irrigation-fed Zaman-Baba villages were abandoned about when the Petrovka miners arrived at Tugai. The steppe tribes quickly appropriated the ore sources of the Zeravshan, and their horses and chariots might have made it impossible for the men of Sarazm to defend themselves.

Central Asian Trade Goods in the Steppes

Did any BMAC products appear in Sintashta or Petrovka settlements? Only a few hints of a return trade can be identified. One intriguing innovation was a new design motif, the stepped pyramid or crenellation. Stepped pyramids or crenellations appeared on the pottery of Sintashta, Potapovka, and Petrovka. The stepped pyramid was the basic element in the decorative artwork on Namazga, Sarazm, and BMAC pottery, jewelry, metalwork, and even in a mural painted on the Proto-Elamite palace wall at Malyan (figure 16.9, bottom). Repeated horizontally, the stepped pyramid became a line of crenellated designs; repeated on four sides, it became a stepped cross. This motif had not appeared in any earlier pottery in the steppes, neither in the Bronze Age nor the Eneolithic. Charts of design motifs are regularly published in Russian archaeological ceramic studies. I have scanned these charts for years and have not found the stepped pyramid in any assemblage earlier than Sintashta. Stepped pyramids appeared for the first time on northern steppe pottery just when northern steppe pottery first showed up in BMAC sites. It was seen first on a small percentage (<5%) of Potapovka pottery on the middle Volga (single vessels in Potapovka kurgans 1, 2, 3, and 5) and at about the same frequency on Sintashta pottery in the Ural-Tobol steppes; later it became a standard design element in Petrovka and Andronovo pottery (but not in Srubnaya pottery, west of the Urals). Although no Sarazm or BMAC pottery has been found in Sintashta contexts, the design could have been conveyed to the northern steppes on textiles—perhaps the commodity exchanged for northern metal. I would guess that Sintashta potters copied the design from imported BMAC textiles.

There are other indications of contact. A lead wire made of two braided strands was found among the metal objects in the Sintashta settlement of Kuisak. Lead had never before appeared in the northern steppes as a pure metal, whereas a single ingot of lead weighing 10 kg was found at Sarazm.

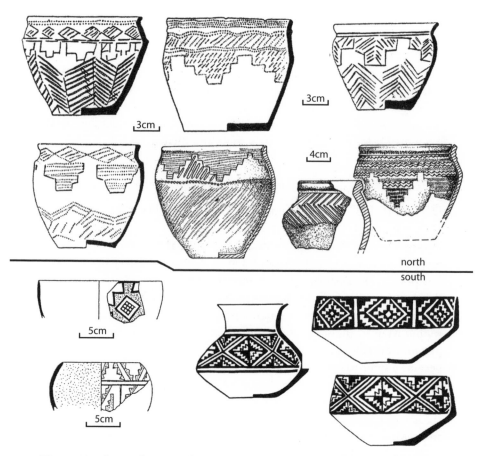

Figure 16.9 Stepped pyramid or crenellation motifs on steppe pottery and on Central Asian pottery: (*top row and left pot in second row*) Potapovka graves, middle Volga region, 2100–1800 BCE, after Vasiliev, Kuznetsov, and Semenova 1994, figures 20 and 22; (*middle row, remaining pots*) Sintashta SII cemetery, grave 1, after Gening, Zdanovich and Gening 1992, figure 172; (*bottom left*) Sarazm, level II, 3000–2500 BCE, after Lyonnet 1996, figures 4 and 12; (*bottom right*) Altyn-Depe, excavation 1, burial 296, after Masson 1988, plate 27.

The Kuishak lead wire probably was an import from the Zeravshan. A lapis lazuli bead from Afghanistan was found at Sintashta. A Bactrian-handled bronze mirror was found in a Sintashta grave at Krasnoe Znamya.[21] Finally, the technique of lost-wax metal casting first appeared in the north during the Sintashta period, in metal objects of Seima-Turbino type (described in more detail below). Lost-wax casting was familiar to BMAC metalsmiths. Southern decorative motifs (stepped pyramids), raw materials (lead and lapis lazuli), one mirror, and metal-working techniques (lost-wax casting) appeared in the north just when northern pottery, chariot-driving cheekpieces, bit wear, and horse bones appeared in the south.

The sudden shift to large-scale copper production that began about 2100–2000 BCE in the earliest Sintashta settlements must have been stimulated by a sharp increase in demand. Central Asia is the most likely source. The increase in metal production deeply affected the internal politics of northern steppe societies, which quickly became accustomed to using and consuming large quantities of bronze. Although the northern steppe producers probably had direct contact with the Central Asian market only for a short time, internal demand in the steppes remained high throughout the LBA. Once the metallurgical pump was primed, so to speak, it continued to flow. The priming happened because of contact with urban markets, but the flow after that raised the usage of metal in the steppes and in the forest zone to the north, starting an internal European cycle of exchange that would lead to a metal boom in the Eurasian steppes after 2100 BCE.

After 1900 BCE a contact zone developed in the Zeravshan valley and extended southward to include the central citadels in the BMAC towns. In the Zeravshan, migrants from the northern steppes mixed with late Kelteminar and BMAC-derived populations. The Old Indic dialects probably evolved and separated from the developing Iranian dialects in this setting. To understand how the Zeravshan-Bactrian contact zone separated itself from the northern steppes, we need to examine what happened in the northern steppes after the end of the Sintashta culture.

THE OPENING OF THE EURASIAN STEPPES

The Srubnaya (or Timber-Grave) culture was the most important LBA culture of the western steppes, from the Urals to the Dnieper (figure 16.10). The Andronovo horizon was the primary LBA complex of the eastern steppes, from the Urals to the Altai and the Tien Shan. Both grew from the Potapovka-Sintashta complex between the middle Volga and the Tobol. With the appearance of Srubnaya and Andronovo between about 1900 and 1800 BCE, for the first time in history a chain of broadly similar cultures extended from the edges of China to the frontiers of Europe. Innovations and raw materials began to move across the continent. The steppe world was not just a conduit, it also became an innovating

Figure 16.10 The Late Bronze Age cultures of the Eurasian steppes, 1900–1500 BCE.

center, particularly in bronze metallurgy and chariot warfare. The chariot-driving Shang kings of China and the Mycenaean princes of Greece, contemporaries at opposite ends of the ancient world at about 1500 BCE, shared a common technological debt to the LBA herders of the Eurasian steppes.

THE SRUBNAYA CULTURE: HERDING AND GATHERING IN THE WESTERN STEPPES

West of the Ural Mountains, the Potapovka and late Abashevo groups of the middle Volga region developed into the Pokrovka complex, dated about 1900–1750 BCE. Pokrovka was a proto-Srubnaya phase that rapidly developed directly into the Srubnaya (or Timber-Grave) culture (1800–1200 BCE). Srubnaya material culture spread as far west as the Dnieper valley. One of the most prominent features of the Srubnaya culture was the appearance of hundreds of small settlement sites, most of them containing just a few houses, across the northern steppe and the southern forest-steppe, from the Urals to the Dnieper. Although settlements had reappeared in a few places east of the Don River during the late Catacomb culture, 2400–2100 BCE, and were even more numerous in Ukraine west of the Don during the Mnogovalikovaya (MVK) period (2100–1800 BCE), the Srubnaya period was the first time since the Eneolithic that settlements appeared across the entire northern steppe zone from the Dnieper to the southern Urals and beyond into northern Kazakhstan.

The reason for this shift back to living in permanent homes is unclear. Most Srubnaya settlements were not fortified or defended. Most were small individual homesteads or extended family ranches rather than nucleated villages. The herding pattern seems to have been localized rather than migratory. During the Samara Valley Project, in 1999–2001, we studied the local Srubnaya herding pattern by excavating a series of Srubnaya herding camps that extended up a tributary stream valley, Peschanyi Dol, from the Srubnaya settlement at Barinovka, near the mouth of the valley on the Samara (figure 16.11). The largest herding camps (PD1 and 2) were those closest to the home settlement, within 4–6 km of Barinovka. Farther upstream the Srubnaya camps were smaller with fewer pottery sherds, and beyond about 10–12 km upstream from Barinovka we found no LBA herding camps at all, not even around the springs that fed the stream at its source, where there was plenty of water and good pastures. So the herding system seems to have been localized, like the new residence pattern.

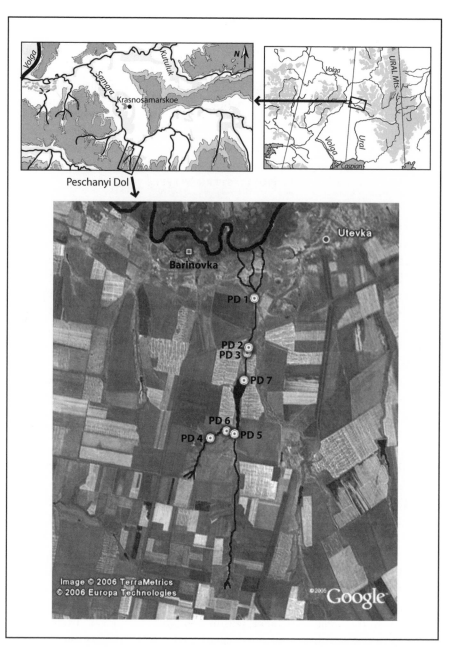

Figure 16.11 The Peschanyi Dol valley, a tributary of the Samara River, surveyed to find ephemeral camps in 1995–96. PD1, 2, and 3, were Srubnaya herding camps excavated in 2000. All numbered sites yielded at least one Srubnaya ceramic sherd. Barinovka was a larger Srubnaya settlement tested in

The Srubnaya economy in the middle Volga steppes does not seem to have required long-distance migrations.

One traditional explanation for the settling-down phenomenon is that this was when agriculture was widely adopted across the northern steppes.[22] But this explanation certainly does not apply everywhere. At the settlement of Krasnosamarskoe in the Samara River valley, where the dog sacrifice was found (chapter 15), a Pokrovka component (radiocarbon dated 1900–1800 BCE) and an early Srubnaya component (dated 1800–1700 BCE) were stratified within a single structure. In the Srubnaya period the structure probably was a well-house and woodshed where a variety of domestic tasks were conducted and food garbage was buried in pits. It was used during all seasons of the year. Anne Pike-Tay's analysis of seasonal bands in the roots of animal teeth established that the cattle and sheep were butchered in all seasons. But there was no agriculture. Laura Popova found no seeds, pollen, or phytoliths of cultivated cereals associated with the LBA occupation, only wild *Chenopodium* and *Amaranthus* seeds. The skeletons of 192 adults from twelve Srubnaya cemeteries in the Samara oblast were examined by Eileen Murray and A. Khokhlov. They showed almost no dental decay. The complete absence of caries usually is associated with a low-starch, low-carbohydrate diet, typical for foragers and quite atypical for bread eaters (figure 16.12). The dental evidence confirmed the botanical evidence. Bread was not eaten much, if at all, in the northern steppes.

In pits at Krasnosamarskoe we found an abundance of carbonized wild seeds, including *Chenopodium album* and *Amaranthus*. Modern wild *Chenopodium* (also known as goosefoot) is a weed that grows in dense stands that can produce seed yields in the range of 500–1000 kg/ha, about the same as einkorn wheat, which yields 645-835 kg/ha.[23] *Amaranthus* is equally prolific. With meat and milk from cattle, sheep, and horses, this was a sufficient diet. Although clear evidence of cereal agriculture has been found in Srubnaya settlements west of the Don in Ukraine, it is possible that agriculture was much less important east of the Don than has often been assumed. Herding and gathering was the basis for the northern steppe economy in at least some regions east of the Don as late as the LBA.[24]

Figure 16.11 (*continued*) 1996 but found to be badly disturbed by a historic settlement. Author's excavation. Bottom image is a Google Earth™ image, © 2006 Terra Metrics, 2006 Europa Technologies.

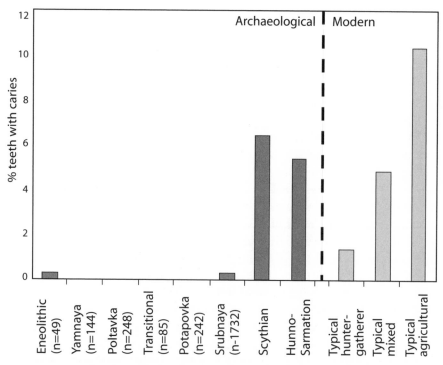

Figure 16.12 Graph of the frequency of dental caries (cavities) in populations with different kinds of food economies (*right*), in Scythian and Sarmatian cemeteries in Tuva (*center*), and in prehistoric populations in the Samara oblast, middle Volga region (*left six bars*). Bread apparently was not part of the diet in the Samara oblast. After Murphy 2003; and Murphy and Khokhlov 2001.

So if agriculture does not provide an answer, then why did people settle down during the MBA/LBA transition in the northern steppes, including the earlier episode at Sintashta? As explained in chapter 15, climate change might have been the principal cause. A cool, arid climate affected the Eurasian steppes between about 2500–2000 BCE. This was the same event that struck Akkadian agriculture and weakened the Harappan civilization. The late MBA/early LBA settling-down phenomenon, including the earliest episodes at Sintashta and Arkaim, can be interpreted as a way to maintain control over the richest winter forage areas for herds, particularly if grazing animals were the principal source of food in an economy that, in many regions, did not include agriculture. Early LBA Krasnosamarskoe overlooked one of the largest marshes on the lower Samara River.

Some permanent settlements also developed near copper mines. Cattle forage was not the only critical resource in the northern steppes. Mining and bronze working became important industries across the steppes during the LBA. A vast Srubnaya mining center operated at Kargaly near Orenburg in the South Urals, and other enormous copper mines operated near Karaganda in central Kazakhstan. Smaller mining camps were established at many small copper outcrops, like the Srubnaya mining camp at Mikhailovka Ovsianka in the southern Samara oblast.[25]

East of the Urals, Phase I: The Petrovka Culture

The first culture of the LBA east of the Urals was the Petrovka culture, an eastern offshoot of Sintashta dated about 1900–1750 BCE. Petrovka was so similar to Sintashta in its material culture and mortuary rituals that many archaeologists (including me) have used the combined term Sintashta-Petrovka to refer to both. But Petrovka ceramics show some distinctive variations in shape and decoration, and are stratified above Sintashta deposits at several sites, so it is clear that Petrovka grew out of and was generally later than Sintashta. The oldest Petrovka sites, like the type site, Petrovka II, were settlements on the Ishim River in the steppes of northern Kazahstan (figure 16.13). The Petrovka culture probably absorbed some people who had roots in the older post-Botai horse-centered cultures of the Ishim steppes, like Sergeivka, but they were materially (and probably linguistically) almost invisible. Petrovka-style pottery then replaced Sintashta ceramics at several Sintashta fortified sites, as at Ust'ye, where the Sintashta settlement was burned and replaced by a Petrovka settlement built on a different plan. Petrovka graves were dug into older Sintashta kurgans at Krivoe Ozero and Kamenny Ambar.[26]

The settlement of Petrovka II was surrounded by a narrow ditch less than 1 m deep, perhaps for drainage. The twenty-four large houses had dug-out floors and measured from 6 by 10 m to about 8 by 18 m. They were built close together on a terrace overlooking the floodplain, a nucleated village pattern quite different from the scattered homesteads of the Srubnaya culture. Petrovka II was reoccupied by people who made classic Andronovo-horizon ceramics of both the Alakul and Federovo types, stratified above the Petrovka layer, and the Andronovo town was succeeded by a "final-LBA" settlement with Sargar ceramics. This stratified sequence made Petrovka II an important yardstick for the LBA chronology of the Kazakh steppes. Chariots continued to be buried in a few early Petrovka graves at Berlyk II and Krivoe Ozero, and many bone disk-shaped cheekpieces have

Petrovka settlement plan

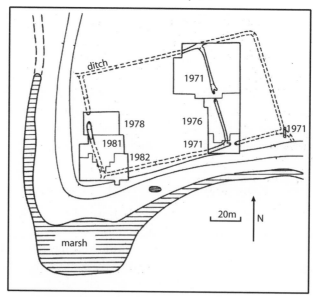

1971 excavation detail

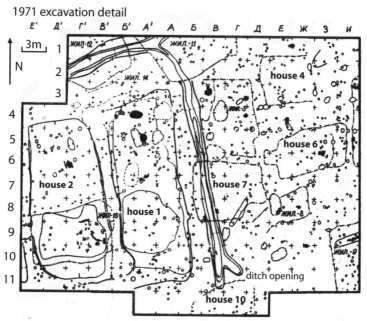

Figure 16.13 The Petrovka settlement, type site for the Petrovka culture, ca. 1900–1750 BCE: (*top*) general plan of the original ditch around the settlement, with a later enlargement at the east end, after Zdanovich 1988, Figure 12; (*bottom*) detail of overlapping rebuilt house floors in the northeast corner of the original settlement, with new houses built over the original eastern ditch, after

come from Petrovka sites. During the Petrovka period, however, chariot burials gradually ceased, the size and number of mortuary animal sacrifices also declined, and large-scale Sintashta-type fortifications were no longer built around settlements in the northern steppes.

Petrovka settlements and kurgan cemeteries spread southward into the arid steppes of central Kazkahstan, and from there to Tugai on the Zeravshan, more than 1,200 km south of central Kazakhstan. Petrovka probably also was in touch with the Okunevo culture in the western Altai, the successor of late Afanasievo. The permanent nucleated settlements of the Petrovka culture do not resemble the temporary camps of nomadic herders, so it is unlikely that the Petrovka economy depended on annual long-distance migrations. Early historic nomads, who did not live in permanent nucleated villages, wintered in the Syr Darya marshes and summered in the north Kazakh steppes, a cycle of annual movements that brought them to the doorstep of Central Asia civilizations each winter. But the Petrovka economy seems to have been less nomadic. If the Petrovka people did *not* engage in long-distance herd migrations, then their movement south to the Zeravshan was not an accidental by-product of annual herding patterns (as is often presumed) but instead was intentional, motivated by the desire for trade, loot, or glory. The later annual migration pattern does at least show that in the spring and fall it was possible to drive herds of animals across the intervening desert and semi-desert.[27]

Petrovka settlements commonly contained two-part furnaces, slag, and abundant evidence of copper smelting, like Sintashta settlements. But, unlike Sintashta, most Petrovka metal objects were made of tin-bronze.[28] A possible source for the tin in Petrovka tin-bronzes, in addition to the Zeravshan valley, was in the western foothills of the Altai Mountains. A remarkable shift occurred in the forest-steppe zone north of the Petrovka territory during the early Petrovka phase.

THE SEIMA-TURBINO HORIZON IN THE FOREST-STEPPE ZONE

The Seima-Turbino horizon marks the entry of the forest-steppe and forest-zone foragers into the cycle of elite competition, trade, and warfare that had erupted earlier in the northern steppes. The tin-bronze spears, daggers, and axes of the Seima-Turbino horizon were among the most

---

Figure 16.13 (continued) Maliutina 1991, Figure 14. The stratigraphic complexity of these settlements contributes to arguments about phases and chronology.

technically and aesthetically refined weapons in the ancient world, but they were made by forest and forest-steppe societies that in some places (Tashkovo II) still depended on hunting and fishing. These very high-quality tin-bronze objects first appeared among the Elunino and Krotovo cultures located on the upper and middle Irtysh and the upper Ob in the western foothills of the Altai Mountains, a surprisingly remote region for such a remarkable exhibition of metallurgical skill. But tin, copper, and gold ores all could be found on the upper Irtysh, near the confluence of the Irtysh and the Bukhtarta rivers about 600 km east of Karaganda. The exploitation of these ore sources apparently was accompanied by an explosion of new metallurgical skills.

One of the earliest and most important Seima-Turbino cemeteries was at Rostovka in the Omsk oblast on the middle Irtysh (figure 16.14). Although skeletal preservation was poor, many of the thirty-eight graves seem to have contained no human bones at all or just a few fragments of a skeleton. In the graves with whole bodies the skeleton was supine with the legs and arms extended. Grave gifts were offered both in the graves and in ritual deposits at the edge of graves. Both kinds of offerings included tin-bronze socketed spearheads, single-edged curved knives with cast figures on the pommel, and hollow-core bronze axes decorated with triangles and lozenges. Grave 21 contained bivalve molds for making all three of these weapon types. Offerings also included stemmed flint projectile points of the same types that appeared in Sintashta graves, bone plates pierced to make plate armor, and nineteen hundred sherds of Krotovo pottery (figure 16.14). One grave (gr. 2) contained a lapis lazuli bead from Afghanistan, probably traded through the BMAC, strung with beads of nephrite, probably from the Baikal region.[29]

Seima-Turbino metalsmiths were, with Petrovka metalsmiths, the first north of Central Asia to regularly use a tin-bronze alloy. But Seima-Turbino metalsmiths were unique in their mastery of lost-wax casting (for decorative figures on dagger handles) and thin-walled hollow-mold casting (for socketed spears and hollow axes). Socketed spearheads were made on Sintashta anvils by bending a bronze sheet around a socket form and then forging the seam (figure 16.15). Seima-Turbino socketed spearheads were made by pouring molten metal into a mold that created a seamless cast socket around a suspended core, making a hollow interior, a much more sophisticated operation, and easier to do with tin-bronze than with arsenical bronze. Axes were made in a similar way, tin-bronze with a hollow interior, cast around a suspended core. Lost-wax and hollow-mold casting methods probably were learned from the BMAC civilization, the only reasonably nearby source (perhaps through a skilled captive?).

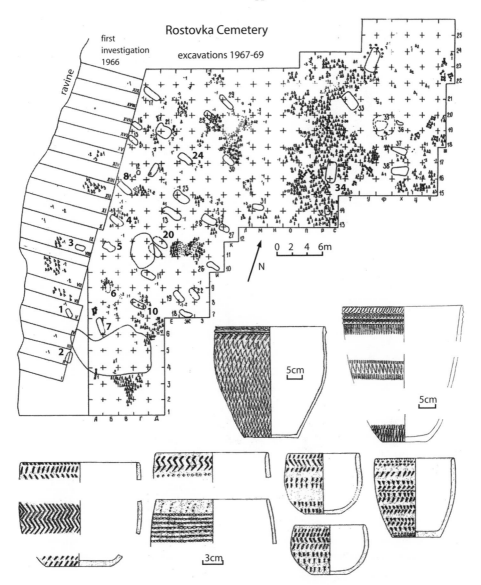

Figure 16.14 The Rostovka cemetery near Omsk, one of the most important sites of the Seima-Turbino culture. Graves are numbered. Black dots represent ceramics, metal objects, and other artifacts deposited above and beside the graves. All the pots conform to the Krotova type. After Matiushchenko and Sinitsyna 1988, figures 4, 81, 82, and 83.

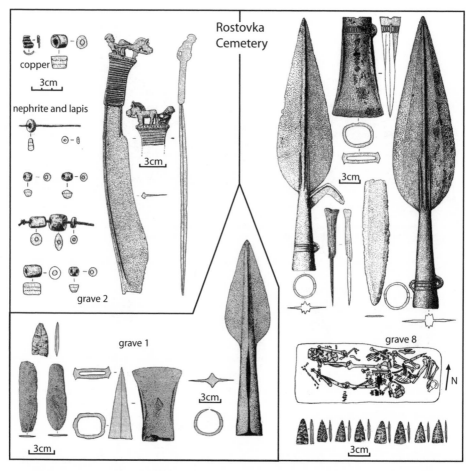

Figure 16.15 Grave lots from the Rostovka cemetery, graves 1, 2, and 8. The lost-wax cast figure of a man roping a horse and the hollow-mold casting of spears and axes were technical innovations probably learned from BMAC metalsmiths. Grave 1 contained beads made of both lapis lazuli from Afghanistan and nephrite probably from the near Lake Baikal. After Matiushchenko and Sinitsyna 1988, figures 6, 7, 17, and 18.

Beyond the western Altai/middle Irtysh core area the Seima-Turbino horizon was not a culture. It did not have a standard ceramic type, settlement type, or even a standard mortuary rite. Rather, Seima-Turbino metal-working techniques were adopted by emerging elites across the southern Siberian forest-steppe zone, perhaps in reaction to and competing with the Sintashta and Petrovka elites in the northern steppes. A series of original and distinctive new metal types quickly diffused through the forest-steppe zone from the east to the west, appearing in late Abashevo and Chirkovskaya cemeteries west of the Urals almost at the same time that they first appeared east of the Urals, beginning about 1900 BCE. The rapidity and reach of this phenomenon in the forest zone is surprising. The new metal styles probably spread more by emulation than by migration, along with fast-moving political changes in the structure of power. Seima-Turbino spearheads, daggers, and axes were displayed at the Turbino cemetery in the forests of the lower Kama, southward up the Oka, and as far south as the Borodino hoard in Moldova, in the East Carpathian foothills. East of the Urals, most Seima-Turbino bronzes were tin-bronzes, and west of the Urals, they were mostly arsenical bronzes. The source of the tin was in the east, but the styles and methods of Seima-Turbino metallurgy were diffused across the forest-steppe and forest zones from the Altai to the Carpathians. The Borodino hoard contained a nephrite axe probably made of stone quarried near Lake Baikal. In the eastern direction, Seima-Turbino metal types (hollow-cast socketed spearheads with a side hook, hollow-cast axes) appeared also in sites on the northwestern edges of the evolving archaic Chinese state, probably through a network of trading trails that passed north of the Tien Shan through Dzungaria.[30]

The dating of the Seima-Turbino horizon has changed significantly in recent years. Similarities between Seima-Turbino socketed spearheads and daggers and parallel objects in Mycenaean tombs were once used to date the Seima-Turbino horizon to a period after 1650 BCE. It is clear now, however, that Mycenaean socketed spearheads, like studded disk cheekpieces, were derived from the east and not the other way around. Seima-Turbino and Sintashta were partly contemporary, so Seima-Turbino probably began before 1900 BCE.[31] Seima-Turbino and Sintasha graves had the same kinds of flint projectile points. Sintashta forged socketed spearheads probably were the simpler predecessors of the more refined hollow-cast Seima-Turbino socketed spearheads. A hollow-cast spearhead of Seima-Turbino type was deposited in a Petrovka-culture chariot grave at Krivoe Ozero (k. 2, gr. 1); and a Sintashta bent and forged spearhead appeared in the Seima-Turbino cemetery at Rostovka (gr. 1) (see figure 16.15).

The metal-working techniques of the northern steppes (Sintashta and Petrovka) and the forest-steppe zone (Seima-Turbino) remained separate and distinct for perhaps one hundred to two hundred years. But by the beginning of the Andronovo period they merged, and some important

Seima-Turbino metal types, such as cast single-edged knives with a ring-pommel, became widely popular in Andronovo communities.

East of the Urals, Phase II: The Andronovo Horizon

The Andronovo horizon was the principal LBA archaeological complex in the steppes east of the Urals, the sister of the Srubnaya horizon west of the Urals, between about 1800 and 1200 BCE. Andronovo sites extended from the Ural steppes eastward to the steppes on the upper Yenisei River in the Altai, and from the southern forest zone southward to the Amu Darya River in Central Asia. Andronovo contained two principal subgroups, Alakul and Federovo. The earliest of these, the Alakul complex, appeared in some places by about 1900–1800 BCE. It grew directly out of the Petrovka culture by small modifications of ceramic decorations and vessel shapes. The Federovo style might have developed from a southern or eastern stylistic variant of Alakul, although some specialists insist that it had completely independent origins. Andronovo continued many of the customs and styles inherited through Sintashta and Petrovka: small family kurgan cemeteries, settlements containing ten to forty houses built close together, similar spear and dagger types, similar ornaments, and even the same decorative motifs on pottery: meanders, hanging triangles, "pine-tree" figures, stepped pyramids, and zig-zags. But chariots were no longer buried.

Alakul and Federovo are described as separate cultures within the Andronovo horizon, but to this observer, admittedly not an expert in the details of LBA ceramic typology, the Alakul and Federovo ceramic styles seem similar. Pot shapes varied only slightly (Federovo pots usually had a more indented, undercut lower profile) and decorative motifs also varied around common themes (some Federovo motifs were "italicized" or forward-slanted versions of Alakul motifs). Pots and potsherds of these two ceramic styles are found in the same sites from the Ural-Tobol steppes southeastward to central Kazkahstan, often in the same house and pit features, and in adjoining kurgans in the same cemeteries. Some pots are described as Alakul with Federovo elements, so the two varieties can appear on the same pot (figure 16.16). Alakul pottery is stratified beneath Federovo pottery in a few key features at some sites (at Novonikol'skoe and Petrovka II in the Ishim steppes and Atasu 1 in central Kazakhstan), but Federovo pottery has never been found stratified beneath Alakul. The earliest Alakul radiocarbon dates (1900–1700 BCE) are a little older than the earliest Federovo dates (1800–1600 BCE), so Alakul

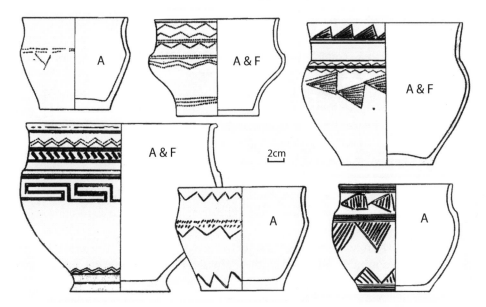

Figure 16.16 Andronovo pots that are described as typical Alkakul (A) or Alakul with Federovo traits (A+F) from the Priplodyi Log kurgan cemetery I on the Ui River, Chelyabinsk oblast, Russia. Traits of both styles can appear on the same pot. After Maliutina 1984, figure 4.

probably began a century or two earlier, although in many settlements the two are thoroughly mixed. Kurgans containing Federovo pots often had larger, more complex stone constructions around the grave and the dead were cremated, whereas kurgans with Alakul pots were simpler and the dead usually were buried in the flesh. Since the two ceramic styles occurred in the same settlements and cemeteries, and even in the same house and pit features, they cannot easily be interpreted as distinct ethnic groups.[32]

The spread of the Andronovo horizon represented the maturation and consolidation of an economy based on cattle and sheep herding almost everywhere in the grasslands east of the Urals. Permanent settlements appeared in every region, occupied by 50 to 250 people who lived in large houses. Wells provided water through the winter. Some settlements had elaborate copper-smelting ovens. Small-scale agriculture might have played a minor role in some places, but there is no direct evidence for it. In the northern steppes cattle were more important than sheep (cattle 40% of bones, sheep/goat 37%, horses 17% in the Ishim steppes), whereas in central Kazakhstan there were more sheep than cattle, and more horses as well (sheep/goat 46%, cattle 29%, horse 24%).[33]

Although it is common in long-established tribal culture areas for a relatively homogeneous material culture to mask multiple languages, the link between language and material culture often is strong among the early generations of long-distance migrants. The source of the Andronovo horizon can be identified in an extraordinary burst of economic, military, and ritual innovations by a single culture—the Sintashta culture. Many of its customs were retained by its eastern daughter, the Petrovka culture. The language spoken in Sintashta strongholds very likely was an older form of the language spoken by the Petrovka and Andronovo people. Indo-Iranian and Proto-Iranian dialects probably spread with Andronovo material culture.

Most Andronovo metals, like Petrovka metals, were tin-bronzes. Andronovo miners mined tin in the Zeravshan and probably on the upper Irtysh. Andronovo copper mines were active in two principal regions: one was south of Karaganda near Uspenskyi, working malachite and azurite oxide ores; and the other was to the west in the southern Ulutau Hills near Dzhezkazgan, working sulfide ores. (Marked on figure 15.9.) One mine of at least seven known in the Dzhezkazgan region was 1,500 m long, 500 m wide, and 15 m deep. Ore was transported from the Uspenskyi mine to copper-smelting settlements such as Atasu 1, where excavation revealed three key-shaped smelting ovens with 4 m-long stone-lined air shafts feeding into two-level circular ovens. The Karaganda-region copper mines are estimated to have produced 30 to 50,000 metric tons of smelted copper during the Bronze Age.[34] The labor and facilities at these places suggest enterprises organized for export.

Trade with and perhaps looting raids into Central Asia left clear evidence surprisingly far north in the steppes. Wheel-made Namamzga VI pottery was found in the Andronovo settlement of Pavlovka, in northern Kazkahstan near Kokchetav, 2,000 km north of Bactria. It was 12% of the pottery on two house floors. The remainder was Andronovo pottery of the Federovo type.[35] The imported Central Asian pots were made with very fine white or red clay fabrics, largely undecorated, and in forms such as pedestaled dishes that were typical of Namazaga VI (figure 16.17). Pavlovka was a settlement of about 5 ha with both Petrovka and Federovo pottery. The Central Asian pottery is said to have been associated with the Federovo component.

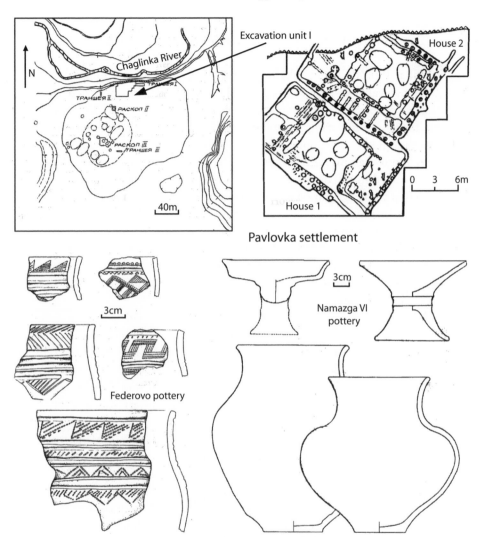

Figure 16.17 Pavlovka, an Alakul-Federovo settlement in the Kokchetav region of northern Kazakhstan, with imported Namazga VI pottery constituting more than 10% of the sherds on two house floors. After Maliutina 1991, figures 4 and 5.

Proto-Vedic Cultures in the Central Asian Contact Zone

By about 1900 BCE Petrovka migrants had started to mine copper in the Zeravshan valley at Tugai. They were followed by larger contingents of Andronovo people who mined tin at Karnab and Mushiston. After 1800 BCE Andronovo mining camps, kurgan cemeteries, and pastoral camps spread into the middle and upper Zeravshan valley. Other Andronovo groups moved into the lower Zeravshan and the delta of the lower Amu Darya (now located in the desert east of the modern delta) and became settled irrigation farmers, known as the Tazabagyab variant of the Andronovo culture. They lived in small settlements of a few large dug-out houses, much like Andronovo houses; used Andronovo pottery and Andronovo-style curved bronze knives and twisted earrings; conducted in-settlement copper smelting as at many Andronovo settlements; but buried their dead in large flat-grave cemeteries like the one at Kokcha 3, with more than 120 graves, rather than in kurgan cemeteries (figure 16.18).[36]

About 1800 BCE the walled BMAC centers decreased sharply in size, each oasis developed its own types of pottery and other objects, and Andronovo-Tazabagyab pottery appeared widely in the Bactrian and Margian countryside. Fred Hiebert termed this the *post-BMAC* period to emphasize the scale of the change, although occupation continued at many BMAC strongholds and Namazga VI–style pottery still was made inside them.[37] But Andronovo-Tazabagyab coarse incised pottery occurred both within post-BMAC fortifications and in occasional pastoral camps located outside the mudbrick walls. Italian survey teams exposed a small Andronovo-Tazabagyab dug-out house southeast of the post-BMAC walled fortress at Takhirbai 3, and American excavations found a similar occupation outside the walls of a partly abandoned Gonur. By this time the people living just outside the crumbling walls and at least some of those now living inside were probably closely related. To the east, in Bactria, people making similar incised coarse ware camped atop the vast ruins (100 ha) of the Djarkutan city. Some walled centers such as Mollali-Tepe continued to be occupied but at a smaller scale. In the highlands above the Bactrian oases in modern Tajikistan, kurgan cemeteries of the Vaksh and Bishkent type appeared with pottery that mixed elements of the late BMAC and Andronovo-Tazabagyab traditions.[38]

Between about 1800 and 1600 BCE, control over the trade in minerals (copper, tin, turquoise) and pastoral products (horses, dairy, leather) gave the Andronovo-Tazabagyab pastoralists great economic power in the old

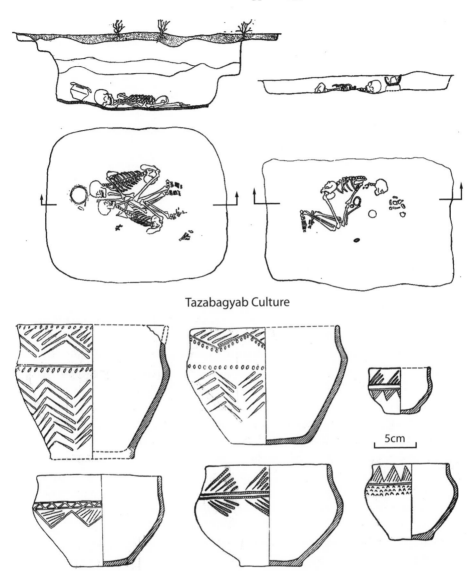

Tazabagyab Culture

Figure 16.18 Graves of the Tazabagyab-Andronovo culture at the Kokcha 3 cemetery on the old course of the lower Amu-Darya River. Pottery like this was widespread in the final phase of occupation in the declining BMAC walled towns of Central Asia, 1700–1500 BCE. After Tolstov and Kes' 1960, figure 55.

BMAC oasis towns and strongholds, and chariot warfare gave them military control. Social, political, and even military integration probably followed. Eventually the simple incised pottery of the steppes gave way to new ceramic traditions, principally gray polished wares in Margiana and the Kopet Dag, and painted wares in Bactria and eastward into Tajikistan.

By 1600 BCE all the old trading towns, cities, and brick-built fortified estates of eastern Iran and the former BMAC region in Central Asia were abandoned. Malyan, the largest city on the Iranian plateau, was reduced to a small walled compound and tower occupied within a vast ruin, where elite administrators, probably representatives of the Elamite kings, still resided atop the former city. Pastoral economies spread across Iran and into Baluchistan, where clay images of riders on horseback appeared at Pirak about 1700 BCE. Chariot corps appeared across the Near East as a new military technology. An Old Indic-speaking group of chariot warriors took control of a Hurrian-speaking kingdom in north Syria about 1500 BCE. Their oaths referred to deites (Indra, Varuna, Mithra, and the Nasatyas) and concepts (*r'ta*) that were the central deities and concepts in the *Rig Veda*, and the language they spoke was a dialect of the Old Indic Sanskrit of the *Rig Veda*.[39] The Mitanni dynasts came from the same ethnolinguistic population as the more famous Old Indic–speakers who simultaneously pushed eastward into the Punjab, where, according to many Vedic scholars, the *Rig Veda* was compiled about 1500–1300 BCE. Both groups probably originated in the hybrid cultures of the Andronovo/ Tazabagyab/ coarse-incised-ware type in Bactria and Margiana.[40]

The language of the *Rig Veda* contained many traces of its syncretic origins. The deity name *Indra* and the drug-deity name *Soma*, the two central elements of the religion of the *Rig Veda*, were non–Indo-Iranian words borrowed in the contact zone. Many of the qualities of the Indo-Iranian god of might/victory, Verethraghna, were transferred to the adopted god Indra, who became the central deity of the developing Old Indic culture.[41] Indra was the subject of 250 hymns, a quarter of the *Rig Veda*. He was associated more than any other deity with *Soma*, a stimulant drug (perhaps derived from *Ephedra*) probably borrowed from the BMAC religion. His rise to prominence was a peculiar trait of the Old Indic speakers. Indra was regarded in later Avestan Iranian texts as a minor demon. Iranian dialects probably developed in the northern steppes among Andronovo and Srubnaya people who had kept their distance from the southern civilizations. Old Indic languages and rituals developed in the contact zone of Central Asia.[42]

Loan Words Borrowed into Indo-Iranian and Vedic Sanskrit

The Old Indic of the *Rig Veda* contained at least 383 non–Indo-European words borrowed from a source belonging to a different language family. Alexander Lubotsky has shown that common Indo-Iranian, the parent of both Old Indic and Iranian, probably had already borrowed words from the *same* non–Indo-European language that later enriched Old Indic. He compiled a list of 55 non–Indo-European words that were borrowed into common Indo-Iranian *before* Old Indic or Avestan evolved, and then later were inherited into one or both of the daughters from common Indo-Iranian. The speakers of common Indo-Iranian were in touch with and borrowed terms from *the same foreign language group* that later was the source from which Old Indic speakers borrowed even more terms. This discovery carries significant implications for the geographic locations of common Indo-Iranian and formative Old Indic—they must have been able to interact with the same foreign-language group.

Among the fifty-five terms borrowed into common Indo-Iranian were the words for bread (**nagna-*), ploughshare (*sphāra*), canal (**iaviā*), brick (**išt(i)a-*, camel (**Huštra-*), ass (**khara-*) sacrificing priest (**ućig-*), soma (**anću-*), and Indra (**indra-*). The BMAC fortresses and cities are an excellent source for the vocabulary related to irrigation agriculture, bricks, camels, and donkeys; and the phonology of the religious terms is the same, so probably came from the same source. The religious loans suggest a close cultural relationship between some people who spoke common Indo-Iranian and the occupants of the BMAC fortresses. These borrowed southern cults might possibly have been one of the features that distinguished the Petrovka culture from Sintashta. Petrovka people were the first to migrate from the northern steppes to Tugai on the northern edge of Central Asia.

Lubotsky suggested that Old Indic developed as a vanguard language south of Indo-Iranian, closer to the source of the loans. The archaeological evidence supports Lubotsky's suggestion. The earliest Old Indic dialects probably developed about 1800–1600 BCE in the contact zone south of the Zeravshan among northern-derived immigrants who were integrated with and perhaps ruled over the declining fortunes of the post-BMAC citadels. They retained a decidedly pastoral set of values. In the *Rig Veda* the clouds were compared to dappled cows full of milk; milk and butter were the symbols of prosperity; milk, butter, cattle, and horses were the proper offerings to the gods; Indra was compared to a mighty bull; and wealth was counted in fat cattle and swift horses. Agricultural products were never offered to the gods. The people of the *Rig Veda* did not live in brick houses and had no cities, although their enemies, the *Dasyus*, did live in walled strongholds. Chariots were used in races and war; the gods drove chariots across the sky. Almost all important deities were masculine. The only important female deity was Dawn, and she was less powerful than Indra, Varuna, Mithra, Agni, or the Divine Twins. Funerals included both cremation (as in Federovo graves) and inhumation (as in Andronovo and Tazabagyab graves). Steppe cultures are an acceptable source for all these details of belief and practice, whereas the culture of the BMAC, with its female deity in a flounced skirt, brick fortresses, and irrigation agriculture, clearly is not.

During the initial phase of contact, the Sintashta or the Petrovka cultures or both borrowed some vocabulary and rituals from the BMAC, accounting for the fifty-five terms in common Indo-Iranian. These included the drug *soma*, which remained in Iranian ritual usage as *haoma*. In the second phase of contact, the speakers of Old Indic borrowed much more heavily from the same language when they lived in the shadows of the old BMAC settlements and began to explore southward into Afghanistan and Iran. Archaeology shows a pattern quite compatible with that suggested by the linguistic evidence.

The Steppes Become a Bridge across Eurasia

The Eurasian steppe is often regarded as a remote and austere place, poor in resources and far from the centers of the civilized world. But during the Late Bronze Age the steppes became a bridge between the civilizations that developed on the edges of the continent in Greece, the Near East, Iran, the Indian subcontinent, and China. Chariot technology, horses and horseback riding, bronze metallurgy, and a strategic location gave steppe societies an importance they never before had possessed. Nephrite from Lake Baikal appeared in the Carpathian foothills in the Borodino hoard; horses and tin from the steppes appeared in Iran; pottery from Bactria appeared in a Federovo settlement in northern Kazakhstan; and chariots appeared across the ancient world from Greece to China. The road from the steppes to China led through the eastern end of the Tarim Basin, where desert-edge cemeteries preserved the dessicated mummies of brown-haired, white-skinned, wool-wearing people dated as early as 1800 BCE. In Gansu, on the border between China and the Tarim Basin, the Qijia culture acquired horses, trumpet-shaped earrings, cast bronze ring-pommel single-edged knives and axes in steppe styles between about 2000 and 1600 BCE.[45] By the time the first Chinese state emerged, beginning about 1800 BCE, it was exchanging innovations with the West. The Srubnaya and Andronovo horizons had transformed the steppes from a series of isolated cultural ponds to a corridor of communication. That transformation permanently altered the dynamics of Eurasian history.

Chapter Seventeen

Words and Deeds

The Indo-European problem can be solved today because archaeological discoveries and advances in linguistics have eaten away at problems that remained insoluble as recently as fifteen years ago. The lifting of the Iron Curtain after 1991 made the results of steppe research more easily available to Western scholars and created new cooperative archaeological projects and radiocarbon dating programs. Linguists like Johanna Nichols, Sarah Thomason, and Terrence Kaufman came up with new ways of understanding language spread and convergence. The publication of the Khvalynsk cemetery and the Sintashta chariot burials revealed unsuspected richness in steppe prehistory. Linguistic and archaeological discoveries now converge on the probability that Proto-Indo-European was spoken in the Pontic-Caspian steppes between 4500 and 2500 BCE, and alternative possibilities are increasingly difficult to square with new evidence. Gimbutas and Mallory preceded me in arguing this case. I began this book by trying to answer questions that still bothered many reasonable observers.

One question was whether prehistoric language borders could be detected in prehistoric material culture. I suggested that they were correlated at persistent frontiers, a generally rare phenomenon that was surprisingly common among the prehistoric cultures of the Pontic-Caspian steppes. Another problem was the reluctance of Western archaeologists and the overenthusiasm of Eastern European archaeologists to use migration as an explanation for prehistoric culture change, a divergence in approach that produced Eastern interpretations that Western archaeologists would not take seriously. I introduced models from demographics, sociology, and anthropology that describe how migration works as a predictable, regular human behavior in an attempt to bring both sides to the middle. The most divisive problem was the absence of convincing evidence indicating when horse domestication and horseback riding began. Bit wear might settle the issue through the presence or absence of a clear riding-related pathology on horse teeth. A separate but related debate swirled around the question of whether pastoral nomadism was possible as early as the Yamnaya horizon, or if it depended on later horseback riding, which in this argument only began in the Iron Age; or perhaps it depended on state economies, which also appeared on the steppe border during the Iron Age. The Samara Valley Project examined the botanical and seasonal aspects of a Bronze Age steppe pastoral economy and found that it did not rely on cultivated grain even in year-round permanent settlements. Steppe pastoralism was entirely self-sustaining and independent in the Bronze Age; wild seed plants were plentiful, and wild seeds were eaten where grain was not cultivated. Pastoral nomadism did not depend for its food supply on Iron Age states. Finally, the narrative culture history of the western steppes was impenetrable to most Western linguists and archaeologists. Much of this book is devoted to my efforts to cut a path through the tangle of arguments about chronology, culture groups, origins, migrations, and influences. I have tried to reduce my areas of ignorance about steppe archaeology, but am mindful of the few years I spent doing federally funded archaeology in Massachusetts, less than half the size of the single Samara oblast on the Volga, and how we all thought it an impossible task to try to learn the archaeology of Massachusetts *and* neighboring Rhode Island—one-tenth the size of Samara oblast. Nevertheless, I have found a path that makes sense through what I have read and seen. Debate will continue on all these subjects, but I sense that a chord is emerging from the different notes.

The Horse and the Wheel

Innovations in transportation technology are among the most powerful causes of change in human social and political life. The introduction of the private automobile created suburbs, malls, and superhighways; transformed heavy industry; generated a vast market for oil; polluted the atmosphere; scattered families across the map; provided a rolling, heated space in which young people could escape and have sex; and fashioned a powerful new way to express personal status and identity. The beginning of horseback riding, the invention of the heavy wagon and cart, and the development of the spoke-wheeled chariot had cumulative effects that unfolded more slowly but eventually were equally profound. One of those effects was to transform Eurasia from a series of unconnected cultures into a single interacting system. How that happened is a principal focus of this book.

Most historians think of war when they begin to list the changes caused by horseback riding and the earliest wheeled vehicles. But horses were first domesticated by people who thought of them as food. They were a cheap source of winter meat; they could feed themselves through the steppe winter, when cattle and sheep needed to be supplied with water and fodder. After people were familiar with horses as domesticated animals, perhaps after a relatively docile male bloodline was established, someone found a particularly submissive horse and rode on it, perhaps as a joke. But riding soon found its first serious use in the management of herds of domesticated cattle, sheep, and horses. In this capacity alone it was an important improvement that enabled fewer people to manage larger herds and move them more efficiently, something that really mattered in a world where domesticated animals were the principal source of food and clothing. By 4800–4600 BCE horses were included with obviously domesticated animals in human funeral rituals at Khvalysnk on the middle Volga.

By about 4200–4000 BCE people living in the Pontic-Caspian steppes probably were beginning to ride horses to advance to and retreat from raids. Once they began to ride, there was nothing to prevent them from riding into tribal conflicts. Organic bits functioned perfectly well, Eneolithic steppe horses were big enough to ride (13–14 hands), and the leaders of steppe tribes began to carry stone maces as soon as they began to keep herds of cattle and sheep, around 5200–4800 BCE. By 4200 BCE people had become more mobile, their single graves emphasized individual status and personal glory unlike the older communal funerals, high-status graves contained stone maces shaped like horse heads and other weapons, and raiding parties migrated hundreds of kilometers to enrich themselves with Balkan copper, which they traded or gifted back to their relatives in the Dnieper-Azov steppes. The collapse of Old Europe about 4200–4000 BCE probably was at least partly their doing.

The relationship between mounted steppe pastoralists and sedentary agricultural societies has usually been seen by historians as either violent, like the Suvorovo confrontation with Old Europe, or parasitic, or both. "Barbaric" pastoral societies, hungry for grain, metals, and wealth, none of which they could produce themselves, preyed upon their "civilized" neighbors, without whom they could not survive. But these ideas are inaccurate and incomplete even for the historical period, as the Soviet ethnographer Sergei Vainshtein, the Western historian Nicola DiCosmo, and our own botanical studies have shown. Pastoralism produced plenty of food—the average nomad probably ate better than the average agricultural peasant in Medieval China or Europe. Steppe miners and craftsmen mined their own abundant ores and made their own metal tools and weapons; in fact, the enormous copper mines of Russia and Kazakhstan and the tin mines of the Zeravshan show that the Bronze Age civilizations of the Near East depended on *them*. For the prehistoric era covered in this book, any model based on relationships between the militarized nomads of the steppes and the medieval civilizations of China or Persia is anachronistic. Although the steppe societies of the Suvorovo-Novodanilovka period did seem to prey upon their neighbors in the lower Danube valley, they were clearly more integrated and apparently had peaceful relationships with their Cucuteni-Tripolye neighbors at the same time. Maikop traders seem to have visited steppe settlements on the lower Don and even perhaps brought weavers there. The institutions that regulated peaceful exchange and cross-cultural relationships were just as important as the institution of the raid.

The reconstructed Proto-Indo-European vocabulary and comparative Indo-European mythology reveal what two of those important integrative institutions were: the oath-bound relationship between patrons and clients, which regulated the reciprocal obligations between the strong and the weak, between gods and humans; and the guest-host relationship, which extended these and other protections to people outside the ordinary social circle. The first institution, legalizing inequality, probably was very old, going back to the initial acceptance of the herding economy, about 5200–5000 BCE, and the first appearance of pronounced differences in wealth. The second might have developed to regulate migrations into unregulated geographic and social space at the beginning of the Yamnaya horizon.

When wheeled vehicles were introduced into the steppes, probably about 3300 BCE, they again found their first use in the herding economy. Early wagons and carts were slow, solid-wheeled vehicles probably pulled by oxen and covered by arched roofs made of reed mats plaited together, perhaps originally attached to a felt backing. Yamnaya-era graves often contain remnants of reed mats with other decayed organic material. On some occasions the mats were painted in red, black, and white stripes and curved designs, certainly at funerals. Wagons permitted herders to migrate with their herds into the deep steppes between the river valleys for weeks or months at a time, relying on the tents, food, and water carried in their wagons. Even if the normal annual range of movement was less than 50 km, which seems likely for Yamnaya herders, the combination of bulk wagon transport with rapid horseback transport revolutionized steppe economies, opening the majority of the Eurasian steppe zone to efficient exploitation. The steppes, largely wild and unused before, were domesticated. The Yamnaya horizon exploded across the Pontic-Caspian steppes about 3300 BCE. With it probably went Proto-Indo-European, its dialects scattering as its speakers moved apart, their migrations sowing the seeds of Germanic, Baltic, Slavic, Italic, Celtic, Armenian, and Phrygian.

The chariot, the first wheeled vehicle designed entirely for speed, first appeared in the graves of the Sintashta culture, in the southern Ural steppes, about 2100 BCE. It was meant to intimidate. A chariot was incredibly difficult to build, a marvel of carpentry and bent-wood joinery. It required a specially trained team of fast, strong horses. To drive it through a turn, you had to rein each horse independently while keeping a backless, bouncing car level by leaning your weight into each bounce. It was even more difficult to throw a javelin accurately at a target while driving a speeding chariot, but the evidence from the Sintashta chariot graves suggests that this is precisely what they did. Only men with a lot of time and resources, as well as balance and courage, could learn to fight from a chariot. When a squadron of javelin-hurling chariot warriors wheeled onto the field of battle, supported by clients and supporters on foot and horseback with axes, spears, and daggers, it was a new, lethal style of fighting that had never been seen before, something that even urban kings soon learned to admire.

This heroic world of chariot-driving warriors was dimly remembered in the poetry of the *Iliad* and the *Rig Veda*. It was introduced to the civilizations of Central Asia and Iran about 2100 BCE, when exotic Sintashta or Petrovka strangers first appeared on the banks of the Zeravshan, probably bouncing along on the backs of the new kinds of equids from the north. At first, this odd way of moving around probably was amusing to the local people of Sarazm and Zaman Baba. Very soon, however, both places were abandoned. Between 2000 and 1800 BCE first Petrovka and then Alakul-Andronovo groups settled in the Zeravshan valley and began mining copper and tin. Horses and chariots appeared across the Near East, and the warfare of cities became dependent, for the first time, on well-trained horses. The Old Indic religion probably emerged among northern-derived immigrants in the contact zone between the Zeravshan and Iran as a syncretic mixture of old Central Asian and new Indo-European elements. From this time forward the people of the Eurasian steppes remained directly connected with the civilizations of Central Asia, South Asia, and Iran, and, through intermediaries, with China. The arid lands that occupied the center of the Eurasian continent began to play a role in transcontinental economies and politics.

Jared Diamond, in *Guns, Germs, and Steel*, suggested that the cultures of Eurasia enjoyed an environmental advantage over those of Africa or the Americas partly because the Eurasian continent is oriented in an east-west direction, making it easier for innovations like farming, herding, and wheeled vehicles to spread rapidly between environments that were basically similar because they were on about the same latitude.[1] But persistent cultural borders like the Ural frontier delayed the transmission of those innovations by thousands of years even within the single ecological zone of the steppes. A herding economy was accepted on the middle Ural River, near the headwaters of the Samara River, by 4800 BCE. Hunters and gatherers in the neighboring steppes of northern Kazakhstan, at the same latitude, refused domesticated cattle and sheep for the next two thousand years (although they did begin to ride horses by 3700–3500 BCE). The potential geographic advantage Diamond described was frustrated for millennia, not a short time, by human distrust of foreign ways of doing things and admiration for the familiar ways. This tendency was hyper-developed when two very different cultures were brought into contact through long-distance migrations or at an ecological border. In the case of the Ural frontier, the Khvalynian Sea separated the populations east and west of the Ural Mountains for millennia, and the saline desert-steppe that replaced it (chapter 8) probably remained a significant ecological barrier for pedestrian foragers. Places like the Ural River frontier became borders where deep-rooted, intransigent traditions of opposition persisted.

These long-lasting, robust kinds of frontiers seem to have been rare in the prehistoric world of tribal politics. We have grown accustomed to them now only because the modern nation-state has made it the standard kind of border everywhere around the world, encouraging patriotism, jingoism, and the suspicion of other nations across sharply defined boundaries. In the tribal past, the long-term survival of sharp, bundled oppositions was unusual. The Pontic-Caspian steppes, however, witnessed an unusual number of persistent tribal frontiers because sharp environmental ecotones ran across it and it had a complex history of long-distance migrations, two important factors in the creation and maintenance of such frontiers.

ARCHAEOLOGY AND LANGUAGE

Indo-European languages replaced non–Indo-European languages in a multi-staged, uneven process that continues today, with the worldwide spread of English. No single factor explains every event in that complicated and drawn-out history—not race, demographics, population pressure, or imagined spiritual qualities. The three most important steps in the spread of Indo-European languages in the last two thousand years were the rise of the Latin-speaking Roman Empire (an event almost prevented by Hannibal); the expansion of Spanish, English, Russian, and French colonial powers in Asia, America, and Africa; and the recent triumph of the English-speaking Western capitalist trade system, in which American-business English has piggybacked onto British-colonial English. No historian would suggest that these events shared a single root cause. If we can draw any lessons about language expansion from them, it is perhaps only that an initial expansion can make later expansions easier (the *lingua franca* effect), and that language generally follows military and economic power (the *elite dominance* effect, so named by Renfrew). The earliest Indo-European expansions described in this book laid a foundation of sorts for later expansions by increasing the territorial extent of the Indo-European languages, but their continued spread never was inevitable, and each expansion had its own local causes and effects. These local events are much more important and meaningful than any imagined spiritual cause.

It is not likely that the initial spread of the Proto-Indo-European dialects into regions outside the Pontic-Caspian steppes was caused primarily by an organized invasion or a series of military conquests. As I suggested in chapter 14, the initial spread of Proto-Indo-European dialects probably was more like a franchising operation than an invasion. At least a few steppe chiefs must have moved into each new region, and their initial arrival might well have been accompanied by cattle raiding and violence. But equally important to their ultimate success were the advantages they enjoyed in institutions (patron-client systems and guest-host agreements that incorporated outsiders as individuals with rights and protections) and perhaps in the public performances associated with Indo-European rituals. Their social system was maintained by myths, rituals, and institutions that were adopted by others, along with the poetic language that conveyed their prayers to the gods and ancestors. Long after the genetic imprint of the original immigrant chiefs faded away, the system of alliances, obligations, myths, and rituals that they introduced was still being passed on from generation to generation. Ultimately the last remnant of this inheritance is the expanding echo of a once-shared language that survives as the Indo-European language family.

Understanding the people who lived before us is difficult, particularly the people who lived in the prehistoric tribal past. Archaeology throws a bright light on some aspects of their lives but leaves much in the dark. Historical linguistics can illuminate a few of those dark corners. But the combination of prehistoric archaeology with historical linguistics has a bad history. The opportunities for imaginative fantasies of many kinds, both innocent and malevolent, seem dangerously increased when these two very different kinds of evidence are mixed. There is no way to stop that from happening—as Eric Hobsbawm once remarked, historians are doomed to provide the raw material for bigotry and nationalism.[2] But he did not let that stop him from doing history.

For Indo-European archaeology, the errors of the past cannot be repeated as easily today. When the nineteenth-century fantasy of the Aryans began there were no material remains, no archaeological findings, to constrain the imagination. The Aryans of Madison Grant were concocted from sparse linguistic evidence (and even that was twisted to his purpose), a large dose of racism, a cover of ideals derived from the Classical literature of Greece and Rome, and the grim zero-sum politics of social Darwinism. Archaeology really played no role. The scattered archaeological discoveries of the first half of the twentieth century could still be forced into this previously established imaginary mold. But that is not so easy today. A convincing narrative about the speakers of Proto-Indo-European must today be pegged to a vast array of archaeological facts, and it must remain un-contradicted by the facts that stand outside the chosen narrative path. I have used a lot of archaeological detail in this account, because the more places a narrative is pegged to the facts, and the more different kinds of facts from different sources are employed as pegs, the less likely it is that the narrative is false. As both the density of the archaeological facts and the quality of the linguistic evidence improve, advances in each field should act as independent checks on the worst abuses. Although I have used linguistic reconstructions for which there is little direct archaeological evidence (importantly patron-client and guest-host relationships), at least both would be compatible with the kinds of societies indicated by the archaeological evidence.

On the positive side, the combination of archaeological evidence and the reconstructed Proto-Indo-European vocabulary can reveal entirely new kinds of information about the prehistoric past. That promise keeps pushing the project forward both for linguists and archaeologists. At many critical points the interpretations presented here have been guided by institutions, rituals, and words that I found in reconstructed Indo-European and applied to archaeological settings. But I have barely scratched the surface of what might be accomplished by pulling material out of Proto-Indo-European and using it as a lens through which to examine archaeological evidence. Reciprocally, archaeological data add real-life complexities and contradictions to the idealized Indo-European social world of the linguists. We might not be able to retrieve the names or the personal accomplishments of the Yamnaya chiefs who migrated into the Danube valley around 3000 BCE, but, with the help of reconstructed

Proto-Indo-European language and mythology, we can say something about their values, religious beliefs, initiation rituals, kinship systems, and the political ideals they admired. Similarly, when we try to understand the personal, human motivation for the enormous animal sacrifices that accompanied the funerals of Sintashta chiefs around 2000 BCE, reading the *Rig Veda* gives us a new way of understanding the value attached to public generosity (RV 10.117):

> That man is no friend who does not give of his own nourishment to his friend, the companion at his side. Let the friend turn away from him; this is not his dwelling-place. Let him find another man who gives freely, even if he be a stranger. Let the stronger man give to the man whose need is greater; let him gaze upon the lengthening path. For riches roll like the wheels of a chariot, turning from one to another.[3]

Archaeologists are conscious of many historical ironies: wooden structures are preserved by burning, garbage pits survive longer than temples and palaces, and the decay of metals leads to the preservation of textiles buried with them. But there is another irony rarely appreciated: that in the invisible and fleeting sounds of our speech we preserve for a future generation of linguists many details of our present world.

Appendix

Author's Note on Radiocarbon Dates

All dates in this book are given as BCE (Before the Common Era) and CE (Common Era), the international equivalent of BC and AD.

All BCE dates in this book are based on calibrated radiocarbon dates. Radiocarbon dates measure the time that has passed since an organic substance (commonly wood or bone) died, by counting the amount of ^{14}C that remains in it. Early radiocarbon scientists thought that the concentration of ^{14}C in the atmosphere, and therefore in all living things, was a constant, and they also knew that the decay rate was a constant; these two factors established the basis for determining how long the ^{14}C in a dead organic substance had been decaying. But later investigations showed that the concentration of ^{14}C in the atmosphere varied, probably with sunspot activity. Organisms that lived at different times had different amounts of ^{14}C in their tissues, so the baseline for counting the amount of ^{14}C in the tissues moved up and down with time. This up-and-down variation in ^{14}C concentrations has been measured in tree rings of known age taken from oaks and bristlecone pines in Europe and North America. The tree-ring sequence is used to calibrate radiocarbon dates or, more precisely, to convert raw radiocarbon dates into real dates by correcting for the initial variation in ^{14}C concentrations as measured in a continuous sequence of annual tree rings. Uncalibrated radiocarbon dates are given here with the designation BP (before present); calibrated dates are given as BCE. Calibrated dates are "real" dates, measured in "real" years. The program used to convert BP to BCE dates is OxCal, which is accessible free for anyone at the website of the Oxford Radiocarbon Accelerator Unit.

Another kind of calibration seems to be necessary for radiocarbon dates taken on human bones, *if the humans ate a lot of fish*. It has long been recognized that in salt-water seas, organic substances like shell or fish bones absorb old carbon that is in solution in the water, which makes radiocarbon dates on shell and fish come out too old. This is called the "reservoir effect" because seas act as a reservoir of old carbon. Recent studies have indicated that the same problem can affect organisms that lived in fresh water, and most important among these were fish. Fish absorb old carbon in solution in fresh water, and people who eat a lot of fish will digest that old carbon

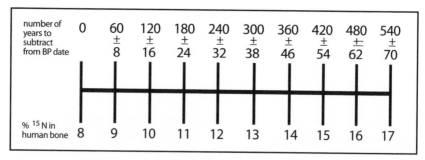

Figure A1. A proposed linear correlation between the % of ^{15}N in dared human bone (*bottom*) and the number of radiocarbon years that should be subtracked from radiocarbon dares (*top*) before they are calibrated.

and use it to build their bones. Radiocarbon dates on their bones will come out too old. Dates measured on charcoal or the bones of horses and sheep are not affected, because wood and grazing animals do not absorb carbon directly from water like fish do, and they do not eat fish. Dates on human bone can come out centuries older than dates measured on animal bone or charcoal *taken from the same grave* (this is how the problem was recognized) if the human ate a lot of fish. The size of the error depends on how much fish the human ate and how much old carbon was in solution in the groundwater where he or she went fishing. Old carbon content in groundwater seems to vary from region to region, although the amount of regional variation is not at all well understood at this time. The amount of fish in the diet can be estimated on the basis of ^{15}N levels in bone. Fish have much higher percentages of ^{15}N in their tissues than does any other animal, so humans with high ^{15}N in their bones probably ate a lot of fish. High ^{15}N in human bones is a signal that radiocarbon dates from those bones probably will yield ages that are too old.

Research to correct for reservoir effects in the steppes is just beginning as I write this, so I cannot solve the problem. But many of the radiocarbon dates from steppe archaeology are from cemeteries, and the dated material often is human bone. Widespread tests of the ^{15}N in human bone from many different steppe cemeteries, from Kazakhstan to Ukraine, indicate that fish was a very important part of most ancient steppe diets, often accounting for 50% of the meat consumed. Because I did not want to introduce dates that were probably wrong, I used an approach discussed by Bonsall, Cook, and others, and described by them as *preliminary* and *speculative*. They studied five graves in the lower Danube valley where human bone and animal bone in the same grave yielded different ages (see chapter 7 for references). Data from these graves suggested a correction method. The average level of ^{15}N in the human skeletons (15.1%) was equated with an average radiocarbon error (425±55) that should be *subtracted* prior to calibrating those dates. These averages could be placed on a scale between the known minimum and maximum levels of ^{15}N found in human bone, and, speculatively, a given level of ^{15}N could be equated with an average error in radiocarbon years. The scale shown in figure A.1 was constructed in this way. It seems to yield results that solve some long-problematic dating offsets in steppe chronology (see ch. 9, notes 4, 16, and 22; and ch. 12, note 30). When I use it—when dates are based principally on human bone—I warn readers in the text. Whatever errors it introduces probably are smaller than those caused by ignoring the problem. All the radiocarbon dates listed in the tables in this book are regular BP and calibrated BCE dates, without any correction for the reservoir effect.

Figure A.1 shows the correction scale I used to revise dates that were measured from human bone in regions where I knew the average ^{15}N levels in human bone. The top number is the number of years that should be

TABLE A.1

The average ^{13}C and ^{15}N% in human bone from seventy-two individuals excavated from graves in the Samara oblast, by time period.

Time Period	Sample Size	C13	N15	Years to Subtract
MESOLITHIC	5	−20.6	13.5	−330±42
NEOLITHIC	8	−22.3	11.8	−228±30
EARLY ENEOL	6	−20.9	14.8	−408±52
LATE ENEOL	6	−21.0	13.1	−306±39
EBA	11	−18.7	11.7	−222±30
MBA	11	−19.0	12.0	−240±32
POTAPOVKA	9	−19.1	11.3	−198±26
EARLY LBA	7	−19.1	11.4	−204±27
LATE LBA	9	−18.9	11.2	−192±26

subtracted from the BP radiocarbon date; the bottom number is the ^{15}N level associated with specific subtraction numbers.

Table A.1, based on our own studies in the Samara oblast, shows the average ^{15}N content in human bone for different periods, taken from measurements on seventy-two individuals.

NOTES

Chapter 1. The Promise and Politics of the Mother Tongue

1. Bloch 1998:109.
2. See Sapir 1912:228.
3. Cannon 1995:28–29.
4. Poliakov 1974:188–214.
5. Veit 1989:38.
6. Grant 1916.
7. For "external origin" passages in the *Rig Veda*, see Witzel 1995. For "indigenous origin" arguments, see N. Kazanas's discussions in the *Journal of Indo-European Studies* 30, nos. 3–4 (2002); and 31, nos. 1–2 (2003).
8. For the Nazi pursuit of Aryan archaeology, see Arnold 1990.
9. For goddesses and Indo-Europeans, see Anthony 1995b; Eisler 1987, 1990; and Gimbutas 1989, 1991. For Aryan-identity politics in Russia, see Shnirelman 1998, 1999.
10. Heidegger 1959:37–51, contrasted to Boaz 1911. For the non-Aryan element in the *Rig Veda*, see Kuiper 1948, 1991.
11. Harding and Sokal 1988.
12. The *American Heritage Dictionary* has thirteen hundred unique Proto-Indo-European roots listed in its appendix. But multiple reconstructed words are derived from the same root morphemes. The number of reconstructed words with distinct meanings is much greater than the number of unique roots.
13. For doubts about proto-languages and tree diagrams, see Lincoln 1991; and Hall 1997. For a more nuanced view of tree diagrams, see Stewart 1976. For "creolization" and convergence creating Proto-Indo-European, see Renfrew 1987:78–86; Robb 1991; and Sherratt and Sherratt 1988.
14. For framing, see Lakoff 1987:328–37.

Chapter 2. How to Reconstruct a Dead Language

1. Here is the text of the tale:
A sheep, shorn of its wool, saw some horses, one moving a heavy cart, another carrying a big load, a third carrying a human speedily. The sheep said to the horses: "It pains me [literally, "the heart narrows itself for me"] to see human driving horses." The horses said: "Listen sheep, it pains us to see that human, the master, makes the wool of the sheep into a warm garment for himself and the sheep no longer has any wool!" On hearing that the sheep ran off into the fields.

It is impossible to construct whole sentences like this with confidence in a language known only in fragments. Proto-Indo-European tense markers in the verbs are debated, the form of the relative pronoun is uncertain, and the exact construction of a Proto-Indo-European complement (sheep saw horse carrying load) is unknown. Linguists still see it as a classic challenge. See Bynon 1977:73–74; and Mallory 1989:16–17.

2. This chapter is generally based on four basic textbooks (Bynon 1977; Beekes 1995; Hock and Joseph 1996; and Fortson 2004), and on various encyclopedia entries in Mallory and Adams 1997.

3. Embleton 1991.

4. Pinker 1994.

5. An example of a change in phonology, or pronunciation, that caused shifts in morphology, or grammar, can be seen in English. German has a complex system of noun and pronoun case endings to identify subjects, objects, and other agents, and verb endings that English lacks. English has lost these features because a particular dialect of Middle English, Old Northumbrian, lost them, and people who spoke the Old Northumbrian dialect, probably rich wool merchants, had a powerful effect on the speech of Medieval London, which happened to give us Modern English. The speakers of Old Northumbrian dropped the Germanic word-final *n* and *m* in most suffixes (*esse'*, not *essen*, for "to eat"). In late Old English the pronunciation of many short vowels (like the final *-e* that resulted here) was already merging into one vowel (the [uh] in sof*a*, called *schwa* by linguists). These two shifts in pronunciation meant that many nouns no longer had distinctive endings, and neither the infinitive nor the subjunctive plural verb had a distinct ending. Later, between 1250 and 1300, the word-final *schwa* began to be dropped from most English speech, which wiped out the distinction between two more grammatical categories. Word order became fixed, as few other guides indicated the difference between subject and object, and auxiliary particles like *to*, *of*, or *by* were employed to distinguish infinitives and other forms. Three shifts in pronunciation were responsible for much of the grammatical simplification of modern English. See Thomason and Kaufman 1988:265–275.

6. For Grimm's Law, see Fortson 2004:300–304.

7. Some linguists argue that the Proto-Indo-European root did not begin with *k* but rather with a palato-velar, a *kh*-type sound, which would require that the first consonant was moved back in the *centum* languages rather than forward in the *satem* languages. See Melchert 1994:251–252. Thanks to Bill Darden for pointing this out.

8. Hock and Joseph 1996:38.

9. For pessimistic views on the "reality" of reconstructed Proto-Indo-European, see Bynon 1977; and Zimmer 1990. For optimistic views, see Hock and Joseph 1996:532–534; and Fortson 2004:12–14.

10. Hall 1950, 1976.

11. Bynon 1977:72. Mycenaean was in a transitional state in 1350 BC, when it was recorded. Some Proto-Indo-European words with k^w had already shifted to k in Mycenaean. The alternation between $*k^w$ and $*p$ probably was already present in some dialects of Proto-Indo-European.

12. For doubts on reconstructed meanings, see Renfrew 1987:80, 82, 260. For the argument that comparing cognates requires that the meanings of the compared terms are subjected to fairly strict limits, see Nichols 1997b.

Chapter 3. Language and Time 1

1. See Swadesh 1952, 1955; and Lees 1953.

2. The replacement rate cited here compares the core vocabulary in Modern English to the core vocabulary in Old English, or Anglo-Saxon. Much of the Old English core vocabulary was replaced by Norse, but, since Norse was another Germanic language, most of the core vocabulary remains Germanic. That is why we can say that 96% of the core vocabulary remains Germanic, and at the same time say that the replacement rate in the core vocabulary was a high 26%.

3. Much of the information in this section came from Embleton 1991, 1986. See also McMahon and McMahon 2003; and Dyen, Kruskal, and Black 1992. Many linguists are hostile to any claim that a cross-cultural core vocabulary can be identified. The Australian aboriginal languages, for example, do not seem to have a core vocabulary—all vocabulary items are equally vulnerable to replacement. We do not understand why. Both sides of the debate are represented in Renfrew, McMahon, and Trask 2000.

4. Meid 1975; Winfred 1989; and Gamkrelidze and Ivanov 1984:267–319.

5. Ivanov derived Hittite (Northern Anatolian) and Luwian (Southern Anatolian) separately and directly from Proto-Indo-European, without an intervening proto-language, making them as different as Celtic and Greek. Most other linguists derive all the Anatolian languages from a common source, Proto-Anatolian; see Melchert 2001 and Diakonoff 1985. Lydian, spoken on the western coast of Anatolia in the Classical era, might have descended from the same dialect group as Hittite. Lycian, spoken on the southwestern coast, probably descended from the same dialects as Luwian. Both became extinct in the Classical era. For all these topics, see Drews 2001.

6. For the Anatolian languages, see Fortson 2004:154–179; Houwink Ten Cate 1995; Veenhof 1995; and Puhvel 1991, 1994. For the glottalic perspective, see Gamkrelidze and Ivanov 1995.

7. Wiluša was a city west of the Hittite realm. It is very possible that Wilusa was Troy and that the Trojans spoke Luwian. See Watkins 1995:145–150; and Latacz 2004.

8. The non–Indo-European substrate effect on Luwian was described by Jaan Puhvel (1994:261–262) as "agglutinative creolization . . . What has happened to Anatolian here is reminiscent of what became of French in places like Haiti." Hittite showed similar non–Indo-European substrate effects and had few speakers, causing Zimmer (1990:325) to note that, "on the whole, the Indo-Europeanization of Anatolia failed."

9. Melchert 2001.

10. Forster 2004; Baldi 1983:156–159.

11. Lehrman 2001. The ten innovations that Lehrman identified as distinctive of Proto-Indo-European included two phonological traits (e.g., loss of the laryngeals), three morphological traits in nouns (e.g., addition of the feminine gender), and five morphological traits in verbs.

12. See Sturtevant 1962 for the Indo-Hittite hypothesis. For Anatolian as a daughter of very early Proto-Indo-European, see Puhvel 1991. Lehrman (2001) pointed out that Anatolian had a different word from Proto-Indo-European for *man*, usually considered part of the core vocabulary. The Anatolian term (*pāsna-*) used a root that also meant "penis," and the Proto-Indo-European term (*wiro-*) used a root that also meant "strength." Proto-Anatolian and Proto-Indo-European did, however, share cognate terms for *grandfather* and *daughter*, so their kinship vocabularies overlapped. Classic Proto-Indo-European and Anatolian probably emerged from different places and different times in the Pre-Proto-Indo-European dialect chain.

13. For Pre-Greek language(s) of Greece, see Hainsworth 1972; and Francis 1992.

14. For the oldest language in the Indic branch I use the term *Old Indic* instead of *Indo-Aryan*. The standard nomenclature today is *Indo-Iranian* for the parent, *Avestan Iranian* for the oldest Iranian daughter, and *Indo-Aryan* for the oldest Indic daughter. But the designation *Aryan* for Indic is unnecessary; they were all Aryan. For the language and history of the *Rig-Veda*, see Erdosy 1995.

15. For Old Indic terms among the Mitanni, see Thieme 1960; Burrow 1973; and Wilhelm 1995. I thank Michael Witzel for his comments on Mitanni names. Any errors are my own.

16. For a date for Zarathustra before 1000 BCE, see Boyce 1975; and Skjærvø 1995. For the "traditional" date promulgated by ancient Greek sources, five hundred years later, see Malandra 1983.

17. Clackson (1994) and Hamp (1998) argued that Pre-Armenian was linked to the Greek-Indo-Iranian block. See also the isogloss map in Antilla 1972, figure 15.2. Many of the shared lexical items are discussed and described in Mallory and Adams 1997. I am grateful to Richard Diebold for his analysis of Greek/Indo-Iranian relations in a long letter of October 1994, where he pointed out that the shared innovations link Greek and Iranian closely, and Greek and Indic somewhat less.

18. See Rijksbaron 1988 and Drinka 1995 for the shared poetic functions of the imperfect. Poetics, shared phrases, and weapon terms are reviewed in Watkins 1995, chap. 2, 435–436.

19. See Ringe et al. 1998; and also Ringe, Warnow, and Taylor 2002. Similar cladistic methods were applied to a purely lexical data set in Rexová, Frynta, and Zrzavý 2003.

CHAPTER 4. LANGUAGE AND TIME 2

1. See Darden 2001, esp. 201–204, for the etymology of the term *wool*. For the actual textiles, see Barber 2001, 1991; and Good 1998.

2. The "unspinnable" quotation is from Barber 2001:2. The mitochondrial DNA in modern domesticated sheep indicates that all are descended from two ancient episodes of domestication. One cluster (B), including all European and Near Eastern sheep, is descended from the wild *Ovis orientalis* of eastern Anatolia or western Iran. The other cluster (A) is descended from another *Ovis orientalis* population, probably in north-central Iran. Other wild Old World ovicaprids, *Ovis ammon* and *Ovis vignei,* did not contribute to the genes of domesticated sheep. See Hiendleder et al. 2002. For a general discussion of sheep domestication, see Davis 1987; and Harris 1996.

3. In the Ianna temple of Uruk IV (3400–3100 BCE) artists depicted women making textiles. The later Sumerian names for some months incorporated the term for plucking sheep. The zoological evidence suggests that the months were named this way during the Late Uruk period or afterward, not before.

4. Zoological evidence for wool production in the Near East is reviewed by Pollack (1999:140–147). For Arslantepe, see Bökönyi 1983. An earlier date for wool sheep could be indicated by a couple of isolated pieces of evidence. The phase A occupation at Hacinebi on the Euphrates, dated 4100–3800 BCE, had spindle-whorls that seemed the right weight for spinning wool, which requires a light spindle; see Keith 1998. A clay sheep figurine from Tepe Sarab in western Iran (Kermanshah) seems to show a wooly fleece, from a level dated about 5000 BCE. For a broader discussion, see Good 2001.

5. For the caprids (sheep and/or goats) at Khvalynsk, see Petrenko 1984. Petrenko did not report the age at death for all the caprids in the Khvalynsk graves, but six of the twelve with reported ages were adults. Sacrificial deposit #11 contained 139 bones of caprids representing four adults and five sub-adults, and the *average* adult withers height was 78 cm, almost 15 cm taller than other European Neolithic caprids. For Svobodnoe sheep, see Nekhaev 1992:81. For sheep in Hungary, see Bökönyi 1979:101–116. For sheep in Poland, see Milisauskas 2002:202.

6. For wool at Novosvobodnaya, see Shishlina, Orfinskaya, and Golikov 2003. For evidence of Catacomb-period wool (dated ca. 2800–2200 BCE) in the North Caucasian steppes, see Shishlina 1999. Sherratt's updated comments on wool are included in the revised text of an older article in Sherratt 1997a.

7. The term for hub or nave, which is often included in other lists, also meant "navel" in Proto-Indo-European, so its exact meaning is unclear. For the wheel-wagon vocabulary, see Specht 1944. Three influential updates were Gamkrelidze and Ivanov 1984:718–738; Meid 1994; and Häusler 1994. I first published on the topic in Anthony and Wailes 1988; and also in Anthony 1991a, 1995a. As with most of the topics covered in this book, there is an excellent review of the Indo-European wheel vocabulary in Mallory and Adams 1997.

8. Don Ringe communicated the argument against *hurki-* to me in a letter in 1997. Bill Darden discussed the Anatolian terms in Darden 2001.

9. I am indebted to Mary Littauer for alerting me to draft experiments carried out in 1838–40 with wagons and carts on different road surfaces, where it was determined that the draft of a wagon was 1.6 times greater than that of a cart of the same weight. See Ryder 1987.

10. For the earliest wheeled vehicles, see Bakker et al. 1999; and Piggott 1983. For European wheels, see Häusler 1992; and Hayen 1989. For Mesopotamia, see Littauer and Crouwel 1979; and Oates 2001. The most comprehensive anlysis of the steppe vehicle burials, still unpublished, is by Izbitser 1993, a thesis for the Institute of the History of Material Culture in

St. Petersburg. Izbitser is working on an English-language update from her post in the New York Metropolitan Museum. Other key steppe accounts are in Mel'nik and Serdiukova 1988, and the section on wagons in Gei 2000:175–192.

11. Sherratt's essays were compiled and amended in Sherratt 1997. He continued to suggest that horseback riding in the steppes was inspired by Near Eastern donkey riding; see 1997:217. An early critical response to the SPR is Chapman 1983.

12. For Neolithic sleds in Russia, see Burov 1997. Most of them were joined with mortice-and-tenon joints, and equipped with bent-wood curved runners. These are the same carpentry skills needed to make wheels and wooden-slat tires.

13. The version of the Renfrew hypothesis I use here was published as Renfrew 2001. For assenting views among archaeologists, see Zvelebil and Zvelebil 1988; Zvelebil 1995; and Robb 1991, 1993. Robert Drews (2001) began in a different place but ended up supporting Renfrew.

14. For the north Syrian origin of the Anatolian Neolithic population, see Bar-Yosef 2002; for the likely Afro-Asiatic linguistic affiliation of these first farmers, see Militarev 2002.

15. See Gray and Atkinson 2003, reviewed by Balter 2003. The linguist L. Trask criticized Gray and Atkinson's methods, and Gray responded on his homepage, updated March 2004, at http:// www.psych.auckland.ac.nz/psych/research/Evolution/GrayRes.htm.

16. Buck 1949:664, with Indo-European terms for *turn*, *turn around*, *wind*, and *roll*. Gray's argument for a natural independent development of the term *wheel* from *to turn* (wheel = the turner) is further complicated by the fact that there are two reconstructed Proto-Indo-European terms for *wheel*, and the other one was based on the Proto-Indo-European verb **reth-* 'run' (wheel = the runner), a different semantic development.

17. Renfrew 2001:40–45; 2000. Renfrew's hypothesis of a very long-lived Proto-Indo-European phase, surviving for many millennia, is supported by some linguists. For a view that Proto-Indo-European was spoken from the Mesolithic through the end of the Corded Ware period, or about 6000–2200 BCE, see Kitson 1997, esp. 198–202.

18. Childe 1957:394.

19. Mallory 1989:145–146; and Anthony 1991a. For Africa, see Nettles 1996.

CHAPTER 5. LANGUAGE AND PLACE

1. For homeland theories, see Mallory 1989, chap. 6. For political uses of the past in the Soviet Union, see Shnirelman 1995, 1999; Chernykh 1995; and Kohl and Tsetskhladze 1995. For the belief in an Aryan-European "race," see Kühl 1994; and Poliakov 1974.

2. The Pontic-Caspian steppe homeland hypothesis was defended in English most clearly by Gimbutas 1970, 1977, 1991; and Mallory 1989, updated in Mallory and Mair 2000. Although I agree with Gimbutas's homeland solution, I disagree with her chronology, her suggested causes for the expansion, and her concept of Kurgan-culture migrations, as I explained in detail in Anthony 1986.

3. See Dixon 1997:43–45. Similarly for Zimmer 1990:312–313, "reconstructions are pure abstracts incapable of being located or dated . . . no philological interpretation of the reconstructed items is possible."

4. The tree model does not exclude or deny some areal convergence. All languages contain elements based on both branching structures and convergence with neighbors. On areal borrowing, see Nichols 1992.

5. See Thomason and Kaufman 1992; Nichols 1992; and Dixon 1997. All support the derivation of the Indo-European languages from Proto-Indo-European. Dixon (1997:31), although a critic of the criteria used to create some family tree models, stated: "The genetic relatedness of the Indo-European languages, in a family tree model, has of course been eminently proved." A good brief review of various approaches to convergence can be found in Hock and Joseph 1996:388–445.

6. Gradual convergence between neighboring languages can result in several different kinds of similarities, depending on the social circumstances. The range of possibilities includes *trade jargons*, crude combinations of words from neighboring languages barely sufficient to communicate for purposes of trade or barter; *pidgins*, which evolve from trade jargons or from a multitude of partially known languages in a colonial encounter where a colonial target language supplies much of the content of the pidgin; and *creoles*, which can evolve from pidgins or can arise abruptly in multiethnic forced labor communities where again a colonial target language supplies much of the content. Unlike pidgins, creoles contain the essential grammatical structures of a natural language, but in a reduced and simple form. They can, of course, be as expressive in song, poetry, and metaphor as any natural language, so the fact that they are grammatically simple is not a value statement. All these ways of speaking pass through a bottleneck of great grammatical simplification. Indo-European grammar is not at all like a creole grammar. See Bickerton 1988; and Thomason and Kaufman 1988.

7. Pulgram, in 1959, suggested that the comparative method, applied to the modern Romance words for *coffee*, would produce a false Latin root for *coffee* in Classical Latin. But Pulgram's claim was rebutted by Hall (1960, 1976). Pulgram's argument was cited in Renfrew (1987:84–86) but corrected in Diakonov (1988: n. 2).

8. For Pre–Indo-European substrate terms in Balto-Slavic, see Andersen 2003. For Greek and pre-Greek place-names, see Hester 1957; Hainsworth 1972; and Renfrew 1998. In northern Europe, at least three different extinct non–Indo-European languages have been identified: (1) the "language of Old European hydronomy," preserved principally in non–Indo-European river names; (2) the "language of bird names," preserved in the names of several kinds of birds, including the blackbird, lark, and heron, and also in other terms borrowed into early Germanic, Celtic, and Latin, including the terms for *ore* and *lightning*; and (3) the "language of geminates," which survives only in a few odd sounds quite atypical for Indo-European, borrowed principally into Germanic but also into a few Celtic words, including doubled final consonants and the word-initial [kn-], as in *knob*. See Schrijver 2001; Venneman 1994; Huld 1990; Polomé 1990; and Krahe 1954.

≠ 9. For *beech* and *salmon* as terms that limited Proto-Indo-European to northern Europe, see Thieme 1958. Friedrich 1970 showed that the *beech* root referred variously to beech, oak, and elder trees in several branches, and that in any case the common beech grew in the Caucasus Mountains, making it useless as a diagnostic northern European tree word. Diebold 1985 summarized the evidence against salmon as a limiting geographic term. For the honeybee argument, see the excellent study by Carpelan and Parpola 2001. See also the articles on salmon and beech in Mallory and Adams 1997.

10. This interpretation of Proto-Indo-European **peku* is that of Benveniste 1973:40–51.

11. This reconstruction of Proto-Indo-European society is based on Benveniste 1973, numerous entries in Mallory and Adams 1997, and Gamkrelidze and Ivanov 1995.

12. For Proto-Uralic linkages with Proto-Indo-European, see Carpelan, Parpola, and Koskikallio 2001, particularly the articles by Koivulehto and Kallio. See also Janhunen 2000; Sinor 1988; and Ringe 1997.

13. For a Yeniseian homeland, see Napol'skikh 1997.

14. Koivulehto 2001.

15. Janhunen (2000) has somewhat different forms for some of the pronouns. Nichols pointed out in a note to me that the *-m* and *-n* shared inflections are not very telling; only a whole paradigm of shared inflections is diagnostic. Also, nasal consonants occur in high frequencies and apparently are prone to occur in grammatical endings, and so it is the pronouns that are really important here.

16. Nichols 1997a.

17. For the glotallic theory, see Gamkrelidze and Ivanov 1973; see also Hopper 1973. For their current views, see Gamkrelidze and Ivanov 1995.

18. For discussions of the glottalic theory, see Diakonov 1985; Salmons 1993; and Szemerényi 1989.

19. For critical discussions of the Semitic-Proto-Indo-European and Kartvelian-Semitic-Proto-Indo-European loan words, see Diakonov 1985:122–140; and Nichols 1997a appendix. On the chronology of the Proto-Kartvelian dispersal or breakup, see Harris 1991.

CHAPTER 6. THE ARCHAEOLOGY OF LANGUAGE

1. My definitions are adapted from Prescott 1987. A different set of definitions was suggested by Parker 2006. He suggested *boundary* as the general term (what I am calling borders) and *border* as a specific term for a political or military boundary (more or less what I am calling a boundary). Parker tried to base his definitions partly on vernacular understandings of how these words are normally used, a noble goal; but I disagree that there is any consistency of usage in the vernacular, and prefer to use established definitions. In their review of the borderland literature, Donnan and Wilson (1999:45–46) followed Prescott in using *border* as the general or unspecialized term. The classic work to which I owe a great deal of my thinking is Barth 1969. For archaeological treatments of ethnic borders, see Shennan 1989, and Stark 1998.

2. For the growth of Medieval European regional identities, see Russell 1972; and Bartlett 1993. For the anthropological deconstruction of tribes and bounded cultures, see Fried 1975; and Wolf 1982, 1984. See also Hill 1992; and Moore 2001. For good archaeological uses of this border-deconstructing approach to ethnicity see Wells 2001; Florin 2001; MacEachern 2000; and James 1999.

3. See Hobsbawm 1990; Giddens 1985; and Gellner 1973. Giddens (1985:120) famously referred to the nation-state as a "bordered power-container." For a different interpretation of ancient tribes and borders, see Smith 1998. He is accused of being a "primordialist"; see his defense in chapter 7. Also see Armstrong 1982.

4. For projectile points and language families in South Africa, see Weissner 1983. For a good review of material culture and ethnicity, see Jones 1997, esp. chap. 6.

5. For New Guineau, see Terrell 2001; see also Terrell, Hunt, and Godsen 1997. For the original argument that biology, culture, and language were separate and independent, see the introduction to Boaz 1911. For California, see Jordan and Shennan 2003. For the other examples, see Silver and Miller 1997:79–98.

6. Persistent frontiers were the subject of a flurry of studies in the 1970s; see Spicer 1971 and a volume dedicated to Spicer by Castile and Kushner 1981. The focus in these papers was the maintenance of stigmatized minority identities. In archaeology, the long-term persistence of prehistoric "culture areas" was discussed long ago in Ehrich 1961. The subject was revisited by Kuna 1991; and Neustupny 1991. My first paper on the subject was Anthony 2001.

7. For the persistence of the Hudson-Valley Iroquoian/Algonkian frontier, see Chilton 1998. For the Linear Pottery frontier, see Zvelebil 2002. For the Jastorf/Halstatt frontier, see Wells 1999.

8. Emberling (1997) used the term *redundant* rather than *robust* for material-culture borders that were marked in multiple categories of material culture, and he recognized that this redundancy suggested that these borders were particularly important socially.

9. For Wales, see Mytum 1994; and John 1972. For the genetic border at the Welsh/English frontier, see Weale et al. 2002. For the border near Basle, see Gallusser 1991. On Breton culture, see Jackson 1994; and Segalen 1991. For the German/Romansh frontier in Italy, see Cole and Wolf 1974.

10. For the Ucayali quotation, see DeBoer 1990:102. For language and genetic correlations, see Jones 2003.

11. For the Iroquois, see Wolf 1982:167; 1984:394; and, in contrast, see Tuck 1978; Snow 1994; and Richter 1992. Moore (2001:43) also used intermarriages between Amerindian tribes as an index of *general* cultural and linguistic mixing: "These [marriage] data show a continual movement of people, and hence their genes, *language, and culture,* from society to society" (emphasis mine).

12. For the borders of functional zones, see Labov 1994. For functional zones, see Chambers and Trudgill 1998; and Britain 2002.

13. See Cole and Wolf 1974:81–282; see also Barth 1969. Cole and Wolf wrote a perceptive analysis of a persistent frontier in Italy, and then in 1982 Wolf published his best-known book, which suggested that tribal borders outside Europe were much more porous and changeable. In making this argument he seems, in my view, to have made some statements contradicted by his own earlier field work.

14. For the billiard-ball analogy, see Wolf 1982:6, 14. On migration processes generally, see Anthony 1990, 1997. Archaeologists of the American Southwest have pushed migration theory further than those of any other region. For a sampling see Spielmann 1998. For migration theory in Iroquoian archaeology, see Sutton 1996.

15. For the four Colonial cultural provinces, see Fischer 1989; Glassie 1965; and Zelinsky 1973. Although anthropology veered away from cultural geography in the 1980s and 1990s, historians and folklorists continued to study it. See Upton and Vlach 1986; and Noble 1992. For a review of the historians' interest in cultural geography in North America, see Nash 1984.

16. Clark 1994.

17. Kopytoff 1987.

18. For the Nuer, see Kelley 1985. For the effect of changes in bride-price currencies on basic subsistence economies, see Cronk 1989.

19. On dialect leveling among colonists, see Siegel 1985; Trudgill 1986; and Britain 2004. The degree of leveling depends on a number of social, economic, and linguistic factors; see Mufwene 2001. For Spanish leveling in the Americas, see Penny 2000. On the history of American English dialects, see Fischer 1989.

20. For charter groups, see Porter 1965; and Breen 1984. On German immigrants in Ohio, see Wilhelm 1992. On Puritan charter groups in new England, see Fischer 1989:57–68. On the Maya, see Fox 1987, although now there are criticisms of Fox's migration-based history; on apex families, see Alvarez 1987; and on the Pueblo, see Schlegel 1992.

21. On leveling and simplification in material culture among colonists, see Noble 1992; and Upton and Vlach 1986. Burmeister (2000) noted that the external form of residential architecture tends to conform to broad norms, whereas ethnicity is expressed in internal details of decoration and ornament.

22. The Boasian approach to borders is reviewed in Bashkow 2004.

23. On the provinces of France, see Chambers and Trudgill 1998:109–123; on the Maasai, see Spear and Waller 1993; on Burma, see Leach 1968, 1960; and for a different interpretation of Burma, see Lehman 1989.

24. On language and ecology, see Hill 1996; and Nettles 1996. Hill's paper was published later in Terrell 2001:257–282. Also see Milroy 1992.

25. The concept of ecologically determined "spread zones" for languages came from Nichols 1992. Similar ideas about arid zones and language expansion can be found in Silver and Miller 1997:79–83. Renfrew (2002) applied the term *spread zone* to any region of rapid language spread, particularly any expansion of pioneer farmers, regardless of ecology. Campbell (2002), however, warned against mixing these definitions.

26. For China, see DiCosmo 2002; and Lattimore 1940.

27. For Acholi origins, see Atkinson 1989, 1994.

28. A similar model for the growth of Bronze Age chiefdoms, described long before Atkinson's case study was published, was by Gilman 1981.

29. For the Pathan-Baluch shift, see Mallory 1992; Barth 1972; and Noelle 1997.

CHAPTER 7. HOW TO RECONSTRUCT A DEAD CULTURE

1. For the history of Christian J. Thomsen's Three-Age System, see Bibby 1956.

2. I generally follow the Eneolithic and Bronze Age chronology of Victor Trifonov at the Institute of the History of Material Culture in St. Petersburg; see Trifonov 2001.

3. For the impact of radiocarbon dating on our understanding of European prehistory, see Renfrew 1973.

4. The old carbon problem in freshwater fish is explained in Cook et al. 2002; and in Bonsall et al. 2004. I used their method to create the correction scale that appears in the appendix.

5. A good historical review of radiocarbon dating in Russian archaeology is in Zaitseva, Timofeev, and Sementsov 1999.

6. For a good example of cultural identity shifting in response to changing historical situations, see Haley and Wilcoxon 2005. For Eric Wolf's and Anthony Smith's comments on situational politics alone being insufficient to explain emotional ties to a cultural identity see Cole and Wolf 1974:281–282; and Smith 1998, chap. 7.

7. For technological style and cultural borders, see Stark 1998.

CHAPTER 8. FIRST FARMERS AND HERDERS

1. The three sky gods named here almost certainly can be ascribed to Proto-Indo-European. *Dyeus Pater*, or Sky/Heaven Father, is the most certain. The Thunder/War god was named differently in different dialects but in each branch was associated with the thunderbolt, the hammer or club, and war. The Divine Twins likewise were named differently in the different branches—the Nāsatyas in Indic, Kastōr and Polydeukēs in Greek, and the Dieva Dēli in Baltic. They were associated with good luck, and often were represented as twin horses, the offspring of a divine mare. For Trita, see Watkins 1995; and Lincoln 1981:103–124. More recently, see Lincoln 1991, chap. 1. For the twins, see Puhvel 1975; and Mallory and Adams 1997:161–165.

2. For the tripartition of Indo-European society, see Dumezil 1958; and Littleton 1982. There is a good review in Mallory 1989:128–142. For an impressive example of the interweaving of three's and two's in Indo-European poetry, see Calvert Watkin's analysis of a traditional Latin poem preserved by Cato in 160 BCE, the "Lustration of the Fields." The structure is tripartite, expressed in a series of doubles. See Watkins 1995:202–204.

3. Przewalkski horses are named after the Polish colonel who first formally described them in 1881. A Russian noble, Frederic von Falz-fein, and a German animal collector, Carl Hagenbeck, captured dozens of them in Mongolia, in 1899 and 1901. All modern Przewalski's are descended from about 15 of these animals. Their wild cousins were hunted to extinction after World War II; the last ones were sighted in Mongolia in 1969. Zoo-bred populations were reintroduced to two preserves in Mongolia in 1992, where once again they are thriving.

4. For differences between east-Ural and west-Ural Upper Paleolithic cultures, see Boriskovskii 1993, and Lisitsyn 1996.

5. For a wide-ranging study of the Ice Age Caspian, the Khvalynian Sea, and the Black Sea, including the "Noah's Flood" hypothesis, see Yanko-Hombach et al. 2006.

6. For the decline of matriliny among cattle herders, see Holden and Mace 2003.

7. For Y-chromosome data on early European cattle, see Gotherstrom et al. 2005. For MtDNA, see Troy et al. 2001; and Bradley et al. 1996.

8. For agricultural frontier demography, see Lefferts 1977; and Simkins and Wernstedt 1971.

9. For the oldest Criş site in the lower Danube valley, see Nica 1977. For a Starcevo settlement in the plains north of Belgrade, see Greenfield 1994.

10. For Criş immigrants in the East Carpathians, see Dergachev, Sherratt, and Larina 1991; Kuzminova, Dergachev, and Larina 1998; Telegin 1996; and Ursulescu 1984. The count of thirty sites refers to excavated sites. Criş pottery is known in unexcavated surface exposures at many more sites listed in Ursulescu 1984. For the Criş economy in eastern Hungary, see Vörös 1980.

11. For Neolithic bread, see Währen 1989. Criş people cultivated gardens containing four varieties of domesticated wheat: *Triticum monococcum, T. dicoccum* Shrank, *T. spelta, T. aestivo-compactum* Schieman; as well as barley (*Hordeum*), millet (*Panicum miliaceum*), and peas (*Pisum*)— all foreign to eastern Europe. On the plant evidence, see Yanushevich 1989; and Pashkevich 1992.

12. Markevich 1974:14.

13. For the possible role of acculturated foragers in the origin of the East Carpathian Criş culture, see Dergachev, Sherratt, and Larina 1991; and, more emphatically, Zvelebil and Lillie 2000.

14. On pioneer farmers and language dispersal, see Bellwood and Renfrew 2002; Bellwood 2001; Renfrew 1996; and Nichols 1994. On the symbolic opposition of wild and domesticated animals, see Hodder 1990.

15. Most archaeologists have accepted the argument made by Perles (2001) that the Greek Neolithic began with a migration of farmers from Anatolia. For the initial spread from Greece into the Balkans, see Fiedel and Anthony 2003. Also see Zvelebil and Lillie 2000; and van Andel and Runnels 1995. The practical logistics of a Neolithic open-boat crossing of the Aegean are discussed in Broodbank and Strasser 1991.

16. For **tawro-s*, see Nichols 1997a: appendixes. For the association of Afro-Asiatic with the initial Neolithic, see Militarev 2003.

17. The classic Russian-language works on the Bug-Dniester culture are in Markevich 1974; and Danilenko 1971; the classic discussion in English is in Tringham 1971. More recently, see Telegin 1977, 1982, and 1996; and Wechler, Dergachev, and Larina 1998.

18. For the Mesolithic groups around the Black Sea, see Telegin 1982; and Kol'tsov 1989. On the Dobrujan Mesolithic, see Paunescu 1987. For zoological analyses, see Benecke 1997.

19. Most of the dates for the earliest Elshanka sites are on shell, which might need correction for old carbon. Corrected, Elshanka dates might come down as low as 6500–6200 BCE. See Mamonov 1995, and other articles in the same edited volume. For radiocarbon dates, see Timofeev and Zaitseva 1997. For the technology and manufacture of this silt/mud/clay pottery, see Bobrinskii and Vasilieva 1998.

20. For the dates from Rakushechni Yar, see Zaitseva, Timofeev, and Sementsov 1999. For the excavations at Rakushechni Yar, see Belanovskaya 1995. Rakushechni Yar was a deeply stratified dune site. Telegin (1981) described sedimentary stratum 14 as the oldest cultural occupation. A series of new radiocarbon dates, which I ignore here, have been taken from organic residues that adhered to pottery vessels said to derive from levels 9 to 20. Levels 15 to 20 would have been beneath the oldest cultural level, so I am unsure about the context of the pottery. These dates were in the calibrated range of 7200–5800 BCE (7930±130 to 6825±100 BP). If they are correct, then this pottery is fifteen hundred years older than the other pottery like it, *and* domesticated sheep appeared in the lower Don valley by 7000 BCE. All domesticated sheep are genetically proven to have come from a maternal gene pool in the mountains of eastern Turkey, northern Syria, and Iraq about 8000–7500 BCE, and no domesticated sheep appeared in the Caucasus, northwestern Anatolia, or anywhere else in Europe in any site dated as early as 7000 BCE. The earliest dates *on charcoal* from Rakushechni Yar (6070+100 BP, 5890+105 BP for level 8) come out about 5200–4800 BCE, in agreement with other dates for the earliest domesticated animals in the steppes. If the dated organic residue was full of boiled fish, it could need a correction of five hundred radiocarbon years, which would bring the earliest dates down to about 6400–6200 BCE—somewhat more reasonable. I think the dates are probably contaminated and the sheep are mixed down from upper levels.

21. For 155 Late Mesolithic and Neolithic radiocarbon dates from Ukraine, see Telegin et al. 2002, 2003.

22. On Bug-Dniester plant foods, see Yanushevich1989; and Kuzminova, Dergachev, and Larina 1998. A report of millet and barley impressions from the middle-phase site of Soroki I/level 1a is contained in Markevich 1965. Yanushevich did not include this site in her 1989 list of Bug-Dniester sites with domesticated seed imprints; it is the only Bug-Dniester site I have seen with reports of barley and millet impressions.

23. The dates here are not on human bones, so they need no correction. The bone percentages are extracted from Table 7 in Markevich 1974; and Benecke 1997. Benecke dismissed the Soviet-era claims that pigs or cattle or both were domesticated independently in the North Pontic region. Telegin (1996:44) agreed. Mullino in the southern Urals produced domesticated sheep bones supposedly dated to 7000 BCE, cited by Matiushin (1986) as evidence for migrations from Central Asia; but like the claimed sheep in deep levels at Rakushechni Yar, these sheep would have been *earlier* than their proposed parent herds at Djeitun, and the wild species was not native to Russia. The sheep bones probably came from later Eneolithic levels. Matiushin's report was criticized for stratigraphic inconsistencies. See Matiushin 1986; and, for his critics, Vasiliev, Vybornov, and Morgunova 1985; and Shorin 1993.

24. Zvelebil and Rowley-Conwy 1984.

25. For captured women and their hyper-correct stylistic behavior, see DeBoer 1986. The archaeological literature on technological style is vast, but a good introduction is in Stark 1998.

26. The Linear Pottery culture in the East Carpathian piedmont overlapped with the Criş culture around 5500–5400 BCE. This is shown at late Criş sites like Grumazeşti and Sakarovka that contained a few Linear Pottery sherds. Sakarovka also had Bug-Dniester sherds, so it shows the brief contemporaneity of all three groups.

27. There is, of course, generosity and sharing among farmers, but farmers also understand that certain potential foods are not food at all but investments. Generosity with food has practical limits in bad times among farmers; these are generally absent among foragers. See Peterson 1993; and Rosenberg 1994.

28. The classic text on the Dnieper-Donets culture is Telegin 1968. For an English-language monograph see Telegin and Potekhina. In this chapter I only discuss the first phase, Dnieper-Donets I.

29. For DDI chipped axes, see Neprina 1970; and Telegin 1968:51–54.

30. Vasilievka V was published as a Dnieper-Donets II cemetery, but its radiocarbon dates suggest that it should have dated to DD I. Vasilievka I and III were published as Late Mesolithic, broadly around 7000–6000 BCE, but have radiocarbon dates of the very Early Mesolithic, closer to 8000 BCE. Vasilievka II and Marievka were published as Neolithic but have no ceramics and Late Mesolithic radiocarbon dates, 6500–6000 BCE, and so are probably Late Mesolithic. Changes in human skeletal morphology that were thought to have occurred between the Late Mesolithic and Neolithic (Jacobs 1993) now appear to have occurred between the Early and Late Mesolithic. These revisions in chronology have not generally been acknowledged. For radiocarbon dates, see Telegin et al. 2002, 2003. See also Jacobs 1993, and my reply in Anthony 1994.

31. For Varfolomievka, see Yudin 1998, 1988.

32. The zoologist Bibikova identified domesticated animals—sheep, cattle, and horses—at Matveev Kurgan in levels dated 6400–6000 BCE. Today neither the German zoologist Benecke nor the Ukrainian archaeologist Telegin give credit to Bibikova's claims for an independent local domestication of animals in Ukraine. Matveev Kurgan (a settlement, not a kurgan) is located in the Mius River valley north of the Sea of Azov, near Mariupol. Two sites were excavated between 1968 and 1973, numbered 1 and 2. Both contained Grebenikov-type microlithic flint tools and were thought to be contemporary. Two radiocarbon dates from MK 1 average about 6400–6000 BCE, but the single date (on bone) from MK 2 was about 4400–4000 BCE.

In the latter period domesticated animals including sheep were common in the region. The artifacts from all depths were analyzed and reported as a single cultural deposit. But at MK 1 the maximum number of flint tools and animal bones was found at a depth of 40–70 cm (Krizhevskaya 1991:8), and the dwelling floor and hearths were at 80–110 cm (Krizhevskaya 1991:16). Most of the animal bones from MK 1 and 2 were from wild animals, principally horses, onagers, and wild pigs, and these probably were associated with the older dates. But the bones identified as domesticated horses, cattle, and sheep probably came from later levels associated with the later date. See Krizhevskaya 1991. Stratigraphic inconsistencies mar the reporting of all three Pontic-Ural sites with claimed very early domesticated animals—Rakushechni Yar, Mullino, and Matveev Kurgan.

CHAPTER 9. COWS, COPPER, AND CHIEFS

1. Benveniste 1973:61–63 for feasts; also see the entry for GIVE in Mallory and Adams 1997:224–225; and the brief recent review by Fortson 2004:19–21.

2. The dates defining the beginning of the Eneolithic in the steppes are principally from human bone, whereas the dates from Old Europe are not. The date of 5200–5000 BCE for the beginning of the Eneolithic Dnieper-Donets II culture incorporates a reduction of −228±30 radiocarbon years prior to recalibration. There is a discussion of this below in note 16.

3. "Old Europe" was a term revived by Marija Gimbutas, perhaps originally to distinguish Neolithic European farming cultures from Near Eastern civilizations, but she also used the term to separate southeastern Europe from all other European Neolithic regions. See Gimbutas 1991, 1974. For chronologies, economy, environment, and site descriptions, see Bailey and Panayotov 1995; and Lichardus 1991. For the origin of the term *Alteuropa* see Schuchhardt 1919.

4. Most of these dates are on charcoal or animal bone and so need no correction. The earliest copper on the Volga is at Khvalynsk, which is dated by human bone that tested high in [15]N (mean 14.8%) and also seemed too old, from about 5200–4700 BCE, older than most of the copper in southeastern Europe, which was the apparent source of the Khvalynsk copper. I have subtracted four hundred radiocarbon years from the original radiocarbon dates to account for reservoir effects, making the Khvalynsk cemetery date 4600–4200 BCE, which accords better with the florescence of the Old European copper age and therefore makes more sense.

5. For the pathologies on cattle bones indicating they were used regularly for heavy draft, see Ghetie and Mateesco 1973; and Marinescu-Bîlcu et al. 1984.

6. For signs and notation, see Gimbutas 1989; and Winn 1981. The best book on female figurines is Pogozheva 1983.

7. Copper tools were found in Early Eneolithic Slatina in southwestern Bulgaria, and copper ornaments and pieces of copper ore (malachite) were found in Late Neolithic Hamangia IIB on the Black Sea coast in the Dobruja hills south of the Danube delta, both probably dated about 5000 BCE. For Old European metals in Bulgaria, see Pernicka et al. 1997. For the middle Danube, see Glumac and Todd 1991. For general overviews of Eneolithic metallurgy, see Chernykh 1992; and Ryndina 1998.

8. For vegetation changes during the Eneolithic, see Willis 1994; Marinescu-Bîlcu, Cârciumaru, and Muraru 1981; and Bailey et al. 2002.

9. Kremenetski et al. 1999; see also Kremenetskii 1997. For those who follow the "beech line" argument in Indo-European origin debates, these pollen studies indicate that Atlantic-period beech forests grew in the Dniester uplands and probably spread as far west as the Dnieper.

10. For the ceramic sequence, see Ellis 1984:48 and n. 3. The Pre-Cucuteni I phase was defined initially on the basis of ceramics from one site, Traian-Dealul Viei; small amounts of

similar ceramics were found later at four other sites, and so the phase probably is valid. For an overview of the Tripolye culture, see Zbenovich 1996.

11. Marinescu-Bilcu et al. 1984.

12. Some Tripolye A settlements in the South Bug valley (Lugach, Gard 3) contained sherds of Bug-Dniester pottery, and others had a few flint microlithic blades like Bug-Dniester forms. These traces suggest that some late Bug-Dniester people were absorbed into Tripolye A villages in the South Bug valley. But late Bug-Dniester pottery was quite different in paste, temper, firing, shape, and decoration from Tripolye pottery, so the shift to using Tripolye wares would have been an obvious and meaningful act. For the absence of Bug-Dniester traits in Tripolye material culture, see Zbenovich 1980:164–167; and for Lugach and Gard 3, see Tovkailo 1990.

13. For Bernashevka, see Zbenovich 1980. For the Tripolye A settlement of Luka-Vrublevetskaya, see Bibikov 1953.

14. For the Karbuna hoard, see Dergachev 1998.

15. The Early Eneolithic cultures I describe in this section are also called Late Neolithic or Neo-Eneolithic. Telegin (1987) called the DDII cemeteries of the Mariupol-Nikol'skoe type Late Neolithic, and Yudin (1988) identified Varfolomievka levels 1 and 2 as Late Neolithic. But in the 1990s Telegin began to use the term "Neo-Eneolithic" for DDII sites, and Yudin (1993) started calling Varfolomievka an Eneolithic site. I have to accept these changes, so sites of Mariupol-Nikol'skoe (DDII) type and all sites contemporary with them, including Khvalynsk and Varfolomievka, are called Early Eneolithic. The Late Neolithic apparently has disappeared. The terminological sequence in this book is Early Neolithic (Surskii), Middle Neolithic (Bug-Dniester–DDI), Early Eneolithic (Tripolye A–DDII–Khvalynsk), and Late Eneolithic (Tripolye B, C1-Sredni Stog-Repin). For key sites in the Dnieper-Azov region, see Telegin and Potekhina 1987; and Telegin 1991. For sites on the middle Volga, see Vasiliev 1981; and Agapov, Vasiliev, and Pestrikova 1990. In the Caspian Depression, see Yudin 1988, 1993.

16. The average level of ^{15}N in DDII human bones is 11.8 percent, which suggests an average offset of about -228 ± 30 BP, according to the method described in the appendix. I subtracted 228 radiocarbon years from the BP dates for the DDII culture and calibrated them again. The unmodified dates from the earliest DDII cemeteries (Dereivka, Yasinovatka) suggested a calibrated earliest range of 5500–5300 BCE (see Table 9.1), but these dates always seemed too early. They would equate DDII with the middle Bug-Dniester and Criş cultures. But DDII came for the most part *after* Bug-Dniester, during the Tripolye A period. The modified radiocarbon dates for Dnieper-Donets II fit better with the stratigraphic data and with the Tripolye A sherds found in Dnieper-Donets II sites. For lists of dates, see Trifonov 2001; Rassamakin 1999; and Telegin et al. 2002, 2003.

17. For lists of fauna, see Benecke 1997:637–638; see also Telegin 1968:205–208. For ^{15}N in the bones, see Lillie and Richards 2000. Western readers might be confused by statements in English that the DDII economy was based on hunting and fishing (Zvelebil and Lillie 2000:77; Telegin, et al. 2003:465; and Levine 1999:33). The DDII people ate cattle and sheep in percentages between 30% and 78% of the animal bones in their garbage pits. Benecke (1997:637), a German zoologist, examined many of the North Pontic bone collections himself and concluded that domesticated animals "first became evident in faunal assemblages that are synchronized with level II of the Dnieper-Donets culture." People who kept domesticated animals were no longer hunter-gatherers.

18. Flint blades 5–14 cm long with sickle gloss are described by Telegin (1968:144). The northwestern DDII settlements with seed impressions are listed in Pashkevich 1992, and Okhrimenko and Telegin 1982. DDII dental caries are described in Lillie 1996.

19. Telegin 1968:87.

20. The Vasilievka II cemetery was recently dated by radiocarbon to the Late Mesolithic, about 7000 BCE. The cemetery was originally assigned to the DDII culture on the basis of

a few details of grave construction and burial pose. Telegin et al. 2002 extended the label "Mariupol culture" back to include Vasilievka II, but it lacks all the artifact types and many of the grave features that define DDII-Mariupol graves. The DDII cemeteries are securely dated to a period after 5400–5200 BCE. Vasilievka II is Late Mesolithic.

21. For funeral feasts, see Telegin and Potekhina 1987:35–37, 113, 130.

22. I have modified Khvalynsk dates on human bone to account for the very high average ^{15}N in human bone from Khvalynsk, which we measured at 14.8%, suggesting that an average −408±52 radiocarbon years should be subtracted from these dates before calibrating them (see Authors Note on Dating, and chapter 7). After doing this I came up with dates for the Khvalynsk cemetery of 4700/4600–4200/4100 BCE, which makes it overlap with Sredni Stog, as many Ukrainian and Russian archaeologists thought it should on stylistic and typological grounds. It also narrows the gap between late Khvalynsk on the lower Volga (now 3600–3400 BCE) and earliest Yamnaya. See Agapov, Vasiliev, and Pestrikova 1990; and Rassamakin 1999.

23. Until Khvalynsk II is published, the figure of forty three graves is conditional. I was given this figure in conversation.

24. For the enhancement of male status with herding economies, see Holden and Mace 2003.

25. In Anthony and Brown (2000) we reported a smaller number of horses, cattle, and sheep from the cemetery at Khvalynsk, based on only the twelve "ritual deposits" placed above the graves. I later compiled the complete animal bone reports from two sources: Petrenko 1984; and Agapov, Vasiliev, and Pestrikova 1990, tables 1, 2. They presented conflicting descriptions of the numbers of sheep in ritual deposits 10 and 11, and this discrepancy resulted in a total count of either fifty-two or seventy sheep MNI.

26. See Ryndina 1998:151–159, for Khvalynsk I and II metals.

27. For ornaments see Vasiliev 2003.

28. For the possibility that the first domesticated animals came across the North Caucasus from the Near East, see Shnirelman 1992; and Jacobs 1993; and, in opposition, see Anthony 1994.

29. Yanushevich 1989.

30. Nalchik is described in Gimbutas 1956:51–53.

31. I found this grave referenced in Gei 2000:193.

32. The bones at Dzhangar were originally reported to contain domesticated cattle, but the zoologist Pavel Kosintsev told me, in 2001, that they were all onager and horse, with no obvious domesticates.

33. The Neolithic cultures of the North Caspian Depression, east of the Volga, were first called the Seroglazivka culture by Melent'ev (1975). Seroglazivka included some Neolithic forager camps similar to Dzhangar and later sites with domesticated animal bones like Varfolomievka. Yudin suggested in 1998 that a new label, "Orlovka culture," should be applied to the Early Eneolithic sites with domesticated animals. On Varfolomievka, see Yudin 1998, 1988. Razdorskoe was described by Kiyashko 1987. Older but still informative is Telegin 1981.

34. The Orlovka site was first described by Mamontov 1974.

35. The Samara Neolithic culture, with the cemetery of S'yezzhe, usually is placed earlier than Khvalynsk, as one S'yezzhe grave contained a boars-tusk plaque exactly like a DDII type. Radiocarbon dates now indicate that early Khvalynsk overlapped with the late Samara Neolithic (and late DDII). The Samara Neolithic settlement of Gundurovka contained Khvalynsk pottery. The Samara culture might have begun before Khvalynsk; see Vasiliev and Ovchinnikova 2000. For S'yezzhe, see Vasiliev and Matveeva 1979. For animal bones, see Petrenko 1984:149; and Kuzmina 2003.

CHAPTER 10. THE DOMESTICATION OF THE HORSE
AND THE ORIGINS OF RIDING

1. See Clayton and Lee 1984; and Clayton 1985. For a recent update, see Manfredi, Clayton, and Rosenstein 2005.

2. For early descriptions of bit wear, see Clutton-Brock 1974; and Azzaroli 1980. Doubts about the causes of this kind of wear had been expressed by Payne (1995) in a study published after long delays.

3. We were provided with horse teeth by Mindy Zeder at the Smithsonian Institution; the Large Mammal Veterinary Facility at Cornell University; the University of Pennsylvania's New Bolton Veterinary Center; the Bureau of Land Management, Winnemucca, Nevada; and Ron Keiper of Pennsylvania State University. We learned mold-making and casting procedures from Sandi Olsen and Pat Shipman, then at Johns Hopkins University. Mary Littauer gave us invaluable advice and the use of her unparalleled library. Our first steps were supported by grants from the Wenner-Gren Foundation and the American Philosophical Society.

4. On horse MtDNA, see Jansen et al. 2002; and Vilà et al. 2001. For horse Y-chromosomes, see Lindgren et al. 2004.

5. For equids in Anatolia, see Summers 2001; and online reports on the Catal Höyuk project. For horses in Europe, see Benecke 1994; and Peške 1986.

6. For Mesolithic and Neolithic Pontic-Caspian horses, see Benecke 1997; Vasiliev, Vybornov, and Komarov 1996; and Vasilev 1998. For horse bones at Ivanovskaya in the Samara Neolithic, see Morgunova 1988. In the same volume, see I. Kuzmina 1988.

7. For Mongol horse keeping, see Sinor 1972; and Smith 1984. For horses and cattle in the blizzard of 1886, see Ryden 1978:160–162. For feral horses see also Berger 1986.

8. For a review of these methods, see Davis 1987. For riding-related pathologies in vertebrae, see Levine 1999b. For crib-biting, see Bahn 1980; and the critique in White 1989.

9. The graphs from Benecke and von den Driesch (2003) are combined and reprinted as figure 10.3 here. See also Bökönyi 1974. For a critical view of Dereivka, see Uerpmann 1990.

10. The ratio of females to males in a harem band, counting immature horses, should be about 2:1, but the *skeletons* of immature males cannot be assigned a sex as the canine teeth do not erupt until about four to five years of age, and the presence of erupted canines is the principal way to identify males. From the bones, a harem band would contain just one *identifiable* male.

11. A horse's age at death can be estimated from a loose molar by measuring the molar crown height, the length of the tooth from the bifurcation between the roots to the occlusal surface. This measurement decreases with age as the tooth wears down. Spinage (1972) was the first to publish crown height-versus-age statistics for equids, based on zebras; Levine (1982) published statistics for a small sample of horses using measurements from X-rays. We largely confirmed Levine's numbers with direct measurements on our larger sample. But we found that estimates based *only* on crown heights have *at best* a±1.5 year degree of uncertainty (a three-year span). The crown height on the right and left P_2s of the same horse can vary by as much as 5 mm, which would normally be interpreted as indicating a difference in age of more than three years. See note 18, below.

12. Bibikova (1967, 1969) noted that fifteen of seventeen sexable mandibles were male. I subtracted the cult stallion, an Iron Age intrusion, making fourteen of sixteen males. Bibikova never published a complete description of the Dereivka horse bones, but she did note that the MNI was fifty-two individuals; 23% of the population was aged one to two years (probably looking at long bone fusion); fifteen of seventeen sexable jaw fragments were from males older than five, as this is when the canine teeth emerge; and there were no very old individuals.

Levine's age-at-death statistics were based on the crown heights of all the teeth kept in 1998, with an MNI of only sixteen—about two-thirds of the original collection had been lost. Only 7% of this remnant population was one to two years of age based on long-bone fusion (1999b:34) and about one-third of the surviving teeth were from the Iron-Age cult stallion. For Levine's age-at-death graphs, see Levine 1990, 1999a, 1999b.

13. The analysis of the equid P_2s from Leisey was conducted by Christian George as part of his MA Thesis in Geosciences at the University of Florida. The 1.5-million-year-old Leisey equids were *Equus "leidyi,"* possibly an eastern variant of *Equus scotti,* a common member of the Rancholabrean fauna, very similar in dentition, diet and stature to true horses. Of the 113 P2s from this site, 39 were eliminated because of age, damage, or pathologies, leaving 74 measurable P_2s from mature equids. See George 2002; Anthony, Brown, and George 2006; and Hulbert, Morgan, and Webb 1995. Our collection of P_2s was assembled through the generosity of the New Bolton Center at the University of Pennsylvania, the Cornell University College of Veterinary Medicine, the Bureau of Land Management in Winnemucca, NE; and Ron Keiper, then at Pennsylvania State University.

14. We are grateful to the National Science Foundation for supporting the riding experiment, and to the State University of New York at Cobleskill for hosting and managing it. Dr. Steve MacKenzie supervised the project, and the riding and recording was done by two students in the Horse Training and Behavior Program, Stephanie Skargensky and Michelle Beleyea. The bone bit and antler cheekpieces were made with flint tools by Paul Trotta. The hemp rope was supplied by Vagn Noeddlund of Randers Ropeworks. Mary Littauer and Sandra Olsen provided valuable suggestions on bits and mold-making. All errors were our own.

15. The pre-experiment, never-bitted mean bevel measurement for the three horses bitted with soft bits was 1.1 mm, the same as the never-bitted Pleistocene Leisey equids. The standard deviation for the three was 0.42 mm. The post-experiment mean was 2.04 mm, more than two standard deviations greater than the pre-experiment mean. Another 300 hours of riding might have created a bevel of 3 mm, our threshold for archaeological specimens.

16. The 74 never-bitted equid teeth from Leisey exhibited a greater range of variation than the 31 never-bitted modern P_2s we collected, not surprising with a larger sample. The distribution of measurements was normal, and a t-Test of the difference between the means for our bitted sample and the Leisey sample showed a significant difference. The threshold of 3 mm for identifying bit wear in archaeological specimens is supported by the Leisey data.

17. Levine outlined six problems with our bit wear studies in 1999b:11–12 and 2004:117–120. She placed it in a category she termed "false direct evidence," with so-called bridle cheekpieces whose forms vary wildly and whose function is entirely speculative. We believe Levine's criticisms are based on factual errors, distortions, and misunderstandings. For our reply to each of her six criticisms, see Anthony, Brown, and George 2006. We remain confident in our analysis of bit wear.

18. Permanent horse P2s become flattened or "tabled" by occlusion with the opposing tooth gradually between two and three years of age. Brown determined that a P_2 with a crown height greater than 5.0 mm *and* an occlusal length-to-width ratio greater than 2.1 is probably from a horse three years old or younger, so should be excluded from studies of bit wear (Brown and Anthony 1998:338–40). Brown was the first to combine the crown height and the occlusal length-width ratio to produce an age-at-death estimate this precise. If she had not done this we would have been forced to discard half of our sample to avoid using 2-3-year-old teeth. Christian George also used Brown's method to eliminate young teeth (≤ 3 yr) from the Leisey sample. It should be noted that George found one P_2 with a bevel of 3.05 mm, but it was probably from a horse less than three years old.

19. Bendrey (2007), as this book went to press, reported new bevel measurements on never-bitted Przewalski horses, from zoos in England and Prague. Bendrey measured 29 P_2s from 15 Przewalksi horses of acceptable age (>3 and <21), and found 3mm bevels on three, or 10%. We found one bevel of *almost* 3mm in 105 never-bitted P_2s, less than 1%. The Przewalski bevels all

were caused by malocclusion with the opposing upper P^2; one 3mm bevel was filed down as a veterinary treatment for underbite. Malocclusion occurred among zoo-kept Przewalskis more frequently than among Pleistocene equids or Nevada mustangs. All zoo Przewalskis are descended from about 15 captured in the wild, and these founders might have had unusually bad occlusion. Also domestic horses were bred with the founders, perhaps mixing genes for different tooth and jaw sizes.

20. Raulwing 2000:61, with references.

21. For Dereivka, see Telegin 1986. For the horse bones, see Bibikova 1967, 1970; Bökönyi 1974, 1978, 1979; and Nobis 1971.

22. For criticisms of the traditional evidence for horse domestication at Dereivka, see Anthony 1986, 1991b; and Levine 1990.

23. Our research at the Institute of Zoology in Kiev was hosted by a generous and thoughtful Natalya Belan; in Samara, Russia, by Igor Vasiliev; and in Petropavlovsk, Kazakhstan, by Victor Zaibert. In Budapest Sandor Bökönyi made us welcome in the gracious manner for which he was widely known and is widely missed. The project was supported by a grant from the National Science Foundation. For reports, see Anthony and Brown 1991; and Anthony, Telegin, and Brown 1991.

24. See Häusler 1994.

25. For the redating of the Dereivka cult stallion, see Anthony and Brown 2000; reiterated in Anthony and Brown 2003.

26. Both Botai and Tersek showed some influence in their ceramics from forager cultures of the forest-steppe zone in the southeastern Urals, known as Ayatskii, Lipchin, and Surtanda. Botai-Tersek might have originated as a southern, steppe-zone offshoot of these cultures. For a description of Botai and Tersek in English, see Kislenko and Tatarintseva 1999; in Russian, see Zaibert 1993. For discussions of the horse remains at Botai and related sites, see Olsen 2003; and Brown and Anthony 1998.

27. Our initial measurements of the horse teeth from Kozhai 1 (made in a hotel room in Petropavlovsk, Kazakhstan) produced one tooth with a 3 mm bevel. This is how we described the Kozhai results before 2006. We remeasured the twelve Kozhai 1 casts for Anthony, Brown, and George 2006, and agreed that a borderline 2.9+ measurement was actually 3 mm, resulting in two teeth with bit wear. Two other P_2s from Kozhai 1 measured 2 mm or more, an unusually high measurement among wild horses.

28. Describing the Botai horses as wild were Levine 1999a, 1999b; Benecke and von den Dreisch 2003; and Ermolova, in Akhinzhalov, Makarova, and Nurumov 1992.

29. See Olsen 2003:98–101.

30. French and Kousoulakou 2003:113.

31. The Atbasar Neolithic preceded Botai in the northern Kazakh steppes; see Kislenko and Tatarintseva 1999. Benecke and von den Dreisch (2003: table 6.3) reported that domesticated sheep and cattle bones were found in Atbasar sites in the Kazakh steppes, dated before Botai. This is true, *but* the Russian and Kazakh authors they cite described the bones of domesticated sheep and cattle as later intrusions in the Neolithic levels; they were less weathered than the bones of the wild animals. The animal bones from Atbasar sites are interpreted by Akhinzhalov, Makarova, and Nurumov as indicating a foraging economy based on wild horses, short-horned bison, saiga antelope, gazelle, red deer, and fish. Domesticated animals appeared at the end of the Botai era. For their comments on differential bone weathering in Atbasar sites, see Akhinzhalov, Makarova, and Nurumov 1992:28–29, 39.

32. Logvin (1992) and Gaiduchenko (1995) interpreted some animal bones in sites of the Eneolithic Tersek culture, centered in the Tugai steppes near Kustenai, Kazakhstan, and dated to the same period as Botai, as domesticated cattle, particularly from Kumkeshu I. Another zoologist, Makarova, had identified the Tersek bovid bones as those of wild bison (Akhinzhalov, Makarova, and Nurumov 1992:38). Some domesticated cattle might have been kept in Tersek sites, which were closer to the Pontic-Caspian herders. None appeared at Botai. For Kumkeshu I, see Logvin, Kalieva, and Gaiduchenko 1989.

33. For horses in the Caucasus I relied on the text of a conference paper by Mezhlumian (1990). A few horses might have passed through the Caucasus into northern Iran before 3000 BCE, indicated by a few probable horse teeth at the site of Qabrestan, west of Teheran (see Mashkour 2003) and a possible horse tooth at Godin Tepe (see Gilbert 1991). No definite horse remains have been identified in eastern Iran, Central Asia, or the Indian subcontinent in deposits dated earlier than 2000 BCE, claims to the contrary notwithstanding. For a review of this debate, see Meadow and Patel 1997.

34. For central European horses, see See Benecke 1994; Bökönyi 1979; and Peške 1986.

35. Khazanov 1994:32.

36. For war and the prestige trade, see Vehik 2002.

37. The American Indian analogy is described in Anthony 1986. The most detailed analysis of the effects of horseback riding and horse keeping on Plains Indian cultures is Ewers 1955.

38. One argument against riding before 1500 BCE was that steppe horses were too small to ride. This is not true. More than 70% of the horses at Dereivka and Botai stood 136–144 cm at the withers, or about 13–14 hands high, and some were 15 hands high. They were the same size as Roman cavalry horses. Another argument is that rope and leather bits were inadequate for controlling horses in battle. This is also not true, as the American Indians demonstrated. Our SUNY students at Cobleskill also had "no problem" controlling horses with rope bits. The third is that riders in the steppes rode sitting back on the rump of the horse, a manner suited only to riding donkeys, which did not exist in the steppes. We have rebutted these doubts about Eneolithic riding in Anthony, Brown, and George 2006. For the arguments against Eneolithic riding, see Sherratt 1997a:217; Drews 2004:42–50; Renfrew 2002; and E. Kuzmina 2003:213.

39. The remains of a bow found in Berezovka kurgan 3, grave 2, on the Volga, in a grave of Pokrovka type probably dated about 1900–1750 BCE, had bone plates reinforcing the shaft and bone tips at the ends—a composite bow. The surviving pieces suggest a length of 1.4–1.5 m, almost five feet from tip to tip. See Shishlina 1990; and Malov 2002. For an overview of early archery and bows, see Zutterman 2003.

40. I am indebted to Dr. Muscarella for some of these ideas about arrow points. For a discussion of the initial appearance and usage of socketed bronze arrowheads, see Derin and Muscarella 2001. For a catalogue and discussion of the early Iron Age socketed arrowheads of the Aral Sea region, see Itina and Yablonskii 1997. Socketed bronze spear points were made in the steppes as early as 2000 BCE, and smaller socketed points began to appear occasionally in steppe sites about the middle of the Late Bronze Age, around 1500 BCE, but their potential was not immediately exploited. The ideal bows, arrows, and arrowheads for mounted archery evolved slowly.

41. For tribal warfare, see Keeley 1996.

CHAPTER 11. THE END OF OLD EUROPE AND THE RISE OF THE STEPPE

1. For the gold at Varna, see Bailey 2000:203–224; Lafontaine and Jordanov 1988; and Eleure 1989.

2. Chapman 1989.

3. For off-tell settlement at Bereket, see Kalchev 1996; at Podgoritsa, see Bailey et al. 1998.

4. The decrease in solar insolation that bottomed out at 4000–3800 BCE is documented in Perry and Hsu 2000; and Bond et al. 2001. For the Piora Oscillation in the Swiss Alps, see Zöller 1977. For indicators of cooling in about 4000 BCE in the Greenland ice cores, see O'Brien et al. 1995. For climate change in Central Europe in the German oak tree rings, see Leuschner et al. 2002. For the Pontic steppes, see Kremenetski, Chichagova, and Shishlina 1999.

5. For the flooding and agricultural shifts, see Bailey et al. 2002. For overgrazing and soil erosion, see Dennell and Webley 1975.

6. For Jilava, see Comsa 1976.

7. The pollen changes are described in Marinova 2003.

8. Cast copper objects began to appear regularly in western Hungary with the Lasinja-Balaton culture at about 4000 BCE; see Bánffy 1995; also Parzinger 1992.

9. Todorova 1995:90; Chernykh 1992:52. The burning of houses might have been an intentional ritual act during the Eneolithic; see Stevanovic 1997. But the final fires that consumed the Eneolithic towns of the lower Danube valley and the Balkans about 4000 BCE were followed by region-wide abandonment and abrupt culture change. Region-wide abandonments of large settlements in the North American Southwest (1100–1400 CE) and in Late Classic Maya sites (700–900 CE) in Mesoamerica were associated with intense warfare; see Cameron and Tomka 1993. The kind of climate shift that struck the lower Danube valley about 4100–3800 BCE would not have made tell settlements uninhabitable. Warfare therefore seems a likely explanation.

10. For evidence of overgrazing and soil erosion at the end of the Karanovo VI period, see Dennell and Webley 1975; for the destruction of Eneolithic Yunatsite, see Merpert 1995; and Nikolova 2000.

11. Todorova 1995.

12. See Ellis 1984 for ceramic workshops, and Popov 1979 for flint workshops. I use the Russian spelling (Tripolye, Tomashovka) rather than the Ukrainian (Tripil'ye, Tomashivka), because many site names such as Tripolye are established in the literature outside Ukraine in their Russian spelling.

13. On the demographics, see Dergachev 2003; and Masson 1979. On the flight of Bolgrad-Aldeni refugees, see Sorokin 1989.

14. On Tripolye B1 warfare generally, see Dergachev 2003, 1998b; and Chapman 1999. On Drutsy 1, see Ryndina and Engovatova 1990. For much of the other information in this section I have relied on the review article by Chernysh 1982.

15. The Cucuteni C designation refers only to a type of shell-tempered pottery. The Cucuteni chronology ends with Cucuteni B_2. Cucuteni C ware appeared first in sites dated to the Cucuteni A_3/Tripolye B1 period and ultimately dominated ceramic assemblages. See Ellis 1984:40–48.

16. The source of the steppe influence on Cucuteni C pottery is usually identified as the early Sredni Stog culture, phase Ib, for Telegin; or the Skelya culture, for Rassamakin.

17. Shell-temper adds to the durability and impact resistance of vessels that are regularly submitted to thermal shock through reheating, and also increases the cooling effect of evaporation, making a shell-tempered pot good for cooking or storing cool drinking water. Cucuteni C ware and fine painted wares were found together both in pit-houses and large two-storied surface houses. Contextual differences in the distribution of Cucuteni C ware and fine ware in settlements have not been described. At some sites the appearance of Cucuteni C wares seems abrupt: Polivanov Yar had traditional grog-tempered coarse wares in the Tripolye B2 occupation but switched to shell-tempered C wares of different shapes and designs in Tripolye C1, whereas the fine painted wares showed clear continuity between the two phases. See Bronitsky and Hamer 1986; Gimbutas 1977; and Marinescu-Bilcu 1981.

18. For the horse-head maces see Telegin et al. 2001; Dergachev 1999; Gheorgiu 1994; and Govedarica and Kaiser 1996.

19. For the skull shapes, see Necrasov 1985; and Marcsik 1971. Gracile "Mediterranean" Tripolye skulls have been found in ritual foundation deposits at Traian (Tripolye B2).

20. For Mirnoe, see Burdo and Stanko 1981.

21. For the eastern migration, see Kruts and Rizhkov 1985.

22. The Iron Age stereotype of nomadic cavalry seems to lie behind some of the writings of Merpert (1974, 1980) and Gimbutas (1977), who were enormously influential.

23. The "awkward seat" hypothesis is based on Near Eastern images that show riders sitting awkwardly on the horse's rump, a seat more suited to donkey riding. Donkeys have low withers

and a high, broad rump. If you sit forward on a donkey and the animal lowers its head, you can easily fall forward to the ground. Donkey riders, therefore, usually sit back on the rump. Horses have high withers, so horse riders sit forward, which also permits them to hang onto the mane. You have to push and lift to get yourself onto a horse's rump, and then there's nothing to hold on to. Artistic images that show riders on horseback sitting back on the rump probably indicate only that many Near Eastern artists before 1000 BCE, particularly in Egypt, were more familiar with riding donkeys than horses. The suggestion that riders in the steppes would adopt and maintain a donkey seat on horses is inherently implausible. See Drews 2004:40–55, for this argument.

24. For mutualism and economic exchanges between Old Europe and the Eneolithic cultures of the Pontic steppe, see Rassamakin 1999:112; see also Manzura, Savva, and Bogotaya 1995; and Nikolova 2005:200. Nikolova has argued that transhumant pastoralism was already part of the Old European economy in Bulgaria, but the Yagodinska cave sites she cited are radiocarbon dated about 3900 BCE, during or just after the collapse. Upland pastoral settlements were a small and comparatively insignificant aspect of the tell economies, and only a serious crisis made them the basis for a new economy.

25. Ewers 1955:10.

26. See Benveniste 1973:53–70, for *Give* and *Take*, esp. 66–67 for the Hittite terms; for the quotation, see 53. Hittite *pai* was derived from the preverb *pe-* with **ai-*, with reflexes meaning "give" in Tocharian *ai-*. Also see the entry for *Give* in Mallory and Adams 1997:224–225.

27. See Keeley 1996. For mutualist models of the Linear Pottery frontier, see Bogucki 1988. An ethnographic case frequently cited in discussions of mutualist food exchange is that of the horticultural Pueblo Indians and the pedestrian buffalo hunters of the Plains. But a recent study by Susan Vehik suggested that the Pueblo Indians and the Plains bison hunters traded prestige commodities—flint arrowheads, painted pottery, and turquoise—not food. And during a period of increasing conflict in the Plains after 1250 CE, trade actually greatly increased; see Vehik 2002.

28. See Kershaw 2000.

29. See "bride-price" in Mallory and Adams 1997:82–83.

30. In East Africa a group of foragers and beekeepers, the Mukogodo, were forced to obtain livestock after they began to interact and intermarry with stock-raising tribes, because it became impossible for Mukogodo men to obtain wives by offering beehives when non-Mukogodo suitors offered cattle. Cattle were just more valuable. The Mukogodo became pastoralists so that they could continue to have children. See Cronk 1989, 1993.

31. Ewers 1955:185–187.

32. The Sredni Stog site had two levels, Sredni Stog 1 and 2. The lower level (Sredni Stog 1) was an Early Eneolithic DDII occupation, and the upper was the type site for the Late Eneolithic Sredni Stog culture. In older publications the Sredni Stog culture is sometimes called Sredni Stog 2 (or II) to differentiate it from Sredni Stog 1 (or I).

33. The Sredni Stog culture is defined in Telegin 1973. The principal settlement site of the Sredni Stog cultre, Dereivka, is described in English in Telegin 1986; for the Sredni Stog origin of Cucuteni C ware, see 111–112. Telegin's chronological outline is described in English in Telegin 1987.

34. The longest and most detailed version of Rassamakin's new model in English is the 123-page article, Rassamakin 1999. Telegin's four phases (Ia, Ib, IIa, IIb) of the Sredni Stog culture represented, for Rassamakin, at least three separate and successive cultures: (1) the Skelya culture, 4500–4000 BCE (named for Strilcha Skelya, a phase Ib Sredni Stog site for Telegin); (2) the Kvityana culture, 3600–3200 BCE (Kvityana was a phase Ia site for Telegin, but Rassamakin moved it to the equivalent of Telegin's *latest* phase IIb); and (3) the Dereivka culture, 3200–3000 BCE (a phase IIa site for Telegin, dated 4200–3700 BCE by radiocarbon). Telegin seemed to stick to the stratigraphy, grave associations, and radiocarbon dates, whereas Rassamakin relied on stylistic arguments.

35. For Sredni Stog ceramics, see Telegin 1986:45–63; 1973:81–101. For skeletal studies, see Potekhina 1999:149–158.

36. For the seeds at Moliukhor Bugor, see Pashkevich 1992:185. For the tools at Dereivka, see Telegin 1973:69, 43. Bibikova actually reported 2,412 horse bones and 52 horse MNI. I have edited out the mandible, skull, and two metacarpals of the "cult stallion."

37. Only four settlement animal bone samples are reported for Sredni Stog. Most of them are worryingly small (a few hundred bones) and screens were not used in excavations (still are not), so bone recovery varied between excavations. For these reasons, the published animal bone percentages can be taken only as rough guides. For an English translation of the faunal reports, see Telegin 1986.

38. Rassamakin (1999:128) assigned the Dereivka cemetery, which he called Dereivka 2, to the Skelya period, before 4000 BCE, and assigned the Dereivka settlement to the Late Eneolithic, around 3300–3000 BCE. Telegin, following the radiocarbon dates from the settlement and the Tripolye B2 bowl found in the cemetery, assigned both to the same period.

39. See Dietz 1992 for the varied interpretations of antler "cheekpieces."

40. For the Suvorovo-Novodanilovka group, see Nechitailo 1996; and Telegin et al. 2001. The metals are analyzed in Ryndina 1998:159–170; for an English summary, see 194–195. English-language discussions of the Suvorovo-Novodanilovka group are few. In addition to Rassamakin's description of the Skelya culture, which incorporates Suvorovo-Novodanilovka, see Dergachev 1999; and Manzura, Savva, and Bogotaya 1995. And there is a useful entry under "Suvorovo" in Mallory and Adams 1997.

41. Telegin 2002, 2001.

42. The physical type in Novodanilovka graves is discussed in Potekhina 1999:149–154. The types of the lower Danube valley are described by Potekhina in Telegin et al. 2001; and in Necrasov and Cristescu 1973.

43. Ryndina (1998:159–170) examined copper objects from graves at Giugiurleşti, Suvorovo, Novodanilovka, Petro-Svistunovo, and Chapli. For the copper of Varna and Gumelnitsa, see Pernicka et al. 1997. They document the end of the Balkan mines and the switch to Carpathian ores at about 4000 BCE.

44. The horse-head examples in the Volga steppes were found at Novoorsk near Orenburg and at Lebyazhinka near Samara. For the polished stone mace heads, see Kriukova 2003.

45. For Old European weapons, see Chapman 1999.

46. *Equus hydruntinus* had a special ritual status in the cemeteries of Varna and Durankulak, but was unimportant in the diet and was on the brink of extinction. Horses (*Equus caballus*) were rare or absent in the Eneolithic settlements and cemeteries of the Danube valley before the Cernavoda I period, except for sites of the Bolgrad variant. The Gumelniţa-related Bolgrad sites had about 8% horse bones. Other Old European sites in the Danube valley had few or no horses. For the Varna and Durankulak equids, see Manhart 1998.

47. See Vehik 2002 on increased warfare and long-distance trade in the Southwest. DiCosmo (1999) observed that increased warfare in the steppes encouraged organizational changes in preexisting institutions, and these changes later made large nomadic armies possible.

48. Contacts between late Tripolye A/early B1 settlements and the Bolgrad culture are summarized in Burdo 2003. Most of the contact is dated to late Tripolye A—Tripolye AIII2 and III3.

49. For Bolgrad sites, see Subbotin 1978, 1990.

50. For the intrusive cemeteries, see Dodd-Opriţescu 1978. For the gold and copper hoards, see Makkay 1976.

51. For the Suvorovo kurgan group, see Alekseeva 1976. The Kopchak kurgan is described in Beilekchi 1985.

52. Giurgiuleşti is described briefly in Haheu and Kurciatov 1993. One radiocarbon date is published from Giurgiuleşti: Ki-7037, 5380 ± 70 BP, or about 4340–4040 BCE, calibrated; I have been told that the date is misprinted in Telegin et al. 2001, 128.

53. The Novodanilovka grave, which was isolated and not in a cemetery, is described in Telegin 1973:113; for Petro-Svistunovo and Chapli, see Bodyans'kii 1968; and Dobrovol'ski 1958.

54. The region-wide abandonment of tells in about 4000–3500 BCE is observed in Coleman 2000. I do not see how this could have been the event that brought Greek speakers into Greece, because Greek shared many traits with the Indo-Iranian language branch (see the end of chapter 3), and Indo-Iranian emerged much later. The crisis of 4000 BCE probably brought Pre-Anatolian speakers into southeastern Europe.

55. See Madgearu 2001 on de-urbanization in post–Roman Bulgaria. Mace (1993) notes that if grain production falls, cattle are insurance against starvation. Cattle can be moved into a protected area during a period of conflict. Under conditions of declining agricultural yields and increasing conflict, a shift to a greater reliance on herding would make good economic sense.

56. For loot, lucre, and booty in Proto-Indo-European, see Benveniste 1973:131–137; for language shift among the Pathan, see Barth 1972.

57. For Cernavoda I, see Morintz and Roman 1968; and Roman 1978; see also Georgieva 1990; Todorova 1995; and Ilčeva 1993. A good recent summary is in Manzura 1999. For the cemetery of Ostrovul Corbului, see Nikolova 2002, 2000.

58. Sherratt 1997b, 1997c. Sherratt suggested that the drinking vessels of the period from 4000 to 2500 BCE were used to serve a beverage that included honey (the basis of mead) and grain (the source of beer), both directly attested in Early Bronze Age Bell Beaker cups. Honey, he suggested, would have been available only in small quantities, and might have been under the control of an elite who apportioned the fermented drink in ceremonies and closed gatherings open to just their inner circle. Proto-Indo-European contained a word for honey (*melit-) and a derivative term for a honey drink (*medhu-).

59. For Cernavoda I-Late Lengyel horses, see Peške 1986; and Bökönyi 1979.

60. For pastoralism, see Greenfield 1999; Bökönyi 1979; and Milisauskas 2002:202.

61. For the prayer to Sius, see Puhvel 1991.

CHAPTER 12. SEEDS OF CHANGE ON THE STEPPE BORDERS

1. Ryndina (1998:170–171) counted 79 copper objects from steppe graves for the Post-Suvorovo period, compared to 362 for Suvorovo-Novodanilovka graves.

2. See Telegin 2002, 1988, 1987; see also Nikolova and Rassamakin 1985; and Rassamakin 1999. Early reports on Mikhailovka are Lagodovskaya, Shaposhnikova, and Makarevich 1959; Shaposhnikova 1961 (this was the article where the division between lower and upper stratum 2 was noticed); and Shevchenko 1957. For the stratigraphic position of Lower Mikhailovka graves, see Cherniakov and Toshchev 1985. Radiocarbon dates for graves with Mikhailovka I pottery are reported in Videiko and Petrenko 2003. Early Mikhailovka II begins about 3500 BCE, in Kotova and Spitsyna 2003.

3. For the Maikop sherd at Mikhailovka I, see Nechitailo 1991:22. For the other pottery exchanges, see Rassamakin 1999:92; and Telegin 2002:36.

4. Pashkevich 2003.

5. The sheep of the Early Bronze Age in southeastern Europe were significantly larger than Eneolithic sheep, which Bökönyi (1987) attributed to a new breed of wool sheep that appeared after about 3500 BCE.

6. At the Cernavoda site three excavation areas yielded three successive archaeological cultures, of which the oldest was Cernavoda I, about 4000–3600 BCE; next was Cernavoda III, about 3600–3000 BCE, contemporary with Baden; and the youngest was Cernavoda II, 3000–2800 BCE. Mikhailovka I probably was contemporary with the end of Cernavoda I and the first half of Cernavoda III. See Manzura, Savva, and Bogatoya 1995.

7. For Mikhailovka I graves at Olaneshti, see Kovapenko and Fomenko 1986; and for Sokolovka, see Sharafutdinova 1980.

8. Potekhina 1999:150–151.

9. "Post-Mariupol" was the label first assigned by Kovaleva in the 1970s. See Nikolova and Rassamakin 1985; Telegin 1987; and Kovaleva 2001.

10. See Ryndina 1998:170–179, for Post-Mariupol metal types.

11. The two graves were Verkhnaya Maevka XII k. 2, gr. 10; and Samarska k.1, gr. 6 in the Orel-Samara region. See Ryndina 1998:172–173.

12. For Razdorske, see Kiyashko 1987, 1994.

13. The percentage of horse bones at Repin is often said to be 80%. Shilov (1985b) reviewed the numbers and came up with 55% horse bones, still a very high number.

14. For Repin/Yamnaya at Cherkasskaya, see Vasiliev and Siniuk 1984:124–125.

15. For Kara Khuduk and Kyzyl-Khak, see Barynkin and Vasiliev 1988; for the fauna, see I. Kuzmina 1988. Also see Ivanov and Vasiliev 1995; and Barynkin, Vasiliev, and Vybornov 1998. For the radiocarbon dates for Kyzyl Khak, see Lavrushin, Spiridonova, and Sulerzhitskii 1998:58–59. For late Khvalynsk graves on the lower Volga, see Dremov and Yudin 1992; and Klepikov 1994.

16. Kruts typed the Chapaevka ceramics as late Tripolye C1, whereas Videiko described Chapaevka as a late Tripolye B2 settlement. See Kruts 1977; and Videiko 2003. Videiko argued that ceramic craft traditions changed at different rates in different settlement groups. Tripolye B2 stylistic habits lingered longer, he suggested, in the Dnieper group (Chapaevka) than they did in the super-settlements of the South Bug group, which shifted to Tripolye C1 styles earlier. Tripolye C2 styles began on the Dniester at Usatovo about 3400–3300 BCE, but Tripolye C2 styles appeared on the Dnieper about 3100 BCE.

17. Kruts 1977:48.

18. For the super-sites, see Videiko 1990, and other articles in the same volume; also see Shmagli and Videiko 1987 and Kohl 2007.

19. At Maidanets'ke, emmer and spelt wheats were the most common cereals recovered; barley and peas also were found in one house. Cattle (35% of domesticates, MNI) were the most important source of meat, with pig (27%) and sheep (26%) as secondary sources; the remaining 11% was equally divided between dogs and horses. About 15% of the animals were red deer, wild boar, bison, hare, and birds. The cattle, pigs, and abundant wild animals indicate substantial forest near the settlement. A forest of about 20 <km^2 would have provided sufficient firewood for the town, figuring about 2.2 ha of hardwood forest per family of five for a sustainable woodlot. Since ecological degradation is not obvious, the abandonment of the town perhaps was caused by warfare. See Shmagli and Videiko 1987:69, and several articles on economy in the volume cited above as Videiko 1990.

20. The Tripolye B1 settlement of Polivanov Yar on the Dniester overlooked outcrops of high-quality flint. One house was engaged heavily in flint working, with all stages of the tool-making process. In the later Tripolye C1 settlement, all six excavated structures were engaged in flint working, the initial shaping occurred elsewhere, and new products were made (heavy flint axes and chisels about 10 cm long). The Tripolye C1 settlement had become a specialized village of flint workers. Maidanets'ke imported finished flint tools of Dniester flint, probably from Polivanov Yar. At Veseli Kut (150 ha), a Tripolye B2 town east of the South Bug valley, two structures were identified as ceramic workshops. Eight buildings dedicated to ceramic production were found at Varvarovka VIII (40 ha and 200 houses—the largest town in its region), and a similar ceramic factory appeared at Petreni on the Dniester, again the largest town in its area. At Maidanets'ke, eight houses in a row contained looms (indicated by clusters of up to seventy ceramic loom weights) and some had two looms, perhaps a specialized weaver's quarter. For Polivanov Yar, see Popova 1979; for ceramic workshops, see Ellis 1984.

21. For the Uruk expansion, see Algaze 1989; Stein 1999; and Rothman 2001. For copper production at Hacinebi, see Özbal, Adriaens, and Earl 2000; for the copper of Iran, see Matthews and Fazeli 2004. For the wool sheep, see Bökönyi 1983; and Pollack 1999.

22. For Sos and Berikldeebi, see Kiguradze and Sagona 2003; and Rothman 2003.

23. The Maikop-like pottery was found in pre-Kura-Araxes levels at Berikldeebi. Early Maikop began before the Early Transcaucasian Culture. See Glonti and Dzhavakhishvili 1987.

24. For pre-Maikop Svobodnoe, see Nekhaev 1992; and Trifonov 1991. For steppe-Svobodnoe exchanges, see Nekhaev 1992; and Rassamakin 2002.

25. The poses of those buried in the Maikop chieftain's grave were not clear. For an English-language description of the Maikop culture, see Chernykh 1992:67–83. Quite dated accounts are Childe 1936; and Gimbutas 1956:56–62. A long, detailed description in Russian is in Munchaev 1994. For the Novosvobodnaya graves, see Rezepkin 2000. For the archaeological culture history in the North Caucasus, see Trifonov 1991.

26. For the silver and gold staff casings with bulls, see Chernopitskii 1987. The 47-cm length of the riveted copper blade is emphasized in Munchaev 1994:199.

27. Rostovtseff (1922:18–32) argued that Maikop was a Copper Age or, in Anatolian terms, a Late Chalcolithic culture. But Maikop became established as a North Caucasian Bronze Age culture, so it begins somewhat earlier than the Anatolian Bronze Age to which it was originally linked. Some Russian archaeologists now suggest an early Maikop phase that would be Late Eneolithic, whereas later Maikop would remain Early Bronze Age. For Maikop chronology, see Trifonov 1991, 2001. For my own mistaken chronology, see Glumac and Anthony 1992. I should have believed Rostovtseff.

28. For the east Anatolian seal, see Nekhaev 1986; and Munchaev 1994:169, table 49:1–4.

29. For Galugai, see Korenevskii 1993, 1995; the fauna is described in 1995:82. Korenevskii considered Galugai a pioneer settlement by migrants from Arslantepe VIA. For Maikop horses, see Chernykh 1992:59.

30. Rezepkin (1991, 2000) argued that Maikop and Novosvobodnaya were separate and contemporary cultures. Similar radiocarbon dates from Galugai (Maikop) and Klady (Novos-vobodnaya) suggested this. But the radiocarbon dates for Galugai are on charcoal and those from Klady are on human bone, which might be affected by old carbon in fish if the Klady people ate a lot of fish. Adjusted for a ^{15}N content of 11%, which would be at the low end of the levels known in the steppes, the *oldest* Klady dates might drop from about 3700–3500 to about 3500–3350 BCE. I follow the traditional view and represent Novosvobodnaya as an outgrowth of Maikop. Rezepkin compared Novosvobodnaya pottery to TRB or Funnel Beaker pottery from Poland, and megalithic porthole graves at Klady to TRB dolmen porthole graves. He suggested that Novosvobodnaya began with a migration from Poland. Sergei Korenevskii (1993) tried to bring the two phases back into a single culture. Black burnished pottery is found in central Anatolia at Late Chalcolithic and at EBI sites such as Kösk Höyük and Pinarbişi, a closer alternative source.

31. Shishlina, Orfinskaya, and Golikov 2003.

32. See Kiguradze and Sagona 2003:89, for the beads at Alikemek Tepesi.

33. The Maikop-Novosvobodnaya connections of the Sé Girdan kurgans were noticed by A. D. Rezepkin and B. A. Trifonov; both published Russian-language articles describing these connections in 2000. These were brought to Muscarella's attention in 2002 by Elena Izbitser at the Metropolitan Museum of Art in New York. Muscarella (2003) reviewed this history.

34. For the symbolic power of long-distance trade, see Helms 1992. For primitive valuables, see Dalton 1977; and Appadurai 1986.

35. For the Novosvobodnaya wagon grave, see Rezepkin and Kondrashov 1988:52.

36. Shilov and Bagautdinov 1998.

37. See Nechitailo 1991, for Maikop-steppe contacts. Rassamakin (2002) suggested that Late Tripolye migrants of the Kasperovka type influenced the formation of the Novosvobod-naya culture.

38. Cannabis might have been traded from the steppes to Mesopotamia. Greek *kánnabis* and Proto-Germanic **hanipiz* seem related to Sumerian *kunibu*. Sumerian was dead as a widely

spoken language by about 1700 BCE, so the connection must have been a very ancient one, and the international trade of the Late Uruk period provides a suitable context; see Sherratt 2003, 1997c. Wine could have been a linked commodity; the Greek, Latin, Armenian, and Hittite roots for "wine" are cognates, and some linguists feel that the root was of Semitic or Afro-Asiatic origin. See Hock and Joseph 1996:513.

39. For Caucasian horses, see Munchaev 1982; Mezhlumian 1990; and Chernykh 1992:59. For Norşuntepe and Anatolia, see Bökönyi 1991.

CHAPTER 13. WAGON DWELLERS OF THE STEPPE

1. For climate change at the beginning of the Yamnaya period, see Kremenetski 1997b, 2002.

2. The *ghos-ti-* root survived only in Italic, Germanic, and Slavic, but the institution was more widespread. See Benveniste 1973:273–288 on *Philos*, and entries in Mallory and Adams 1997 on *guest* and *friend*. Ivanov suggested that Luwian *kaši-* 'visit' might possibly be cognate with Proto-Indo-European *ghos-ti-*, but the relationship was unclear. See Gamkrelidze and Ivanov 1995:657–658, for their discussion of *hospitality*. In later Indo-European societies, this institution was critical for the protection of merchants and visiting elites or nobles; see Kristiansen and Larsson 2005:236–240. See also Rowlands 1980.

3. As Mallory has noted, the eastern Indo-European branches did have some agricultural vocabulary. The eastern Indo-Europeans talked about plowed fields, grain, and chaff. The archaeological contrast between east and west is more extreme than the linguistic one, which perhaps reflects the difference between what people knew and could talk about (language) and how they actually behaved most of the time (archaeology). See entries on *agriculture, field*, and *plow* in Mallory and Adams 1997.

4. For the feminine gender as one of the ten innovations distinguishing classic Proto-Indo-European from the archaic form preserved in Anatolian, see Lehrman 2001. For the Afro-Asiatic loans in western Indo-European, see Hock and Joseph 1996:513. For Rudra's female consorts, see Kershaw 2000:212

5. Gimbutas 1956:70ff. I would never have thought it possible to penetrate the archaeology of Eastern Europe had it not been for this pioneering English-language synthesis, which opened the door. Nevertheless, I soon began to disagree with her; see Anthony 1986. I was very pleased to spend a few days with her in 1991 at a the National Endowment for the Humanities conference in Austin, Texas, organized by Edgar Polomé.

6. The hundred-year anniversary of Gorodtsov's 1903 archaeological expedition on the Northern Donets River was celebrated by three conferences on the Bronze Age (or at least three were planned). The first conference was in Samara in 2001, and the proceedings make a valuable primer on the Bronze Age cultures of the steppes. See Kolev et al. 2001.

7. See Merpert 1974:123–146, for the Yamnaya "cultural-historical community."

8. This steppe-pine-forest vegetation community is designated number 19 in the Atlas SSSR, 1962, edited by S. N. Teplova, 88–89. It occurs both in the lowland and mountain steppe environments.

9. Afanasievo radiocarbon dates are listed in table 13.3. Most of the Afanasievo dates appear to be on wood from the graves, but some are on human bone. Although I have not seen ^{15}N measurements for Afanasievo individuals, later skeletons from graves in the Altai had ^{15}N levels of 10.2 to 14.3%. Applying the correction scale I am using in this book, the Afanasievo dates taken on bone might be too old by 130 to 375 radiocarbon years. I have not corrected them, because, as I said, most appear to have been measured on samples of wood taken from graves, not human bone.

10. V. N. Logvin (1995) noted that some undated flat-grave cemeteries in northern Kazakhstan might represent a short-lived mixture of early Yamnaya or Repin and Botai-Tersek people. For the Karagash kurgan, see Evdokimov and Loman 1989.

11. The pottery in the earliest Yamnaya graves in the Volga-Ural region (Pokrovka cemetery I, k. 15, gr. 2; Lopatino k. 1, gr. 31; Gerasimovka II, k. 4, gr. 2) was Repin-influenced; and the pottery in the earliest Afanasievo kurgans (Bertek 33, Karakol) in the Gorny-Altai region also looks Repin-influenced.

12. For Afanasievo, see Molodin 1997; and Kubarev 1988. On the craniometrics, see Hemphill and Mallory 2003; and Hemphill, Christensen, and Mustafakulov 1997. For the faunal remains from Balyktyul, see Alekhin and Gal'chenko 1995.

13. On the local cultures, see Weber, Link, and Katzenberg 2002; also Bobrov 1988.

14. Chernykh 1992:88; Chernykh, Kuz'minykh, and Orlovskaya 2004.

15. For Tocharian linkages to Afanasievo, see Mallory and Mair 2000.

16. See Gei 2000:176, for the count of all steppe vehicle graves, and for the wagons of the Novotitorovskaya culture. For the Yamnaya wagon grave at Balki kurgan, see Lyashko and Otroshchenko 1988. For the Yamnaya vehicle at Lukyanovka, see Mel'nik and Serdyukova 1988. For the Yamnaya vehicle graves north of the Danube delta, see Gudkova and Chernyakov 1981. The Yamnaya vehicle graves at Shumaevo cemetery II, kurgans 2 and 6, were the first wagon graves found in the Volga-Ural region in decades, excavated by M. A. Turetskii and N. L. Morgunova in 2001–2002. One wheel was recognized in kurgans 6 and three in kurgan 2; see Morgunova and Turetskii 2003. For early wheeled vehicles in general, see Bakker, et al. 1999.

17. Mel'nik and Serdiukova (1988:123) suggested that Yamnaya wagons had no practical use but were purely ritual imitations of vehicles used in the cults of Near Eastern kings. This ascribes to the Yamnaya people more veneration of distant Near Eastern symbols and less practical sense than seems likely to me. It also leaves unexplained the Yamnaya shift to an economy based on mobility. Even if some of the wagons placed in graves *were* lightly built funeral objects, that does not mean that sturdier originals did not exist.

18. Izbitser (1993) asserted that all these steppe vehicles, including those in graves where only two wheels were found, were four-wheeled wagons. Her opinion has been cited in arguments over the origin of the chariot to suggest that the steppe cultures perhaps had no experience making two-wheeled vehicles; see Littauer and Crouwel 1996:936. But many graves contain just two wheels, including Bal'ki kurgan, grave 57. The image on the Novosvobodnaya cauldron at Evdik looks like a cart. Ceramic cart models associated with the Catacomb culture (2800–2200 BCE) and in the North Caucasus at the Badaani site of the ETC or Kura-Araxes culture (3500–2500 BCE) are interpreted by Izbitser as portraying something other than vehicles. Gei, on the other hand, sees evidence for both carts and wagons, as do I. See Gei 2000:186.

19. The Dnieper region of Merpret 1974 was divided into no fewer than six microregions by Syvolap 2001.

20. Telegin, Pustalov, and Kovalyukh 2003.

21. See Sinitsyn 1959; Merpert 1974; and Mallory 1977. For reconsiderations of Merpert's scheme in the light of the discovery of the Khvalynsk culture, see Dremov and Yudin 1992; and Klepikov 1994. For a review of all the early Yamnaya variants in the Volga-Don-Caucasus region, and their chronology, see Vasiliev, Kuznetsov, and Turetskii 2000.

22. Whereas Mikhailovka I produced 1,166 animal bones, Mikhailovka II and III together yielded 52,540 bones.

23. For Yamnaya seed imprints, see Pashkevich 2003. Pashkevich identifies Mikhailovka II as a settlement of the Repin culture, reflecting the debate about its ceramic affiliation referred to in the text; see also Kotova and Spitsyna 2003.

24. For Yamnaya and Catacomb chronology, see Trifonov 2001; Gei 2000; and Telegin, Pustalov, and Kovalyukh 2003. For western Yamnaya and Catacomb dates, see Kośko and Klochko 2003.

25. These views were well stated by Khazanov (1994) and Barfield (1989).

26. For grain cultivation by steppe nomads, see Vainshtein 1980; and DiCosmo 1994. For modern nomads who ate very little grain, see Shakhanova 1989. For the growth of bodyguards into armies, see DiCosmo 1999, 2002.

27. See Shilov 1985b.

28. For a study of seasonal indicators in kurgans in the Kalmyk steppes, see Shishlina 2000. For comments on the Yamnaya herding pattern in the Dnieper steppes, see Bunyatyan 2003.

29. For Samsonova, see Gei 1979. For Liventsovka, see Bratchenko 1969. The predominance of cattle at these places is mentioned in Shilov 1985b:30.

30. Surface scatters of Yamnaya lithics and ceramics in the Manych Depression in Kalmykia are mentioned by Shishlina and Bulatov 2000; and in the lower Volga and North Caspian steppes by Sinitsyn 1959:184. Desert or semi-desert conditions in these places make surface sites more visible than they are in the northern steppes, where the sod hides the ground. In the Samara oblast we found LBA occupations 20–30 cm beneath the modern ground surface; see Anthony et al. 2006. The winter camps of the Blackfeet are described in Ewers 1955:124–126: "Green Grass Bull said that bands whose members owned large horse herds had to move camp several times each winter. . . . However, a short journey of less than a day's march might bring them to a new site possessing adequate resources for another winter camp . . . Demands on fuel and grass were too great to allow all the members of a tribe to winter in one large village." This kind of behavior might make Yamnaya camps hard to find.

31. The Tsa-Tsa grave is described in Shilov 1985a.

32. Yamnaya dental pathologies in the middle Volga region with comparative data from Hsiung-Nu and other cemeteries were studied by Eileen Murphy at Queen's University Belfast as part of the Samara Valley Project. The unpublished internal report is in Murphy and Khokhlov 2004; see also Anthony et al. 2006. For caries in different populations, see Lukacs 1989.

33. For phytoliths in Yamnaya graves, see Shishlina 2000. The yields of *Chenopodium* and einkorn wheat were compared by Smith 1989. *Amaranthus* has 22% more protein (g/kg) than bread wheat, and *Chenopodium* has 34% more; wheat is higher in carbohydrates than either. For nutrient comparisons, see Gremillion 2004.

34. For the high incidence of curbitra orbitalis among Yamnaya skeletons, see Murphy and Khokhlov 2004; and Anthony et al. 2006.

35. For lactose tolerance, see Enattah 2005.

36. See Vainshtein 1980:59, 72, for comments on cows, milk foods, and poverty.

37. Mallory 1990.

38. On genders in Yamnaya graves, see Murphy and Khokhlov 2004; Gei 1990; Häusler 1974; and Mallory 1990.

39. On "Amazon" graves, see Davis-Kimball 1997; and Guliaev 2003.

40. Alexander Gei (1990) estimated a population density of 8–12 people per 100 km^2 in the EBA Novotitorovskaya and 12–14 per 100 km^2 in the MBA Catacomb periods in the Kuban steppes. But kurgans were erected only for a small percentage of those who died, so Gei's figures undercount the actual population density by an order of magnitude. At ten times his grave-based estimate, or about 120 people per 100 km^2, the population density would have been like that of modern Mongolia, where pastoralism is the dominant element in the economy.

41. Golyeva 2000.

42. For the equation between the status and man-days invested in the funeral, see Binford 1971. See also Dovchenko and Rychkov 1988; Mallory's analysis of their study in Mallory 1990; and Morgunova 1995.

43. The granulated decoration on the two golden rings from Utyevka I, kurgan 1, grave 1, is surprising, since the technique of making and applying golden granulation requires very specific skills that first appeared about 2500 BCE (Troy II, Early Dynastic III). The middle Volga was apparently connected with the Troad through some kind of network at this time. The axe in the Utyevka grave is an early type, similar to the axes of Novosvobodnaya and Yamnaya, and that implies a very early Poltavka date. The grave form and artifact assemblage taken together suggested to Vasiliev a date at the late Yamnaya–early Poltavka transition, so probably about 2800 BCE. The grave has not been dated by radiocarbon. For Utyevka I and its analogies, see

Vasiliev 1980. For the Kutuluk grave with the mace, see Kuznetsov 1991, 2005. For an overview, see Chernykh 1992:83–92.

44. Chernykh 1992:83–92.

45. For the Yamnaya grave at Pershin, see Chernykh; and Isto 2002. For the "clean" copper on the Volga, see Korenevskii 1980.

46. For the Post-Mariupol graves, see Ryndina 1998:170–179; for Lebedi, see Chernykh 1992:79–83; and for Voroshilovgrad, see Berezanskaya 1979.

47. For the iron blade, see Shramko and Mashkarov 1993.

48. Oared longboats are not actually portrayed in surviving art until Early Cycladic II, after 2900–2800 BCE, but the number of settled Cycladic Islands jumped from 10% to 90% for the first time in Early Cycladic I, beginning about 3300 BCE. This was possible only with a reliable form of seagoing transport. Longboats capable of holding twenty to forty oarsmen probably appeared earlier than ECII. See Broodbank 1989.

49. For Kemi-Oba graves in the Odessa oblast, see Subbotin 1995. For stone stelae in the North Pontic steppes generally, see Telegin and Mallory 1994.

CHAPTER 14. THE WESTERN INDO-EUROPEAN LANGUAGES

1. For a good essay on the subject of language shift, see the introduction in Kulick 1992. For Scots Gaelic, see Dorian 1981; see also Gal 1978.

2. For the Galgenberg site of the Cham culture, see Ottaway 1999. Bökönyi saw the statistical source of the larger horses that appeared in Central Europe in the horse population at Dereivka; Benecke suggested that the horses of Late Mesolithic Mirnoe in the steppes north of the Danube delta were a closer match. But both agreed that the source of the new larger breeds was in the steppes. See Benecke 1994:73–74; and Bökönyi 1974.

3. For the Bukhara horse trade, see Levi 2002. I am indebted to Peter Golden and Ranabir Chakravarti for calling my attention to it.

4. Polomé 1991. For the translation of the *Rig Veda* passage, see O'Flaherty 1981:92.

5. See Kristiansen and Larsson 2005:238.

6. See Benveniste 1973:61–63 for feasts; also see the entry for GIVE in Mallory and Adams 1997:224–225; and Markey 1990. For poets, see Watkins 1995:73–84. For the general importance of feasting in tribal societies, see Dietler and Hayden 2001. For an ethnographic parallel where chiefs and poets were mutually dependent, see Lehman 1989.

7. Mallory (1998) referred to this process using the wry metaphor of the *Kulturkugel*, a bullet of language and culture that acquired a new cultural skin after penetrating a target culture, but retained its linguistic core.

8. A broad scatter of kurgan graves in the steppes contained imported Tripolye C2 pots (among other imported pot types) and a few, like Serezlievka, also contained Tripolye-like schematic rod-headed figurines. The Serezlievka-type graves in the South Bug valley probably were contemporary with Yamnaya graves of the Zhivotilovka-Volchansk group in the Dnieper-Azov steppes that also contained imported Tripolye C2 pots, dated by radiocarbon about 2900–2800 BCE. Rassamakin (1999, 2002) thought that Zhivotilovka-Volchansk graves represented a migration of Tripolye C2 people from the forested upper Dniester deep into the steppes east of the Dnieper. But a Tripolye pot in a Yamnaya grave is most simply interpreted as a souvenir, gift, or acquisition rather than as a migrant Tripolye person. Yamnaya graves rarely contained any pots. Cotsofeni pots filled that customary void in the Yamnaya graves of the Danube valley, just as pottery of the Tripolye C2, late Maikop, and Globular Amphorae types did in the Ukrainian steppes.

9. For the Usatovo culture see Zbenovich 1974; Dergachev 1980; Chernysh 1982; and Patovka et al. 1989. For a history of excavations at Usatovo, see Patovka 1976. The Cernavoda I affiliations of pre-Usatovo coastal steppe kurgans are discussed in Manzura, Savva and Boga-

toya 1995. A Cernavoda I feature in Usatovo is described in Boltenko 1957:42. Recent radiocarbon dates are discussed in Videiko 1999.

10. For Usatovo fauna see Zbenovich 1974: 111–115.

11. For spindle whorls, see Dergachev 1980:106.

12. See Kuz'minova 1990, for Usatovo paleobotany.

13. For Usatovo ceramics, see Zbenovich 1968, with a brief notice of the orange-slipped grey wares on page 54.

14. For trade between Usatovo, late Cernavoda III, and late Maikop, see Zbenovich 1974:103, 141. The single glass bead at Usatovo was colored white by the inclusion of phosphorus. It was in a grave pit covered by a stone lid, a stone cairn, and then by the kurgan. The pear-shaped bead measured 9 mm in diameter, had a hole 5 mm in diameter, and had slightly darker spiraling on its surface. Two cylindrical glass beads, colored with copper (green-blue) were recovered from the Tripolye C2 grave 125 at Sofievka on the Dnieper near Kiev, dated a century or two later, about 3000–2800 BCE (4320+70 BP, 4270+90 BP, 4300+45 BP, from three other graves at Sofievka). Two other glass beads were found on the surface near this grave but certainly were not from it. The glass in both Sofievka and Usatovo was made with ash as an alkali, not soda. An ash recipe was used in the Near East. For analyses, see Ostroverkhov 1985. For the radiocarbon dates from Sofievka and the amber beads from Zavalovka, see Videiko 1999.

15. For the daggers, see Anthony 1996. For oared longboats, see the end of the last chapter of this volume, and Broodbank 1989.

16. For the ochre-painted skulls, see Zin'kovskii and Petrenko 1987.

17. For Zimnea, see Bronicki, Kadrow, and Zakościelna 2003; see also Movsha 1985; and Kośko 1999.

18. For fortifications, see Chernysh 1982:222.

19. See Boyadziev 1995, for the dating of the migration.

20. For the large cluster in Hungary, see Ecsedy 1979, 1994. For the cluster in Oltenia, see Dumitrescu 1980. For the cluster in northern Serbia, see Jovanovich 1975. For Bulgaria, see Panayotov 1989. For overviews see, Nikolova 2000, 1994. For relative chronologies at the time of the migration event in southeastern Europe generally, see Parzinger 1993. For the wagon grave at Plachidol, see Sherratt 1986. For the stone stelae, see Telegin and Mallory 1994. Ecsedy mentions that undecorated stone stelae were found near Yamnaya kurgans in Hungary.

21. The graves in Hungary could possibly have been the result of a separate migration stream that passed directly over the Carpathians through Late Tripolye territory rather than being a continuation of the lower Danube valley stream.

22. Most of the radiocarbon dates for Yamnaya graves in the Odessa oblast, the heart of the Dniester steppes, are quite late, beginning about 2800–2600 BCE, by which time the Usatovo culture was gone. There are a few earlier radiocarbon dates (Semenovskii, k.11, 14; Liman, k.2; Novoseltsy, k.19), but in both of the Semenovskii kurgans the primary grave for which the kurgan was raised was an Usatovo grave, and all the Yamnaya graves were secondary. The stratigraphy makes me wonder about the early radiocarbon dates. Yamnaya seems to have taken over the Odessa oblast steppes after the Usatovo culture. See Gudkova and Chernyakov 1981; and Subbotin 1985.

23. Kershaw 2000; see also entries on *korios* and warfare in Mallory and Adams 1997. The cattle raid, a related institution, is discussed in Walcot 1979.

24. For Yamnaya dog-tooth ornaments on the Ingul, see Bondar and Nechitailo 1980.

25. For the stelae of the steppes, see Telegin and Mallory 1994. For the symbolic importance of belts, see Kershaw 2000:202–203; and Falk 1986:22–23.

26. Kalchev 1996.

27. Nikolova 1996.

28. Alexandrov 1995.

29. Panayotov 1989:84–93.

30. Barth 1965:69.

31. Bell Beaker decorated cup styles, domestic pot types, and grave and dagger types from the middle Danube were adopted about 2600 BCE in Moravia and Southern Germany. This material network could have been the bridge through which pre-Celtic dialects spread into Germany. See Heyd, Husty, and Kreiner 2004, especially the final section by Volker Heyd.

32. See Hamp 1998; and Schmidt 1991, for connections between Italic and Celtic.

33. For the effects of wheeled vehicles, see Maran 2001.

34. See Szmyt 1999, esp. 178–188.

35. On the Slavic homeland, see Darden 2004.

36. Coleman (2000) argued that Greek speakers entered Greece during the Final Neolithic/Bronze Age transition, about 3200 BCE. If an Indo-European language spread into Greece this early I think it was more likely an Anatolian-type language. For a northern steppe origin for Greek, but in a later era more amenable to my scenario, see Lichardus and Vladar 1996; and Penner 1998. The same evidence is marshaled for another purpose in Makkay 2000, and in detail by Kristiansen and Larsson 2005. Another argument for a northern connection of the Shaft Grave princes is presented in Davis 1983. Connections between southeastern Europe and Greece are outlined in Hänsel 1982. Robert Drews (1988) also argued that the Shaft Grave princes were an immigrant dynasty from the north, although he derived them from Anatolia.

37. Mallory 1998:180.

CHAPTER 15. CHARIOT WARRIORS OF THE NORTHERN STEPPES

1. See Gening, Zdanovich, and Gening 1992, for the original report on Sintashta.

2. The Sintashta culture remained unrecognized as recently as 1992. Chernykh (1992:210–234) discussed Sintashta-type metals as part of the "Andronovo historico-cultural community," assigning it to about 1600–1500 BCE. Dorcas Brown and I visited Nikolai Vinogradov in 1992, and I was permitted to take bone samples from the chariot grave at Krivoe Ozero for radiocarbon dating. This resulted in two articles: Anthony 1995a; and Anthony and Vinogradov 1995. See Vinogradov 2003, for the complete report on the Krivoe Ozero cemetery. For the settlement and cemeteries at Arkaim, see Zdanovich 1995; and Kovaleva and Zdanovich 2002. For the Sintashta cemetery at Kammeny Ambar, see Epimakhov 2002. For a wide-ranging overview, see Grigoriev 2002, marred by the assumption that the Sintashta culture and many other steppe cultures originated from a series of south-to-north folk migrations from Anatolia and Syria, where he argued that the Indo-European homeland was located. See Lamberg-Karlovsky 2002, for connections to Central Asia. For conference proceedings, see Jones-Bley and Zdanovich 2002; Boyle, Renfrew, and Levine 2002; and Levine, Renfrew, and Boyle 2003.

3. I use the term *Aryan* here as it is defined it in chapter 1, as the self-designation of the people who composed the hymns and poems of the *Rig Veda* and *Avesta* and their immediate Indo-Iranian ancestors.

4. For the contact zone between Corded Ware, Globular Amphorae, and Yamnaya at about 2800–2600 BCE, see Szmyt 1999, esp. pp. 178–188. Also see Machnik 1999; and Klochko, Kośko, and Szmyt 2003. A classic review of the archaeological evidence for mixed Yamnaya, late Tripolye (Chapaevka), and Corded Ware elements in Middle Dnieper origins is Bondar 1974. A recent review emphasizes the Yamnaya influence on the Middle Dnieper culture, in Telegin 2005.

5. For Middle Dnieper chronology, see Kryvaltsevich and Kovalyukh 1999; and Yazepenka and Kośko 2003.

6. Machnik 1999.

7. Before the Middle Dnieper culture appeared, the east side of the river near Kiev had been occupied between about 3000 and 2800 BCE by the mixed-origin late Tripolye C2 Sofievka group, which cremated its dead, used riveted daggers like those at Usatovo, and made

pottery that showed both cord-impressed steppe elements and late Tripolye elements. For the Sofievka settlement, see Kruts 1977:109–138; for radiocarbon dates, see Videiko 1999.

8. See Carpelan and Parpola 2001. This almost monograph-length article covers much of the subject matter discussed in this chapter. For Corded Ware migrations from the genetic point of view, see Kasperavičiūtė, Kučinskas, and Stoneking 2004.

9. For Balanovo, Abashevo, and Volosovo, see Bol'shov 1995. For Abashevo ceramics, see Kuzmina 1999. The classic work on Abashevo is Pryakhin 1976, updated in Pryakhin 1980. For an English account, in addition to Carpelan and Parpola 2001, see Chernykh 1992:200–204 and Koryakova and Epimakhov 2007.

10. For the Volosovo culture, see Korolev 1999; Vybornov and Tretyakov 1991; and Bakharev and Obchinnikova 1991.

11. For Abashevo and Indo-Iranian linkages, see Carpelan and Parpola 2001; and Pryakhin 1980.

12. For the headbands, see Bol'shov 1995.

13. See Keeley 1996, on tribal war.

14. See Koivulehto 2001; and Carpelan and Parpola 2001.

15. See Ivanova 1995:175–176, for the Aleksandrovska IV kurgan cemetery.

16. For Kuisak settlement, see Maliutina and Zdanovich 1995.

17. In Table 1, sample AA 47803, dated ca. 2900–2600 BCE, was from a human skeleton of the Poltavka period that was later cut through and decapitated by a much deeper Potapovka grave pit. A horse sacrifice above the Potapovka grave is dated by sample AA 47802 to about 1900–1800 BCE. Although they were almost a thousand years apart, they looked, on excavation, like they were deposited together, with the Potapovka horse skull lying above the shoulders of the decapitated Poltavka human. Before dates were obtained on both the horse and the skeleton this deposit was interpreted as a "centaur"—a decapitated human with his head replaced by the head of a horse, an important combination in Indo-Iranian mythology. But Nerissa Russell and Eileen Murphy found that both the horse and the human were female, and the dates show that they were buried a thousand years apart. Similarly sample AA-12569 was from an older Poltavka-period dog sacrifice found on the ancient ground surface at the edge of Potapovka grave 6 under kurgan 5 at the same cemetery. Older Poltavka sacrifices and graves were discovered under both kurgans 3 and 5 at Potapovka cemetery I. The Poltavka funeral deposits were so disturbed by the Potapovka grave diggers that they remained unrecognized until the radiocarbon dates made us take a second look. The "centaur" possibility was mentioned in Anthony and Vinogradov 1995, five or six years before the two pieces were dated. Of course, it now must be abandoned.

18. For Sarazm, see Isakov 1994.

19. For Kelteminar, see Dolukhanov 1986; and Kohl, Francfort, and Gardin 1984. The classic work on Kelteminar is Vinogradov 1981.

20. For a radiocarbon date from Sergeivka, see Levine and Kislenko 2002, but note that their discussion mistakenly assigns it to the Andronovo period, 1900–1700 BCE. See also Kislenko and Tatarintseva 1990. Another transitional forager-herder group influenced by Poltavka was the Vishnevka 1 pottery group in the forest-steppe on the northern Ishim; see Tatarintseva 1984. For Sergeivka sherds at the Poltavka cemetery of Aleksandrovka, see Maliutina and Zdanovich 1995:105.

21. For climate deterioration, see Blyakharchuk et al. 2004; and Kremenetski 2002, 1997a, 1997b.

22. Rosenberg 1998.

23. For the Mesopotamian metal trade, see Muhly 1995; Potts 1999:168–171, 186.

24. For metals and mining, see Grigoriev 2002:84; and Zaikov, Zdanovich, and Yuminov 1995. See also Kovaleva and Zdanovich 2002. Grigoriev suggested that the amount of slag found in each house was so small that it could represent household production. However, slag is often found in small amounts even at industrial sites, and that all houses contained slag and

production facilities (ovens with attached wells that aided in the updraft) shows an intensity of metal production that was unprecedented in the steppes.

25. See DiCosmo 1999, 2002; and Vehik 2002.

26. Ust'e, like Chernorech'e III, was excavated by Nikolai Vinogrado. Vinogradov was kind enough to show me his plans and photographs from Ust'e, where Sintashta houses are clearly stratified beneath a Petrovka occupation.

27. See Epimakhov 2002:124–132 for the artifact catalogue.

28. For the ballistics of flint projectile points, see Knecht 1997; and Van Buren 1974. For javelins in Greek chariot warfare, see Littauer 1972; and Littauer and Crouwel 1983.

29. For the chariot petroglyphs, see Littauer 1977; Samashev 1993; and Jacobsen-Tepfer 1993. On the derivation of steppe cheekpieces from Mycenaean cheekpieces, see E. Kuzmina 1980. For a review of European cheekpieces, see Hüttel 1992. Littauer and Crouwel (1979) argued persuasively for the Near Eastern origin of the chariot, overthrowing pre-World War II suggestions that the chariot was a super-weapon of the steppe Aryans. Piggott (1983, 1992) began to challenge the Near Eastern origin hypothesis almost immediately. Moorey (1986) also supported a multiregional invention of the various elements combined in the chariot.

30. See Epimakhov 2002:124–132 for a grave inventory that totals sixteen chariot graves; see Kuzmina 2001:12 for an estimate of twenty. The sites Kuzmina lists include Sintashta (seven chariot graves), Kamenny Ambar (two), Solntse II (three), Krivoe Ozero (three), and, in northern Kazakhstan, in Petrovka graves, Ulybai (one), Kenes (one), Berlyk II (two), and Satan (one).

31. For arguments against the functionality of steppe chariots, see Littauer and Crouwel 1996; Jones-Bley 2000; and Vinogradov 2003:264, 274. For arguments in favor of the steppe chariots as effective instruments of war, see Anthony and Vinogradov 1995; and Nefedkin 2001.

32. For English descriptions of the narrow-gauge chariots, see Gening 1979; Anthony and Vinogradov 1995; and Anthony 1995a. For two critical replies, see Littauer and Crouwel 1996; and Jones-Bley 2000. For the limitations of the chariot in battle, see Littauer 1972; and Littauer and Crouwel 1983.

33. For Bronze Age steppe bows, see Grigoriev 2002:59–60; Shishlina 1990; Malov 2002; and Bratchenko 2003:199. For ancient bows of the Near East and Iran, see Zutterman 2003.

34. See Littauer 1968.

35. For the disk cheekpieces, see Priakhin and Besedin 1999; Usachuk 2002; and Kuzmina 2003, 1980. For left and right side differences, see Priakhin and Besedin 1999:43–44. For chariots in the *Rig Veda*, see Sparreboom 1985. For the metal examples in the Levant, see Littauer and Crouwel 1986, 2001. This type of cheekpiece probably spread into Mycenaean Greece from southeastern Europe, where it appeared in Otomani, Monteoru, and Vatin contexts. For radiocarbon dates for these cultures, see Forenbaher 1993, and for disk-shaped cheekpieces in those contexts, see Boroffka 1998, and Hüttel 1994. The European origin of Mycenaean chariotry might explain why Mycenaean chariot warriors, like the early charioteers of the northern steppes, sometimes carried spears or javelins. For chariots in Greece, see Crouwel 1981.

36. For a review of the Near Eastern evidence for chariots, see Oates 2003; for older studies, see Moorey 1986, and Littauer and Crouwel 1979. For vehicles at Tell Brak, see Oates 2001:141–154. If we were to accept the "low" chronology, which seems increasingly likely, the date for the end of Ur III and the earliest proto-chariots would shift down from 2000 to 1900 BCE. See Reade 2001.

37. See Stillman and Tallis 1984:25 for Mitanni chariot squadrons; for Chinese chariot squadrons, see Sawyer 1993:5.

38. See Appuradai 1986:21 for the "tournament of values."

39. For human pathologies, see Lindstrom 2002, who notes the complete absence of dental caries, even in the oldest individuals (161). Lindstrom was the first Western archaeologist to participate in excavations at a Sintashta site.

40. Igor Ivanov, a geomorphologist at Arkaim, told me in 2000 that the reports of irrigation channels at Arkaim were mistaken, that these were natural features.

41. See Gening, Zdanovich, and Gening 1992:234–235 for Sacrificial Complex 1, and page 370 for the man-days for the SB kurgan.

42. For feasting in tribal societies, see Hayden 2001.

43. For the fauna, see Kosintsev 2001; and Gaiduchenko 1995. For $N15$ isotopes in human and animal bones, see Privat 2002.

44. For doubts about social hierarchy in Sintashta society, see Epimakhov 2000:57–60.

45. Witzel 1995:109, citing Kuiper 1991.

46. For various theories on how to link Sintashta and the Indo-Iranians, see Parpola 1988, 2004–2005; E. Kuzmina 1994, 2001; and Witzel 2003.

47. All quotations are from O'Flaherty 1981.

48. For the Indo-European dog sacrifice and New Year initiation ceremony, see Kershaw 2000; and Kuiper 1991, 1960.

49. Epimakhov 2002; and Anthony et al. 2005.

Chapter 16. The Opening of the Eurasian Steppes

1. For exotic knowledge and power, see Helms 1992.

2. For Indic terms among the Mitanni, see chapter 3; Thieme 1960; and Burrow 1973.

3. Elamite was a non–Indo-European language of uncertain affiliations. As Dan Potts stressed, the people of the western Iranian highlands never used this or any other common term as a blanket ethnic designation for themselves. They did not even all speak Elamite. See Potts 1999:2–4. For the appearance of horses, see Oates 2003.

4. See Weiss 2000; also Perry and Hsu 2000.

5. At Godin Tepe, onagers were 94% of the equid bones. A cheektooth and a metacarpal from Godin IV, dated about 3000–2800 BCE, might be horse. The first clear and unambiguous horse bones at Godin appeared in period III, dated 2100–1900 BCE; see Gilbert 1991. On horses and mules at Malyan, see Zeder 1986. The bit wear at Malyan is the earliest unambiguous bit wear in the Near East. Copper stains reported on the P_2s of asses from Tell Brak, dated 2300–2000 BCE, might have had another cause (perhaps corroded lip rings). See Clutton-Brock 2003.

6. Owen 1991.

7. The phrase *Fahren und Reiten*, or "To drive and to ride," appeared between 1939 and 1968 in the titles of three influential publications by Joseph Weisner, and the order of terms in this phrase—driving *before* riding—has become a form of shorthand referring to the historical priority of the chariot over the ridden horse in the Bronze Age civilizations of the Near East. Certainly wheeled vehicles preceded horseback riding in the Near East, and horse-drawn chariots dominated Near Eastern warfare long before cavalry, but this was not because riding was invented after chariotry (see chapter 10). If images of horseback riding can now be dated before 1800 BCE, as seems to be the case, they preceded the appearance of horses with chariots in Near Eastern art. See Weisner 1939, 1968; Drews 2004:33–41, 52; and Oates 2003.

8. For Zimri-Lim's adviser's advice, see Owen 1991; n. 12.

9. For tin sources, see Muhly 1995:1501–1519; Yener 1995; and Potts 1999:168–171, 186. For Eneolithic Serbian tin-copper alloys, see Glumac and Todd 1991. For the possible mistranslation of the Gudea inscription I am indebted to Chris Thornton, and, through him, to Greg Possehl and Steven Tinney. For the seaborne tin trade in the Arabian Gulf, see Weeks 1999; and for the Bactrian comb at Umm-al-Nar, see Potts 2000:126. For Harappan metals, see Agrawal 1984.

10. The polymetallic ores of the Zeravshan probably produced the metals of Ilgynly-Depe, near Anau, during the fourth millennium BCE. At Ilgynly, among sixty-two copper artifacts, primarily tanged knives, one object contained traces of tin; see Solovyova et al. 1994. For tin

bronzes in early third-millennium Namazga IV, see Salvatori et al. 2002. For Sarazm, see Isakov 1994; for its radiocarbon dates and metals, see Isakov, et al. 1987.

11. For the tin mines of the Zeravshan, see Boroffka et al. 2002; and Parzinger and Boroffka 2003.

12. Zaman Baba graves have been seen as a hybrid between Kelteminar and Namazga V/VI-type cultures, see Vinogradov 1960:80–81; and as a hybrid with Catacomb cultures on the supposition that Catacomb-culture people migrated to Central Asia, see Klejn 1984. I support the former. For recent debates over Zaman Baba, see E. Kuzmina 2003:215–216.

13. Lyonnet (1996) sees Sarazm IV ending during Namazga IV, or during the middle of the third millennium BCE. I see Sarazm ending in late Namazga V/early VI, based on the co-occurrence of Petrovka and late Sarazm pottery at Tugai, and on radiocarbon dates indicating that Sarazm III was occupied in 2400–2000 BCE, so Sarazm IV had to be later.

14. For skull type affiliations, see Christensen, Hemphill, and Mustafakulov 1996.

15. For BMAC, see Hiebert 1994, 2002. Salvatori (2000) disagreed with Hiebert, suggesting that BMAC began much earlier than 2100 BCE, and grew from local roots, not from an intrusion from the south, making the growth of BMAC more gradual. For the BMAC graves at Mehrgarh VIII, see Jarrige 1994. For BMAC materials in the Arabian Gulf, see Potts 2000, During Caspers 1998; and Winckelmann 2000.

16. For tin-bronzes in Bactria and lead-copper alloys in Margiana, see Chernykh 1992:176–182; and Salvatori et al. 2002. For the lead ingot at Sarazm, see Isakov 1994:8. For the Iranian background, see Thornton and Lamberg-Karlovsky 2004.

17. For horse bones in BMAC, see Salvatori 2003; and Sarianidi 2002. For the BMAC seal with the rider, see Sarianidi 1986. A few horses might have passed through the Caucasus into western Iran before 3000 BCE, indicated by a few probable horse teeth at the site of Qabrestan, west of Teheran; see Mashkour 2003. No definite horse remains have been identified in eastern Iran or the Indian subcontinent dated earlier than 2000 BCE. See Meadow and Patel 1997.

18. For the steppe sherds in BMAC sites, see Hiebert 2002. For the "Abashevo-like"sherds at Karnab, see Parzinger and Boroffka 2003:72, and Figure 49.

19. For Tugai, see Hiebert 2002; E. Kuzmina 2003; and the original report, Avanessova 1996. The talc temper in two pots, an indication that they were made in the South Ural steppes, is described in Avanessova 1996:122.

20. For Zardcha Khalifa, see Bobomulloev 1997; and E. Kuzmina 2001, 2003:224–225.

21. For the lead wires at Kuisak, see Maliutina and Zdanovich 1995:103. For the lapis bead and the grave at Krasnoe Znamya, see E. Kuzmina 2001:20.

22. For Srubnaya subsistence, see Bunyatyan 2003; and Ostroshchenko 2003.

23. For Chenopodium yields, see Smith 1989:1569.

24. For the Samara Valley Project, see Anthony et al. 2006. The results obtained here were replicated at Kibit, another Srubnaya settlement in Samara Oblast, excavated by L. Popova and D. Peterson, where there was no cultivated grain and many seeds of *Chenopodium*.

25. For the enormous Srubnaya mining center at Kargaly, see Chernykh 1997, 2004. For the mining center in Kazakhstan near Atasu, see Kadyrbaev and Kurmankulov 1992.

26. For stratigraphic relationships between Sintashta and Petrovka, see Vinogradov 2003; and Kuzmina 2001:9. The Petrovka culture was a transitional culture marking the beginning of the LBA. For Petrovka and its stratigraphic relationships to Alakul and Federovo, see Maliutina 1991. I would like to acknowledge the difficulty of keeping all these P-k cultures straight: on the middle Volga the MBA Poltavka culture evolved into final MBA Potapovka and then into early LBA Pokrovka, which was contemporary with early LBA Petrovka in Kazakhstan.

27. For the north-south movements of nomads in Kazakhstan, see Gorbunova 1993/94.

28. See Grigoriev 2002:78–84, for Petrovka metals.

29. For the Rostovka cemetery, see Matiushchenko and Sinitsyna 1988. For general discussions in English, see Chernykh 1992:215–234; and Grigoriev 2002:192–205.

30. For Seima-Turbino hollow-cast bronze casting and its influence on early China through the Qijia culture of Gansu province, see Mei 2003a, 2003b; and Li 2002. See also Fitzgerald-Huber 1995 and Linduff, Han, and Sun 2000.

31. See Epimakhov, Hanks, and Renfrew 2005 for dates. Seima-Turbino might possibly have begun west of the Urals and spread eastward. Sintashta fortifications might then be seen as a reaction to the emergence of Seima-Turbino warrior bands in the forest zone, but this is a minority position; see Kuznetsov 2001.

32. For Alakul and Federovo elements on the same pot, see Maliutina 1984; for the stratigraphic relations between the two, see Maliutina 1991. For radiocarbon dates, see Parzinger and Boroffka 2003:228.

33. E. Kuzmina 1994:207–208.

34. For Andronovo mines near Karaganda, see Kadyrbaev and Kurmankulov 1992; for mines near Dzhezkazgan, see Zhauymbaev 1984. For the estimate of copper production, see Chernykh 1992:212

35. For the Namazga VI pottery at Pavlovka, see Maliutina 1991:151–159.

36. For Andronovo sites in the Zeravshan, see Boroffka et al. 2002. For Tazabagyab sites on the former Amu-Darya delta, see Tolstov and Kes' 1960:89–132.

37. Hiebert 2002.

38. For the post-BMAC pastoral groups who made coarse incised ware, see Salvatori 2003:13; also Salvatori 2002. For the Vaksh and Bishkent groups, see Litvinsky and P'yankova 1992.

39. See Witzel 1995.

40. Books 2 and 4 of the *Rig Veda* referred to places in eastern Iran and Afghanistan. Book 6 described two clans who claimed they had come from far away, crossed many rivers, and gone through narrow passages, fighting indigenous people referred to as *Dasyus*. These details suggest that the Aryans fought their way into the Indian subcontinent from eastern Iran and Afghanistan. Although some new elements such as horses can be seen moving from Central Asia into the Indian subcontinent at this time, and intrusive pottery styles can be identified here or there, no single material culture spread with the Old Indic languages. For discussions, see Parpola 2002; Mallory 1998; and Witzel 1995:315–319.

41. For *Indra* and *Soma* as loan words, see Lubotsky 2001. Indra combined attributes that originally were separate: the mace was Mithra's; some of his epithets, his martial power, and perhaps his ability to change form were Verethraghna's; and the slaying of the serpent was the feat of the hero Thrataona, the Third One. The Old Indic poets gave these Indo-Iranian traits to Indra. The most prominent aspect of Indo-Iranian Verethraghna, the god of might/victory, was his shape-shifting ability, especially his form as the Boar. See Malandra 1983:80–81.

42. V. Sarianidi proposed that the people of the BMAC spoke Iranian. Sarianidi suggested that "white rooms" inside the walled buildings at Togolok 21, Togolok 1, and Gonur were fire temples like those of the Zoroastrians, with vessels containing *Ephedra*, *Cannabis*, and poppy seeds, which he equated with *Soma* (RV) or *Haoma* (AV). But examinations of the seed and stem impressions from the "white rooms" at Gonur and Togolok 21 by paleobotanists at Helsinki and Leiden Universities proved that the vessels contained no *Cannabis* or *Ephedra*. Instead the impressions probably were made by millet seeds and stems (*Panicum miliaceum*); see Bakels 2003. The BMAC culture makes a poor match with Indo-Iranian. The BMAC people lived in brick-built fortified walled towns, depended on irrigation agriculture, worshiped a female deity who was prominent in their iconography (a goddess with a flounced skirt), had few horses, no chariots, did not build kurgan cemeteries, and did not place carefully cut horse limbs in their graves.

43. Li 2002; and Mei 2003a.

Chapter 17. Words and Deeds

1. See Diamond 1997.

2. Hobsbawm 1997:5–6: "For history is the raw material for nationalist or ethnic or fundamentalist ideologies, as poppies are the raw material for heroin addiction. . . . This state of affairs affects us in two ways. We have a responsibility for historical facts in general and for criticizing the politico-ideological abuse of history in particular."

3. O'Flaherty 1981:69.

REFERENCES

Agapov, S. A., I. B. Vasiliev, and V. I. Pestrikova. 1990. *Khvalynskii Eneoliticheskii Mogil'nik.* Saratov: Saratovskogo universiteta.

Agrawal, D. P. 1984. Metal technology of the Harappans. In *Frontiers of the Indus Civilization,* ed. B. B. Lal and S. P. Gupta, pp. 163–167. New Delhi: Books and Books, Indian Archaeological Society.

Akhinzhalov, S. M., L. A. Makarova, and T. N. Nurumov. 1992. *K Istorii Skotovodstva i Okhoty v Kazakhstane.* Alma-Ata: Akademiya nauk Kazakhskoi SSR.

Alekhin, U. P., and A. V. Gal'chenko. 1995. K voprosu o drevneishem skotovodstve Altaya. In *Rossiya i Vostok: Problemy Vzaimodeistviya,* pt. 5, bk. 1: *Kul'tury Eneolita-Bronzy Stepnoi Evrazii,* pp. 22–26. Chelyabinsk: 3-ya Mezhdunarodnaya nauchnaya konferentsiya.

Alekseeva, I. L. 1976. O drevneishhikh Eneoliticheskikh pogrebeniyakh severo-zapadnogo prichernomor'ya. In *Materialy po arkheologii severnogo prichernomor'ya* (Kiev) 8:176–186.

Alexandrov, Stefan. 1995. The early Bronze Age in western Bulgaria: Periodization and cultural definition. In *Prehistoric Bulgaria,* ed. Douglass W. Bailey and Ivan Panayotov, pp. 253–270. Monographs in World Archaeology 22. Madison, Wis.: Prehistory Press.

Algaze, G. 1989. The Uruk Expansion: Cross-cultural exchange in Early Mesopotamian civilization. *Current Anthropology* 30:571–608.

Alvarez, Robert R., Jr. 1987. *Familia: Migration and Adaptation in Baja and Alta California, 1800–1975.* Berkeley: University of California Press.

Amiet, Pierre. 1986. *L'Âge des Échanges Inter-Iraniens 3500–1700 Avant J-C.* Paris: Editions de la Réunion des Musées Nationaux.

Andersen, Henning. 2003. Slavic and the Indo-European migrations. In *Language Contacts in Prehistory: Studies in Stratigraphy,* ed. Henning Andersen, pp. 45–76. Amsterdam and Philadelphia: Benjamins.

Antilla, R. 1972. *An Introduction to Historical and Comparative Linguistics.* New York: Macmillan.

Anthony, David W. 2001. Persistent identity and Indo-European archaeology in the western steppes. In *Early Contacts between Uralic and Indo-European: Linguistic and Archaeological Considerations,* ed. Christian Carpelan, Asko Parpola, and Petteri Koskikallio, pp. 11–35. Memoires de la Société Finno-Ugrienne 242. Helsinki: Suomalais-Ugrilainen Seura.

———. 1997. "Prehistoric migration as social process." In *Migrations and Invasions in Archaeological Explanation,* ed. John Chapman and Helena Hamerow, pp. 21–32. British Archaeological Reports International Series 664. Oxford: Archeopress.

———. 1996. V. G. Childe's world system and the daggers of the Early Bronze Age. In *Craft Specialization and Social Evolution: In Memory of V. Gordon Childe,* ed. Bernard Wailes, pp. 47–66. Philadelphia: University of Pennsylvania Museum Press.

———. 1995a. Horse, wagon, and chariot: Indo-European languages and archaeology. *Antiquity* 69 (264): 554–565.

———. 1995b. Nazi and Ecofeminist prehistories: ideology and empiricism in Indo-European archaeology. In *Nationalism, Politics, and the Practice of Archaeology,* ed. Philip Kohl and Clare Fawcett, pp. 82–96. Cambridge: Cambridge University Press.

———. 1994. On subsistence change at the Mesolithic-Neolithic transition in Ukraine. *Current Anthropology* 35 (1): 49–52.

———. 1991a. The archaeology of Indo-European origins. *Journal of Indo-European Studies* 19 (3–4): 193–222.

————. 1991b. The domestication of the horse. In *Equids in the Ancient World*, vol. 2, ed. Richard H. Meadow and Hans-Peter Uerpmann, pp. 250–277. Weisbaden: Verlag.

————. 1990. Migration in archaeology: The baby and the bathwater. *American Anthropologist* 92 (4): 23–42.

————. 1986. The "Kurgan Culture," Indo-European origins, and the domestication of the horse: A reconsideration. *Current Anthropology* 27:291–313.

Anthony, David W., and Dorcas Brown. 2003. Eneolithic horse rituals and riding in the steppes: New evidence. In *Prehistoric Steppe Adaptation and the Horse*, ed. Marsha Levine, Colin Renfrew, and Katie Boyle, pp. 55–68. Cambridge: McDonald Institute for Archaeological Research.

————. 2000. Eneolithic horse exploitation in the Eurasian steppes: Diet, ritual, and riding. *Antiquity* 74:75–86.

————. 1991. The origins of horseback riding. *Antiquity* 65:22–38.

Anthony, David W., D. Brown, E. Brown, A. Goodman, A. Kokhlov, P. Kosintsev, P. Kuznetsov, O. Mochalov, E. Murphy, D. Peterson, A. Pike-Tay, L. Popova, A. Rosen, N. Russel, and A. Weisskopf. 2005. The Samara Valley Project: Late Bronze Age economy and ritual in the Russian steppes. *Eurasia Antiqua* 11:395–417.

Anthony, David W., Dorcas R. Brown, and Christian George. 2006. Early horseback riding and warfare: The importance of the magpie around the neck. In *Horses and Humans: The Evolution of the Equine-Human Relationship*, ed. Sandra Olsen, Susan Grant, Alice Choyke, and László Bartosiewicz. pp. 137–156. British Archaeological Reports International Series 1560. Oxford: Archeopress.

Anthony, David W., Dimitri Telegin, and Dorcas Brown. 1991. The origin of horseback riding. *Scientific American* 265:94–100.

Anthony, David W., and Nikolai Vinogradov. 1995. The birth of the chariot. *Archaeology* 48 (2): 36–41.

Anthony, David W., and B. Wailes. 1988. CA review of *Archaeology and Language* by Colin Renfrew. *Current Anthropology* 29 (3): 441–445.

Appadurai, Arjun. 1986. Introduction: Commodities and the politics of value. In *The Social Life of Things: Commodities in Cultural Perspective*, ed. Arjun Appadurai, pp. 3–63. Cambridge: Cambridge University Press.

Armstrong, J. A. 1982. *Nations before Nationalism*. Chapel Hill: University of North Carolina Press.

Arnold, Bettina. 1990. The past as propaganda: Totalitarian archaeology in Nazi Germany. *Antiquity* 64:464–478.

Aruz, Joan. 1998. Images of the supernatural world: Bactria-Margiana seals and relations with the Near East and the Indus. *Ancient Civilizations from Scythia to Siberia* 5 (1): 12–30.

Atkinson, R. R. 1994. *The Roots of Ethnicity: The Origins of the Acholi of Uganda before 1800.* Philadelphia: University of Pennsylvania Press.

————. 1989. The evolution of ethnicity among the Acholi of Uganda: The precolonial phase. *Ethnohistory* 36 (1): 19–43.

Avanessova, N. A. 1996. Pasteurs et agriculteurs de la vallée du Zeravshan (Ouzbekistan) au début de l'age du Bronze: relations et influences mutuelles. In B. Lyonnet, *Sarazm (Tadjikistan) Céramiques (Chalcolithique et Bronze Ancien)*, pp. 117–131. Paris: Mémoires de la Mission Archéologique Française en Asie Centrale Tome 7.

Azzaroli, Augusto. 1980. Venetic horses from Iron Age burials at Padova. *Rivista di Scienze Preistoriche* 35 (1–2): 282–308.

Bahn, Paul G. 1980. "Crib-biting: Tethered horses in the Palaeolithic?" *World Archaeology* 12:212–217.

Bailey, Douglass W. 2000. *Balkan Prehistory: Exclusion, Incorporation, and Identity*. London: Routledge.

Bailey, Douglass W., R. Andreescu, A. J. Howard, M. G. Macklin, and S. Mills. 2002. Alluvial landscapes in the temperate Balkan Neolithic: Transitions to tells. *Antiquity* 76:349–355.

Bailey, Douglass W., and Ivan Panayotov, eds. 1995. Monographs in World Archaeology 22. *Prehistoric Bulgaria*. Madison, Wis.: Prehistory Press.

Bailey, Douglass W., Ruth Tringham, Jason Bass, Mirjana Stefanović, Mike Hamilton, Heike Neumann, Ilke Angelova, and Ana Raduncheva. 1998. Expanding the dimensions of early agricultural tells: The Podgoritsa archaeological project, Bulgaria. *Journal of Field Archaeology* 25:373–396.

Bakels, C. C. 2003. The contents of ceramic vessels in the Bactria-Margiana Archaeological Complex, Turkmenistan. *Electronic Journal of Vedic Studies* 9 (1).

Bakharev, S. S., and N. V. Obchinnikova. 1991. Chesnokovskaya stoiankana na reke Sok. In *Drevnosti Vostochno-Evropeiskoi Lesotepi*, ed. V. V. Nikitin, pp. 72–93. Samara: Samarskii gosudartsvennyi pedagogicheskii institut.

Bakker, Jan Albert, Janusz Kruk, A. L. Lanting, and Sarunas Milisauskas. 1999. The earliest evidence of wheeled vehicles in Europe and the Near East. *Antiquity* 73:778–790.

Baldi, Philip. 1983. *An Introduction to the Indo-European Languages*. Carbondale: Southern Illinois University Press.

Balter, Michael. 2003. Early date for the birth of Indo-European languages. *Science* 302 (5650): 1490–1491.

Bánffy, Ester. 1995. South-west Transdanubia as a mediating area: on the cultural history of the early and middle Chalcolithic. In *Archaeology and Settlement History in the Hahót Basin, South-West Hungary*, ed. Béla Miklós Szőke. Antaeus 22. Budapest: Archaeological Institute of the Hungarian Academy of Sciences.

Bar-Yosef, Ofer. 2002. The Natufian Culture and the Early Neolithic: Social and Economic Trends in Southwestern Asia. In *Examining the Farming/Language Dispersal Hypothesis*, ed. Peter Bellwood and Colin Renfrew, pp. 113–126. Cambridge: McDonald Institute for Archaeological Research.

Barber, Elizabeth J. W. 2001. The clues in the clothes: Some independent evidence for the movement of families. In *Greater Anatolia and the Indo-Hittite Language Family*, ed. Robert Drews, pp. 1–14. Journal of Indo-European Studies Monograph 38. Washington, D.C.: Institute for the Study of Man.

———. 1991. *Prehistoric Textiles*. Princeton, N. J.: Princeton University Press.

Barfield, Thomas. 1989. *The Perilous Frontier*. Cambridge: Blackwell.

Barth, Frederik. 1972 [1964]. "Ethnic processes on the Pathan-Baluch boundary." In *Directions in Sociolinguistics: The Ethnography of Communication*, ed. John J. Gumperz and Dell Hymes, pp. 454–464. New York: Holt Rinehart.

———. 1965 [1959]. *Political Leadership among Swat Pathans*. Rev. ed. London: Athalone.

Barth, Fredrik. 1969. *Ethnic Groups and Boundaries: The Social Organization of Culture Difference*. Repr. ed. Prospect Heights: Waveland.

Bartlett, Robert. 1993. *The Making of Europe: Conquest, Colonization, and Cultural Change, 950–1350*. Princeton, N. J.: Princeton University Press.

Barynkin, P. P., and E. V. Kozin. 1998. Prirodno-kilmaticheskie i kul'turno-demograficheskie protsessy v severnom priKaspii v rannem i srednem Golotsene. In *Arkheologicheskie Kul'tury Severnogo Prikaspiya*, ed. R. S. Bagautdinov, pp. 66–83. Kuibyshev: Kuibyshevskii gosudarts-vennyi pedagogicheskii institut.

Barynkin, P. P., and I. B. Vasiliev. 1988. Stoianka Khvalynskoi eneoliticheskoi kulturi Kara-Khuduk v severnom Prikaspii. In *Arkheologicheskie Kul'tury Severnogo Prikaspiya*, ed. R. S. Bagautdinov, pp. 123–142, Kuibyshev: Kuibyshevskii gosudarstvennyi pedagogicheskii institut.

Barynkin, P. P., I. B. Vasiliev, and A. A. Vybornov. 1998. Stoianka Kyzyl-Khak II: pamyatnik epokhi rannei Bronzy severnogo prikaspiya. In *Problemy Drevnei Istorii Severnogo Prikaspiya*, ed. V. S. Gorbunov, pp. 179–192, Samara: Samarskogo gosudarstvennogo pedagogicheskogo universiteta.

Bashkow, Ira. 2004. A neo-Boasian conception of cultural boundaries. *American Anthropologist* 106 (3): 443–458.

Beekes, Robert S. P. 1995. *Comparative Indo-European Linguistics: An Introduction.* Amsterdam: John Benjamins.

Beilekchi, V. S. 1985. Raskopki kurgana 3 u s. Kopchak. *Arkheologicheskie Issledovaniya v Moldavii v 1985 g.*, pp. 34–49. Kishinev: Shtiintsa.

Belanovskaya, T. D. 1995. *Iz drevneishego proshlogo nizhnego po Don'ya.* St. Petersburg: IIMK.

Bellwood, Peter. 2001. Early agriculturalist population diasporas? Farming, language, and genes. *Annual Review of Anthropology* 30:181–207.

Bellwood, Peter, and Colin Renfrew, eds. 2002. *Examining the Farming/Language Dispersal Hypothesis.* Cambridge: McDonald Institute for Archaeological Research.

Bendrey, Robin. 2007. New methods for the identification of evidence for bitting on horse remains from archaeological sites. *Journal of Archaeological Science* 34:1036–1050.

Benecke, Norbert. 1997. Archaeozoological studies on the transition from the Mesolithic to the Neolithic in the North Pontic region. *Anthropozoologica* 25–26:631–641.

———. 1994. *Archäologische Studien zur Entwicklung der Haustierhaltung in Mitteleuropa und Sødskandinavien von Anfängen bis zum Ausgehenden Mittelalter.* Berlin: Akademie Verlag.

Benecke, Norbert, and Angela von den Dreisch. 2003. Horse exploitation in the Kazakh steppes during the Eneolithic and Bronze Age. In *Prehistoric Steppe Adaptation and the Horse*, ed. Marsha Levine, Colin Renfrew, and Katie Boyle, pp. 69–82. Cambridge: McDonald Institute for Archaeological Research.

Benveniste, Emile. 1973 [1969]. *Indo-European Language and Society.* Translated by Elizabeth Palmer. Coral Gables, Fla.: University of Miami Press.

Berger, Joel. 1986. *Wild Horses of the Great Basin: Social Competition and Population Size.* Chicago: University of Chicago Press.

Berezanskaya, S. S. 1979. Pervye mastera-metallurgi na territorii Ukrainy. In *Pervobytnaya arkheologiya: poiski i nakhodki*, ed. N. N. Bondar and D. Y. Telegin, pp. 243–256. Kiev: Naukova Dumka.

Bibby, Geoffrey. 1956. *The Testimony of the Spade.* New York: Knopf.

Bibikov, S. N. 1953. *Rannetripol'skoe Poselenie Luka-Vrublevetskaya na Dnestre.* Materialy i issledovaniya po arkheologii SSR 38. Moscow: Akademii Nauk SSSR.

Bibikova, V. I. 1970. K izucheniyu drevneishikh domashnikh loshadei vostochnoi Evropy, soobshchenie 2. *Biulleten moskovskogo obshchestva ispytatlei prirodi otdel biologicheskii* 75 (5): 118–126.

———. 1967. K izucheniyu drevneishikh domashnikh loshadei vostochnoi Evropy. *Biulleten moskovskogo obshchestva ispytatelei prirodi Otdel Biologicheskii* 72 (3): 106–117.

Bickerton, D. 1988. Creole languages and the bioprogram. In *Linguistics: The Cambridge Survey*, vol. 2 ed. F. J. Newmeyer, pp. 267–284. Cambridge: Cambridge University Press.

Binford, Lewis. 1971. Mortuary practices: Their study and their potential. In *Approaches to the Social Dimensions of Mortuary Practices*, ed. James A. Brown, pp. 92–112. Memoirs No. 25. Washington, D.C.: Society for American Archaeology.

Blyakharchuk, T. A., H. E. Wright, P. S. Borodavko, W. O. van der Knaap, and B. Ammann. 2004. Late Glacial and Holocene vegetational changes on the Ulagan high-mountain plateau, Altai Mts., southern Siberia. *Palaeogeography, Paleoclimatology, and Paleoecology* 209:259–279.

Bloch, Maurice E. F. 1998. Time, narratives, and the multiplicity of representations of the past. In *How We Think They Think*, ed. Maurice E. F. Bloch, 100–113. Boulder, CO: Westview Press.

Boaz, Franz. 1911. Introduction. In *Handbook of American Indian Languages*, pt. 1, pp. 1–82. Bulletin 40. Washington, D.C.: Bureau of American Ethnology.

Bobomulloev, Saidmurad. 1997. Ein bronzezeitliches Grab aus Zardča Chalifa bei Pendžikent (Zeravšan-Tal). *Archäologische Mitteilungen aus Iran und Turan* 29:122–134.

Bobrinskii, A. A., and I. N. Vasilieva. 1998. O nekotorykh osobennostiakh plasticheskogo syr'ya v istorii goncharstva. In *Problemy drevnei istorii severnogo prikaspiya*, pp. 193–217. Samara: Institut istorii i arkheologii povolzh'ya.

Bobrov, V. V. 1988. On the problem of interethnic relations in South Siberia in the third and second millennia BC. *Arctic Anthropology* 25 (2): 30–46.

Bodyans'kii, O. V. 1968. Eneolitichnii mogil'nik bilya s. Petyro-Svistunovo. *Arkheologiya* (Kiev) 21:117–125.

Bogucki, Peter. 1988. *Forest Farmers and Stockherders*. Cambridge: Cambridge University Press.

Bökönyi, Sandor. 1991. Late Chalcolithic horses in Anatolia. In *Equids in the Ancient World*, ed., Richard Meadow and Hans-Peter Uerpmann, vol. 2, pp. 123–131. Wiesbaden: Ludwig Reichert.

———. 1987. Horses and sheep in East Europe. In *Proto-Indo-European: The Archaeology of a Linguistic Problem*, ed. Susan Skomal, pp. 136–144. Washington, D.C.: Institute for the Study of Man.

———. 1983. Late Chalcolithic and Early Bronze I animal remains from Arslantepe (Malatya), Turkey: A preliminary report. *Origini* 12 (2): 581–598.

———. 1979. Copper age vertebrate fauna from Kétegyháza. In *The People of the Pit-Grave Kurgans in Eastern Hungary*, ed. Istvan Ecsedy, pp. 101–116. Budapest: Akademiai Kiado.

———. 1978. The earliest waves of domestic horses in East Europe. *Journal of Indo-European Studies* 6 (1/2): 17–76.

———. 1974. *History of Domestic Animals in Central and Eastern Europe*. Budapest: Akademiai Kiado.

Bol'shov, S. V. 1995. Problemy kulturogeneza v lesnoi polose srednego povolzh'ya v Abashevskoe vremya. In *Drevnie IndoIranskie Kul'tury Volgo-Ural'ya*, ed. I. B. Vasilev and O. V. Kuz'mina, pp. 141–156. Samara: Samara Gosudarstvennogo Pedagogicheskogo Universiteta.

Boltenko, M. F. 1957. Stratigrafiya i khronologiya Bol'shogo Kulial'nika. *Materiali i issledovaniya po arkheologii severnogo prichernomoriya (Kiev)* 1:21–46.

Bond, G., Kromer, B., Beer, J., Muscheler, R., Evans, M. N., Showers, W., Hoffmann, S., Lotti-Bond, R., Hajdas, I. and Bonani, G., 2001. Persistent solar influence on North Atlantic climate during the Holocene. *Science* 294:2130–2136.

Bondar, N. N. and Nechitailo, A. L., eds. 1980. *Arkheologicheskie pamyatniki po ingul'ya*. Kiev: Naukova Dumka.

Bondar, N. N. 1974. K voprosu o proiskhozhdenii serdnedneprovskoi kul'tury. *Zborník Filozofickej Fakulty Univerzity Komenského Musaica (Bratislava)* 14:37–53.

Bonsall, C., G. T. Cook, R. E. M. Hedges, T. F. G. Higham, C. Pickard, and I. Radovanovic. 2004. Radiocarbon and stable isotope evidence of dietary change from the Mesolithic to the Middle Ages in the Iron Gates: New results from Lepenski Vir. *Radiocarbon* 46 (1): 293–300.

Boriskovskii, Pavel I. 1993. Determining Upper Paleolithic historico-cultural regions. In *From Kostienki to Clovis, Upper Paleolithic: Paleo-Indian Adaptations*, ed. Olga Soffer and N. D. Praslov, pp. 143–147. New York: Plenum.

Boroffka, Nikolaus. 1998. Bronze- und früheizenzeitliche Geweihtrensenknebel aus Rumänien und ihre Beziehungen. *Eurasia Antiqua* (Berlin) 4:81–135.

Boroffka, Nikolaus, Jan Cierny, Joachim Lutz, Hermann Parzinger, Ernst Pernicka, and Gerd Weisberger, 2002. Bronze Age tin from central Asia: Preliminary notes. In *Ancient Interactions: East and West in Eurasia*, ed. Katie Boyle, Colin Renfrew, and Marsha Levine, pp. 135–159, Cambridge: McDonald Institute for Archaeological Research.

Boyadziev, Yavor D. 1995. Chronology of the prehistoric cultures in Bulgaria. In *Prehistoric Bulgaria*, ed. Douglass W. Bailey and Ivan Panayotov, pp. 149–191. Monographs in World Archaeology 22. Madison, Wis.: Prehistory Press.

Boyce, Mary. 1975. *A History of Zoroastrianism*. Vol. 1. Leiden: Brill.

Britain, David. 2002. Space and spatial diffusion. In *The Handbook of Language Variation and Change*, ed. J. Chambers, P. Trudgill, and N. Schilling-Estes, pp. 603–637. Oxford: Blackwell.

Boyle, Katie, Colin Renfrew, and Marsha Levine, eds. 2002. *Ancient Interactions: East and West in Eurasia*. Cambridge: McDonald Institute for Archaeological Research.

Bradley D. G., D. E. MacHugh, P. Cunningham, and R. T. Loftus. 1996. Mitochondrial diversity and the origins of African and European cattle. *Proceedings of the National Academy of Sciences* 93 (10): 5131–5135.

Bratchenko, S. N. 2003. Radiocarbon chronology of the Early Bronze Age of the Middle Don, Svatove, Luhansk region. *Baltic-Pontic Studies* 12:185–208.

———. 1976. *Nizhnee Podone v Epokhu Srednei Bronzy*. Kiev: Naukovo Dumka.

———. 1969. Bagatosha rove poselennya Liventsivka I na Donu. *Arkheologiia* (Kiev) 22:210–231.

Breen, T. H. 1984. Creative adaptations: Peoples and cultures. In *Colonial British America*, ed. Jack P. Green and J. R. Pole, pp. 195–232. Baltimore, Md.: Johns Hopkins University Press.

Britain, David. 2004. Geolinguistics—Diffusion of Language. In *Sociolinguistics: International Handbook of the Science of Language and Society* vol. 1, ed. Ulrich Ammon, Norbert Dittmar, Klaus J. Mattheier, and Peter Trudgill, pp. 34–48, Berlin: Mouton de Gruyter.

Bronicki, Andrzej, Sławomir Kadrow, and Anna Zakościelna. 2003. Radiocarbon dating of the Neolithic settlement in Zimne, Volhynia. *Baltic-Pontic Studies* 12:22–66.

Bronitsky, G., and R. Hamer. 1986. Experiments in ceramic technology: The effects of various tempering material on impact and thermal-shock resistance. *American Antiquity* 51 (1): 89–101.

Broodbank, Cyprian. 1989. The longboat and society in the Cyclades in the Keros-Syros culture. *American Journal of Archaeology* 85:318–337.

Broodbank, Cyprian, and T. F. Strasser. 1991. Migrant farmers and the colonization of Crete. *Antiquity* 65:233–245.

Brown, D. R., and David W. Anthony. 1998. Bit wear, horseback riding, and the Botai site in Kazakstan. *Journal of Archaeological Science* 25:331–347.

Bryce, T. 1998. *The Kingdom of the Hittites*. Oxford: Clarendon.

Buchanan, Briggs. 1966. *Catalogue of Ancient Near Eastern Seals in the Ashmolean Museum*. Vol. 1, *Cylinder Seals*. Oxford: Clarendon.

Buck, Carl Darling. 1949. *A Dictionary of Selected Synonyms in the Principal Indo-European Languages*. Chicago: University of Chicago Press.

Bunyatyan, Katerina P. 2003. Correlations between agriculture and pastoralism in the northern Pontic steppe area during the Bronze Age. In *Prehistoric Steppe Adaptation and the Horse*, ed. Marsha Levine, Colin Renfrew, and Katie Boyle, pp. 269–286. Cambridge: McDonald Institute for Archaeological Research.

Burdo, Natalia B. 2003. Cultural contacts of early Tripolye tribes. Paper delivered at the Ninth Annual Conference of the European Association of Archaeologists. St Petersburg, Russia.

Burdo, Natalia B., and V. N. Stanko. 1981. Eneoliticheskie nakhodki na stoianke Mirnoe. In *Drevnosti severo-zapadnogo prichernomor'ya*, pp. 17–22. Kiev: Naukovo Dumka.

Burmeister, Stefan. 2000. Archaeology and migration: Approaches to an archaeological proof of migration. *Current Anthropology* 41 (4): 554–555.

Burov, G. M. 1997. Zimnii transport severnoi Evropy i Zaural'ya v epokhu Neolita i rannego metalla. *Rossiskaya arkheologiya* 4:42–53.

Burrow, T. 1973. The Proto-Indoaryans. *Journal of the Royal Asiatic Society* (n. 5.) 2:123–40.

Bynon, Theodora. 1977. *Historical Linguistics*. Cambridge: Cambridge University Press.

Cameron, Catherine, and Steve A. Tomka, eds. 1993. *Abandonment of Settlements and Regions: Ethnoarchaeological and Archaeological Approaches*. Cambridge: Cambridge University Press.

Campbell, Lyle. 2002. What drives linguistic diversification and language spread? In *Examining the Farming/Language Dispersal Hypothesis*, ed. Peter Bellwood and Colin Renfrew, pp. 49–63. Cambridge: McDonald Institute for Archaeological Research.

Cannon, Garland. 1995. "Oriental Jones: Scholarship, Literature, Multiculturalism, and Humankind." In *Objects of Enquiry: The Life, Contributions, and Influences of Sir William Jones*, pp. 25–50. New York: New York University Press.

Carpelan, Christian, and Asko Parpola. 2001. Emergence, contacts and dispersal of Proto-Indo-European, proto-Uralic and proto-Aryan in archaeological perspective. In *Early Contacts between Uralic and Indo-European: Linguistic and Archaeological Considerations*, ed. Christian Carpelan, Asko Parpola, and Petteri Koskikallio, pp. 55–150. Memoires de la Société Finno-Ugrienne 242. Helsinki: Suomalais-Ugrilainen Seura.

Castile, George Pierre, and Gilbert Kushner, eds. 1981. *Persistent Peoples: Cultural Enclaves in Perspective*. Tucson: University of Arizona Press.

Chambers, Jack, and Peter Trudgill. 1998. *Dialectology*. Cambridge: Cambridge University Press.

Chapman, John C. 1999. The origins of warfare in the prehistory of Eastern and central Europe. In *Ancient Warfare: Archaeological Perspectives*, ed. John Carman and Anthony Harding, pp. 101–142. Phoenix Mill: Sutton.

———. 1989. The early Balkan village. In *The Neolithic of Southeastern Europe and Its Near Eastern Connections*, ed. Sándor Bökönyi, pp. 33–53. Budapest: Varia Archaeologica Hungarica II.

———. 1983. The Secondary Products Revolution and the limitations of the Neolithic. *Bulletin of the Institute of Archaeology* (London) 19:107–122.

Cherniakov, I. T., and G. N. Toshchev. 1985. Kul'turno-khronologicheskie osobennosti kurgannykh pogrebenii epokhi Bronzy nizhnego Dunaya. In *Novye Materialy po Arkheologii Severnogo-Zapadnogo Prichernomor'ya*, ed. V. N. Stanko, pp. 5–45, Kiev: Naukovo Dumka.

Chernopitskii, M. P. 1987. Maikopskii "baldachin." *Kratkie soobshcheniya institut arkheologii* 192:33–40.

Chernykh, E. N., ed. 2004. *Kargaly*. Vol. 3, *Arkheologicheskie materialy, tekhnologiya gornometallurgicheskogo proizvodstva, arkheobiologicheskie issledovaniya*. Moscow: Yaziki slavyanskoi kul'tury.

———. 1997. *Kargaly: Zabytyi Mir*. Moscow: NOX.

———. 1995. Postscript: Russian archaeology after the collapse of the USSR: Infrastructural crisis and the resurgence of old and new nationalisms. In *Nationalism, Politics, and the Practice of Archaeology*, ed. Philip L. Kohl and Clare Fawcett, pp. 139–148, Cambridge: Cambridge University Press.

———. 1992. *Ancient Metallurgy in the USSR*. Cambridge: Cambridge University Press.

Chernykh, E. N., and K. D. Isto. 2002. Nachalo ekspluatsii Kargalov: Radiouglerodnyi daty. *Rossiiskaya arkheologiya* 2: 44–55.

Chernykh, E.N., E. V. Kuz'minykh, and L. B. Orlovskaya. 2004. Ancient metallurgy of northeast Asia: From the Urals to the Saiano-Altai. In *Metallurgy in Ancient Eastern Eurasia from the Urals to the Yellow River*, ed. Katheryn M. Linduff, pp. 15–36. Lewiston, Me.: Edwin Mellen.

Chernysh, E. K. 1982. Eneolit pravoberezhnoi Ukrainy i Moldavii. In *Eneolit SSSR*, ed. V. M. Masson and N. Y. Merpert, pp. 165–320. Moscow: Nauka.

Childe, V. Gordon. 1957. *The Dawn of European Civilization*. 6th ed. London: Routledge Kegan Paul.

———. 1936. The axes from Maikop and Caucasian metallurgy. *Annals of Archaeology and Anthropology* (Liverpool) 23:113–119.

Chilton, Elizabeth S. 1998. The cultural origins of technical choice: Unraveling Algonquian and Iroquoian ceramic traditions in the Northeast. In *The Archaeology of Social Boundaries*, ed. Miriam Stark, pp 132–160. Washington, D.C.: Smithsonian Institution Press.

Chretien, C. D. 1962. The mathematical models of glottochronology. *Language* 38:11–37.

Christensen, A. F., Brian E. Hemphill, and Samar I. Mustafakulov. 1996. Bactrian relationships to Russian and Central Asian populations: A craniometric assessment. *American Journal of Physical Anthropology* 22:84–85.

Clackson, James. 1994. *The Linguistic Relationship between Greek and Armenian.* Oxford: Blackwell.

Clark, Geoffry. 1994. Migration as an explanatory concept in Paleolithic archaeology. *Journal of Archaeological Method and Theory* 1 (4): 305–343.

Clark, Grahame. 1941. Horses and battle-axes. *Antiquity* 15 (57): 50–69.

Clayton, Hilary. 1985. A fluoroscopic study of the position and action of different bits in the horse's mouth. *Equine Veterinary Science* 5 (2): 68–77.

Clayton, Hilary M., and R. Lee. 1984. A fluoroscopic study of the position and action of the jointed snaffle bit in the horse's mouth. *Equine Veterinary Science* 4 (5): 193–196.

Clutton-Brock, Juliet. 2003. Were the donkeys of Tell Brak harnessed with a bit? In *Prehistoric Steppe Adaptation and the Horse*, ed. Marsha Levine, Colin Renfrew, and Katie Boyle, pp. 126–127. Cambridge: McDonald Institute for Archaeological Research.

———. 1974. The Buhen horse. *Journal of Archaeological Science* 1:89–100.

Cole, John W., and Eric Wolf. 1974. *The Hidden Frontier: Ecology and Ethnicity in an Alpine Valley.* New York: Academic Press.

Coleman, John. 2000. An archaeological scenario for the "Coming of the Greeks" ca. 3200 BC." *Journal of Indo-European Studies* 28 (1–2): 101–153.

Comsa, Eugen. 1976. Quelques considerations sur la culture Gumelnitsa. *Dacia* 20:105–127.

Cook, G. T., C. Bonsall, R. E. M. Hedges, K. McSweeney, V. Boroneanţ, L. Bartosiewicz, and P. B. Pettitt, 2002. Problems of dating human bones from the Iron Gates. *Antiquity* 76:77–85.

Cronk, Lee. 1993. CA comment on transitions between cultivation and pastoralism in Sub-Saharan Africa. *Current Anthropology* 34 (4): 374.

———. 1989. From hunters to herders: Subsistence change as a reproductive strategy. *Current Anthropology* 30:224–34.

Crouwel, Joost H. 1981. *Chariots and Other Means of Land Transport in Bronze Age Greece.* Allard Pierson Series 3. Amsterdam: Allard Pierson Museum.

Dalton, G. 1977. Aboriginal economies in stateless societies. In *Exchange Systems in Prehistory*, ed. Timothy Earle and J. Ericson, pp. 191–212, New York: Academic Press.

Danilenko, V. M. 1971. *Bugo-Dnistrovs'ka Kul'tura.* Kiev: Dumka.

Darden, Bill J. 2001. On the question of the Anatolian origin of Indo-Hittite. In *Greater Anatolia and the Indo-Hittite Language Family*, ed. Robert Drews, pp. 184–228. Journal of Indo-European Studies Monograph 38. Washington, D.C.: Institute for the Study of Man.

———. 2004. Who were the Sclaveni and where did they come from? *Byzantinische Forschungen* 28:133–157.

Davis, E. M. 1983. The gold of the shaft graves: The Transylvanian connection. *Temple University Aegean Symposium* 8:32–38.

Davis, Simon J. M. 1987. *The Archaeology of Animals.* New Haven: Yale University Press.

Davis-Kimball, Jeannine. 1997. Warrior women of the Eurasian steppes. *Archaeology* 50 (1): 44–49.

DeBoer, Warren. 1990. Interaction, imitation, and communication as expressed in style: The Ucayali experience. In *The Uses of Style in Archaeology*, ed. M. Conkey and Christine Hastorf, pp. 82–104. Cambridge: Cambridge University Press.

———. 1986. Pillage and production in the Amazon: A view through the Conibo of the Ucayali Basin, eastern Peru. *World Archaeology* 18 (2): 231–246.

Dennell, R. W., and D. Webley. 1975. Prehistoric settlement and land use in southern Bulgaria. In *Palaeoeconomy*, ed. E. S. Higgs, pp. 97–110. Cambridge: Cambridge University Press.

Dergachev, Valentin A. 2003. Two studies in defense of the migration concept. In *Ancient Interactions: East and West in Eurasia*, ed. Katie Boyle, Colin Renfrew, and Marsha Levine, pp. 93–112. McDonald Institute Monographs. Cambridge: University of Cambridge Press.

————. 1999. Cultural-historical dialogue between the Balkans and Eastern Europe (Neolithic-Bronze Age). *Thraco-Dacica* 20 (1–2): 33–78.

————. 1998a. *Karbunskii Klad*. Kishinev: Academiei Ştiinţe.

————. 1998b. Kulturell und historische Entwicklungen im Raum zwischen Karpaten und Dnepr. In *Das Karpatenbecken und Die Osteuropäische Steppe*, ed. Bernhard Hänsel and Jan Machnik, pp. 27–64. München: Südosteuropa-Schriften Band 20, Verlag Marie Leidorf GmbH.

————. 1980. *Pamyatniki Pozdnego Tripol'ya*. Kishinev: Shtiintsa.

Dergachev, V., A. Sherratt, and O. Larina. 1991. Recent results of Neolithic research in Moldavia (USSR). *Oxford Journal of Prehistory* 10 (1): 1–16.

Derin, Z., and Oscar W. Muscarella. 2001. Iron and bronze arrows. In *Ayanis I. Ten Years' Excavations at Rusahinili Eiduru-kai 1989–1998*, ed. A. Çilingiroğlu and M. Salvini, pp. 189–217. Roma: Documenta Asiana VI ISMEA.

Diakonov, I. M. 1988. Review of *Archaeology and Language*. *Annual of Armenian Linguistics* 9:79–87.

————. 1985. On the original home of the speakers of Indo-European. *Journal of Indo-European Studies* 13 (1–2): 93–173.

Diamond, Jared. 1997. *Guns, Germs, and Steel: The Fates of Human Societies*. New York: Norton.

DiCosmo, Nicola. 2002. *Ancient China and Its Enemies: The Rise of Nomadic Power in East Asian History*. Cambridge: Cambridge University Press.

————. 1999. State Formation and periodization in Inner Asian prehistory. *Journal of World History* 10 (1): 1–40.

————. 1994. Ancient Inner Asian Nomads: Their Economic basis and its significance in Chinese history. *Journal of Asian Studies* 53 (4): 1092–1126.

Diebold, Richard. 1985. *The Evolution of the Nomenclature for the Salmonid Fish: The Case of "huchen" (Hucho spp.)*. Journal of Indo-European Studies Monograph 5. Washington, D.C.: Institute for the Study of Man.

Dietler, Michael, and Brian Hayden, eds. 2001. *Feasts*. Washington, D.C.: Smithsonian Institution Press.

Dietz, Ute Luise. 1992. Zur frage vorbronzezeitlicher Trensenbelege in Europa. *Germania* 70 (1): 17–36.

Dixon, R. M. W. 1997. *The Rise and Fall of Languages*. Cambridge: Cambridge University Press.

Dobrovol'skii, A. V. 1958. Mogil'nik vs. Chapli. *Arkheologiya* (Kiev) 9:106–118.

Dodd-Opriţescu, 1978, Les elements steppiques dans l'Énéolithique de Transylvanie. *Dacia* 22:87–97.

Dolukhanov, P. M. 1986. Foragers and farmers in west-Central Asia. In *Hunters in Transition*, ed. Marek Zvelebil, pp. 121–132. Cambridge: Cambridge University Press.

Donnan, Hastings, and Thomas M. Wilson. 1999. *Borders: Frontiers of Identity, Nation, and State*. Oxford: Berg.

Dorian, N. 1981. *Language Death: The Life Cycle of a Scottish Gaelic Dialect*. Philadelphia: University of Pennsylvania Press.

Dovchenko, N. D., and N. A. Rychkov. 1988. K probleme sotsial'noi stratigrafikatsii plemen Yamnoi kul'turno-istoricheskoi obshchnosti. In *Novye Pamyatniki Yamnoi Kul'tury Stepnoi Zony Ukrainy*, pp. 27–40. Kiev: Naukova Dumka.

Dremov, I. I., and A. I. Yudin. 1992. Drevneishie podkurgannye zakhoroneniya stepnogo zaVolzh'ya. *Rossiskaya arkheologiya* 4:18–31.

Drews, Robert. 2004. *Early Riders*. London: Routledge.

————, ed. 2001. *Greater Anatolia and the Indo-Hittite Language Family*. Journal of Indo-European Studies Monograph 38. Washington, D.C.: Institute for the Study of Man.

————. 1988. *The Coming of the Greeks: Indo-European Conquests in the Aegean and the Ancient Near East*. Princeton, N. J.: Princeton University Press.

Drinka, Bridget. 1995. Areal linguistics in prehistory: Evidence from Indo-European aspect. In *Historical Linguistics 1993*, ed. Henning Andersen, pp. 143–158. Amsterdam: John Benjamins.

Dumezil, Georges. 1958. *L'Idéologie Tripartie des Indo-Européens*. Brussels: Latomus.

Dumitrescu, Vladimir. 1980. Tumuli from the period of transition from the Eneolithic to the Bronze Age excavated near Rast. In *The Neolithic Settlement at Rast*, appendix 3, pp. 126–133. British Archaeological Reports International Series 72. Oxford: Archaeopress.

During Caspers, E. C. L. 1998. The MBAC and the Harappan script. *Ancient Civilizations from Scythia to Siberia* 5 (1): 40–58.

Dyen, I., J. B. Kruskal, and P. Black. 1992. An Indo-European classification: A lexicostatistical experiment. *Transactions of the American Philosophical Society* 82 (5): 1–132.

Ecsedy, István. 1994. "Camps for eternal rest: Some aspects of the burials by the earliest nomads of the steppes." In *The Archaeology of the Steppes: Methods and Strategies*, ed. Bruno Genito, pp. 167–176. Napo: Instituto universitario oreintale series minor 44.

———, ed. 1979. *The People of the Pit-Grave Kurgans in Eastern Hungary*. Budapest: Akadémia Kiadó.

Ehrich, Robert W. 1961. On the persistence and recurrences of culture areas and culture boundaries during the course of European prehistory, protohistory, and history. In *Berichte über den V Internationalen Kongress für Vor- und Frühgeschichte*, pp. 253–257. Berlin: Gebrüder Mann.

Eisler, Riane. 1990. The Gaia tradition and the partnership future: An ecofeminist manifesto. In *Reweaving the World*, ed. Irene Diamond and G. F. Orenstein, pp. 23–34. San Francisco: Sierra Club Books.

———. 1987. *The Chalice and the Blade*. San Francisco: Harper and Row.

Eleure, C., ed. 1989. *Le Premier Or de l'Humanité en Bulgarie 5e millénaire*. Paris: Musées Nationaux.

Ellis, Linda. 1984. *The Cucuteni-Tripolye Culture: A Study in Technology and the Origins of Complex Society*. British Archaeological Reports International Series 217. Oxford: Archaeopress.

Emberling, Geoff. 1997. Ethnicity in complex societies: Archaeological perspectives. *Journal of Archaeological Research* 5 (4): 295–344.

Embleton, Sheila. 1991. Mathematical methods of genetic classification. In *Sprung from Some Common Source: Investigations into the Prehistory of Languages*, ed. Sidney Lamb and E. Douglass Mitchell, pp. 365–388. Stanford: Stanford University Press.

———. 1986. *Statistics in Historical Linguistics*. Bochum: Brockmeyer.

Enattah, Nabil Sabri. 2005. *Molecular Genetics of Lactase Persistence*. Ph.D. dissertation, Department of Medical Genetics, Faculty of Medicine, University of Helsinki, Finland.

Epimakhov, A. V. 2002. *Iuzhnoe zaural'e v epokhu srednei bronzy*. Chelyabinsk: YUrGU.

Epimakhov, A., B. Hanks, and A. C. Renfrew. 2005. Radiocarbon dating chronology for the Bronze Age monuments in the Transurals, Russia. *Rossiiskaia Arkheologiia* 4:92–102.

Erdosy, George, ed. 1995. *The Indo-Aryans of Ancient South Asia: Language, Material Culture and Ethnicity*. Indian Philology and South Asian Studies 1. Berlin: Walter de Gruyter.

Euler, Wolfram. 1979. *Indoiranisch-griechische Gemeinsamkeiten der Nominalbildung und deren Indogermanische Grundlagen*. Innsbruck: Institut für Sprachwissenschaft der Universität Innsbruck, vol. 30.

Evdokimov, V. V., and V. G. Loman. 1989. Raskopi Yamnogo kurgana v Karagandinskoi Oblasti. In *Voprosy arkheologii tsestral'nogo i severnogo Kazakhstana*, ed. K.M. Baipakov, pp. 34–46. Karaganda: Karagandinskii gosudarstvennyi universitet.

Ewers, John C. 1955. *The Horse in Blackfoot Indian Culture*. Washington, D.C.: Smithsonian Institution Press.

Falk, Harry. 1986. *Bruderschaft und Würfelspiel*. Freiburg: Hedwig Falk.

Fiedel, Stuart, and David W. Anthony. 2003. Deerslayers, pathfinders, and icemen: Origins of the European Neolithic as seen from the frontier. In *The Colonization of Unfamiliar Landscapes*, ed. Marcy Rockman and James Steele, pp. 144–168. London: Routledge.

Fischer, David Hackett. 1989. *Albion's Seed: Four British Folkways in America*. New York: Oxford University Press.

Fitzgerald-Huber, Louise G. 1995. Qijia and Erlitou: The question of contacts with distant cultures. *Early China* 20:17–67.

Florin, Curta. 2001. *The Making of the Slavs*. Oxford: Oxford University Press.

Forenbaher, S. 1993. Radiocarbon dates and absolute chronology of the central European Early Bronze Age. *Antiquity* 67:218–256.

Forsén, J. 1992. *The Twilight of the Early Helladics: A Study of the Disturbances in East-Central and Southern Greece toward the End of the Early Bronze Age*. Jonsered, Sweden: P. Åströms Förlag.

Fortson, Benjamin W., IV. 2004. *Indo-European Language and Culture: An Introduction*. Oxford: Blackwell.

Fox, John W. 1987. *Maya Postclassic State Formation*. Cambridge: Cambridge University Press.

Francis, E. D. 1992. The impact of non-Indo-European languages on Greek and Mycenaean. In *Reconstructing Languages and Cultures*, ed. E. Polome and W. Winter, pp. 469–506. Trends in Linguistics: Studies and Monographs 58. Berlin: Mouton de Gruyter.

French, Charly, and Maria Kousoulakou. 2003. Geomorphological and micro-morphological investigations of paleosols, valley sediments and a sunken-floored dwelling at Botai, Kazakstan. In *Prehistoric Steppe Adaptation and the Horse*, ed. Marsha Levine, Colin Renfrew, and Katie Boyle, pp. 105–114. Cambridge: McDonald Institute for Archaeological Research.

Fried, Morton H. 1975. *The Notion of Tribe*. Menlo Park, Calif.: Cummings.

Friedrich, Paul. 1970. *Proto-Indo-European Trees*. Chicago: University of Chicago Press.

Gaiduchenko, L. L. 1995. Mesto i znachenie Iuzhnogo Urala v eksportno-importnikh operatsiyakh po napravleniu vostok-zapad v eopkhu bronzy. In *Rossiya i vostok: Problemy vzaimodeistviya*, pt. 5, bk. 1: *Kul'tury eneolita-bronzy stepnoi evrazii*, pp. 110–115. Chelyabinsk: 3-ya Mezhdunarodnaya nauchnaya konferentsiya.

Gal, S. 1978. *Language Shift: Social Determinants of Linguistic Change in Bilingual Austria*. New York: Academic Press.

Gallusser, W. A. 1991. Geographical investigations in boundary areas of the Basle region ("Regio"). In *The Geography of Border Landscapes*, ed. D. Rumley and J. V. Minghi, pp. 32–42. London: Routledge.

Gamkrelidze, Thomas V., and Vyacheslav Ivanov. 1995. *Indo-European and the Indo-Europeans: A Reconstruction and Historical Analysis of a Proto-Language and a Proto-Culture*. Vol. 1. Translated by Johanna Nichols. Edited by Werner Winter. Trends in Linguistics: Studies and Monographs 80. Berlin: Mouton de Gruyter.

————. 1984. *Indoevropeiskii iazyk i indoevropeitsy*. Tiflis: Tbilisskogo Universiteta.

————. 1973. Sprachtypologie und die Rekonstruktion der gemeinindogermanischen Verschlüsse. *Phonetica* 27:150–156.

Gei, A. N. 2000. *Novotitorovskaya kul'tura*. Moscow: Institut Arkheologii.

————. 1990. Poyt paleodemograficheskogo analiza obshchestva stepnykh skotovodov epokhi bronzy: po pogrebal'nym pamyatkikam prikuban'ya. *Kratkie Soobshcheniya Institut Arkheologii* 201:78–87.

————. 1986. Pogrebenie liteishchika Novotitorovskoi kul'tury iz nizhnego pri kuban'ya. In *Arkheologicheskie Otkrytiya na Novostroikakh: Drevnosti severnogo kavkaza* (Moscow) 1:13–32.

————. 1979. Samsonovskoe mnogosloinoe poselenie na Donu. *Sovietskaya arkheologiya* (2): 119–131.

Gellner, Ernest. 1973. *Nations and Nationalism*. Ithaca, N.Y.: Cornell University Press.

Gening, V. F. 1979. The cemetery at Sintashta and the early Indo-Iranian peoples. *Journal of Indo-European Studies* 7:1–29.

Gening, V. F., G. B. Zdanovich, and V. V. Gening. 1992. *Sintashta*. Chelyabinsk: Iuzhno-ural'skoe knizhnoe izdatel'stvo.

George, Christian. 2002. *Quantification of Wear in Equus Teeth from Florida*. MA thesis, Department of Geological Sciences, University of Florida, Gainesville.

Georgieva, P. 1990. Ethnocultural and socio-economic changes during the transitional period from Eneolithic to Bronze Age in the region of the lower Danube. *Glasnik Centara za Balkanoloških Ispitavanja* 26:123–154.

Gheorgiu, Drago. 1994. Horse-head scepters: First images of yoked horses. *Journal of Indo-European Studies* 22 (3–4): 221–250.

Ghetie, B., and C. N. Mateesco. 1973. L'utilisation des bovines a la tracation dans le Neolithique Moyen. *International Conference of Prehistoric and Protohistoric Sciences* (Belgrade) 10:454–461.

Giddens, Anthony. 1985. *The Nation-state and Violence*, Cambridge: Polity.

Gilbert, Allan S. 1991. Equid remains from Godin Tepe, western Iran: An interim summary and interpretation, with notes on the introduction of the horse into Southwest Asia. In *Equids in the Ancient World*, vol. 2, ed. Richard H. Meadow and Hans-Peter Uerpmann, pp. 75–122. Wiesbaden: Reichert.

Gilman, Antonio. 1981. The development of social stratification in Bronze Age Europe. *Current Anthropology* 22 (1): 1–23.

Gimbutas, Marija. 1991. *The Civilization of the Goddess*. San Francisco: Harper.

———. 1989a. *The Language of the Goddess*. London: Thames and Hudson.

———. 1989b. Women and culture in Goddess-oriented Old Europe. In *Weaving the Visions*, ed. Judith Plaskow, and C. C. Christ, pp. 63–71. San Francisco: Harper and Row.

———. 1977. The first wave of Eurasian steppe pastoralists into Copper Age Europe. *Journal of Indo-European Studies* 5 (4): 277–338.

———. 1974. *The Goddesses and Gods of Old Europe: Myths and Cult Images (6500–3500 B.C.)*, London: Thames and Hudson.

———. 1970. Proto-Indo-European culture: The Kurgan Culture during the fifth, fourth, and third millennia B.C. In *Indo-European and the Indo-Europeans*, ed. George Cardona, Henry Hoenigswald, and Alfred Senn, pp. 155–198. Philadelphia: University of Pennsylvania Press.

———. 1956. *The Prehistory of Eastern Europe, Part 1*. Cambridge: American School of Prehistoric Research Bulletin 20.

Glassie, Henry. 1965. *Pattern in the Material Folk Culture of the Eastern United States*. Philadelphia: University of Pennsylvania Press.

Glonti, L. I. and A. I. Dzhavakhishvili. 1987. Novye dannye o mnogosloinom pamyatniki epokhi Eneolita-Pozdnei Bronzy v shida Kartli-Berkldeebi. *Kratkie Soobshcheniya Institut Arkheologii* 192:80–87.

Glumac, P. D., and J. A. Todd. 1991. Eneolithic copper smelting slags from the Middle Danube basin. In *Archaeometry '90*, ed. Ernst Pernicka and Günther A. Wagner, pp. 155–164. Basel: Birkhäuser Verlag.

Glumac, Petar, and David W. Anthony. 1992. Culture and environment in the prehistoric Caucasus, Neolithic to Early Bronze Age. In *Chronologies in Old World Archaeology, 3rd ed.*, ed. Robert Ehrich, pp. 196–206. Chicago: Aldine.

Golyeva, A. A. 2000. Vzaimodeistvie cheloveka i prirody v severo-zapadnom Prikaspii v epokhu Bronzy. In *Sezonnyi ekonomicheskii tsikl naseleniya severo-zapadnogo Prikaspiya v Bronzovom Veke*, vol. 120, ed. N. I. Shishlina, pp. 10–29. Moscow: Trudy gosudarstvennogo istoricheskogo muzeya.

Good, Irene. 2001. Archaeological textiles: A review of current research. *Annual Review of Anthropology* 30:209–226.

———. 1998. Bronze Age cloth and clothing of the Tarim Basin: The Chärchän evidence. In *The Bronze Age and Early Iron Age Peoples of Eastern Central Asia*, ed. Victor Mair, vol. 2, pp. 656–668. Journal of Indo European Studies Monograph 26. Washington, D.C.: Institute for the Study of Man.

Gorbunova, Natalya G. 1993/94. Traditional movements of nomadic pastoralists and the role of seasonal migrations in the formation of ancient trade routes in Central Asia. *Silk Road Art and Archaeology* 3:1–10.

Gotherstrom, A., C. Anderung, L. Hellborg, R. Elburg, C. Smith, D. G. Bradley, H. Ellegren 2005. Cattle domestication in the Near East was followed by hybridization with aurochs bulls in Europe. *Proceedings of Biological Sciences* 272 (1579): 2337–44.

Govedarica, B., and E. Kaiser. 1996. Die äneolithischen abstrakten und zoomorphen Steinzepter Südostund Europas. *Eurasia Antiqua* 2:59–103.

Grant, Madison. 1916. *The Passing of the Great Race; or, The Racial Basis of European History.* New York: Scribner's.

Gray, Russell D., and Quentin D. Atkinson. 2003. Language-tree divergence times support the Anatolian theory of Indo-European origin. *Nature* 426 (6965): 435–439.

Greenfield, Haskell. 1994. Preliminary report on the 1992 excavations at Foeni-Sălaş: An early Neolithic Starčevo-Criş settlement in the Romanian Banat. *Analele Banatului* 3:45–93.

———. 1999. The advent of transhumant pastoralism in temperate southeast Europe: A zooarchaeological perspective from the central Balkans. In *Transhumant Pastoralism in Southern Europe*, ed. L. Bartosiewicz and Haskell Greenfield, pp. 15–36. Budapest: Archaeolingua.

Gremillion, Kristen J. 2004. Seed processing and the origins of food production in eastern North America. *American Antiquity* 69 (2): 215–233.

Grigoriev, Stanislav A. 2002. *Ancient Indo-Europeans.* Chelyabinsk: RIFEI.

Gudkova, A. V., and I. T. Chernyakov. 1981. Yamnye pogebeniya s kolesami u s. Kholmskoe. In *Drevnosti severo-zapanogo prichernomor'ya*, pp. 38–50. Kiev: Naukovo Dumka.

Guliaev, V. I. 2003. Amazons in the Scythia: New finds at the Middle Don, Southern Russia. *World Archaeology* 35 (1): 112–125.

Haheu, Vasile, and Serghei Kurciatov. 1993. Cimitirul plan Eneolitic de lingă satul Giurgiuleşti. *Revista Arkheologică* (Kishinev) 1:101–114.

Hainsworth, J. B. 1972. Some observations on the Indo-European placenames of Greece. In *Acta of the 2nd International Colloquium on Aegean Prehistory*, pp. 39–42. Athens: Ministry of Culture and Sciences.

Haley, Brian D., and Larry R. Wilcoxon. 2005. How Spaniards became Chumash and other tales of ethnogenesis. *American Anthropologist* 107 (3): 432–445.

Hall, Jonathan M. 1997. *Ethnic Identity in Greek Antiquity.* Cambridge: Cambridge University Press.

Hall, Robert A., Jr. 1976. *Proto-Romance Phonology.* New York: Elsevier.

———. 1960. On realism in reconstruction. *Language* 36:203–206.

———. 1950. The reconstruction of Proto-Romance. *Language* 26:6–27.

Hamp, Eric. 1998. Whose were the Tocharians? In *The Bronze Age and Early Iron Age Peoples of Eastern Central Asia*, ed. Victor H. Mair, vol. 1, pp. 307–346. Journal of Indo-European Studies Monograph 26. Washington, D.C.: Institute for the Study of Man.

Hänsel, B. 1982. Südosteuropa zwischen 1600 und 1000 V. Chr. In *Südosteuropa zwischen 1600 und 1000 V. Chr.*, ed. B. Hänsel, pp. 1–38. Berlin: Moreland Editions.

Harding, R. M., and R. R. Sokal. 1988. Classification of the European language families by genetic distance. *Proceedings of the National Academy of Sciences* 85:9370–9372.

Harris, Alice C. 1991. Overview on the history of the Kartvelian languages. In *The Indigenous Languages of the Caucasus*, vol. 1, *The Kartvelian Languages*, ed. Alice C. Harris, pp. 7–83. Delmar, N.Y.: Caravan Books.

Harris, D. R., ed. 1996. *The Origins and Spread of Agriculture and Pastoralism in Eurasia.* London: University College.

Häusler, A. 1994. Archäologische Zeugnisse für Pferd und Wagen in Ost- und Mitteleuropa. In *Die Indogermanen und das Pferd: Festschrift für Bernfried Schlerath*, ed. B. Hänsel and S. Zimmer, pp. 217–257. Budapest: Archaeolingua.

———. 1992. "Der ursprung der Wagens in der Diskussion der gegenwart." *Archäologische Mitteilungen aus Nordwestdeutschland* 15:179–190.

———. 1974. *Die Gräber der älteren Ockergrabkultur zwischen Dnepr und Karpaten*. Berlin: Akadmie-Verlag.

Hayden, Brian. 2001. Fabulous feasts: A prolegomenon to the importance of feasting. In *Feasts*, ed. M. Dietler, and Brian Hayden, pp.23–64. Washington, D.C.: Smithsonian Institution Press.

Hayen, Hajo. 1989. Früheste Nachweise des Wagens und die Entwicklung der Transport-Hilfsmittel. *Mitteilungen der Berliner Gesellschaft für Anthropologie, Ethnologie und Urgeschichte* 10:31–49.

Heidegger, Martin. 1959. *An Introduction to Metaphysics*. 1953 [1935]. Translated by Ralph Manheim. New Haven: Yale University Press.

Helms, Mary. 1992. Long-distance contacts, elite aspirations, and the age of discovery. In *Resources, Power, and Inter-regional Interaction*, ed. Edward M. Schortman and Patricia A. Urban, pp. 157–174. New York: Plenum.

Hemphill, Brian E., A. F. Christensen, and Samar I. Mustafakulov. 1997. Trade or travel: An assessment of interpopulational dynamics among Bronze-Age Indo-Iranian populations. *South Asian Archaeology, 1995: Proceedings of the 13th Meeting of the South Asian Archaeologists of Europe, Cambridge, UK*, ed. Bridget Allchin, pp. 863–879, Oxford: IBH.

Hemphill, Brian E., and J. P. Mallory. 2003. Horse-mounted invaders from the Russo-Kazakh steppe or agricultural colonists from western Central Asia? A craniometric investigation of the Bronze Age settlements of Xinjiang. *American Journal of Physical Anthropology* 124 (3): 199–222.

Hester, D. A. 1957. Pre-Greek placenames in Greece and Asia Minor. *Revue Hittite et Asianique* 15:107–119.

Heyd, V., L. Husty, and L. Kreiner. 2004. *Siedlungen der Glockenbecherkultur in Süddeutschland und Mitteleuropa*. Büchenbach: Arbeiten zur Archäologie Süddeutschlands 17 (Dr. Faustus Verlag).

Hiebert, Frederik T. 2002. Bronze age interaction between the Eurasian steppe and Central Asia. In *Ancient Interactions: East and West in Eurasia*, ed. Katie Boyle, Colin Renfrew, and Marsha Levine, pp. 237–248, Cambridge: McDonald Institute for Archaeological Research.

———. 1994. *Origins of the Bronze Age Oasis Civilizations of Central Asia*. Bulletin of the American School of Prehistoric Research 42. Cambridge, Mass.: Peabody Museum of Archaeology and Ethnology, Harvard University.

Hiendleder, Stefan, Bernhard Kaupe, Rudolf Wassmuth, and Axel Janke. 2002. Molecular analysis of wild and domestic sheep. *Proceedings of the Royal Society of London* 269:893–904.

Hill, Jane. 1996. Languages on the land: Toward an anthropological dialectology. In *David Skomp Distinguished Lectures in Dialectology*. Bloomington: Indiana University Press.

Hill, Jonathon D. 1992. Contested pasts and the practice of archaeology: Overview. *American Anthropologist* 94 (4): 809–815.

Hobsbawm, Eric. 1997. *On History*. New York: New Press.

———. 1990. *Nations and Nationalism since 1780*. Cambridge: Cambridge University Press.

Hock, Hans Henrich, and Brian D. Joseph. 1996. *Language History, Language Change, and Language Relationship: An Introduction to Historical and Comparative Linguistics*. Berlin: Mouton de Gruyter.

Hodder, Ian. 1990. *The Domestication of Europe: Structure and Contingency in Neolithic Societies*. Cambridge: Cambridge University Press.

Holden, Clare, and Ruth Mace. 2003. Spread of cattle led to the loss of matriliny in Africa: A co-evolutionary analysis. *Proceedings of the Royal Society B* 270:2425–2433.

Hopper, Paul. 1973. Glottalized and murmured occlusives in Indo-European. *Glossa* 7:141–166.

Houwink Ten Cate, P. H. J. 1995. Ethnic diversity and population movement in Anatolia. In *Civilizations of the Ancient Near East*, ed. Jack M. Sasson, John Baines, Gary Beckman, and Karen R. Rubinson, vol. 1, pp. 259–270, New York: Scribner's.

Hulbert, R. C., G. S. Morgan, and S. D. Webb, eds. 1995. Paleontology and Geology of the Leisey Shell Pits, Early Pleistocene of Florida. *Bulletin of the Florida Museum of Natural History* 37 (1–10).

Huld, Martin E. 2002. "Linguistic science, truth, and the Indocentric hypothesis." *Journal of Indo-European Studies* 30 (3–4): 353–364.

———. 1990. "The linguistic typology of Old European substrata in north central Europe." *Journal of Indo-European Studies* 18:389–417.

Hüttel, Hans-Georg. 1992. "Zur archäologischen Evidenz der Pfredenutzung in der Kupfer- und Bronzezeit." In *Die Indogermanen und das Pferd: Festschrift für Bernfried Schlerath*, ed. B. Hänsel and S. Zimmer, pp. 197–215. Archaeolingua 4. Budapest: Archaeolingua Foundation.

Ilčeva, V. 1993. Localités de periode de transition de l'énéolithique a l'âdu bronze dans la region de Veliko Tîrnovo. In *The Fourth Millennium B.C.*, ed. Petya Georgieva, pp. 82–98. Sofia: New Bulgarian University.

Isakov, A. I. 1994. Sarazm: An agricultural center of ancient Sogdiana. *Bulletin of the Asia Institute* (n. s.) 8:1–12.

Isakov, A. I., Philip L. Kohl, C. C. Lamberg-Karlovsky, and R. Maddin. 1987. Metallurgical analysis from Sarazm, Tadjikistan SSR. *Archaeometry* 29 (1): 90–102.

Itina, M. A., and L. T. Yablonskii. 1997. *Saki Nizhnei Syrdar'i*. Moscow: Rosspen.

Ivanov, I. V., and I. B. Vasiliev. 1995. *Chelovek, Priroda i Pochvy Ryn-Peskov Volgo-Ural'skogo Mezhdurech'ya v Golotsene*. Moscow: Intellekt.

Ivanova, N. O. 1995. Arkheologicheskaya karta zapovednika Arkaim: Istotiya izucheniya arkheologicheskikh pamyatnikov. In *Arkaim*, ed. G. B. Zdanovich, pp. 159–195. Chelyabinsk: "Kammennyi Poyas."

Izbitser, Elena. 1993. Wheeled vehicle burials of the steppe zone of Eastern Europe and the Northern Caucasus, 3rd to 2nd millennium B.C. Doctoral Thesis, Institute of the History of Material Culture, St. Petersburg, Russia.

Jackson, Kenneth H. 1994. *Language and History in Early Britain*. Dublin: Four Courts.

Jacobs, Kenneth. 1993. Human postcranial variation in the Ukrainian Mesolithic-Neolithic. *Current Anthropology* 34 (3): 311–324.

Jacobsen-Tepfer, Esther. 1993. *The Deer-Goddess of Ancient Siberia: A Study in the Ecology of Belief.* Leiden: Brill.

James, Simon. 1999. *The Atlantic Celts: Ancient People or Modern Invention?* London: British Musem Press.

Janhunen, Juha. 2001. "Indo-Uralic and Ural-Altaic: On the diachronic implications of areal typology." In *Early Contacts between Uralic and Indo-European: Linguistic and Archaeological Considerations*, ed. Christian Carpelan, Asko Parpola, and Petteri Koskikallio, pp. 207–220. Memoires de la Société Finno-Ugrienne 242. Helsinki: Suomalais-Ugrilainen Seura.

———. 2000. Reconstructing Pre-Proto-Uralic typology: Spanning the millennia of linguistic evolution. In *Congressus Nonus Internationalis Fenno-Ugristarum*, pt. 1: *Orationes Plenariae & Orationes Publicae*, ed. Anu Nurk, Triinu Palo, and Tõnu Seilenthal, pp. 59–76. Tartu: CIFU.

Jansen, Thomas, Peter Forster, Marsha A. Levine, Hardy Oelke, Matthew Hurles, Colin Renfrew, Jürgen Weber, and Klaus Olek. 2002. Mitochondrial DNA and the origins of the domestic horse. *Proceedings of the National Academy of Sciences* 99:10905–10910.

Jarrige, Jean-Francois. 1994. The final phase of the Indus occupation at Nausharo and its connection with the following cultural complex of Mehrgarh VIII. *South Asian Archaeology* 1993 (1): 295–313.

John, B. S. 1972. The linguistic significance of the Pembrokeshire Landsker. *The Pembrokeshire Historian* 4:7–29.

Jones, Doug. 2003. Kinship and deep history: Exploring connections between culture areas, genes, and languages. *American Anthropologist* 105 (3): 501–514.

Jones, Siân. 1997. *The Archaeology of Ethnicity: Constructing Identities in the Past and Present.* London: Routledge.

Jones-Bley, Karlene. 2000. The Sintashta "chariots." In *Kurgans, Ritual Sites, and Settlements: Eurasian Bronze and Iron Age,* ed. Jeannine Davis-Kimball, Eileen Murphy, Ludmila Koryakova, and Leonid Yablonsky, pp. 135–140. BAR International Series 89. Oxford: Archeopress.

Jones-Bley, Karlene, and D. G. Zdanovich, eds. 2002. *Complex Societies of Central Eurasia from the 3rd to the 1st Millennium BC.* Vols. 1 and 2. Journal of Indo-European Studies Monograph 45. Washington, D.C.: Institute for the Study of Man.

Jordan, Peter, and Stephen Shennan. 2003. Cultural transmission, language, and basketry traditions amongst the California Indians. *Journal of Anthropological Archaeology* 22:42–74.

Jovanovich, B. 1975. Tumuli stepske culture grobova jama u Padunavlu," *Starinar* 26:9–24.

Kadyrbaev, M. K., and Z. Kurmankulov. 1992. *Kul'tura Drevnikh Skotobodov i Metallurgov Sary-Arki.* Alma-Ata: Gylym.

Kalchev, Petar. 1996. Funeral rites of the Early Bronze Age flat necropolis near the Bereket tell, Stara Zagora." In *Early Bronze Age Settlement Patterns in the Balkans,* pt. 2. Reports of Prehistoric Research Projects 1 (2–4): 215–225. Sofia: Agatho Publishers, Prehistory Foundation.

Kallio, Petri. 2001. Phonetic Uralisms in Indo-European? In *Early Contacts between Uralic and Indo-European: Linguistic and Archaeological Considerations,* ed. Christian Carpelan, Asko Parpola, and Petteri Koskikallio, pp. 221–234. Memoires de la Société Finno-Ugrienne 242. Helsinki: Suomalais-Ugrilainen Seura.

Kasperavičiūtė, D., V. Kučinskas, and M. Stoneking. 2004. Y chromosome and mitochondrial DNA variation in Lithuanians. *Annals of Human Genetics* 68:438–452.

Keeley, Lawrence, H. 1996. *War before Civilization.* New York: Oxford University Press.

Keith, Kathryn. 1998. Spindle whorls, gender, and ethnicity at Late Chalcolithic Hacinebi Tepe. *Journal of Field Archaeology* 25:497–515.

Kelley, Raymond C. 1985. *The Nuer Conquest.* Ann Arbor: University of Michigan Press.

Kershaw, Kris. 2000. *The One-Eyed God: Odin and the Indo-Germanic Männerbünde.* Journal of Indo-European Studies Monograph 36. Washington, D.C.: Institute for the Study of Man.

Khazanov, Anatoly. 1994 [1983]. *Nomads and the Outside World.* Rev. ed. Madison: University of Wisconsin Press.

Kiguradze, Tamaz, and Antonio Sagona. 2003. On the origins of the Kura-Araxes cultural complex. In *Archaeology in the Borderlands,* ed. Adam T. Smith and Karen Rubinson, pp. 38–94. Los Angeles: Cotsen Institute.

Kislenko, Aleksandr, and N. Tatarintseva. 1999. The eastern Ural steppe at the end of the Stone Age. In *Late Prehistoric Exploitation of the Eurasian Steppe,* ed. Marsha Levine, Yuri Rassamakin, A. Kislenko, and N. Tatarintseva, pp. 183–216. Cambridge: McDonald Institute for Archaeological Research.

Kitson, Peter R. 1997. Reconstruction, typology, and the "original homeland" of the Indo-Europeans. In *Linguistic Reconstruction and Typology,* ed. Jacek Fisiak, pp. 183–239, esp. pp. 198–202. Berlin: Mouton de Gruyter.

Kiyashko, V. Y. 1994. *Mezhdu Kamnem i Bronzoi.* Vol. 3. Azov: Donskie drevnosti.

———. 1987. Mnogosloinoe poselenie Razdorskoe i na Nizhnem Donu. *Kratkie soobschcheniya institut arkheologii* 192:73–79.

Klejn, L. 1984. The coming of the Aryans: Who and whence? *Bulletin of the Deccan College Research Institute* 43:57–69.

Klepikov, V. M. 1994. Pogrebeniya pozdneneoliticheskkogo vremeni u Khutora Shlyakhovskii v nizhnem Povolzh'e. *Rossiskaya arkheologiya* (3): 97–102.

Klochko, Viktor I., Aleksandr Kośko, and Marzena Szmyt. 2003. A comparative chronology of the prehistory of the area between the Vistula and the Dnieper: 4000–1000 BC. *Baltic-Pontic Studies* 12:396–414.

Knecht, Heidi, ed. 1997. *Projectile Technology*. New York: Plenum.

Kniffen, F. B. 1986. Folk housing: Key to diffusion. In *Common Places: Readings in American Vernacular Architecture*, ed. Dell V. Upton and John M. Vlach, pp. 3–23. Athens: University of Georgia Press.

Kohl, Philip, 2007. *The Making of Bronze Age Eurasia*. Cambridge: Cambridge University Press.

Kohl, Philip L., and Gocha R. Tsetskhladze. 1995. Nationalism, politics, and the practice of archaeology in the Caucasus. In *Nationalism, Politics, and the Practice of Archaeology*, ed. Philip L. Kohl and Clare Fawcett, pp. 149–174. Cambridge: Cambridge University Press.

Kohl, Philip L., Henri-Paul Francfort, and Jean-Claude Gardin. 1984. *Central Asia Palaeolithic Beginnings to the Iron Age*. Paris: Editions recherche sur les civilisations.

Koivulehto, Jorma. 2001. The earliest contacts between Indo-European and Uralic speakers in the light of lexical loans. In *Early Contacts between Uralic and Indo-European: Linguistic and Archaeological Considerations*, ed. Christian Carpelan, Asko Parpola, and Petteri Koskikallio, pp. 235–263. Memoires de la Société Finno-Ugrienne 242. Helsinki: Suomalais-Ugrilainen Seura.

Kolev, U. I., Kuznetsov, P. F., Kuz'mina, O. V., Semenova, A. P., Turetskii, M. A., and Aguzarov, B. A., eds. 2001. *Bronzovyi Vek Vostochnoi Evropy: Kharaketristika Kul'tur, Khronologiia i Periodizatsiya*. Samara: Samarskii gosudarstvennyi pedagogicheskii universitet.

Kol'tsov, L. V., ed. 1989. *Mezolit SSSR*. Moscow: Nauka.

Kopytoff, Igor. 1987. The internal African frontier: The making of African political culture. In *The African Frontier: The Reproduction of Traditional African Societies*, ed. Igor Kopytoff, pp. 3–84, Bloomington: Indiana University Press.

Korenevskii, S. N. 1995. *Galiugai I, poselenie Maikopskoi kul'tury*. Moscow: Biblioteka rossiskogo etnografa.

———. 1993. *Drevneishee osedloe naselenie na srednem Tereke*. Moscow: Stemi.

———. 1980. O metallicheskikh veshchakh i Utyevskogo mogil'nika. In *Arkheologiya Vostochno-Evropeiskoi Lesostepi*, ed. A. D. Pryakhin, pp. 59–66. Voronezh: Vorenezhskogo universiteta.

Korpusova, V. N., and S. N. Lyashko. 1990. Katakombnoe porgebenie s pshenitsei v Krimu. *Sovietskaya Arkheologiia* 3:166–175.

Korolev, A. I. 1999. Materialy po khronologii Eneolita pri Mokshan'ya. In *Voprosy Arkheologii Povolzh'ya, Sbornik Statei*, Vol. 1, ed. A. A. Vybornov and V. N. Myshkin, pp. 106–115. Samara: Samarskii gosudarstvennyi pedagogicheskii universitet.

Koryakova, L., and A. D. Epimakhov, 2007. *The Urals and Western Siberia in the Bronze and Iron Ages*. Cambridge: Cambridge University Press.

Kosintsev, Pavel. 2001. Kompleks kostnykh ostatkov domashnikh zhivotnykh iz poselenii i mogilnikov epokhi Bronzy Volgo-Ural'ya i ZaUral'ya. In *Bronzovyi Vek Vostochnoi Evropy: Kharakteristika Kul'tur, Khronologiya i Periodizatsiya*, ed. Y. I. Kolev, pp. 363–367. Samara: Samarskii gosudarstvennyi pedagogicheskii universitet.

Kośko, Aleksander, ed. 1999. *The Western Border Area of the Tripolye Culture*. Baltic-Pontic Studies 9. Poznań: Adam Mickiewicz University.

Kośko, Aleksandr, and Viktor I. Klochko, eds. 2003. *The Foundations of Radiocarbon Chronology of Cultures between the Vistula and Dnieper, 4000–1000 BC*. Baltic-Pontic Studies 12. Poznán: Adam Mickiewicz University.

Kotova, Nadezhda, and L. A. Spitsyna. 2003. Radiocarbon chronology of the middle layer of the Mikhailivka settlement. *Baltic-Pontic Studies* 12:121–131.

Kovaleva, I. F. 2001. "Vityanutye" pogrebeniya iz raskopok V. A. Gorodtsovym kurganov Donetchiny v kontekste Postmariupol'skoi kul'tury. In *Bronzovy Vek v Vostochnoi Evropy: Kharakteristika Kul'tur', Khronologiya i Periodizatsiya*, ed. Y. U. Kolev, pp. 20–24. Samara: Samara gosudarstvennyi pedagogicheskii universitet.

Kovaleva, V. T., and Zdanovich, G. B., eds. 2002. *Arkaim: Nekropol (po materialam kurgana 25 Bol'she Karaganskoe Mogil'nika)*. Chelyabinsk: Yuzhno-Ural'skoe knizhnoe izdatel'stvo.

Kovapenko, G. T., and V. M. Fomenko. 1986. Pokhovannya dobi Eneolitu-ranni Bronzi na pravoberezhzhi Pivdennogo Bugu. *Arkheologiya* (Kiev) 55:10–25.

Krahe, Hans. 1954. *Sprach und Vorzeit*. Heidelberg: Quelle und Meyer.

Kremenetski, C. V. 2002. Steppe and forest-steppe belt of Eurasia: Holocene environmental history. In *Prehistoric Steppe Adaptation and the Horse*, ed. M. Levine, C. Renfrew, and K. Boyle, pp. 11–27. Cambridge: Cambridge University Press.

———. 1997a. Human impact on the Holocene vegetation of the South Russian plain. In *Landscapes in Flux: Central and Eastern Europe in Antiquity*, ed. John Chapman and Pavel Dolukhanov, pp. 275–287. London: Oxbow Books.

———. 1997b. The Late Holocene environment and climate shift in Russia and surrounding lands. In *Climate Change in the Third Millennium BC*, ed. H. Dalfes, G. Kukla, and H. Weiss, pp. 351–370. New York: Springer.

Kremenetski, C. V., T. Böttger, F. W. Junge, A. G. Tarasov. 1999. Late- and postglacial environment of the Buzuluk area, middle Volga region, Russia. *Quaternary Science Reviews* 18:1185–1203.

Kremenetski, C. V., O. A. Chichagova, and N. I. Shishlina. 1999. Palaeoecological evidence for Holocene vegetation, climate and land-use change in the low Don basin and Kalmuk area, southern Russia. *Vegetation History and Archaeology* 8 (4): 233–246.

Kristiansen, Kristian, and Thomas Larsson. 2005. *The Rise of Bronze Age Society: Travels, Transmissions, and Transformations*. Cambridge: Cambrideg University Press.

Kriukova, E. A. 2003. Obraz loshadi v iskusstve stepnogo naseleniya epokhi Eneolita-Rannei Bronzy. In *Voprosy Arkheologii Povolzh'ya*, pp. 134–143. Samara: Samarskii nauchnyi tsentr RAN.

Krizhevskaya, L. Y. 1991. *Nachalo Neolita v stepyakh severnogo Priochernomor'ya*. St. Petersburg: Institut istorii material'noi kul'tury Akademii Nauk SSSR.

Kruts, V. O. 1977. *Pozdnetripol'skie pamyatniki srednego Podneprov'ya*. Kiev: Naukovo Dumka.

Kruts, V. O., and S. M. Rizhkov, 1985, Fazi rozvitku pam'yatok Tomashivs'ko-Syshkivs'koi grupi. *Arkheologiya* (Kiev) 51:45–56.

Kryvaltsevich, Mikola M., and Nikolai Kovalyukh. 1999. Radiocarbon dating of the Middle Dnieper culture from Belarus. *Baltic Pontic Studies* 7:151–162.

Kubarev, V. D. 1988. *Drevnie rospisi Karakola*. Novosibirsk: Nauka.

Kühl, Stefan. 1994. *The Nazi Connection: Eugenics, American Racism, and German National Socialism*. New York: Oxford University Press.

Kuiper, F. B. J. 1991. *Aryans in the Rig-Veda*. Amsterdam: Rodopi.

———. 1960. The ancient Aryan verbal contest. *Indo-Iranian Journal* 4:217–281.

———. 1955. Rig Vedic Loanwords. *Studia Indologica* (Festschrift für Willibaldkirfel), pp. 137–185. Bonn: Selbst Verlag des Orientalishen Seminars des Universität.

———. 1948. *Proto-Munda Words in Sanskrit*. Amsterdam: Noord-Hollandische Uitgevers Maatschappij.

Kulick, Don. 1992. *Language Shift and Cultural Reproduction: Socialization, Self, and Syncretism in a Papuan New Guineau Village*. Cambridge: Cambridge University Press.

Kuna, Martin. 1991. The structuring of the prehistoric landscape. *Antiquity* 65:332–347.

Kuzmina, I. E. 1988. Mlekopitayushchie severnogo pri Kaspiya v Golotsene. *Arkheologocheskie Kul'tury Severnogo Prikaspiya*, ed. R. S. Bagautdinov, pp. 173–188. Kuibyshev: Samarskii gosudarstvennyi pedagogicheskii universitet.

Kuzmina, Elena E. 2003. Origins of pastoralism in the Eurasian steppes. In *Prehistoric Steppe Adaptation and the Horse*, ed. Marsha Levine, Colin Renfrew, and Katie Boyle, pp. 203–232. Cambridge: McDonald Institute for Archaeological Research.

———. 2001. The first migration wave of Indo-Iranians to the south. *Journal of Indo-European Studies* 29 (1–2): 1–40.

———. 1994. *Otkuda prishli indoarii?* Moscow: MGP "Kalina" VINITI RAN.

———. 1980. Eshche raz o diskovidniykh psaliakh Evraziiskikh stepei. *Kratkie Soobshcheniya Institut Arkheologii* 161:8–21.

Kuzmina, O. V. 1999. Keramika Abashevskoi kul'tury. In *Voprosy Arkheologii Povolzh'ya, Sbornik Statei*, vol. 1, ed. A. A. Vybornov and V. N. Myshkin, pp. 154–205. Samara: Samarskii gosudarstvennyi pedagogicheskii universitet.

Kuzminova, N. N. 1990. Paleoetnobotanicheskii i palinologicheskii analizy materialov iz kurganov nizhnego podnestrov'ya. In *Kurgany Eneolita-Eopkhi Bronzy Nizhnego Podnestrov'ya*, ed. E. V. Yarovoi, pp. 259–267. Kishinev: Shtiintsa.

Kuzminova, N. N., V. A. Dergachev, and O. V. Larina. 1998. Paleoetnobotanicheskie issledovaniya na poselenii Sakarovka I. *Revista Arheologică* (Kishinev) (2): 166–182.

Kuznetsov, Pavel. 2005. An Indo-European symbol of power in the earliest steppe kurgans. *Journal of Indo-European Studies* 33 (3–4): 325–338.

———. 2001. Territorial'nye osobennosti i vremennye ramki perekhodnogo perioda k epokhe Pozdnei Bronzy Vostochnoi Evropy. In *Bronzovyi Vek Vostochnoi Evropy: Kharakteristika Kul'tur, Khronologiya i Periodizatsiya*, ed. Y. I. Kolev et al., pp. 71–82. Samara: Samarskii gosudarstvennyi pedagogicheskii universitet.

———. 1991. Unikalnoe pogrebenie epokhi rannei Bronzy na r. Kutuluk. In *Drevnosti Vostochno-Evropeiskoi Lesostepi*, ed. V. V. Nikitin, pp. 137–139. Samara: Samarskii gosudarstvennyi pedagogicheskii institut. Labov, William. 1994. *Principles of Linguistic Change: Internal Factors*. Oxford: Blackwell.

Lafontaine, Oskar, and Georgi Jordanov, eds. 1988. *Macht, Herrschaft und Gold: Das Gräberfeld von Varna (Bulgarien) und die Anfänge Einer Neuen Europäischen Zivilisation*. Saarbrücken: Moderne Galerie des Saarland-Museums.

Lagodovskaya, E. F., O. G. Shaposhnikova, and M. L. Makarevich. 1959. Osnovnye itogi issledovaniya Mikhailovskogo poseleniya. *Kratkie soobshcheniya institut arkheologii* 9:21–28.

Lakoff, George. 1987. *Women, Fire and Dangerous Things: What Categories Reveal about the Mind*. Chicago: University of Chicago Press.

Lam, Andrew. 2006. *Learning a Language, Inventing a Future*. Commentary on National Public Radio, May 1, 2006.

———. 2005. *Perfume Dreams: Reflections on the Vietnamese Diaspora*. Foreword by Richard Rodriguez. Berkeley: Heyday Books.

Lamberg-Karlovsky, C. C. 2002. Archaeology and language: The Indo-Iranians. *Current Anthropology* 43 (1): 63–88.

Latacz, Joachim. 2004. *Troy and Homer: Toward a Solution of an Old Mystery*. Oxford: Oxford University Press.

Lattimore, Owen. 1940. *Inner Asian Frontiers of China*. Boston: Beacon.

Lavrushin, Y. A., E. A. Spiridonova, and L. L. Sulerzhitskii. 1998. Geologo-paleoekologocheskie sobytiya severa aridnoi zony v poslednie 10- tys. let. In *Problemy Drevnei Istorii Severnogo Prikaspiya*, ed. V. S. Gorbunov, pp. 40–65. Samara: Samarskogo gosudarstvennogo pedagogicheskogo universiteta.

Leach, Edmund R. 1968. *Political Systems of Highland Burma*. Boston: Beacon.

———. 1960. The frontiers of Burma. *Comparative Studies in Society and History* 3 (1): 49–68.

Lees, Robert. 1953. The basis of glottochronology. *Language* 29 (2): 113–127.

Lefferts, H. L., Jr. 1977. Frontier demography: An introduction. In *The Frontier, Comparative Studies*, ed. D. H. Miller and J. O. Steffen, pp. 33–56. Norman: University of Oklahoma Press.

Legge, Tony. 1996. The beginning of caprine domestication in southwest Asia. In *The Origins and Spread of Agriculture and Pastoralism in Eurasia*, ed. David R. Harris, pp. 238–262. London: University College London Press.

Lehman, F. K. 1989. Internal inflationary pressures in the prestige economy of the Feast of Merit complex: The Chin and Kachin cases from upper Burma. in *Ritual, Power and Economy:*

Upland-Lowland Contrasts in Mainland Southeast Asia, ed. Susan D. Russell, pp. 89–101. Occasional Paper 14. DeKalb, Ill.: Center for Southeast Asian Studies.

Lehmann, Winfred. 1989. Earlier stages of Proto-Indo-European. In *Indogermanica Europaea*, ed. K. Heller, O. Panagi, and J. Tischler, pp. 109–131. Grazer Linguistische Monographien 4. Graz: Institut für Sprachwissenschaft der Universität Graz.

Lehrman, Alexander. 2001. Reconstructing Proto-Hittite. In *Greater Anatolia and the Indo-Hittite Language Family*, ed. Robert Drews, pp. 106–130. Journal of Indo-European Studies Monograph 38. Washington, D.C.: Institute for the Study of Man.

Leuschner, Hans Hubert, Ute Sass-Klaassen, Esther Jansma, Michael Baillie, and Marco Spurk. 2002. Subfossil European bog oaks: Population dynamics and long-term growth depressions as indicators of changes in the Holocene hydro-regime and climate. *The Holocene* 12 (6): 695–706.

Levi, Scott C. 2002. *The Indian Diaspora in Central Asia and Its Trade, 1550–1900*. Leiden: Brill.

Levine, Marsha. 2004. Exploring the criteria for early horse domestication. In *Traces of Ancestry: Studies in Honor of Colin Renfrew*, ed. Martin Jones, pp. 115–126. Cambridge: McDonald Institute for Archaeological Research.

———. 2003. Focusing on Central Eurasian archaeology: East meets west. In *Prehistoric Steppe Adaptation and the Horse*, ed. Marsha Levine, Colin Renfrew, and Katie Boyle, pp. 1–7. Cambridge: McDonald Institute for Archaeological Research.

———. 1999a. Botai and the origins of horse domestication. *Journal of Anthropological Archaeology* 18:29–78.

———. 1999b. The origins of horse husbandry on the Eurasian steppe. In *Late Prehistoric Exploitation of the Eurasian Steppe*, ed. Marsha Levine, Yuri Rassamakin, Aleksandr Kislenko, and Nataliya Tatarintseva, pp. 5–58. Cambridge: McDonald Institute for Archaeological Research.

———. 1990. Dereivka and the problem of horse domestication. *Antiquity* 64:727–740.

———. 1982. The use of crown height measurements and eruption-wear sequences to age horse teeth. In *Ageing and Sexing Animal Bones from Archaeological Sites*, ed. B. Wilson, C. Grigson, and S. Payne, pp. 223–250. British Archaeological Reports, British Series 109. Oxford: Archaeopress.

Levine, Marsha, and A. M. Kislenko. 2002. New Eneolithic and Early Bronze Age radiocarbon dates for northern Kazakhstan and south Siberia. In *Ancient Interactions: East and West in Eurasia*, ed. Katie Boyle, Colin Renfrew, and Marsha Levine, pp. 131–134, Cambridge: McDonald Institute for Archaeological Research.

Levine, Marsha, Colin Renfrew, and Katie Boyle, eds. 2003. *Prehistoric Steppe Adaptation and the Horse*. Cambridge: McDonald Institute for Archaeological Research.

Li, Shuicheng. 2002. The interaction between northwest China and Central Asia during the second millennium BC: An archaeological perspective. In *Ancient Interactions: East and West in Eurasia*, ed. Katie Boyle, Colin Renfrew, and Marsha Levine, pp. 171–182. Cambridge: McDonald Institute for Archaeological Research.

Lichardus, Jan, ed. 1991. *Die Kupferzeit als historische Epoche*. Bonn: Dr. Rudolf Hebelt GMBH.

Lichardus, Jan, and Josef Vladar. 1996. Karpatenbecken-Sintashta-Mykene: ein Beitrag zur Definition der Bronzezeit als Historischer Epoche. *Slovenska Archeologia* 44 (1): 25–93.

Lillie, Malcolm C. 1996. Mesolithic and Neolithic populations in Ukraine: Indications of diet from dental pathology. *Current Anthropology* 37 (1): 135–142.

Lillie, Malcolm C., and M. P. Richards. 2000. Stable isotope analysis and dental evidence of diet at the Mesolithic-Neolithic transition in Ukraine. *Journal of Archaeological Science* 27:965–972.

Lincoln, Bruce. 1981. *Priests, Warriors, and Cattle: A Study in the Ecology of Religions*. Berkeley: University of California Press.

———. 1991. *Death, War and Sacrifice: Studies in Ideology and Practice*. Chicago: University of Chicago Press.

Lindgren, G., N. Backström, J. Swinburne, L. Hellborg, A. Einarsson, K. Sandberg, G. Co-
thran, Carles Vilà, M. Binns, and H. Ellegren. 2004. Limited number of patrilines in horse
domestication. *Nature Genetics* 36 (3): 335–336.

Lindstrom, Richard W. 2002. Anthropological characteristics of the population of the Bol-
shekaragansky cemetery, kurgan 25. In *Arkaim: Nekropol (po materialam kurgana 25 Bol'she
Karaganskoe Mogil'nika)*, ed. V. T. Kovaleva and G.B. Zdanovich, pp. 159–166, Chelyabinsk:
Yuzhno-Ural'skoe knizhnoe izdatel'stvo.

Linduff, Katheryn M., Han Rubin, and Sun Shuyun, eds. 2000. *The Beginnings of Metallurgy in
China*. New York: Edwin Mellen Press.

Lisitsyn, N. F. 1996. Srednii etap pozdnego Paleolita Sibiri. *Rossiskaya arkheologiya* (4): 5–17.

Littauer, Mary A. 1977. Rock carvings of chariots in Transcaucasia, Central Asia, and Outer
Mongolia. *Proceedings of the Prehistoric Society* 43:243–262.

———. 1972. The military use of the chariot in the Aegean in the Late Bronze Age. *American
Journal of Archaeology* 76:145–157.

———. 1968. A 19th and 20th dynasty heroic motif on Attic black-figured vases? *American
Journal of Archaeology* 72:150–152.

Littauer, Mary A., and Joost H. Crouwel. 1996. The origin of the true chariot. *Antiquity*
70:934–939.

———. 1986. A Near Eastern bridle bit of the second millennium BC in New York. *Levant*
18:163–167.

———. 1983. Chariots in Late Bronze Age Greece. *Antiquity* 57:187–192.

———. 1979. *Wheeled Vehicles and Ridden Animals in the Ancient Near East*. Leiden: Brill.

Littleton, C. S. 1982. *The New Comparative Mythology*. Berkeley: University of California Press.

Litvinsky, B. A., and L. T. P'yankova. 1992. Pastoral tribes of the Bronze Age in the Oxus
valley (Bactria). In *History of the Civilizations of Central Asia*, ed. A. H. Dani and V. M.
Masson, vol. 1, pp. 379–394. Paris: UNESCO.

Logvin, V. N. 1995. K probleme stanovleniya Sintashtinsko-Petrovskikh drevnostei. In *Rossiya i Vostok: Problemy Vzaimodeistviya*, pt. 5, bk. 1: *Kul'tury Eneolita-Bronzy Stepnoi Evrazii*,
pp. 88–95. Chelyabinsk: 3-ya Mezhdunarodnaya nauchnaya konferentsiya.

———. 1992. Poseleniya Tersekskogo tipa Solenoe Ozero I. *Rossiskaya arkheologiya* (1): 110–120.

Logvin, V. N., S. S. Kalieva, and L. L. Gaiduchenko. 1989. O nomadizme v stepyakh Kazakhstana v III tys. do n. e. In *Margulanovskie chteniya*, pp. 78–81. Alma-Ata: Akademie Nauk
Kazakhskoi SSR.

Lubotsky, Alexsander. 2001. The Indo-Iranian substratum. In *Early Contacts between Uralic
and Indo-European: Linguistic and Archaeological Considerations*, ed. Christian Carpelan,
Asko Parpola, and Petteri Koskikallio, pp. 301–317. Helsinki: Suomalais-Ugrilainen
Seura.

Lukacs, J. R. 1989. Dental paleopathology: Methods for reconstructing dietary patterns. In
Reconstruction of Life From the Skeleton, ed. M. Y. Iscan, and K. A. R. Kennedy, pp. 261–286.
New York: Alan Liss.

Lyashko, S. N., and V. V. Otroshchenko. 1988. Balkovskii kurgan. In *Novye pamyatniki yamnoi kul'tury stepnoi zony Ukrainy*, ed. A. A. Zolotareva, pp. 40–63. Kiev: Naukovo Dumka.

Lyonnet, B., ed. 1996. *Sarazm (Tajikistan). Céramiques (Chalcolithiques et Bronze Ancien)*.
Mémoire de la Mission Archéologique Française en Asie Centrale 7. Paris: De Boccard.

Mace, Ruth. 1993. Transitions between cultivation and pastoralism in sub-Saharan Africa.
Current Anthropology 34 (4): 363–382.

MacEachern, Scott. 2000. Genes, tribes, and African history. *Current Anthropology* 41 (3):
357–384.

Machnik, Jan. 1999. Radiocarbon chronology of the Corded Ware culture on Grzeda Sokalska:
A Middle Dnieper traits perspective. *Baltic-Pontic Studies* 7:221–250.

Madgearu, Alexandru. 2001. The end of town life in Scythia Minor. *Oxford Journal of Archaeology* 20 (2): 207–217.

Makkay, Janos. 2000. *The Early Mycenaean Rulers and the Contemporary Early Iranians of the Northeast.* Tractata Miniscula 22. Budapest: szerzo kiadása.

———. 1976. Problems concerning Copper Age chronology in the Carpathian Basin: Copper Age gold pendants and gold discs in central and south-east Europe. *Acta Archaeologica Hungarica* 28:251–300.

Malandra, William. 1983. *An Introduction to Ancient Iranian Religion.* Minneapolis: University of Minnesota Press.

Maliutina, T. S. 1991. Stratigraficheskaya pozitsiya materilaov Fedeorovskoi kul'tury na mnogosloinikh poseleniyakh Kazakhstanskikh stepei. In *Drevnosti Vostochno-Evropeiskoi Lesostepi,* ed. V. V. Nikitin, pp. 141–162. Samara: Samarskii gosudarstvennyi pedagogicheskii institut.

———. 1984. Mogil'nik Priplodnyi Log 1. In *Bronzovyi Vek Uralo-Irtyshskogo Mezhdurech'ya,* pp. 58–79. Chelyabinsk: Chelyabinskii gosudarstvennyi universitet.

Maliutina, T. S., and G. B. Zdanovich. 1995. Kuisak—ukreplennoe poselenie protogorodskoi tsivilizatsii iuzhnogo zaUral'ya. In *Rossiya i Vostok: Problemy Vzaimodeistviya,* pt. 5, bk. 1: *Kul'tury Eneolita-Bronzy Stepnoi Evrazii,* pp. 100–106. Chelyabinsk: 3-ya Mezhdunarodnaya nauchnaya konferentsiya.

Mallory, J. P. 1998. A European perspective on Indo-Europeans in Asia. In *The Bronze Age and Early Iron Age Peoples of Eastern Central Asia,* ed. Victor H. Mair, vol. 1, pp. 175–201. Philadelphia: University of Pennsylvania Press.

———. 1992. Migration and language change. *Peregrinatio Gothica III, Universitetets Oldsaksamlings Skrifter Ny Rekke* (Oslo) 14:145–153.

———. 1990. Social structure in the Pontic-Caspian Eneolithic: A preliminary review. *Journal of Indo-European Studies* 18 (1–2): 15–57.

———. 1989. *In Search of the Indo-Europeans.* London: Thames and Hudson.

———. 1977. The chronology of the early Kurgan tradition. *Journal of Indo-European Studies* 5:339–368.

Mallory, J. P., and Douglas Q. Adams. 1997. *Encyclopedia of Indo-European Culture.* London: Fitzroy Dearborn.

Mallory, J. P., and Victor H. Mair. 2000. *The Tarim Mummies: Ancient China and the Mystery of the Earliest Peoples from the West.* London: Thames and Hudson.

Malov, N. M. 2002. Spears: Signs of archaic leaders of the Pokrovsk archaeological cultures. In *Complex Societies of Central Eurasia from the 3rd to the 1st Millennium BC,* vols. 1 and 2, ed. Karlene Jones-Bley and D. G. Zdanovich, pp. 314–336. Journal of Indo-European Studies Monograph 45. Washington, D.C.: Institute for the Study of Man.

Mamonov, A. E. 1995. Elshanskii kompleks stoianki Chekalino IV. In *Drevnie kul'tury lesostepnogo povolzh'ya,* pp. 3–25. Samara: Samarskogo gosudarstvennogo pedagogicheskogo universiteta.

Mamontov, V. I. 1974. Pozdneneoliticheskaya stoianka Orlovka. *Sovietskaya arkheologiya* (4): 254–258.

Manfredi, J., Hilary M. Clayton, and D. Rosenstein. 2005. Radiographic study of bit position within the horse's oral cavity. *Equine and Comparative Exercise Physiology* 2 (3): 195–201.

Manhart, H. 1998. Die vorgeschichtliche Tierwelt von Koprivec und Durankulak und anderen prähistorischen Fundplätzen in Bulgarien aufgrund von Knochenfunden aus archäologischen Ausgrabungen. *Documenta Naturae* (München) 116:1–353.

Manzura, I. 1999. The Cernavoda I culture. In *The Balkans in Later Prehistory,* ed. Lolita Nikolova, pp. 95–174. British Archaeological Reports, International Series 791. Oxford: Archaeopress.

Manzura, I., E. Savva, and L. Bogotaya. 1995. East-west interactions in the Eneolithic and Bronze Age cultures of the north-west Pontic region. *Journal of Indo-European Studies* 23 (1–2): 1–51.

Maran, Joseph. 2001. Zur Westausbreitung von Boleráz-Elementen in Mitteleuropa. In *Cernavoda III-Boleráz, Ein vorgeschichtliches Phänomen zwischen dem Oberrhein und der unteren Donau,* ed. P. Roman, and S. Diamandi, pp. 733–752. Bucharest: Studia Danubiana.

————. 1998. *Kulturwandel auf dem Griechischen Festland und den Kykladen im späten 3. Jahrtausend v. Chr.* Bonn: Habelt.

Marcsik, Antónia. 1971. Data of the Copper Age anthropological find of Bárdos-Farmstead at Csongrád-Kettöshalom. *A Móra Ferenc Múzeum Évkönyve* (2): 19–27.

Marinescu-Bîlcu, S. 1981. Tîrpeşti: From prehistory to history in Eastern Romania. British Archaeological Reports, International Series 107. Oxford: Archeopress.

Marinescu-Bîlcu, S., Alexandra Bolomey, Marin Cârciumâru, and Adrian Muraru. 1984. Ecological, economic and behavioral aspects of the Cucuteni A4 community at Draguşeni. *Dacia* 28 (1–2): 41–46.

Marinesu-Bîlcu, Silvia, M. Cârciumaru, and A. Muraru. 1981. Contributions to the ecology of pre- and proto-historic habitations at Tîrpeşti. *Dacia* 25:7–31.

Marinova, Elena. 2003. The new pollen core Lake Durankulak-3: The vegetation history and human impact in Northeastern Bulgaria. In *Aspects of Palynology and Paleontology*, ed. S. Tonkov, pp. 279–288. Sofia: Pensoft.

Markevich, V. I. 1974. *Bugo-Dnestrovskaya kul'tura na territorii Moldavii.* Kishinev: Shtintsa.

————. 1965. Issledovaniia Neolita na srednem Dnestre. *Kratkie soobshcheniya institut arkheologii* 105:85–90.

Markey, T. L. 1990. Gift, payment, and reward revisited. In *When Worlds Collide: The Indo-Europeans and the Pre-Indo-Europeans*, ed. T. L. Markey and John Grippin, pp. 345–362. Ann Arbor, Mich.: Karoma.

Markovin, V. I. 1980. O nekotorykh voprosakh interpretatsii dol'mennykh i drugikh arkheologicheskikh pamyatnikov Kavkaza. *Kratkie soobshchenniya institut arkheologii* 161:36–45.

Mashkour, Marjan. 2003. Equids in the northern part of the Iranian central plateau from the Neolithic to the Iron Age: New zoogeographic evidence. In *Prehistoric Steppe Adaptation and the Horse*, ed. Marsha Levine, Colin Renfrew, and Katie Boyle, pp. 129–138. Cambridge: McDonald Institute for Archaeological Research.

Masson, V. M. 1988. *Altyn-Depe.* Translated by Henry N. Michael. University Museum Monograph 55. Philadelphia: University of Pennsylvania Press.

————. 1979. Dinamika razvitiya Tripol'skogo obshchestva v svete paleo-demograficheskikh otsenok. In *Pervobytnaya Arkheologiya, Poiski i Nakhodki*, ed. N. N. Bondar and D. Y. Telegin, pp. 204–212. Kiev: Naukovo Dumka.

Matiushchenko, V. I., and G. V. Sinitsyna. 1988. *Mogil'nik u d. Rostovka Vblizi Omska.* Tomsk: Tomskogo universiteta.

Matiushin, G. N. 1986. The Mesolithic and Neolithic in the southern Urals and Central Asia. In *Hunters in Transition: Mesolithic Societies of Temperate Eurasia and Their Transition to Farming*, ed. M. Zvelebil, pp. 133–150. Cambridge: Cambridge University Press.

Matthews, Roger, and Hassan Fazeli. 2004. Copper and complexity: Iran and Mesopotamia in the fourth millennium BC. *Iran* 42:61–75.

McMahon, April, and Robert McMahon. 2003. Finding families: Quantitative methods in language classification. *Transactions of the Philological Society* 10:7–55.

Meadow, Richard H., and Ajita Patel. 1997. A comment on "Horse Remains from Surkotada" by Sándor Bökönyi. *South Asian Studies* 13:308–315.

Mei, Jianjun. 2003a. Cultural interaction between China and Central Asia during the Bronze Age. *Proceedings of the British Academy* 121:1–39.

————. 2003b. Qijia and Seima-Turbino: The question of early contacts between northwest China and the Eurasian steppe. *Bulletin of the Museum of Far Eastern Antiquities* 75: 31–54.

Meid, Wolfgang. 1994. Die Terminologie von Pferd und Wagen im Indogermanischen. In *Die Indogermanen und das Pferd*, ed. B. Hänsel and S. Zimmer, pp. 53–65. Budapest: Archaeolingua.

————. 1975. Probleme der räumlichen und zeitlichen Gliederung des Indogermanischen. In *Flexion und Wortbildung*, ed. Helmut Rix, pp. 204–219. Weisbaden: Reichert.

Melchert, Craig. 2001. Critical responses. In *Greater Anatolia and the Indo-Hittite Language Family*, ed. Robert Drews, pp. 229–235. Journal of Indo-European Studies Monograph 38. Washington, D.C.: Institute for the Study of Man.

———. 1994. *Anatolian Historical Phonology*. Amsterdam: Rodopi.

Melent'ev, A. N. 1975. Pamyatniki seroglazivskoi kul'tury (neolit Severnogo Prikaspiya). *Kratkie soobshcheniya institut arkheologii* (Moscow) 141:112–118.

Mel'nik, A. A., and I. L. Serdiukova. 1988. Rekonstruktsiya pogrebal'noi povozki Yamnoi kul'tury. In *Novye pamyatniki yamnoi kul'tury stepnoi zony Ukrainy*, ed. N. N. Bondar and D. Y. Telegin, pp. 118–124. Kiev: Dumka.

Merpert, N. Y. 1995. Bulgaro-Russian archaeological investigations in the Balkans. *Ancient Civilizations from Scythia to Siberia* 2 (3): 364–383.

———. 1980. Rannie skotovody vostochnoi Evropy i sudby drevneishikh tsivilizatsii. *Studia Praehistorica* 3:65–90.

———. 1974. *Drevneishie Skotovody Volzhsko-Uralskogo Mezhdurechya*. Moscow: Nauka.

Mezhlumian, S. K. 1990. Domestic horse in Armenia. Paper delivered at the International Conference on Archaeozoology, Washington, D.C.

Milisauskas, Sarunas. 2002. *European Prehistory: A Survey*. New York: Kluwer.

Militarev, Alexander. 2002. The prehistory of a dispersal: The Proto-Afrasian (Afroasiatic) farming lexicon. In *Examining the Farming/Language Dispersal Hypothesis*, ed. Peter Bell-wood and Colin Renfrew, pp. 135–150. Cambridge: McDonald Institute for Archaeological Research.

Milroy, James. 1992. *Linguistic Variation and Change*. Oxford: Blackwell.

Molleson, Theya, and Joel Blondiaux. 1994. Riders' bones from Kish, Iraq. *Cambridge Archaeological Journal* 4 (2): 312–316.

Molodin, V. I. 1997. Nekotoriye itogi arkheologicheskikh isseldovanii na Iuge Gornogo Altaya. *Rossiiskaya arkheologiya* (1): 37–49.

Moore, John. 2001. Ethnogenetic patterns in Native North America. In *Archaeology, Language, and History*, ed. John E. Terrell, pp. 31–56. Westport, Conn.: Bergin and Garvey.

Moorey, P. R. S. 1986. The emergence of the light, horse-drawn chariot in the Near East, c. 2000–1500 BC. *World Archaeology* 18 (2): 196–215.

Morgunova, N. L. 1995. Elitnye kurgany eopkhi rannei I srednei bronzy v stepnom Orenburzh'e. In *Rossiya i Vostok: Problemy Vzaimodeistviya*, pt. 5, bk. 1, *Kul'tury Eneolita-Bronzy Stepnoi Evrazii*, pp. 120–123. Chelyabinsk: 3-ya Mezhdunarodnaya nauchnaya konferentsiya.

———. 1988. Ivanovskaya stoianka v Orenburgskoi oblasti. In *Arkheologocheskie kul'tury severnogo prikaspiya*, ed. R. S. Bagautdinov, pp. 106–122. Kuibyshev: Samarskii gosudarstvennyi pedagogicheskii universitet.

Morgunova, N. L., and M. A. Turetskii. 2003. Yamnye pamyatniki u s. Shumaevo: novye dannye o kolesnom transporte u naseleniya zapadnogo Orenburzh'ya v epokha rannego metalla. In *Voprosy arkheologii povozh'ya*, vol. 3, pp. 144–159. Samara: Samarskii nauchnyi tsentr RAN.

Morintz, Sebastian, and Petre Roman. 1968. Aspekte des Ausgangs des Äneolithikums und der Übergangsstufe zur Bronzezeit im Raum der Niederdonau. *Dacia* 12:45–128.

Movsha, T. G. 1985. Bzaemovidnosini Tripillya-Kukuteni z sinkhronimi kul'turami Tsentral'noi Evropi. *Arkheologiia* (Kiev) 51:22–31.

Mufwene, Salikoko. 2001. *The Ecology of Language Evolution*. Cambridge: Cambridge University Press.

Muhly, J. D. 1995. Mining and Metalwork in Ancient Western Asia. In *Civilizations of the Ancient Near East*, ed. Jack M. Sasson, John Baines, Gary Beckman, and Karen R. Rubinson, vol. 3, pp. 1501–1519. New York: Scribner's.

Munchaev, R. M. 1994. Maikopskaya kul'tura. In *Epokha Bronzy Kavkaza i Srednei Azii: Rannyaya i Srednyaya Bronza Kavkaza*, ed. K. X. Kushnareva and V. I. Markovin, pp. 158–225. Moscow: Nauka.

————. 1982. Voprosy khozyaistva i obshchestvennogo stroya Eneoliticheskikh plemen Kavkaza. In *Eneolit SSSR*, ed. V. M. Masson and N. Y. Merpert, pp. 132–137. Moscow: Akademiya nauk.

Murphy, Eileen. 2003. *Iron Age Archaeology and Trauma from Aymyrlyg, South Siberia*. British Archaeological Reports International Series 1152. Oxford: Archeopress.

Murphy, Eileen, and Aleksandr Kokhlov. 2004. Osteological and paleopathological analysis of Volga populations from the Eneolithic to the Srubnaya periods. Samara Valley Project Interim Reports, private manuscript.

Muscarella, Oscar W. 2003. The chronology and culture of Se Girdan: Phase III. *Ancient Civilizations* 9 (1–2): 117–131.

Mytum, Harold. 1994. Language as symbol in churchyard monuments: the use of Welsh in nineteenth and twentieth-century Pembrokeshire. *World Archaeology* 26 (2): 252–267.

Napol'skikh, V. V. 1997. *Vvedenie v Istoricheskuiu Uralistiku*. Izhevsk: Udmurtskii institut istorii, yazika i literatury.

Nash, Gary. 1984. Social development. In *Colonial British America*, ed. Jack P. Green and J. R. Pole, pp. 233–261. Baltimore, Md.: Johns Hopkins University Press.

Nechitailo, A. P. 1996. Evropeiskaya stepnaya obshchnost' v epokhu Eneolita. *Rossiiskaya arkheologiya* (4): 18–30.

————. 1991. *Svyazi naseleniya stepnoi Ukrainy i severnogo Kavkaza v epokhy Bronzy*. Kiev: Nauknovo Dumka.

Necrasov, Olga. 1985. Doneés anthropologiques concernant la population du complexe culturel Cucuteni-Ariuşd-Tripolié: Phases Cucuteni et Ariuşd. *Annuaire Roumain D'Anthropologie (Bucarest)* 22:17–23.

Necrasov, Olga, and M. Cristescu. 1973. Structure anthropologique des tribus Neo-Eneolithiques et de l'age du Bronze de la Roumanie. In *Die Anfänge des Neolithikums vom Orient bis Nordeuropa VIIIa, Fundamenta*, vol. 3, pp. 137–152. Cologne: Institut für Ur-und Frügeschichte der Universität zu Köln.

Nefedkin, A. K. and E. D. Frolov. 2001. *Boevye kolesnitsy i kolesnichie drevnikh Grekov (XVI–I vv. do n.e.)*. St. Petersburg: Peterburgskoe Vostokovedenie.

Nekhaev, A. A. 1992. Domakiopskaya kul'tura severnogo Kavkaza. *Arkheologicheskie vesti* 1:76–96.

————. 1986. Pogrebenie Maikopskoi kul'tury iz kurgana u s. Krasnogvardeiskoe. *Sovietskaya arkheologiya* (1): 244–248.

Neprina, V. I. 1970. Neolitichne poselenniya v Girli r. Gnilop'yati. *Arkheologiya* (Kiev) 24:100–111.

Nettles, Daniel. 1996. Language diversity in West Africa: An ecological approach. *Journal of Anthropological Archaeology* 15:403–438.

Neustupny, E. 1991. Community areas of prehistoric farmers in Bohemia. *Antiquity* 65:326–331.

Nica, Marin. 1977. Cîrcea, cea mai veche aşezare neolită de la sud de carpaţi. *Studii si Cercetări de Istore Veche şi Arheologie* 27 (4): 4, 435–463.

Nichols, Johanna. 1997a. The epicentre of the Indo-European linguistic spread. In *Archaeology and Language, I vol. 1, Theoretical and Methodological Orientations*, ed. Roger Blench, and Matthew Spriggs, pp. 122–148. London: Routledge.

————. 1997b. Modeling ancient population structures and movement in linguistics. *Annual Review of Anthropology* 26:359–384.

————. 1994. The spread of languages around the Pacific rim. *Evolutionary Anthropology* 3:206–215.

————. 1992. *Linguistic Diversity in Space and Time*. Chicago: University of Chicago Press.

Nikolova, A. V., and Y. Y. Rassamakin. 1985. O pozdneeneoliticheskie pamyatnikakh pravoberezh'ya Dnepra. *Sovietskaya arkheologiya* (3):37–56.

Nikolova, Lolita. 2005. Social changes and cultural interactions in later Balkan prehistory (later fifth and fourth millennia calBC). *Reports of Prehistoric Research Projects* 6–7:87-96. Salt Lake City, Utah: International Institute of Anthropology.

―――. 2002. Diversity of prehistoric burial customs. In *Material Evidence and Cultural Pattern in Prehistory*, ed. L. Nikolova, pp. 53–87. Salt Lake City: International Institute of Anthropology.

―――. 2000. Social transformations and evolution in the Balkans in the fourth and third millennia BC. In *Analyzing the Bronze Age*, ed. L. Nikolova, pp. 1–8. Sofia: Prehistory Foundation.

―――. 1996. Settlements and ceramics: The experience of Early Bronze Age in Bulgaria. In *Early Bronze Age Settlement Patterns in the Balkans,* pt. 2, ed. Lolita Nikolova, pp. 145–186. Sofia: Reports of Prehistoric Research Projects 1 (2–4).

―――. 1994. On the Pit-Grave culture in northeastern Bulgaria. *Helis* (Sofia) 3:27–42.

Nobis, G. 1971. *Vom Wildpferd zum Hauspferd*. Fundamenta Reihe B, vol. 6. Cologne: Bohlau-Verlag.

Noble, Allen G. 1992. Migration to North America: Before, during, and after the nineteenth century. In *To Build in a New Land: Ethnic Landscapes in North America*, ed. Allen G. Noble, pp. 3–24. Baltimore, Md.: Johns Hopkins University Press.

Noelle, Christine. 1997. *State and Tribe in Nineteenth-Century Afghanistan: The reign of Amir Dost Muhammad Khan (1826–1863)*. Richmond, Surrey: Curzon.

Oates, Joan. 2003. A note on the early evidence for horse and the riding of equids in Western Asia. In *Prehistoric Steppe Adaptation and the Horse*, ed. Marsha Levine, Colin Renfrew, and Katie Boyle, pp. 115–125. Cambridge: McDonald Institute for Archaeological Research.

―――. 2001. Equid figurines and "chariot" models. In *Excavations at Tell Brak*, ed. David Oates, Joan Oates, and Helen McDonald, vol. 2, pp. 279–293. Cambridge: McDonald Institute for Archaeological Research.

O'Brien, S. R., P. A. Mayewski, L. D. Meeker, D. A. Meese, M. S. Twickler, and S. I. Whitlow. 1995. Complexity of Holocene climate as reconstructed from a Greenland ice core. *Science* 270:1962–1964.

O'Flaherty, Wendy Doniger. 1981. *The Rig Veda: An Anthology*. London: Penguin.

Olsen, Sandra. 2003. The exploitation of horses at Botai, Kazakhstan. In *Prehistoric Steppe Adaptation and the Horse*, ed. Marsha Levine, Colin Renfrew, and Katie Boyle, pp. 83–104. Cambridge: McDonald Institute for Archaeological Research.

Okhrimenko, G. V., and D. Y. Telegin. 1982. Novi pam'yatki mezolitu ta neolitu Volini. *Arkheologiya* (Kiev) 39:64–77.

Ostroshchenko, V. V. 2003. The economic preculiarities of the Srubnaya cultural-historical entity. In *Prehistoric Steppe Adaptation and the Horse*, ed. Marsha Levine, Colin Renfrew, and Katie Boyle, pp. 319–328. Cambridge: McDonald Institute for Archaeological Research.

Ostroverkhov, A. S. 1985. Steklyannye busy v pamyatnikakh pozdnego Tripolya. In *Novye materialy po arkheologii severo-zapadnogo prichernomorya*, ed. V. N. Stanko, pp. 174–180. Kiev: Naukovo Dumka.

Ottaway, Barbara S., ed. 1999. *A Changing Place: The Galgenberg in Lower Bavaria from the Fifth to the First Millennium BC*. British Archaeological Reports, n.s. 752. Oxford: Archeopress.

Owen, David I. 1991. The first equestrian: An UrIII glyptic scene. *Acta Sumerologica* 13:259–273.

Özbal, H., A. Adriaens, and B. Earl. 2000. Hacinebi metal production and exchange. *Paleorient* 25 (1): 57–65.

Panayotov, Ivan. 1989. *Yamnata Kultuyra v B'lgarskite Zemi*. Vol. 21. Sofia: Razkopki i Prouchvaniya.

Parker, Bradley. 2006. Toward an understanding of borderland processes. *American Antiquity* 71 (1): 77–100.

Parpola, Asko. 2004–2005. The Nāsatyas, the chariot, and Proto-Aryan religion. *Journal of Indological Studies* 16, 17:1–63.

―――. 2002. From the dialects of Old Indo-Aryan to Proto-Indo-Aryan and Proto-Iranian. In *Indo-Iranian Languages and Peoples*, ed. N. Sims-Williams, pp. 43–102. London: Oxford University Press.

————. 1988. The coming of the Aryans to Iran and India and the cultural and ethnic identity of the Dāsas. *Studia Orientalia* (Helsinki) 64:195–302.

Parzinger, H. 2002. Germanskii Arkheologogicheskii Institut: zadachi i perspektivy arkheologicheskogo izucheniya Evrazii. *Rossiiskaya arkheologiya* (3): 59–78.

————. 1993. *Studien zur Chronologie und Kulturgeschichte der Jungstein, Kupfer- und Frühbronzezeit Zwischen Karpaten und Mittelerem Taurus.* Mainz am Rhein: Römish-Germanische Forschungen B 52.

————. 1992. Hornstaad-Hlinskoe-Stollhof: Zur absoluten datierung eines vor-Badenzeitlichen Horizontes. *Germania* 70:241–250.

Parzinger, Hermann, and Nikolaus Boroffka. 2003. *Das Zinn der Bronzezeit in Mittelasien.* Vol. 1, *Die siedlungsarchäologischen Forschgungen im Umfeld der Zinnlagerstätten.* Archäologie in Iran und Turan, Band 5. Mainz am Rhein: Philipp von Zabern.

Pashkevich. G. O. 2003. Paleoethnobotanical evidence of agriculture in the steppe and the forest-steppe of east Europe in the late Neolithic and the Bronze Age. In *Prehistoric Steppe Adaptation and the Horse*, ed. Marsha Levine, Colin Renfrew, and Katie Boyle, pp. 287–297. Cambridge: McDonald Institute for Archaeological Research.

————. 1992. Do rekonstruktsii asortmentu kul'turnikh roslin epokhi Neolitu-Bronzi na territorii Ukraini. In *Starodavne Vibornitstvo na Teritorii Ukraini*, ed. S. V. Pan'kov and G. O. Voznesens'ka, pp. 179–194. Kiev: Naukovo Dumka.

Patovka, E. F. 1976. Usatovo: iz istorii issledovaniya. *Materiali i issledovaniya po arkheologii severnogo prichernomoriya* (Kiev) 8:49–60.

Patovka, E. F., et al. 1989. *Pamyatniki tripol'skoi kul'tury v severo-zapadnom prichernomor'ye.* Kiev: Naukovo Dumka.

Payne, Sebastian. 1995. Appendix B. In *The Gordion Excavations (1950–1973) Final Reports*, vol. 2, pt. 1, *The Lesser Phrygian Tumuli: The Inhumations*, ed. Ellen L. Kohler. Philadelphia: University Museum Press.

Paunescu, Alexandru. 1987. Tardenoasianul din Dobrogea. *Studii și Cercetări de Istorie Veche și Arheologie* 38 (1): 3–22.

Penner, Sylvia. *Schliemanns Schachtgräberund und der Europäische Nordosten: Studien zur Herkunft der frühmykenischen Streitwagenausstattung.* Vol. 60. Bonn: Saarbrücker Beiträge zur Alterumskunde.

Penny, Ralph. 2000. *Variation and Change in Spanish.* Cambridge: Cambridge University Press.

Perles, Catherine. 2001. *Early Neolithic Greece.* Cambridge: Cambridge University Press.

Pernicka, Ernst, et al. 1997. Prehistoric copper in Bulgaria. *Eurasia Antiqua* 3:41–179.

Perry, C. A., and K. J. Hsu. 2000. Geophysical, archaeological, and historical evidence support a solar-output model for climate change. *Proceedings of the National Academy of Sciences* 7 (23): 12,433–12,438.

Peške, Lubomir. 1986. Domesticated horses in the Lengyel culture? In *Internationales Symposium Über die Lengyel-Kultur*, pp. 221–226. Nitra-Wien: Archäologisches Institut der Slowakischen Akademie der Wissenschaften in Nitra.

Peterson, Nicholas. 1993. Demand sharing: Reciprocity and the pressure for generosity among foragers. *American Anthropologist* 95 (4): 860–874.

Petrenko, A. G. 1984. *Drevnee i srednevekovoe zhivotnovodstvo srednego povolzh'ya i predural'ya.* Moscow: Nauka.

Piggott, Stuart. 1992. *Wagon, Chariot and Carriage: Symbol and Status in the History of Transport.* London: Thames and Hudson.

————. 1983. *The Earliest Wheeled Transport: From the Atlantic Coast to the Caspian Sea.* New York: Cornell University Press.

————. 1974. Chariots in the Caucasus and China. *Antiquity* 48:16–24.

————. 1962. Heads and hoofs. *Antiquity* 36 (142): 110–118.

Pinker, Steven. 1994. *The Language Instinct.* New York: William Morrow.

Pogozheva, A. P. 1983. *Antropomorfnaya Plastika Tripol'ya*. Novosibirsk: Akademiia nauk, Sibirskoe otdelenie.

Poliakov, Leon. 1974. *The Aryan Myth: A History of Racist and Nationalist Ideas in Europe*. Translated by Edmund Howard. New York: Basic Books.

Pollack, Susan. 1999. *Ancient Mesopotamia*. Cambridge: Cambridge University Press.

Polomé, Edgar C. 1991. Indo-European religion and the Indo-European religious vocabulary. In *Sprung from Some Common Source: Investigations into the Prehistory of Languages*, ed. S. M. Lamb and E. D. Mitchell, pp. 67–88. Stanford: Stanford University Press.

———. 1990. Types of linguistic evidence for early contact: Indo-Europeans and non-Indo-Europeans. In *When Worlds Collide: Indo-Europeans and the Pre-Indo-Europeans*, ed. T. L. Markey, and John A. C. Greppin, pp. 267–289. Ann Arbor, Mich.: Karoma.

Popova, T. A. 1979. Kremneobrabatyvaiushchee proizvodstvo Tripol'skikh plemen. In *Pervobytnaya Arkheologiya, Poiski i Nakhodki*, ed. N. N. Bondar and D. Y. Telegin, pp. 145–163. Kiev: Nauknovo Dumka.

Porter, John. 1965. *The Vertical Mosaic: An Analysis of Social Class and Power in Canada*. Toronto: University of Toronto Press.

Potekhina, I. D. 1999. *Naselenie Ukrainy v Epokhi Neolita i Rannego Eneolita*. Kiev: Insitut arkheologii.

Potts, Dan T. 2000. *Ancient Magan: The Secrets of Tell Abraq*. London: Trident.

———. 1999. *The Archaeology of Elam*. Cambridge: Cambridge University Press.

Prescott, J. R. V. 1987. *Political Frontiers and Boundaries*. London: Unwin Hyman.

Privat, Karen. 2002. Preliminary report of paleodietary analysis of human and faunal remains from Bolshekaragansky kurgan 25. In *Arkaim: Nekropol (po materialam kurgana 25 Bol'she Karaganskoe Mogil'nika)*, ed. V. T. Kovaleva, and G. B. Zdanovich, pp. 166–171. Chelyabinsk: Yuzhno-Ural'skoe knizhnoe izdatel'stvo.

Pryakhin, A. D., 1980. Abashevskaya kul'turno-istoricheskaya obshchnost' epokhi bronzy i lesostepe. In *Arkheologiya Vostochno-Evropeiskoi Lesostepi*, ed. A. D. Pryakhin, pp. 7–32. Voronezh: Voronezhskogo universiteta.

———. 1976. *Poseleniya Abashevskoi Obshchnosti*. Voronezh: Voronezhskogo universiteta.

Pryakhin, A. D., and V. I. Besedin. 1999. The horse bridle of the Middle Bronze Age in the East European forest-steppe and the steppe. *Anthropology and Archaeology of Eurasia* 38 (1): 39–59.

Puhvel, Jaan. 1994. Anatolian: Autochthonous or interloper? *Journal of Indo-European Studies* 22:251–263.

———. 1991. Whence the Hittite, whither the Jonesian vision? In *Sprung from Some Common Source*, ed. Sydney M. Lamb and E. D. Mitchell, pp. 52–66. Stanford: Stanford University Press.

———. 1975. Remus et Frater. *History of Religions* 15:146–157.

Pulgram, E. 1959. Proto-Indo-European reality and reconstruction. *Language* 35:421–426

Rassamakin, Yuri. 2002. Aspects of Pontic steppe development (4550–3000 BC) in the light of the new cultural-chronological model. In *Ancient Interactions: East and West in Eurasia*, ed. Katie Boyle, Colin Renfrew, and Marsha Levine, pp. 49–74. Cambridge: McDonald Institute for Archaeological Research.

———. 1999. The Eneolithic of the Black Sea steppe: dynamics of cultural and economic development, 4500–2300 BC. In *Late Prehistoric Exploitation of the Eurasian Steppe*, ed. Marsha Levine, Yuri Rassamakin, Aleksandr Kislenko, and Nataliya Tatarintseva, pp. 59–182. Cambridge: McDonald Institute for Archaeological Research.

Raulwing, Peter. 2000. *Horses, Chariots and Indo-Europeans*. Archaeolingua Series Minor 13. Budapest: Archaeolingua Foundation.

Reade, Julian. 2001. Assyrian king-lists, the royal tombs of Ur, and Indus Origins. *Journal of Near Eastern Studies* 60 (1): 1–29.

Renfrew, Colin. 2002a. Pastoralism and interaction: Some introductory questions. In *Ancient Interactions: East and West in Eurasia*, ed. Katie Boyle, Colin Renfrew, and Marsha Levine, pp. 1–12. Cambridge: McDonald Institute for Archaeological Research.

———. 2002b. The emerging synthesis: The archaeogenetics of farming/language dispersals and other spread zones. In *Examining the Farming/Language Dispersal Hypothesis*, ed. Peter Bellwood and Colin Renfrew, pp. 3–16. Cambridge: McDonald Institute for Archaeological Research.

———. 2001. The Anatolian origins of Proto-Indo-European and the autochthony of the Hittites. In *Greater Anatolia and the Indo-Hittite Language Family*, ed. Robert Drews, pp. 36–63. Journal of Indo-European Studies Monograph 38. Washington, D.C.: Institute for the Study of Man.

———. 2000. At the edge of knowability: Towards a prehistory of languages. *Cambridge Archaeological Journal* 10 (1): 7–34.

———. 1998. Word of Minos: The Minoan contribution to Mycenaean Greek and the linguistic geography of the Bronze Age Aegean. *Cambridge Archaeological Journal* 8 (2): 239–264.

———. 1996. Language families and the spread of farming. In *The Origins and Spread of Agriculture and Pastoralism in Eurasia*, ed. David Harris, pp. 70–92. Washington, D.C.: Smithsonian Institution Press.

———. 1987. *Archaeology and Language: The Puzzle of Indo-European Origins*. London: Jonathon Cape.

———. 1973. *Before Civilization: The Radiocarbon Revolution and Prehistoric Europe*. London: Jonathon Cape.

Renfrew, Colin, April McMahon, and Larry Trask, eds. 2000. *Time Depth in Historical Linguistics*. Cambridge: McDonald Institute for Archaeological Research.

Rexová, Katerina, Daniel Frynta, and Jan Zrzavý. 2003. Cladistic analysis of languages: Indo-European classification based on lexicostatistical data. *Cladistics* 19 (2): 120–127.

Rezepkin, A. D. 2000. *Das Frühbronzezeitliche Gräberfeld von Klady und die Majkop-Kultur in Nordwestkaukasien*. Archäologie in Eurasien 10. Rahden: Verlag Marie Leidorf.

———. 1991. Kurgan 31 mogil'nika Klady problemy genezisa i khronologii Maikopskoi kul'tury. In *Drevnie kul'tury prikuban'ya*, ed. V. M. Masson, pp. 167–197. Leningrad: Nauka.

Rezepkin, A. D. and A. V. Kondrashov. 1988. Novosvobodnenskoe pogrebenie s povozkoy. *Kratkie soobshcheniya instituta arkheologii AN SSSR* 193:91–97.

Richter, Daniel K. 1992. *The Ordeal of the Longhouse: The Peoples of the Iroquois League in the Era of European Colonization*. Chapel Hill: University of North Carolina Press.

Rijksbaron, A. 1988. The discourse function of the imperfect. In *In the Footsteps of Raphael Kühner*, ed. A. Rijksbaron, H. A. Mulder, and G. C. Wakker, pp. 237–254. Amsterdam: J. C. Geiben,

Ringe, Don. 1997. A probabilistic evaluation of Indo-Uralic. In *Nostratic: Sifting the Evidence*, ed. B. Joseph and J. Salmons, pp. 153–197. Philadelphia: Benjamins.

Ringe, Don, Tandy Warnow, and Ann Taylor. 2002. Indo-European and computational cladistics. *Transactions of the Philological Society* 100:59–129.

Ringe, Don, Tandy Warnow, Ann Taylor, A. Michailov, and Libby Levison. 1998. Computational cladistics and the position of Tocharian. In *The Bronze Age and Early Iron Age Peoples of Eastern Central Asia*, ed. Victor Mair, pp. 391–414. Washington, D.C.: Institute for the Study of Man.

Robb, J. 1993. A social prehistory of European languages. *Antiquity* 67:747–760.

———. 1991. Random causes with directed effects: The Indo-European language spread and the stochastic loss of lineages. *Antiquity* 65:287–291.

Roman, Petre. 1978. Modificări în tabelul sincronismelor privind eneoliticul Tîrziu. *Studii si Cercetări de Istorie Veche și Arheologie* (Bucharest) 29 (2): 215–221.

Rosenberg, Michael. 1998. Cheating at musical chairs: Territoriality and sedentism in an evolutionary context. *Current Anthropology* 39 (5): 653–681.

———. 1994. Agricultural origins in the American Midwest: A reply to Charles. *American Anthropologist* 96 (1): 161–164.

Rostovtseff, M. 1922. *Iranians and Greeks in South Russia.* Oxford: Clarendon.

Rothman, Mitchell S. 2003. Ripples in the stream: Transcaucasia-Anatolian interaction in the Murat/Euphrates basin at the beginning of the third millennium BC. In *Archaeology in the Borderlands*, ed. Adam T. Smith and Karen Rubinson, pp. 95–110. Los Angeles: Cotsen Institute.

———. 2001. *Uruk Mesopotamia and Its Neighbors: Cross-cultural Interactions in the Era of State Formation.* Santa Fe: SAR.

Russell, Josiah Cox. 1972. *Medieval Regions and Their Cities.* Bloomington: Indiana University Press.

Rutter, Jeremy. 1993. Review of Aegean prehistory II: The prepalatial Bronze Age of the southern and central Greek mainland. *American Journal of Archaeology* 97:745–797.

Ryden, Hope. 1978. *America's Last Wild Horses.* New York: Dutton.

Ryder, Tom. 1987. Questions and Answers. *The Carriage Journal* 24 (4): 200–201.

Ryndina, N. V. 1998. *Dreneishee Metallo-obrabatyvaiushchee Proizvodstvo Iugo-Vostochnoi Evropy.* Moscow: Editorial.

Ryndina, N. V. and A. V. Engovatova. 1990. Opyt planigraficheskogo analiza kremnevykh orudii Tripol'skogo poseleniya Drutsy 1. In *Rannezemledel'cheskie Poseleniya-Giganty Tripol'skoi Kul'tury na Ukraine*, ed. I. T. Chernyakov, pp. 108–114. Tal'yanki: Institut arkheologii akademii nauk USSR.

Salminen, Tapani. 2001. The rise of the Finno-Ugric language family. In *Early Contacts between Uralic and Indo-European: Linguistic and Archaeological Considerations*, ed. Christian Carpelan, Asko Parpola, and Petteri Koskikallio, pp. 385–395. Memoires de la Société Finno-Ugrienne 242. Helsinki: Suomalais-Ugrilainen Seura.

Salmons, Joe. 1993. *The Glottalic Theory: Survey and Synthesis.* Journal of Indo-European Studies Monograph 10. Washington, D.C.: Institute for the Study of Man.

Salvatori, Sandro. 2003. Pots and peoples: The "Pandora's Jar" of Central Asian archaeological research: On two recent books on Gonur graveyard excavations. *Rivista di Archeologia* 27:5–20.

———. 2002. Project "Archaeological map of the Murghab Delta" (Turkmenistan): Test trenches at the sites of Adzhi Kui 1 and 9. *Ancient Civilizations from Scythia to Siberia* 8 (1–2): 107–178.

———. 2000. Bactria and Margiana seals: A new assessment of their chronological position and a typological survey. *East and West* 50 (1–4): 97–145.

Salvatori, Sandro, Massimo Vidale, Giuseppe Guida, and Giovanni Gigante. 2002. A glimpse on copper and lead metalworking at Altyn-Depe (Turkmenistan) in the 3rd millennium BC. *Ancient Civilizations from Scythia to Siberia* 8:69–101.

Samashev, Z. 1993. *Petroglyphs of the East Kazakhstan as a Historical Source.* Almaty: Rakurs.

Sapir, Edward, 1912. Language and environment. *American Anthropologist* 14(2): 226–42.

Sarianidi, V. I. 2002. *Margush: Drevnevostochnoe tsarstvo v staroi del'te reki Murgab.* Ashgabat: Turkmendöwlethebarlary.

———. 1995. New discoveries at ancient Gonur. *Ancient Civilizations from Scythia to Siberia* 2 (3): 289–310.

———. 1987. Southwest Asia: Migrations, the Aryans, and Zoroastrians. *Information Bulletin, International Association for the Study of the Cultures of Central Asia* (Moscow) 13:44–56.

———. 1986. Mesopotamiia i Baktriia vo ii tys. do n.e. *Sovietskaia Arkheologiia* (2): 34–46.

———. 1977. *Drevnie Zemledel'tsy Afganistana: Materialy Sovetsko-Afganskoi Ekspeditsii 1969–1974 gg.* Moscow: Akademiia Nauka.

Sawyer, Ralph D. 1993. *The Seven Military Classics of Ancient China.* Boulder, Colo.: Westview.

Schlegel, Alice. 1992. African political models in the American Southwest: Hopi as an internal frontier society. *American Anthropologist* 94 (2): 376–97.

Schmidt, Karl Horst. 1991. Latin and Celtic: Genetic relationship and areal contacts. *Bulletin of the Board of Celtic Studies* 38:1–19.

Schrijver, Peter. 2001. Lost languages in northern Europe. In *Early Contacts between Uralic and Indo-European: Linguistic and Archaeological Considerations*, ed. Christian Carpelan, Asko Parpola, and Petteri Koskikallio, pp. 417–425. Memories de la Société Finno-Ugrienne 242. Helsinki: Suomalais-Ugrilainen Seura.

Schuchhardt, C. 1919. *Alteuropa in seiner Kultur- und Stilentwicklung.* Berlin: Walter de Gruyter.

Segalen, Martine. 1991. *Fifteen Generations of Bretons: Kinship and Society in Lower Brittany, 1720–1980.* Cambridge: Cambridge University Press.

Shakhanova, N. 1989. The system of nourishment among the Eurasian nomads: The Kazakh example. In *Ecology and Empire: Nomads in the Cultural Evolution of the Old World*, pp. 111–117. Los Angeles: University of Southern California Ethnographics Press.

Shaposhnikova, O. G. 1961. Novye dannye o Mikhailovskom poselenii. *Kratkie soobshcheniya institut arkheologii* 11:38–42.

Sharafutdinova, I. N. 1980. Severnaya kurgannaya grupa u s. Sokolovka. In *Arkheologicheskie pamyatniki poingul'ya*, pp. 71–123. Kiev: Naukovo Dumka.

Shaughnessy, Edward L. 1988. Historical perspectives on the introduction of the chariot into China. *Harvard Journal of Asian Studies* 48:189–237.

Shennan, Stephen J., ed. 1989. *Archaeological Approaches to Cultural Identity.* London: Routledge.

Sherratt, Andrew. 2003. The horse and the wheel: The dialectics of change in the circum-Pontic and adjacent areas, 4500–1500 BC. In *Prehistoric Steppe Adaptation and the Horse*, ed. Marsha Levine, C. Renfrew, and K. Boyle, pp. 233–252. McDonald Institute Monographs. Cambridge: University of Cambridge Press.

———. 1997a [1983]. The secondary exploitation of animals in the Old World. In *Economy and Society in Prehistoric Europe: Changing Perspectives*, rev. ed., ed. Andrew Sherratt, pp.199–228. Princeton, N.J.: Princeton University Press.

———. 1997b. The introduction of alcohol to prehistoric Europe. In *Economy and Society in Prehistoric Europe*, ed. Andrew Sherratt, pp. 376–402. Princeton, N.J.: Princeton University Press.

———. 1997c [1991]. Sacred and profane substances: The ritual use of narcotics in later Neolithic Europe. In *Economy and Society in Prehistoric Europe*, ed. Andrew Sherratt, rev. ed. pp. 403–430. Princeton, N.J.: Princeton University Press.

———. 1986. Two new finds of wooden wheels from Later Neolithic and Early Bronze Age Europe. *Oxford Journal of Archaeology* 5:243–248.

Sherratt, Andrew, and E. S. Sherratt. 1988. The archaeology of Indo-European: An alternative view. *Antiquity* 62 (236): 584–595.

Shevchenko, A. I., 1957. Fauna poseleniya epokhi bronzy v s. Mikhailovke na nizhnem Dnepre. *Kratkie soobshcheniya institut arkheologii* 7:36–37.

Shilov, V. P. 1985a. Kurgannyi mogil'nik y s. Tsatsa. In *Drevnosti Kalmykii*, pp. 94–157. Elista: Kalmytskii nauchno-issledovatel'skii institut istorii, filogii i ekonomiki.

———. 1985b. Problemy proiskhozhdeniya kochevogo skotovodstva v vostochnoi Evrope. In *Drevnosti kalmykii*, pp. 23–33. Elista: Kalmytskii nauchno-issledovatel'skii institut istorii, filogii i ekonomiki.

Shilov, V. P., and R. S. Bagautdinov. 1998. Pogebeniya Eneolita-rannei Bronzy mogil'nika Evdyk. In *Problemy drevnei istorii severnogo prikaspiya*, ed. I. B. Vasiliev, pp. 160–178. Samara: Samarskii gosudarstvenyi pedagogicheskii universitet.

Shishlina, N. I., ed. 2000. *Sezonnyi ekonomicheskii tsikl naseleniya severo-zapadnogo Prikaspiya v Bronzovom Veke.* Vol. 120. Moscow: Trudy gosudarstvennogo istoricheskogo muzeya.

———, ed. 1999. *Tekstil' epokhi Bronzy Evraziiskikh stepei.* Vol. 109. Moscow: Trudy gosudarstvennogo istoricheskogo muzeya.

―――. 1990. O slozhnom luke Srubnoi kul'tury. In *Problemy arkheologii evrazii*, ed. S. V. Studzitskaya, vol. 74, pp. 23–37. Moscow: Trudy gosudarstvennogo oedena Lenina istoricheskogo muzeya.

Shishlina, N. I., and V. E. Bulatov. 2000. K voprosu o sezonnoi sisteme ispol'zovaniya pastbishch nositelyami Yamnoi kul'tury Prikaspiiskikh stepei v III tys. do n.e. In *Sezonnyi Ekonomicheskii Tsikl Naseleniya Severo-Zapadnogo Prikaspiya v Bronzovom Veke*, ed. N. I. Shishlina, vol. 120, pp. 43–53. Moscow: Trudy gosudarstvennogo istoricheskogo muzeya.

Shishlina, N. I., O. V. Orfinskaya, and V. P. Golikov. 2003. Bronze Age textiles from the North Caucasus: New evidence of fourth millennium BC fibres and fabrics. *Oxford Journal of Archaeology* 22 (4): 331–344.

Shmagli, M. M., and M. Y. Videiko. 1987. Piznotripil'ske poseleniya poblizu s. Maidanets'kogo na Cherkashchini. *Arkheologiya* (Kiev) 60:58–71.

Shnirelman, Victor, A. 1999. Passions about Arkaim: Russian nationalism, the Aryans, and the politics of archaeology. *Inner Asia* 1:267–282.

―――. 1998. Archaeology and ethnic politics: The discovery of Arkaim. *Museum International* 50 (2): 33–39.

―――. 1995. Soviet archaeology in the 1940s. In *Nationalism, Politics, and the Practice of Archaeology*, ed. Philip L. Kohl and Clare Fawcett, pp. 120–138. Cambridge: Cambridge University Press.

―――. 1992. The emergence of food-producing economy in the steppe and forest-steppe zones of Eastern Europe. *Journal of Indo-European Studies* 20:123–143.

Shorin, A. F. 1993. O za Uralskoi oblasti areala lesnikh Eneoliticheskikh kul'tur grebenchatoi keramiki. In *Voprosy arkheologii Urala*, pp. 84–93. Ekaterinburg: Uralskii gosudarstvenyi universitet.

Shramko, B. A., and Y. A. Mashkarov. 1993. Issledovanie bimetallicheskogo nozha iz pogrebeniya Katakombnoi kul'tury. *Rossiskaya arkheologiya* (2): 163–170.

Siegel, Jeff. 1985. Koines and koineisation. *Language in Society* 14:357–378.

Silver, Shirley, and Wick R. Miller. 1997. *American Indian Languages: Cultural and Social Contexts*. Tucson: University of Arizona Press.

Simkins, P. D., and F. L. Wernstedt. 1971. *Philippines Migration: Settlement of the Digos-Padada Valley, Padao Province*. Southeast Asia Studies 16. New Haven: Yale University Press.

Sinitsyn, I. V. 1959. Arkheologicheskie issledovaniya Zavolzhskogo otriada (1951–1953). *Materialy i issledovaniya Institut arkheologii* (Moscow) 60:39–205.

Siniuk, A. T., and I. A. Kozmirchuk. 1995. Nekotorye aspekti izucheniya Abashevskoi kul'tury v basseine Dona. In *Drevnie IndoIranskie Kul'tury Volgo-Ural'ya*, ed. V. S. Gorbunov, pp. 37–72. Samara: Samarskogo gosudarstvennogo pedagogicheskogo universiteta.

Sinor, Dennis, ed. 1988. *The Uralic Languages*. Leiden: Brill.

―――. 1972. Horse and pasture in Inner Asian history. *Oriens Extremus* 19:171–183.

Skjærvø, P. Oktor. 1995. The Avesta as a source for the early history of the Iranians. In *The Indo-Aryans of Ancient South Asia: Language, Material Culture and Ethnicity*, ed. George Erdosy, pp. 155–176. Indian Philology and South Asian Studies 1. Berlin: Walter de Gruyter.

Smith, Anthony D. 1998. *Nationalism and Modernism*. London: Routledge.

Smith, Bruce. 1989. Origins of agriculture in eastern North America. *Science* 246 (4937): 1,566–1,571.

Smith, John Masson. 1984. Mongol campaign rations: Milk, marmots, and blood? In *Turks, Hungarians, and Kipchaks: A Festschrift in Honor of Tibor Halasi-Kun*, ed. şinasi Tekin and Gönül Alpay Tekin. Journal of Turkish Studies 8:223–228. Cambridge, Mass.: Harvard University Print Office.

Snow, Dean. 1994. *The Iroquois*. Oxford: Blackwell.

Solovyova, N. F., A. N. Yegor'kov, V. A. Galibin, and Y. E. Berezkin. 1994. Metal artifacts from Ilgynly-Depe, Turkmenistan. In *New Archaeological Discoveries in Asiatic Russia and*

Central Asia, ed. A. G. Kozintsev, V. M. Masson, N. F. Solovyova, and V. Y. Zuyev, pp. 31–35. Archaeological Studies 16. St. Petersburg: Institute of the History of Material Culture.

Sorokin, V. Y. 1989. Kulturno-istoricheski problemy plemen srednogo Triploya Dnestrovsko-Prutskogo mezhdurechya. *Izvestiya Akademii Nauk Moldavskoi SSR* 3:45–54.

Southworth, Franklin. 1995. Reconstructing social context from language: Indo-Aryan and Dravidian prehistory. In *The Indo-Aryans of Ancient South Asia: Language, Material Culture and Ethnicity*, ed. George Erdosy, pp. 258–277. Indian Philology and South Asian Studies 1. Berlin: Walter de Gruyter.

Sparreboom, M. 1985. *Chariots in the Vedas*. Edited by J. C. Heesterman and E. J. M. Witzel. Memoirs of the Kern Institute 3. Leiden: Brill.

Spear, Thomas, and Richard Waller, eds. 1993. *Being Maasai: Ethnicity and Identity in East Africa*. Oxford: James Currey.

Specht, F. 1944. *Der Ursprung der Indogermanischen Deklination*. Göttingen: Vandenhoeck and Ruprecht.

Spicer, Edward. 1971. Persistent cultural systems: A comparative study of identity systems that can adapt to contrasting environments. *Science* 174:795–800.

Spielmann, Katherine A., ed. 1998. *Migration and Reorganization: The Pueblo IV Period in the American Southwest*. Anthropological Research Papers 51. Tempe: Arizona State University Press.

Spinage, C. A. 1972. Age estimation of zebra. *East African Wildlife Journal* 10:273–277.

Stark, Miriam T., ed. 1998. *The Archaeology of Social Boundaries*. Washington, D.C.: Smithsonian Institution Press.

Stein, Gil. 1999. *Rethinking World Systems: Diasporas, Colonies, and Interaction in Uruk Mesopotamia*. Tucson: University of Arizona Press.

Stevanovic, Mirjana. 1997. The Age of Clay: The Social Dynamics of House Destruction. *Journal of Anthropological Archaeology* 16:334–395.

Stewart, Ann H. 1976. *Graphic Representation of Models in Linguistic Theory*. Bloomington: Indiana University Press.

Stillman, Nigel, and Nigel Tallis. 1984. *Armies of the Ancient Near East*. Worthing, Sussex: Flexiprint.

Sturtevant, William. 1962. The Indo-Hittite hypothesis. *Language* 38:105–110.

Subbotin, L.V. 1995. Grobniki Kemi-Obinskogo tipa severo-zapadnogo Prichernomor'ya. *Rossiskaya arkheologiya* (3): 193–197.

———. 1990. Uglubennye zhilishcha kul'tury Gumelnitsa v nizhnem podunav'e. In *Rannezemledel'cheski poseleniya-giganty Tripol'skoi kul'tury na Ukraine*, ed. I. T. Chenyakov, pp. 177–182. Tal'yanki: Institut arkheologii AN USSR.

———. 1985. Semenovskii mogil'nik epokhi Eneolita-Bronzy. In *Novye material'i po arkheologii severo-zapadnogo prichernomor'ya*, ed. V. N. Stanko, pp. 45–95. Kiev: Naukovo Dumka.

———. 1978. O sinkhronizatsii pamyatnikov kul'tury Gumelnitsa v nizhnem Podunav'e. In *Arkheologicheskie issledovaniya severo-zapadnogo prichernomor'ya*, ed. V. N. Stanko, pp. 29–41. Kiev: Naukovo Dumka.

Summers, Geoffrey D. 2001. Questions raised by the identification of the Neolithic, Chalcolithic, and Early Bronze Age horse bones in Anatolia. In *Greater Anatolia and the Indo-Hittite Language Family*, ed. Robert Drews, pp. 285–292. Journal of Indo-European Studies Monograph 38. Washington, D.C.: Institute for the Study of Man.

Sutton, Richard E. 1996. The Middle Iroquoian colonization of Huronia. Ph.D. dissertation. McMaster University, Hamilton, Ontario.

Swadesh, M. 1955. Towards greater accuracy in lexicostatistic dating. *International Journal of American Linguistics* 21:121–37.

———. 1952. Lexico-statistic dating of prehistoric ethnic contacts. *Proceedings of the American Philosophical Society* 96:452–463.

Syvolap, M. P. 2001. Kratkaya kharakteristika pamyatnikov Yamnoi kul'tury srednego podneprov'ya. In *Bronzovyi vek vostochnoi Evropy: Kharakteristika kul'tur, khronologiya i periodizatsiya*, ed. Y. I. Kolev, P. F. Kuznetsov, O. V. Kuzmina, A. P. Semenova, M. A. Turetskii, and B. A. Aguzarov, pp. 109–117. Samara: Samarskii Gosudarstvennyi Pedagogicheskii Universitet.

Szemerényi, Oswald. 1989. The new sound of Indo-European. *Diachronica* 6:237–269.

Szmyt, Marzena. 1999. *Between West and East: People of the Globular Amphorae Culture in Eastern Europe, 2950–2350 BC.* Baltic-Pontic Studies 8. Poznań: Adam Mickiewicz University.

Tatarintseva, N. S. 1984. Keramika poseleniya Vishnevka 1 v lesostepnom pri Ishim'e. In *Bronzovyi Vek Uralo-Irtyshskogo Mezhdurech'ya*, pp. 104–113. Chelyabinsk: Chelyabinskii gosudarstvennyi universitet.

Telegin, D. Y. 2005. The Yamna culture and the Indo-European homeland problem. *Journal of Indo-European Studies* 33 (3–4): 339–358.

———. 2002. A discussion on some of the problems arising from the study of Neolithic and Eneolithic cultures in the Azov-Black Sea region. In *Ancient Interactions: East and West in Eurasia*, ed. Katie Boyle, Colin Renfrew, and Marsha Levine, pp. 25–47. Cambridge: McDonald Institute for Archaeological Research.

———. 1996. Yugo-zapad vostochnoi Evropy; and Yug vostochnoi Evropy. In *Neolit severnoi evrazii*, ed. S. V. Oshibkina, pp. 19–86. Moscow: Nauka.

———. 1991. *Neoliticheskie mogil'niki mariupol'skogo tipa.* Kiev: Naukovo Dumka.

———. 1988. Keramika rannogo Eneolitu tipu Zasukhi v lisostepovomu liboberezhzhi Ukriani. *Arkheologiya* (Kiev) 64:73–84.

———. 1987. Neolithic cultures of the Ukraine and adjacent areas and their chronology. *Journal of World Prehistory* 1 (3): 307–331.

———. 1986. *Dereivka: A Settlement and Cemetery of Copper Age Horse Keepers on the Middle Dnieper.* Edited by J. P. Mallory. Translated by V. K. Pyatkovskiy. British Archaeological Reports International Series 287. Oxford: Archeopress.

———. 1982. *Mezolitichni pam'yatki Ukraini.* Kiev: Naukovo Dumka.

———. 1981. Pro neolitichni pam'yatki Podonnya i steponogo Povolzhya. *Arkheologiya* (Kiev) 36:3–19.

———. 1977. Review of Markevich, V. I., 1974. *Bugo-Dnestrovskaya kul'tura na territorii Moldavii. Arkheologiia* (Kiev) 23:88–91.

———. 1973. *Seredno-Stogivs'ka kul'tura Epokha Midi.* Kiev: Naukovo Dumka.

———. 1968. *Dnipro-Donets'ka kul'tura.* Kiev: Naukovo Dumka.

Telegin, D. Y., and James P. Mallory. 1994. *The Anthropomorphic Stelae of the Ukraine: The Early Iconography of the Indo-Europeans.* Journal of Indo-European Studies Monograph 11. Washington D.C.: Institute for the Study of Man.

Telegin, D. Y., A. L. Nechitailo, I. D. Potekhina, and Y. V. Panchenko. 2001. *Srednestogovskaya i novodanilovskaya kul'tury Eneolita Azovo-Chernomorskogo regiona.* Lugansk: Shlyakh.

Telegin, D. Y., and I. D. Potekhina. 1987. *Neolithic Cemeteries and Populations in the Dnieper Basin*, ed. J. P. Mallory. British Archaeological Reports International Series 383. Oxford: Archeopress.

Telegin, D. Y., I. D. Potekhina, M. Lillie, and M. M. Kovaliukh. 2003. Settlement and economy in Neolithic Ukraine: A new chronology. *Antiquity* 77 (296): 456–470.

———. 2002. The chronology of the Mariupol-type cemeteries of Ukraine revisited. *Antiquity* 76:356–363.

Telegin, D. Y., Sergei Z. Pustalov, and N. N. Kovalyukh. 2003. Relative and absolute chronology of Yamnaya and Catacomb monuments: The issue of co-existence. *Baltic-Pontic Studies* 12:132–184.

Teplova, S. N. 1962. *Atlas SSSR.* Moscow: Ministerstva geologii i okhrany nedr SSSR.

Terrell, John Edward, ed. 2001. *Archaeology, Language and History: Essays on Culture and Ethnicity.* Westport, Conn.: Bergin and Garvey.

Terrell, John Edward, T. L. Hunt, and Chris Godsen. 1997. The dimensions of social life in the Pacific: Human diversity and the myth of the primitive isolate. *Current Anthropology* 38:155–195.

Thieme, Paul. 1960. The Aryan gods of the Mitanni treaties. *Journal of the American Oriental Society* 80:310–317.

——. 1958. The Indo-European language. *Scientific American* 199 (4): 63–74.

Thomason, Sarah Gray, and Terrence Kaufman. 1988. *Language Contact, Creolization, and Genetic Linguistics.* Los Angeles: University of California Press.

Thornton, C. P., and C. C. Lamberg-Karlovsky. 2004. A new look at the prehistoric metallurgy of southeastern Iran. *Iran* 42:47–59.

Timofeev, V. I., and G. I. Zaitseva. 1997. K probleme radiouglerodnoi khronologii Neolita stepnoi i iuga lesnoi zony Evropeiskoi chasti Rossii i Sibiri. *Radiouglerod i arkheologiya* (St. Petersburg) 2:98–108.

Todorova, Henrietta. 1995. The Neolithic, Eneolithic, and Transitional in Bulgarian Prehistory. In *Prehistoric Bulgaria*, ed. Douglass W. Bailey and Ivan Panayotov, pp. 79–98. Monographs in World Archaeology 22. Madison, Wis.: Prehistory Press.

Tolstov, S. P., and A. S. Kes'. 1960. *Nizov'ya Amu-Dar'i, Sarykamysh, Uzboi: Istoriya formirovaniya i zaseleniya.* Vol. 3. Moscow: Materialy khorezmskoi ekspeditsii.

Tovkailo, M. T. 1990. Do pitannya pro vzaemini naseleniya Bugo-Dnitrovskoi ta ranne Triplil'skoi kul'tur u stepovomu po Buzhi. In *Rannezemledel'cheskie poseleniya-Giganty Tripol'skoi Kul'tury na Ukraine*, ed. V. G. Zbenovich, and I. T. Chernyakov, pp. 191–194. Tal'yanki: Institut arkheologii akademiya nauk.

Trifonov, V. A. 2001. Popravki absoliutnoi khronologii kultur epokha Eneolita-Srednei Bronzy Kavkaza, stepnoi i lesostepnoi zon vostochnoi Evropy (po dannym radiouglerodnogo datirovaniya). In *Bronzovyi vek Vostochnoi Evropy: Kharakteristika kul'tur, Khronologiya i Periodizatsiya*, ed. Y. I. Kolev, P. F. Kuznetsov, O. V. Kuzmina, A. P. Semenova, M. A. Turetskii, and B. A. Aguzarov, pp.71–82, Samara: Samarskii gosudarstvennyi pedagogicheskii universitet.

——. 1991. Stepnoe prikuban'e v epokhu Eneolita: Srednei Bronzy (periodizatsiya). In *Drevnie kul'tury Prikuban'ya*, ed. V. M. Masson, pp. 92–166. Leningrad: Nauka.

Tringham, Ruth. 1971. *Hunters, Fishers and Farmers of Eastern Europe, 6000–3000 BC.* London: Hutchinson.

Troy, C. S., D. E. MacHugh, J. F. Bailey, D. A. Magee, R. T. Loftus, P. Cunningham, A. T. Chamberlain, B. C. Sykes, and D. G. Bradley. 2001. Genetic Evidence for Near-Eastern Origins of European Cattle. *Nature* 410:1088–1091.

Trudgill, Peter. 1986. *Dialects in Contact.* Oxford: Blackwell.

Tuck, J. A. 1978. Northern Iroquoian prehistory. In *Northeast Handbook of North American Indians*, ed. Bruce G. Trigger, vol. 15, pp. 322–333. Washington, D.C.: Smithsonian Institution Press.

Uerpmann, Hans-Peter. 1990. Die Domestikation des Pferdes im Chalcolithikum West- und Mitteleuropas. *Madrider Mitteilungen* 31:109–153.

Upton, Dell, and J. M. Vlach, eds. 1986. *Common Places: Readings in American Vernacular Architecture.* Athens: University of Georgia Press.

Ursulescu, Nicolae. 1984. *Evoluția Culturii Starčevo-Criş Pe Teritoriul Moldovei.* Suceava: Muzeul Județean Suceava.

Vainshtein, Sevyan. 1980. *Nomads of South Siberia: The Pastoral Economies of Tuva.* Edited by Caroline Humphrey. Translated by M. Colenso. Cambridge: Cambridge University Press.

Van Andel, T. H., and C. N. Runnels. 1995. The earliest farmers in Europe. *Antiquity* 69:481–500.

Van Buren, G. E. 1974. *Arrowheads and Projectile Points.* Garden Grove, Calif.: Arrowhead.

Vasiliev, I. B. 2003. Khvalynskaya Eneoliticheskaya kul'tura Volgo-Ural'skoi stepi i lesostepi (nekotorye itogi issledovaniya). *Voprosy Arkeologii Povolzh'ya* v.3: 61–99. Samara: Samarskii Gosudarstvennyi Redagogieheskii Univerditet.

Vasiliev, I. B., ed. 1998. *Problemy drevnei istorii severnogo prikaspiya*. Samara: Samarskii gosu-darstvennyi pedagogicheskii universitet.

———. 1981. *Eneolit Povolzh'ya*. Kuibyshev: Kuibyshevskii gosudarstvenyi pedagogicheskii institut.

———. 1980. Mogil'nik Yamno-Poltavkinskogo veremeni u s. Utyevka v srednem Povolzh'e. In *Arkheologiya Vostochno-Evropeiskoi Lesostepi*, pp. 32–58. Voronezh: Voronezhskogo universiteta.

Vasiliev, I. B., and G. I. Matveeva. 1979. Mogil'nik u s. S'yezhee na R. Samare. *Sovietskaya arkheologiia* (4): 147–166.

Vasiliev, I. B., P. F. Kuznetsov, and A. P. Semenova. 1994. *Potapovskii Kurgannyi Mogil'nik Indoiranskikh Plemen na Volge*. Samara: Samarskii universitet.

Vasiliev, I. B., P. F. Kuznetsov, and M. A. Turetskii. 2000. Yamnaya i Poltavkinskaya kul'tura. In *Istoriya samarskogo po volzh'ya s drevneishikh vremen do nashikh dnei: Bronzovyi Vek*, ed. Y. I. Kolev, A. E. Mamontov, and M. A. Turetskii, pp. 6–64. Samara: Samarskogo nauchnogo tsentra RAN.

Vasiliev, I. B., and N. V. Ovchinnikova. 2000. Eneolit. In *Istoriya samarskogo povolzh'ya s drevneishikh vremen do nashikh dnei*, ed. A. A. Vybornov, Y. I. Kolev, and A. E. Mamonov, pp. 216–277. Samara: Integratsiya.

Vasiliev, I. B., and Siniuk, A. T. 1984. Cherkasskaya stoiyanka na Srednem Donu. In *Epokha Medi Iuga Vostochnoi Evropy*, ed. S. G. Basina and G. I. Matveeva, pp. 102–129. Kuibyshev: Kuibyshevskii gosudarstvennyi pedagogicheskii institut.

Vasiliev, I. B., A. Vybornov, and A. Komarov. 1996. *The Mesolithic of the North Caspian Sea Area*. Samara: Samara State Pedagogical University.

Vasiliev, I. B., A. A. Vybornov, and N. L. Morgunova. 1985. Review of *Eneolit iuzhnogo Urala* by G. N. Matiushin. *Sovetskaia arkheologiia* (2): 280–289.

Veenhof, Klaas R. 1995. Kanesh: An Assyrian Colony in Anatolia. In *Civilizations of the Ancient Near East*, ed. Jack M. Sasson, John Baines, Gary Beckman, and Karen R. Rubinson, vol. 1, pp. 859–871. New York: Scribner's.

Vehik, Susan. 2002. Conflict, trade, and political development on the southern Plains. *American Antiquity* 67 (1): 37–64.

Veit, Ulrich. 1989. Ethnic concepts in German prehistory: A case study on the relationship between cultural identity and archaeological objectivity." In *Archaeological Approaches to Ethnic Identity*, ed. S. J. Shennan, pp. 35–56. London: Unwin Hyman.

Venneman, Theo. 1994. Linguistic reconstruction in the context of European prehistory. *Transactions of the Philological Society* 92:215–284.

Videiko, Mihailo Y. 2003. Radiocarbon chronology of settlements of BII and CI stages of the Tripolye culture at the middle Dnieper. *Baltic-Pontic Studies* 12:7–21.

———. 1999. Radiocarbon dating chronology of the late Tripolye culture. *Baltic-Pontic Studies* 7:34–71.

———. 1990. Zhilishchno-khozyaistvennye kompleksy poseleniya Maidanetskoe i voprosy ikh interpretatsii. In *Rannezemledel'cheskie Poseleniya-Giganty Tripol'skoi kul'tury na Ukraine*, ed. I. T. Cherniakhov, pp. 115–120. Tal'yanki: Vinnitskii pedagogicheskii institut.

Videiko, Mihailo Y., and Vladislav H. Petrenko. 2003. Radiocarbon chronology of complexes of the Eneolithic–Early Bronze Age in the North Pontic region, a preliminary report. *Baltic-Pontic Studies* 12:113–120.

Vilà, Carles, J. A. Leonard, A. Götherdtröm, S. Marklund, K. Sandberg, K. Lidén, R. K. Wayne, and Hans Ellegren. 2001. Widespread origins of domestic horse lineages. *Science* 291 (5503): 474–477.

Vinogradov, A. V. 1981. *Drevnie okhotniki i rybolovy sredneaziatskogo mezhdorechya*. Vol. 10. Moscow: Materialy khorezmskoi ekspeditsii.

———. 1960. Novye Neoliticheskie nakhodki Korezmskoi ekspeditsii AN SSSR 1957 g. In *Polevye issledovaniya khorezmskoi ekspeditsii v 1957 g.*, ed. S. P. Tolstova, vol. 4, pp. 63–81. Moscow: Materialy khorezmskoi ekspeditsii.

Vinogradov, Nikolai. 2003. *Mogil'nik Bronzovogo Beka: Krivoe ozero v yuzhnom Zaural'e*. Chelyabinsk: Yuzhno-Ural'skoe knizhnoe izdatel'stvo.

Vörös, Istvan. 1980. Zoological and paleoeconomical investigations on the archaeozoological material of the Early Neolithic Körös culture. *Folia Archaeologica* 31:35–64.

Vybornov, A. A., and V. P. Tretyakov. 1991. Stoyanka Imerka VII v Primokshan. In *Drevnosti Vostochno-Evropeiskoi Lesotepi*, ed. V. V. Nikitin, pp. 42–55. Samara: Samarskii gosudartsvennyi pedagogicheskii institut.

Währen, M. 1989. Brot und Gebäck von der Jungsteinzeit bis zur Römerzeit. *Helvetia Archaeologica* 20:82–116.

Walcot, Peter. 1979. Cattle raiding, heroic tradition, and ritual: The Greek evidence. *History of Religions* 18:326–351.

Watkins, Calvert. 1995. *How to Kill a Dragon: Aspects of Indo-European Poetics*. Oxford: Oxford University Press.

Weale, Michael E., Deborah A. Weiss, Rolf F. Jager, Neil Bradman, and Mark G. Thomas. 2002. Y Chromosome Evidence for Anglo-Saxon Mass Migration. *Molecular Biology and Evolution* 19:1008–1021.

Weber, Andrzej, David W. Link, and M. Anne Katzenberg. 2002. Hunter-gatherer culture change and continuity in the Middle Holocene of the Cis-Baikal, Siberia. *Journal of Anthropological Archaeology* 21:230–299.

Wechler, Klaus-Peter, V. Dergachev, and O. Larina. 1998. Neue Forschungen zum Neolithikum Osteuropas: Ergebnisse der Moldawisch-Deutschen Geländearbeiten 1996 und 1997. *Praehistorische Zeitschrift* 73 (2): 151–166.

Weeks, L. 1999. Lead isotope analyses from Tell Abraq, United Arab Emirates: New data regarding the "tin problem" in Western Asia. *Antiquity* 73:49–64.

Weisner, Joseph. 1968. *Fahren und Reiten*. Göttingen: Vandenhoeck and Ruprecht, Archaeologia Homerica.

———. 1939. Fahren und Reiten in Alteuropa und im alten Orient. In *Der Alte Orient* Bd. 38, fascicles 2–4. Leipzig: Heinrichs Verlag.

Weiss, Harvey. 2000. Beyond the Younger Dryas: Collapse as adaptation to abrupt climate change in ancient West Asia and the Eastern Mediterranean. In *Environmental Disaster and the Archaeology of Human Response*, ed. Garth Bawden and Richard M. Reycraft, pp. 75–98. Anthropological Papers no. 7. Albuquerque: Maxwell Museum of Anthropology.

Weissner, Polly. 1983. Style and social information in Kalahari San projectile points. *American Antiquity* 48 (2): 253–275.

Wells, Peter S. 2001. *Beyond Celts, Germans and Scythians: Archaeology and Identity in Iron Age Europe*. London: Duckworth.

———. 1999. *The Barbarians Speak*. Princeton, N. J.: Princeton University Press.

White, Randall. 1989. Husbandry and herd control in the Upper Paleolithic: A critical review of the evidence. *Current Anthropology* 30 (5): 609–632.

Wilhelm, Gernot. 1995. The Kingdom of Mitanni in Second-Millennium Upper Mesopotamia. In *Civilizations of the Ancient Near East*, vol. 2, ed. Jack M. Sasson, John Baines, G. Beckman, and Karen S. Rubinson, pp. 1243–1254. New York: Scribner's.

Wilhelm, Hubert G. H. 1992. Germans in Ohio. In *To Build in a New Land: Ethnic Landscapes in North America*, ed. Allen G. Noble, pp. 60–78. Baltimore, Md.: Johns Hopkins University Press.

Willis, K. J. 1994. The vegetational history of the Balkans. *Quaternary Science Reviews* 13: 769–788.

Winckelmann, Sylvia. 2000. Intercultural relations between Iran, the Murghabo-Bactrian Archaeological Complex (BMAC), northwest India, and Falaika in the field of seals. *East and West* 50 (1–4): 43–96.

Winn, S.M.M., 1981. *Pre-Writing in Southeastern Europe: The Sign System of the Vinča Culture ca. 4000 B.C.* Calgary: Western.

Witzel, Michael. 2003. *Linguistic Evidence for Cultural Exchange in Prehistoric Western Central Asia*. Sino-Platonic Papers 129:1–70. Philadelphia: Department of Asian and Middle Eastern Languages, University of Pennsylvania.

———. 1995. Rgvedic history: Poets, chieftans, and polities. In *The Indo-Aryans of Ancient South Asia: Language, Material Culture and Ethnicity*, ed. George Erdösy, pp. 307–352. Indian Philology and South Asian Studies 1. Berlin: Walter de Gruyter.

Wolf, Eric. 1984. Culture: Panacea or problem? *American Antiquity* 49 (2): 393–400.

———. 1982. *Europe and the People without History*. Berkeley: University of California Press.

Wylie, Alison. 1995. Unification and convergence in archaeological explanation: The agricultural "wave of advance" and the origins of Indo-European languages. In *Explanation in the Human Sciences*, ed. David K. Henderson, pp. 1–30. Southern Journal of Philosophy Supplement 34. Memphis: Department of Philosophy, University of Memphis.

Yanko-Hombach, Valentina, Allan S. Gilbert, Nicolae Panin, and Pavel M. Dolukhanov. 2006. *The Black Sea Flood Question: Changes in Coastline, Climate, and Human Settlement*. NATO Science Series. Dordrecht: Springer.

Yanushevich, Zoya V. 1989. Agricultural evolution north of the Black Sea from the Neolithic to the Iron Age. In *Foraging and Farming: The Evolution of Plant Expoitation*, ed. David R. Harris and Gordon C. Hillman, pp. 607–619. London: Unwin Hyman.

Yarovoy, E. V. 1990. *Kurgany Eneolita-epokhi Bronzy nizhnego poDnestrov'ya*. Kishinev: Shtiintsa.

Yazepenka, Igor, and Aleksandr Kośko. 2003. Radiocarbon chronology of the beakers with short-wave moulding component in the development of the Middle Dnieper culture. *Baltic-Pontic Studies* 12:247–252.

Yener, A. 1995. Early Bronze Age tin processing at Göltepe and Kestel, Turkey. In *Civilizations of the Ancient Near East*, ed. Jack M. Sasson, John Baines, Gary Beckman, and Karen R. Rubinson, vol. 3, pp. 1519–1521. New York: Scribner's.

Yudin, A. I. 1998. Orlovskaya kul'tura i istoki formirovaniya stepnego Eneolita za Volzh'ya. In *Problemy Drevnei Istorii Severnogo Prikaspiya*, pp. 83–105. Samara: Samarskii gosudarstvennyi pedagogicheskii universitet.

———. 1988. Varfolomievka Neoliticheskaya stoianka. In *Arkheologicheskie kul'tury severnogo Prikaspiya*, pp. 142–172. Kuibyshev: Kuibyshevskii gosudarstvenii pedagogicheskii institut.

Zaibert, V. F. 1993. *Eneolit Uralo-Irtyshskogo Mezhdurech'ya*. Petropavlovsk: Nauka.

Zaikov, V. V., G. B. Zdanovich, and A. M. Yuminov. 1995. Mednyi rudnik Bronzogo veka "Vorovskaya Yama." In *Rossiya i Vostok: Problemy Vzaimodeistviya*, pt. 5, bk. 1: *Kul'tury Eneolita-Bronzy Stepnoi Evrazii*, pp. 157–162. Chelyabinsk: 3-ya Mezhdunarodnaya nauchnaya konferentsiya.

Zaitseva, G. I., V. I. Timofeev, and A. A. Sementsov. 1999. Radiouglerodnoe datirovanie v IIMK RAN: istoriya, sostoyanie, rezul'taty, perspektivy. *Rossiiskaya arkheologiia* (3): 5–22.

Zbenovich, V. G. 1996. The Tripolye culture: Centenary of research. *Journal of World Prehistory* 10 (2): 199–241.

———. 1980. *Poselenie Bernashevka na Dnestre (K Proiskhozhdenniu Tripol'skoi Kul'tury)*. Kiev: Naukovo Dumka.

———. 1974. *Posdnetriplos'kie plemena severnogo Prichernomor'ya*. Kiev: Naukovo Dumka.

———. 1968. Keramika usativs'kogo tipu. *Arkheologiya* (Kiev) 21:50–78.

Zdanovich, G. B., ed. 1995. *Arkaim: Issledovaniya, Poiski, Otkrytiya*. Chelyabinsk: "Kammennyi Poyas."

———. 1988. *Bronzovyi Vek Uralo-Kazakhstanskikh Stepei*. Sverdlovsk: Ural'skogo universiteta, for Berlyk II.

Zeder, Melinda. 1986. The equid remains from Tal-e Malyan, southern Iran. In *Equids in the Ancient World*, vol. 1, ed. Richard Meadow and Hans-Peter Uerpmann, pp. 366–412. Weisbaden: Reichert.

Zelinsky, W. 1973. *The Cultural Geography of the United States.* Englewood Cliffs, N.J.: Prentice-Hall.

Zhauymbaev, S. U. 1984. Drevnie mednye rudniki tsentral'nogo Kazakhstana. In *Bronzovyi Vek Uralo-Irtyshskogo Mezhdurech'ya*, pp. 113–120. Chelyabinsk: Chelyabinskii gosudarstvennyi universitet.

Zimmer, Stefan. 1990. The investigation of Proto-Indo-European history: Methods, problems, limitations. In *When Worlds Collide: Indo-Europeans and the Pre-Indo-Europeans*, ed. T. L. Markey, and John A. C. Greppin, pp. 311–344. Ann Arbor, Mich.: Karoma.

Zin'kovskii, K. V., and V. G. Petrenko. 1987. Pogrebeniya s okhroi v Usatovskikh mogil'nikakh. *Sovietskaya arkheologiya* (4): 24–39.

Zöller, H. 1977. Alter und Ausmass postgläzialer Klimaschwankungen in der Schweizer Alpen. In *Dendrochronologie und Postgläziale Klimaschwangungen in Europa*, ed. B. Frenzel, pp. 271–281. Wiesbaden: Franz Steiner Verlag.

Zutterman, Christophe. 2003. The bow in the ancient Near East, a re-evaluation of archery from the late 2nd millennium to the end of the Achaemenid empire. *Iranica Antiqua* 38: 119–165.

Zvelebil, Marek. 2002. Demography and dispersal of early farming populations at the Mesolithic/Neolithic transition: Linguistic and demographic implications. In *Examining the Farming/Language Dispersal Hypothesis*, ed. Peter Bellwood and Colin Renfrew, pp. 379–394. Cambridge: McDonald Institute for Archaeological Research.

———. 1995. Indo-European origins and the agricultural transition in Europe. *Journal of European Archaeology* 3:33–70.

Zvelebil, Marek, and Malcolm Lillie. 2000. Transition to agriculture in eastern Europe. In *Europe's First Farmers*, ed. T. Douglas Price, pp. 57–92. Cambridge: Cambridge University Press.

Zvelebil, Marek, and Peter Rowley-Conwy. 1984. Transition to farming in northern Europe: A hunter-gatherer perspective. *Norwegian Archaeological Review* 17:104–128.

Zvelebil, Marek, and K. Zvelebil. 1988. Agricultural transition and Indo-European dispersals. *Antiquity* 62:574–583.

INDEX